Advances in Science, Technology & Innovation

IEREK Interdisciplinary Series for Sustainable Development

Advances in Science, Technology & Innovation (ASTI) is a series of peer-reviewed books based on important emerging research that redefines the current disciplinary boundaries in science, technology and innovation (STI) in order to develop integrated concepts for sustainable development. It not only discusses the progress made towards securing more resources, allocating smarter solutions, and rebalancing the relationship between nature and people, but also provides in-depth insights from comprehensive research that addresses the **17 sustainable development goals (SDGs)** as set out by the UN for 2030.

The series draws on the best research papers from various IEREK and other international conferences to promote the creation and development of viable solutions for a **sustainable future and a positive societal** transformation with the help of integrated and innovative science-based approaches. Including interdisciplinary contributions, it presents innovative approaches and highlights how they can best support both economic and sustainable development, through better use of data, more effective institutions, and global, local and individual action, for the welfare of all societies.

The series particularly features conceptual and empirical contributions from various interrelated fields of science, technology and innovation, with an emphasis on digital transformation, that focus on providing practical solutions to **ensure food, water and energy security to achieve the SDGs.** It also presents new case studies offering concrete examples of how to resolve sustainable urbanization and environmental issues in different regions of the world.

The series is intended for professionals in research and teaching, consultancies and industry, and government and international organizations. Published in collaboration with IEREK, the Springer ASTI series will acquaint readers with essential new studies in STI for sustainable development.

ASTI series has now been accepted for Scopus (September 2020). All content published in this series will start appearing on the Scopus site in early 2021.

Shahfahad · Pir Mohammad ·
Abraham R. Matamanda · Jesús Rodrigo-Comino
Editors

Geographic Approaches to Climate Change and Mitigation: Urban and Rural Perspectives (Volume 2)

A culmination of selected research papers from the International conference on Geographic Perspectives on Climate Change Mitigation in Urban and Rural Environment—1st Edition—University of Salento, Italy

Springer

Editors
Shahfahad
Department of Geography
Institute of Science
Banaras Hindu University
Varanasi, Uttar Pradesh, India

Pir Mohammad
Department of Land Surveying
and Geo-Informatics
The Hong Kong Polytechnic University
Hung Hom, Hong Kong

Abraham R. Matamanda
Department of Urban and Regional Planning
University of the Free State
Bloemfontein, South Africa

Jesús Rodrigo-Comino
Department of Regional Geographical Analysis
and Physical Geography
University Campus of Cartuja
Granada, Spain

ISSN 2522-8714 ISSN 2522-8722 (electronic)
Advances in Science, Technology & Innovation
IEREK Interdisciplinary Series for Sustainable Development
ISBN 978-3-031-92114-8 ISBN 978-3-031-92115-5 (eBook)
https://doi.org/10.1007/978-3-031-92115-5

This Springer imprint is published by the registered company Springer Nature Switzerland AG
The registered company address is: Gewerbestrasse 11, 6330 Cham, Switzerland

Scientific Committee

Nitin Bisht, Shri Mata Vaishno Devi University, Jammu and Kashmir, India
Saraah Imran, Indian Institute of Technology Roorkee, India
Johannes Bhanye, University of Cape Town, African Climate and Development Initiative (ACDI)
Maomao Zhang, Huazhong University of Science and Technology, China
Swapan Talukdar, Asutosh College, Kolkata, India
Mohd Rihan, Jamia Millia Islamia, India
Md Nawaj Sarif, Rashtriya Raksha University, Lucknow, India
Gomal Amin, The Hong Kong Polytechnic University, Hong Kong
Aqil Tariq, Mississippi State University, USA
Naeem Shahzad, The Hong Kong Polytechnic University, Hong Kong
Siqi Jia, The Hong Kong Polytechnic University, Hong Kong
Farhan Khan, Rungta college of Engineering and Technology, India
Retsepile C. Kalaoane, University of Johannesburg, South Africa
Darya Gachkar, Universidad De Sevilla, Spain
Saraah Imran, Indian Institute of Technology, Roorkee, India
Gomal Amin, The Hong Kong Polytechnic University, Hong Kong
Bhumika Das, Lakhmi Chand Institute of Technology, India
Charles Mukumba, University of the Free State, South Africa
Natal Buthelezi, University of the Free State, South Africa
Salvatore G.P. Virdis, Asian Institute of Technology, Thailand
Thi Phuoc Lai Nguyen, Asian Institute of Technology, Thailand
Md Sarfaraz Asgher, University of Jammu, India
Intejar Ansari, Faculty of Sciences, Jamia Millia Islamia. New Delhi, India
Md Rejaul Islam, Faculty of Sciences, Jamia Millia Islamia. New Delhi, India
Mohd Waseem Naikoo, University of Kashmir, India
Md Babor Ali, Aligarh Muslim University, India
Pankaj Singha, IIT Delhi, India
Swapan Talukdar, Asutosh College, University of Calcutta, Kolkata, India
Ishita Afreen Ahmed, IIT Delhi, India
Md Masroor, Centre for Distance and Online Education (CDOE), Jamia Millia Islamia, India
Indrajit Roy Chowdhury, University of North Bengal, Darjeeling, West Bengal, India

The Editors warmly thank all the Reviewers who have contributed their authority to the double-blind review process, to ensure the quality of this publication.

Preface

The impact of climate change is pervasive and far-reaching manifesting in diverse forms that differ significantly between urban and rural environments. Climate change is not merely an environmental concern; it represents a comprehensive challenge and impact that reverberates through every aspect of human existence. In the rapid urbanization world that exacerbates this issue, there is an urgent need for developing innovative solutions and mitigation strategies that can enhance resilience in both urban and rural environments and promote sustainable development.

This book, *Geographic Approaches to Climate Change and Mitigation: Urban and Rural Perspectives (Volume 2)*, is the result of selected papers presented at the 1st International Conference on Geographic Perspectives on Climate Change Mitigation in Urban and Rural Environments (GCUE 2024), held from June 25 to 27, 2024, at the University of Salento, Italy. The conference reflects a shared commitment to exploring these diverse impacts and emphasizes innovative strategies that can be applied across various settings. Our goal is to bridge the gap between urban and rural approaches, fostering an understanding of how climate solutions can be tailored to the unique characteristics of each environment. This second volume of the book is divided into eight different topics, consisting of 27 chapters. Each chapter addresses distinct yet interconnected aspects of climate change mitigation, contributing to a comprehensive understanding of the subject.

In this book, various aspects of climate change mitigation in urban and rural environments are discussed that offer valuable knowledge and insights through case studies. It highlights the impactful initiative and the knowledge gained through the process. The exploration of climate change mitigation requires collaboration across various sectors and disciplines. By bridging the gap between urban and rural perspective, we can harness the strength of each environment to develop holistic solutions that pave the way for a sustainable future.

As a reader of this book, whether you are researcher, planner, or public, I encourage you to consider your own role in this crucial issue of urban and rural perspective and think how you can contribute for the better sustainable development. Your role, thinking, and actions play a vital role in shaping our future planets. Finally, I want to extend my heartfelt thanks to all the contributors, reviewers, and supporters who have made this book possible. Together, we can chart a course toward a sustainable and resilient future that recognizes the essential contributions of both urban and rural environments in combating climate change.

Varanasi, India — Shahfahad
Hung Hom, Hong Kong — Pir Mohammad
Bloemfontein, South Africa — Abraham R. Matamanda
Granada, Spain — Jesús Rodrigo-Comino

Acknowledgments

The completion of *Geographic Approaches to Climate Change and Mitigation: Urban and Rural Perspectives-(Volume 2)* is the result of a collaborative effort involving numerous individuals and organizations whose contributions have been crucial. We extend our heartfelt gratitude to the contributing authors; your expertise and insights have enriched this volume significantly. The dedication you have shown to advancing our understanding of climate change mitigation across various contexts is evident in every chapter. We truly appreciate the time, effort, and passion you have devoted to your research and writing.

We are also deeply thankful to the reviewers for their constructive feedback and critical evaluations, which have greatly improved the quality of this book. Your meticulous attention to detail and commitment to excellence have made the content both rigorous and accessible. Our appreciation goes to IEREK for organizing the conference in collaboration with the University of Salento, Italy, providing a platform for researchers worldwide to convene and discuss the challenges of urban and rural environments in relation to climate change. We recognize the efforts of the hosts and co-hosts of the conference and thank all supporting organizations for their contributions.

A special acknowledgment is due to our editorial team and publishing staff, whose professionalism and guidance have been invaluable in bringing this book to life. Your expertise in navigating the publishing process has been immensely appreciated. We also want to thank our families and friends for their understanding and encouragement during the long hours dedicated to this project; your support has been a source of motivation.

Lastly, we wish to recognize the urban and rural communities that are at the forefront of climate change impacts. Your resilience, creativity, and commitment to sustainability inspire us all. This book is dedicated to you and to those who are working tirelessly to create a more equitable and sustainable future for our planet.

Introduction

As the world faces the multifaceted challenges posed by climate change and global warming, the need for sustainable solutions for better climate conditions has never been more urgent. Its impacts are widespread from land to ocean, witnessing an increase frequency of extreme weather events, agricultural production, uneven distribution of precipitation, and extreme flooding affecting millions of people worldwide. These profound and far-reaching reamplifications call for action to tackle the climate change issue and provide sustainable innovative solutions.

This book *Geographic Approaches to Climate Change and Mitigation: Urban and Rural Perspectives (Volume 2)* brings together a diverse range of topics that explore the interplay between environmental sustainability, governmental policy, urban and rural planning, and the socioeconomic impacts of climate phenomena.

Urban environments hosting more than half of the world's population will face adverse impacts of future climate change, facing unique challenges and opportunities in the quest for sustainability. Cities play a pivotal role in global response to climate change, characterized by high population density and extensive infrastructure making it more vulnerable. As a significant contribution to greenhouse gas emissions, urban centers often serve as hot spots for climate impacts. Understanding the role of cities in the global climate will be crucial for formulating mitigation and adaptation strategies. The urban planners, stakeholders, and policymakers should work together for better sustainable urban planning and development. The need for sustainable governmental policies and leadership is essential in urban planning.

The livelihood of rural communities is closely related to the naturel resources like agricultural, forestry etc., which makes them most vulnerable to climate change fluctuations. The rural area, despite their strategic significant importance in terms of living and climate condition, often receives less attention as compared to urban area in climate discussions. However, rural areas play a significant role in climate change mitigation through sustainable land management practices. With the help of sustainable reforestation, agroecology, and crop farming, the rural areas contribute to carbon sequestration and biodiversity preservation. Understanding the unique challenges of each rural area and communities is essential for the development of effective climate strategies.

This book brings together a diverse range of topics that explore the relationship between the environmental sustainability, urban and rural planning, government policy development, and the socioeconomic injustice of climate phenomena among urban and rural environments. The book volume is divided into eight different topics. The first topic of the book, Sustainable Governmental Policy and Leadership, emphasizes the essential role of different urban policies in addressing climate changes issues and promoting sustainable development. It highlights the need of visionary leadership and policies to prioritize sustainable development targets at all government levels. The second topic, Urban Planning and Climate Change Mitigation, examines the effectiveness and challenges in reducing carbon emissions in urban areas for better quality of life for urban people. Complementing this is the third topic on Rural Land Use and Climate Change Mitigation, which highlights the importance and strategies of sustainable agriculture practices and proper land management to increase rural resilience to climate change. The fourth topic is the Food Crisis in Climate Change Phenomena, where the pressing issue of food security in the context of climate change is discussed with the aim of providing solutions to ensure the future generations have access to adequate nutrition. Moreover, this book also explores the Heritage Conservation with Climate Change Hazards in fifth topic, focusing on the importance of preserving ancient cultural heritage sites in the face of environmental degradation and highlights the importance of adaptive strategies to protect these sites, while acknowledging their significance to community identity. In the sixth topic, Sustainable Transportation and Transit Systems, the innovative approach to reduce carbon emissions is presented and provides justification for its vital need for building resilient urban infrastructures. Additionally, the seventh topic, Climatic Impacts and Response Methods in Rural Environments provides insight into how rural areas can adapt to climate change through community-driven initiatives and resilience-building measures. Lastly, in eight topics, Environmental Impacts on Human Geography and Culture, the impact of climate change influences on cultural practices, population dynamics, and social structures are explored.

This book provides critical insights into effective climate change initiatives across diverse geographic settings in various urban and rural environments from green space in urban areas to sustainable agriculture practices in rural areas. In includes theoretical frameworks to contextualize these studies with the help of various case studies aim to foster innovative thinking and collaboration efforts by highlighting the distinct challenges and opportunities in urban and rural environment. The goal of this book is to build resilient communities that can confront climate change while advancing sustainable development for both urban and rural environments.

Pir Mohammed

Contents

Human Geography, Culture, and Education

Integrating Computational Architecture in Sustainable Schools Design: A Case Study of Al-Mreigha Mixed Basic School

Zaid M. Al-Zrigat and Majed Abu-Zreig

Abstract

Sustainability is a central focus across various fields and is recognized as a key driver for achieving sustainable development. Education, in particular, plays a vital role in this effort, as many sustainable development initiatives emphasize its importance. This research aims to promote innovative strategies for sustainable school design, specifically targeting the environment of Al-Mreigha Mixed Basic School in Ma'an, Jordan, through the lens of computational architecture. The study is divided into two main areas of focus. First, it identifies and addresses concerns related to the comfort of school occupants—thermal, visual, and acoustic comfort—using both questionnaires and interviews. Second, it evaluates the impact of proposed sustainability measures on the school's operational efficiency through the use of simulation software, such as Design Builder, as a computational architectural tool. The research explores four design alternatives: upgrading wall insulation, adding highly efficient ceiling insulation, modifying glazing types, and implementing shading devices. Simulations reveal varying effects on temperature regulation, heat gains, and energy consumption, highlighting how computational approaches can optimize design decisions. The discussion emphasizes the importance of optimal insulation, shading solutions, and energy-efficient lighting, showcasing how computational architecture can facilitate sustainable design outcomes.

Z. M. Al-Zrigat (✉)
Department of Architecture, Faculty of Engineering, Al al-Bayt University, Mafraq, Jordan
e-mail: zadez6040@yahoo.com

M. Abu-Zreig
Department of Civil Engineering, Water Diplomacy Center, Jordan University of Science and Technology, Irbid, Jordan

Keywords

Building efficiency · Design-builder software · Educational institutions · Thermal well-being · Visual satisfaction

1 Introduction

The concept of achieving green schools is widely debated both internationally and locally, with a central focus on how schools meet the basic needs of their users (students, teachers, and staff), while enhancing living standards in alignment with cultural, social, economic, and environmental contexts. As noted by Gough (2006), finding school design solutions that consider social, economic, and demographic factors is crucial, particularly when the target group represents a significant portion of the population.

Given Jordan's economic reality as a middle-income country, challenges such as limited resources and high educational costs necessitate innovative approaches, including computational architecture, to improve sustainability in school design. Although a building system law was introduced in 2018, gaps remain in local legislation concerning school design, leading to inadequate national guidance for this sector. These challenges underscore the urgent need to enhance the quality and performance of local schools. Water and energy scarcity further highlight the importance of creating sustainable indoor environments that maximize students' physiological and physical comfort. This research aims to explore sustainable design strategies for schools through computational architecture, focusing on land, water, air, and natural energy resources. Additionally, it seeks to provide a comprehensive evaluation of the effectiveness of sustainable educational facilities, particularly concerning the unique challenges posed by Jordan's resources and economic conditions.

P. Mohammad et al. (eds.), *Geographic Approaches to Climate Change and Mitigation: Urban and Rural Perspectives (Volume 2)*, Advances in Science, Technology & Innovation, https://doi.org/10.1007/978-3-031-92115-5_1

A key issue in Jordan's education sector is the ongoing challenge of limited resources, raising concerns about the quality and sustainability of educational environments. Currently, there is a lack of universally recognized metrics to assess the sustainability of educational structures, particularly in schools. Therefore, developing appropriate strategies that incorporate computational architecture to improve the indoor environment of schools is crucial, focusing on the impact these environments have on students' health and well-being. To achieve maximum comfort, it is essential to adopt sustainable construction standards that reduce energy and water consumption while enhancing the overall learning environment.

School buildings are vital in creating learning environments that meet the necessary technical standards for student spaces. Simultaneously, they must contribute to improving the overall school environment by incorporating sustainable design strategies and standards, supported by computational methods.

This dual focus ensures that schools not only provide functional and comfortable spaces but also promote environmental responsibility and long-term resource efficiency.

1.1 Research Justification

The research challenge arises from the limited implementation of sustainability standards and regulations for buildings in Jordan, particularly in educational institutions. This study aims to assess issues related to occupant comfort—thermal, visual, and acoustic—and to gauge the satisfaction levels of students utilizing educational facilities. The ultimate goal is to foster a safe and conducive environment that supports student achievement. Furthermore, the research seeks to evaluate the impact of proposed sustainability strategies on building performance through the integration of computational architecture. By emphasizing the necessary standards for designing a sustainable school, this study specifically focuses on Al-Mreigha Mixed School, aiming to create a model for enhancing educational environments across Jordan.

1.2 Research Question

1. How might we enhance the operational efficiency of educational settings, particularly schools, through computational design approaches that address sustainability concerns and the holistic well-being of students, encompassing both physical and mental health aspects?
   1.1. How can computational design tools be utilized to promote thermal comfort in local Jordanian schools?
   1.2. In what ways can computational design enhance visual comfort in local Jordanian schools?
   1.3. How can computational design strategies contribute to the reduction of energy consumption in schools?

1.3 Research Background and Context

The Al-Mreigha area, located south of Ma'an city in Jordan, faces significant sustainability challenges, including weak infrastructure, inadequate health care, and limited transportation. Within this region, Al-Mreigha Mixed Basic School was established in 2011 with royal endorsement. However, structural issues, such as chronic cracks and fissures, led to the school's closure by the Ministry of Education in 2022, displacing 430 students (Omar 2023). Addressing these challenges is critical to ensuring a safe learning environment that promotes student growth and emphasizes the role of education in sustainable development.

Serving a diverse student population in Ma'an's arid climate—characterized by hot summers and mild winters—the school plays a key role in the local community. Ma'an, an important administrative and economic hub in southern Jordan, is home to a culturally diverse population with varying socioeconomic backgrounds. The school provides essential primary and possibly lower secondary education for both male and female students of different ages, acting as a foundation for the community's educational development.

Additionally, Al-Mreigha is rich in historical and cultural heritage, with roots dating back to ancient civilizations such as the Nabateans, Romans, and Byzantines. The region's desert landscape, marked by deep valleys and rocky plateaus, includes notable landmarks like Mreigha Castle, a medieval Islamic structure that underscores the area's historical significance. Mreigha also serves as a center for traditional Bedouin life, with agriculture and animal husbandry forming the backbone of the local economy (Saidu and Yeom 2020).

2 Literature Review

2.1 Sustainable Buildings

Sustainable building fosters an interactive relationship between users and the structure, where the key criterion for sustainability is enhancing the social, economic, and cultural environment. This is accomplished by optimizing the internal environment's efficiency while minimizing damage to natural resources and energy consumption.

Effective strategies include leveraging the original building design and adapting it to climate change conditions, thereby reducing the environmental impact on users' living spaces (Adamkiewicz et al. 2011). Sustainable construction requires a holistic approach that addresses the entire lifespan of a building, focusing on environmental and functional quality, as well as long-term value.

To achieve this, sustainable building design must integrate architectural, electrical, mechanical, and structural engineering expertise, ensuring that all systems work cohesively to promote energy efficiency, durability, and overall sustainability (Ayeni et al. 2017).

Prudent utilization of natural resources and effective management of building stock are pivotal in conserving resources, minimizing energy consumption, and enhancing environmental quality. Extensive literature indicates five primary objectives attained through sustainable building practices (Ramli et al. 2012), including:

a. Efficient use of resources.
b. Energy efficiency (including mitigating greenhouse gas emissions).
c. Pollution mitigation (covering indoor air quality and noise reduction).
d. Compatibility with the environment.
e. Integrated and green approach.

2.2 Sustainable Schools

Sustainable schools serve as essential tools that meet the needs of environmental education and community participation, while also providing a healthy environment conducive to learning. These schools create greater opportunities for students to engage with concepts of productivity, problem-solving, and critical thinking skills. By integrating these processes, sustainable schools cultivate a culture within the school community that fosters learning, personal growth, and sustainable environmental practices. This holistic approach not only enhances educational outcomes but also empowers students to become active stewards of their environment.

Schools are considered sustainable primarily because students spend a significant portion of their day in the school environment, particularly during crucial stages of physical and mental development. As a result, educational programs play a vital role in integrating physical, cultural, and organizational elements that foster a sense of belonging among students. Establishing a healthy, sustainable learning environment is the first step toward achieving this objective.

A review of extensive literature and studies indicates that society benefits in numerous ways from sustainable schools, including reduced environmental impact, enhanced comfort, improved health, and an overall better quality of life as shown in Table 1.

To effectively implement the sustainable school framework, the Green School associated with the US Green Building Council has outlined a series of standards and attributes aimed at achieving sustainability objectives. These encompass energy conservation, enhancement of indoor air quality, elimination of hazardous materials from learning and recreational areas, adoption of daylighting strategies, enhancement of classroom acoustics, and wastewater recycling. Furthermore, considerations include selecting school sites near public transportation to minimize pollution, situating buildings on green sites to mitigate environmental impact, and opting for renewable materials conducive to easy recycling (Rodriguez et al. 2019).

Table 1 Illustrates the advantages of implementing sustainability strategies and standards in school educational programs

No.	Students	Teachers	Schools
1	Implementing sustainability standards increases student concentration and happiness due to environmental education	The energy efficiency of a sustainable school building is 35% higher than that of a traditional school	Sustainable building design relies on renewable building materials, which reduce the emission of carbon dioxide gases
2	The sustainable school relies on natural light sources, which has a positive impact on students' performance in exams	Sustainable schools are characterized by their ability to save operating costs	The life span of green roofs is longer than traditional roofs, which has a positive impact in reducing the amount of waste generated
3	The literature review confirmed the ability of students to participate effectively and equally in environmental education regardless of their level of mental abilities	Sustainable schools improve thermal, optical, and acoustic comfort, which has the effect of supporting the health of students and staff	A sustainable school helps convey awareness to future generations regarding environmental issues
4	Student engagement increases in classes located in sustainable environments	The Sustainable School has proven its ability to increase teacher retention	

2.3 Performance of the Schools

2.3.1 Comfort in Thermal Conditions

Thermal comfort is intricately linked to sustainable architecture, representing a fundamental aspect that architects must address through careful considerations such as building orientation, material selection, and the utilization of advanced technological systems. Therefore, designers must analyze the thermal characteristics of each building individually, taking into account factors such as the design of internal spaces, the relationships between mass elements, height ratios, and the size of openings (Bataineh and Ali 2010). Processes to enhance thermal comfort include reducing fossil fuel consumption, optimizing air conditioning and heating systems, and selecting appropriate lighting types. Given that students spend more than 60% of their time in school, sustainable school projects that focus on thermal comfort are of significant interest to all specialists, as they align with the primary goal of ensuring comfort for both students and teachers (Atzeri 2017).

Most studies specialized in thermal comfort in schools indicated users' dissatisfaction with aspects of heat and humidity, according to the study conducted by researchers Bogota, Colombia (2019), which represents a large proportion of green school designs in Latin America. This study recorded the most prominent problems related to thermal comfort in classrooms through post-occupancy evaluation, including lack of insulation of the external walls, and leakage of cold air (Bataineh and Ali 2010).

2.3.2 Comfort in Visual Conditions

Our ability to control visual comfort depends on the amount of light present in any given space, which involves balancing large or small levels of light to achieve psychological comfort in the initial phase. Variations in light levels or sudden changes can lead to fatigue, exhaustion, and a lack of concentration among users. The concept of visual comfort encompasses multiple aspects, including aesthetic efficiency, quality of lighting, the effective use of natural sunlight, views associated with the natural environment, and the absence of glare (Asfour and Al Shurafa 2016).

There are three factors based on which visual comfort is evaluated, including the lighting source (natural or artificial), the possibility of distributing lighting within spaces (degree of intensity, color, and glare), and how the user perceives the light source. Hence, scientists realized the effect of light on the user's mental health, such as sleep, mood, wakefulness, concentration, well-being, and activity (Maachi et al. 2019).

The study conducted by researcher Maachi et al. (2019), evaluating the lighting levels in schools within the framework of the new Algerian school building designated by the Housing Improvement and Development Agency (AADL), revealed the dissatisfaction of students and teachers with the levels of visual comfort due to the inappropriate size and locations of openings, and poor selection of the site is surrounded by tall adjacent buildings that prevent sunlight from entering (Zhang et al. 2021).

Consequently, the degree of daylight incorporation in school design emerges as a significant determinant influencing students' visual comfort, encompassing lighting levels, spatial orientation, window-to-wall ratios, reflective materials on interior surfaces, window shading, and orientation toward the south to optimize lighting levels.

2.3.3 Comfort in Sound Conditions

Acoustics has emerged as a significant focus in the field of sustainability, highlighting the importance of sound management within building design. There are two primary approaches to addressing acoustic concerns: sound isolation and sound treatment.

To enhance acoustic comfort, the walls of a building play a critical role as the first line of defense against noise from surrounding environments. Properly designed walls can significantly reduce external noise intrusion, contributing to a quieter and more conducive learning environment for occupants.

Accordingly, there are differences in the properties of soundproof walls in terms of building materials, the design of the wall itself is either reflective or a transmitter of sound to reduce the amount of echo produced. Conversely, the second aspect pertains to sound absorption via windows, as sound passing through them is not reflected (Labia et al. 2020).

Many studies have revealed the dangerous effects of unwanted sounds, especially on school students, in terms of poor hearing, headaches, high blood pressure, heart disease, aggression, and feelings of resentment, and frustration. The effect of noise can be reduced by eliminating the causative source, using electrical appliances that do not emit sound, and increasing the diameters of the air ducts in the installations to reduce the air velocity. The designer's knowledge of the acoustic properties and their impact on materials contributes to creating better behavior in spaces designed to meet the user's needs and acoustic comfort (Sulaiman and Beithou 2011).

2.4 Material Used in the Schools

External conditions, heating methods, materials used for thermal insulation, and natural ventilation are essentially the keys to controlling the internal environment of schools. The high cost of sustainable materials in Jordan is the reason why user's orientation to using poor-quality materials in construction operations, which creates unhealthy environments in terms of increased humidity and carbon dioxide consumption (Creswell 2008).

Many schools in Jordan feature exteriors composed of white stone or concrete blocks for walls, windows utilizing non-insulating glass, and insulating materials like rock wool. Hence, designers must select locally available materials that are cost-effective to maintain, durable throughout the building's lifespan, easy to manage, and quick to install, aiming to minimize expenses in low-income nations such as Jordan (Sulaiman and Beithou 2011).

To achieving sustainable building practices requires focusing on enhancing the internal environment of schools that promotes student health. Giving priority to economic, cultural, and social factors achieves the vital interaction between users and the building.

When the designer considers the building's lifespan in interaction with the environmental and functional quality in terms of achieving thermal, visual, and acoustic comfort, it ensures an effective positive impact on the user spaces in the school. Given these considerations, prioritizing resource efficiency and employing judicious material selection practices not only decreases energy consumption but also enhances the environmental quality of the building.

The existing literature provides valuable insights into sustainable school design, but significant gaps remain in proactive design strategies, empirical evaluations of interventions, and consideration of socioeconomic factors in material selection. There is a need for interdisciplinary research efforts to address these gaps and inform effective design practices for sustainable education environments. Specifically, exploring the trade-offs between different design options in terms of cost, feasibility, and long-term performance, such as changes in insulation and glazing types, could enhance decision-making processes in sustainable school design. Further research in these areas is crucial to advancing knowledge and promoting sustainability in educational facilities.

3 Research Methodology

The research relied on combining quantitative and qualitative methods to obtain the largest possible amount of data. The focus was on the basic principles of each of them, in addition to their strengths and weaknesses. After that, the researcher checked the reliability and validity of the results (Allesina et al. 2018) as shown in Fig. 1.

3.1 Project Description

The area targeted in this research is Mreigha, located south of the city of Ma'an. It comprises many residential buildings and only one school, Mreigha, which was constructed by royal decree. The majority of educational buildings in this area utilize locally sourced brick materials and single-pane, non-insulated glass.

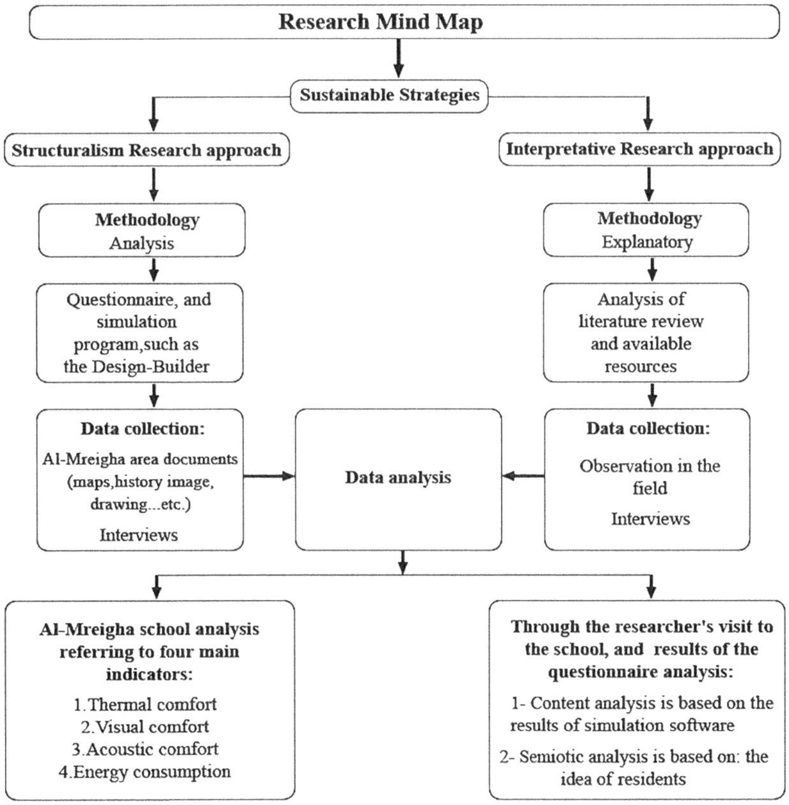

Fig. 1 Research mind map (by authors)

Table 2 Building summary

Title	Description
Location	Base case Al-Mreigha School
Spaces type	(8) classroom type (a), space of each room (8.48 * 5.40 m^2) (3) classroom type (b), space of each room (5.50 * 8.55 m^2) Sewing production room (5.40 * 12.35 m^2) (2) Bathrooms for each bathroom (1.95 * 1.75 m^2) (2) staircases, space of each one (3.80 * 5.20 m^2)
Orientation	The school building is oriented toward the northwest
Use	Educational facilities
Stories	Three typical floors
Building area	878 m^2 (area of one floor)
Structure	Reinforced concrete
Floor Height	3.30 m

Recently, many Jordanian news channels have focused on the problems that the school suffers from and affect the comfort of students and teachers (health, psychological, and thermal). This was reflected in the performance of students and their interaction with the educational environment due to the weakness of the building materials used, which affected the structural system, and contributed to raising the level of humidity and rot. In addition to poor sanitary and wastewater disposal conditions in the bathrooms (Omar 2023). As shown in Table 2.

3.2 Data Collection Procedures Used in This Research Included: (Inquiry Strategy)

1. Building Simulation.
2. Field Observation and Occupants Interview.
3. Questionnaires.

3.2.1 Building Simulation

The research employed the Design Builder simulation program, a methodology that utilizes computer software to analyze, synthesize, formulate, and assess hierarchical design activities. Design Builder integrates rapid three-dimensional building modeling with dynamic energy simulations, offering exceptional user-friendliness. It features specialized modules applicable at various design stages, facilitating the evaluation of facade options regarding overheating, energy consumption, and shading parameters. Additionally, it enables the assessment of optimal daylight utilization, modeling of lighting control systems, and determination of electricity savings rates.

The researcher utilized the Design Builder simulation program to explore various alternatives aimed at enhancing the building's performance regarding thermal comfort, indoor air quality, and energy consumption. Variables were systematically altered in each experiment, with a single change implemented at a time. The calculations will focus on measuring the temperatures, regarding gain, loss, and energy consumption in the school, on the hottest days of heat in the summer from August 8 to August 14, and on the coldest winter days from January 14 to January 25, and measuring the degree of cooling on the highest day on temperatures throughout the year it is July 15 (Table 3).

3.2.2 On-Site Observation and Occupants Interviews

Field observation or surveying plays a crucial role as the initial step in identifying the physical attributes of buildings, which aid in determining variables based on site visualization, building area, utilization of heating, ventilation, and air conditioning systems, as well as activity patterns. In this study, two methods were employed: interviews and questionnaires, serving as self-reported data sources. These techniques, commonly used as alternatives to direct monitoring, enable the assessment of user-perceived comfort (thermal, acoustic, and visual), helping determine when users feel comfortable. Personal interviews further inform the design of key strategies for enhancing energy efficiency. Through these approaches, preliminary data quality is established, providing statistical insights into user satisfaction levels (students and teachers) with the school's interior environment and attainment of thermal comfort standards (Creswell 2008).

- ***Occupant's interview***

The way the conductive interview by dialogue is the target groups and they are the officials (owners of authorities—mayor—project manager—architects), and directing a group of important questions to them based on the literature review. The most important of these questions:

a. Why did they not use sustainable solutions in project design?
b. What are the most prominent challenges that prevented the application of the sustainability rules outlined in the Jordanian codes in the project?
c. Were the user's opinions and discussion of the aspects of sustainable design taken before starting the implementation of the project?

Table 3 Illustrates the physical characteristics of the school's building materials alongside proposed alternatives for enhancing school performance through the utilization of the DesignBuilder simulation program

Thickness/materials	Basic case (Abu Aland)	Alternative (1)	Alternative (2)	Alternative (3)	Alternative (4)
Walls	– (10) cm brick	– (10) cm brick	– (10) cm brick	– (10) cm brick	– (10) cm brick
	– (5) cm vacuum insulation	– (5) cm rock wool insulation	– (5) cm vacuum insulation	– (5) cm vacuum insulation	– (5) cm vacuum insulation
	– (10) cm brick	– (10) cm brick	– (10) cm brick	– (10) cm brick	– (10) cm brick
	– (3) cm plaster/paint	– (3) cm plaster/paint	– (3) cm plaster/paint	– (3) cm plaster/paint	– (3) cm plaster/paint
Roof/floor	– (25) cm concrete reinforcement	– (25) cm concrete reinforcement	– (25) cm concrete reinforcement	– (25) cm concrete reinforcement	– (25) cm concrete reinforcement
	– (10) cm aggregate	– (10) cm aggregate	– (10) cm aggregate	– (10) cm aggregate	– (10) cm aggregate
	– (3) cm cement	Non	– (10) mm phenolic foam insulation	– (3) cm cement	– (3) cm cement
	– (3) cm tiles	– (3) cm cement	– (3) cm cement	– (3) cm tiles	– (3) cm tiles
Window	– (10) cm sliding type glass (aluminum frame surrounded glass)	– (3) cm tiles	– (3) cm tiles	– (10) cm double glass type (aluminum frame surrounded glass)	Adding shedding device
Doors	– (5) cm wooden door (includes door thickness with frame)	– (10) cm sliding type glass (aluminum frame surrounded glass)	– (10) cm sliding type glass (aluminum frame surrounded glass)	– (5) cm wooden door (includes door thickness with frame)	– (5) cm wooden door (includes door thickness with frame)
		– (5) cm wooden door (includes door thickness with frame)	– (5) cm wooden door (includes door thickness with frame)		

Source Authors (2024)

Interviews will be two days a week to meet the largest possible number of architects supervising the project and authorities and to ensure that accurate answers are obtained regarding the application of sustainability standards.

- ***Researcher observation and subjective surveys***

During the school survey, assessments were conducted on acoustic quality, ventilation, temperature, lighting, and humidity levels. The subjective survey questionnaires comprised two sections: a general survey targeting school users (students and teachers) and a second section focused specifically on the research objectives. Detailed contents for each section are available in supplementary documents. Participants were assured that their responses would be treated confidentially and used solely for research purposes. The questionnaires were formulated using straightforward language and terminology, and participants were encouraged to complete them while on school premises. It was estimated that participants would need approximately 10–15 min to provide feedback on their current environment.

The general survey questionnaire addressed three aspects: thermal comfort, visual comfort, and acoustic comfort. Questions were designed with multiple response options based on various Likert scales, while others offered binary choices such as acceptance or refusal. The researcher targeted a sample of (240) people, including (students, teachers, and other workers). observations were done (4) days a week by explaining the questionnaire to (15) people the users distributed and explained the questionnaire to other users, as the questionnaire covers several important questions represented in five main sections, which are basic information, economic information, health information, Thermal, visual and audio comfort, and last part other connections.

The primary emphasis lies on the fourth section, which assesses the satisfaction of school users with thermal, visual, and acoustic comfort. Thermal comfort is gauged by users' contentment with temperature, air movement, and humidity

Fig. 2 The images explain the location of Al-Mreigha School in the south of the city of Ma'an. *Source* Authors (2024)

levels within the school premises. Visual comfort is evaluated based on the intensity of lighting. Sound comfort is determined by identifying the extent and sources of external sounds, including those from adjacent buildings. Furthermore, users were given the opportunity to propose measures aimed at mitigating adverse health impacts within the school environment. Refer to Fig. 1 for illustration. As shown in Fig. 2.

3.3 Analysis of Data

The questionnaire was distributed in paper format, featuring questions and options formulated in straightforward language to ensure clarity across all age groups. Responses were collected over nine evenings, specifically on the 23rd of October, between 8:55 AM and 7:30 PM. Analysis revealed that individuals arriving from outside the school perceived warmer thermal conditions compared to those inside the building. Different areas within the school premises were surveyed, with a total of 57 respondents completing the questionnaire. Among them, 59% were male, and 40% were under 24 years old. A small proportion of respondents (36 out of 40) chose not to answer certain questions, constituting less than 1% of the total responses. Statistical analysis was conducted using Microsoft Excel.

4 Results

4.1 Questionnaire Result

In the survey questionnaire, comfort variables were assessed using Likert scales to gauge the school's suitability as an optimal educational environment. Comfort satisfaction was measured based on responses from users (students and teachers), including:

- ***Thermal Satisfaction***

Based on the questionnaire responses, the findings indicated that 60% of users perceive the school's temperature as moderate, while 20% of students find it comfortable, and 35% feel somewhat comfortable. Additionally, 22% are unsure about the comfort level. Among teachers, 25% observe significant temperature fluctuations throughout the day, while 45% notice slight variations, and 17% report no changes.

Regarding temperature preferences, 14% of students prefer warmer classrooms, while 20% prefer cooler ones. Concerning air quality, 25% of students perceive it as stable, 12% find air movement moderate, and 47% feel slight air movement. Meanwhile, 26% of teachers perceive the classrooms as humid, with 35% preferring reduced humidity levels. Furthermore, 56% of users expressed dissatisfaction with classroom locations due to their south-facing orientation, which exacerbates temperatures during summer.

- ***Visual Satisfaction***

Based on the questionnaire responses, the findings revealed that 27% of users find the lighting intensity in the school very comfortable, while 22% perceive it as somewhat comfortable, and 33% find it slightly annoying. Concerning classroom lighting, 25% of students and teachers find it uncomfortable, with 55% indicating a preference for increased lighting intensity, while 13% prefer the current intensity. Furthermore, 43% of all classroom responses suggest that their location in the south affects illumination intensity.

- ***Acoustic Satisfaction***

Based on the questionnaire responses, the findings indicated sources of noise were categorized into those originating from neighboring buildings and those from outside the school. 31% of students and teachers do not find sounds from outside the school bothersome, while 16% find sounds from neighboring apartments somewhat annoying, and 38% find external sounds extremely disturbing. Additionally, 28% of respondents perceive sounds from the schoolyard as somewhat bothersome, and 32% find sounds from adjacent buildings very annoying.

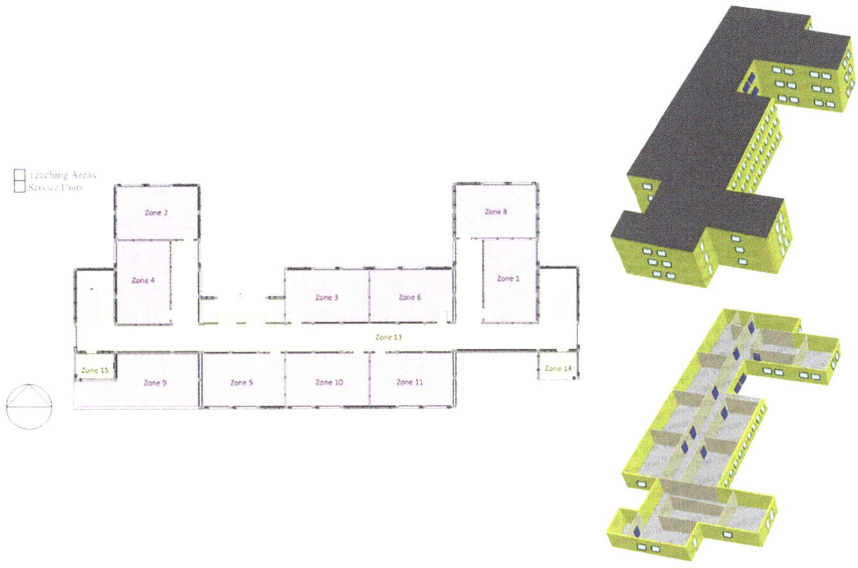

Fig. 3 The building geometry and building materials. *Source* Authors (2024)

4.2 Design-Builder/Energy Plus Program Result

The building simulation began with modeling its structure. Subsequently, construction materials for interior and exterior elements such as walls, floors, roofs, doors, and window glazing were selected based on the architectural drawing. As shown in Fig. 3.

Calculations of the Original School Building, Using the Design-Builder Program

As shown in Fig. 4, the original building's temperature distribution, surface heat transfer, and outdoor environment impact on indoor conditions were examined from August 8th to 14th (summer week). It was noted that indoor temperatures fluctuated in response to outdoor weather conditions. The temperature peaked at 31.30 °C on August 14th and reached a low of 30.07 °C on August 9th.

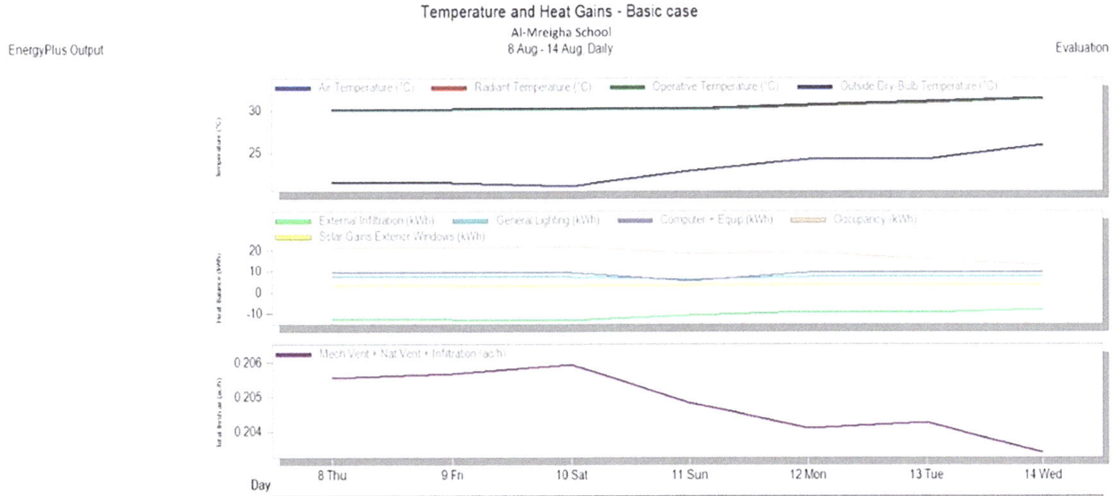

Day	8 Thu	9 Fri	10 Sat	11 Sun	12 Mon	13 Tue	14 Wed
Air Temperature (°C)	30.07	30.16	30.16	30.15	30.65	31.02	31.46
Radiant Temperature (°C)	29.96	30.07	30.07	30.07	30.46	30.88	31.30
Operative Temperature (°C)	30.01	30.11	30.11	30.11	30.56	30.95	31.38
Outside Dry-Bulb Temperature (°C)	21.56	21.46	21.08	22.85	24.21	24.24	25.92
External Infiltration (kWh)	-12.35	-12.61	-13.14	-10.57	-9.29	-9.76	-7.95
General Lighting (kWh)	7.22	7.23	7.24	6.04	7.24	7.25	7.25
Computer + Equip (kWh)	9.37	9.37	9.37	5.49	9.37	9.37	9.37
Occupancy (kWh)	21.24	21.10	21.86	18.45	18.91	15.65	12.77
Solar Gains Exterior Windows (kWh)	3.27	3.23	3.13	3.30	3.31	3.34	3.29
Mech Vent + Nat Vent + Infiltration (ac/h)	0.21	0.21	0.21	0.20	0.20	0.20	0.20

Fig. 4 The accompanying table illustrate temperature distribution, surface heat transfer, and environmental factors (basic case)

4.2.1 In Summer Week from 8 to 14 August

Temperature, Heat Gains, and Energy Consumption-in the city of Ma'an, the base case as shown in Fig. 4.

4.2.2 During the Winter Period, Spanning from January 19th to January 25th

As shown in Fig. 5, temperature distribution, surface heat transfer, and the influence of outdoor conditions on enclosed spaces were analyzed from January 19th to January 25th (winter). It was observed that indoor temperature fluctuated in response to external weather conditions. The temperature ranged from 19.65 °C on January 19th to 20.65 °C on January 25th.

4.2.3 Cooling Design

The heat transfer gain/loss between the interior and exterior walls, roofs, and window glazing of the school building in the base case scenario.

In Summer Week from 15 July-Sub Hourly

As shown in Fig. 6, in the base scenario, temperature distribution and heat transfer gain/loss between the outer and inner walls were analyzed on July 15th (Summer Week). It was noted that the internal temperature fluctuated in response to the external weather conditions of the building, with a recorded temperature of 36.76 °C.

4.2.4 Daylighting

The quantity of daylight in classrooms is notable, with each classroom featuring two windows, as shown in Fig. 7. However, a concern arises regarding the design's longitudinal windows, resulting in limited daylight penetration and heightened reliance on artificial lighting.

Below is the comparison between the original and redesigned school buildings.

4.2.5 The First Alternative: Changing the Type of Wall Insulation to Rock Wool with a Thickness of (5) cm

In Summer Week from 8 to 14 August

The result is in Fig. 8 and compared to the results of the basic case calculations in Fig. 4. It was noted that the internal temperature varies depending on the type of insulation used. The temperature rose to 31.83 on 14 August and reached

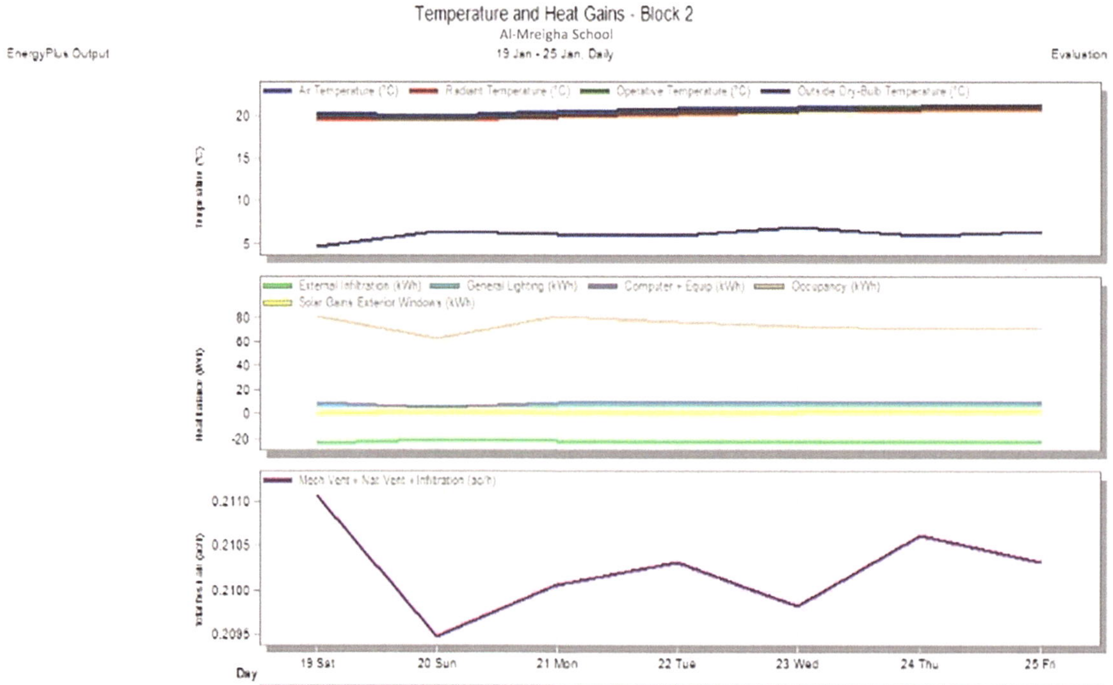

Day	19 Sat	20 Sun	21 Mon	22 Tue	23 Wed	24 Thu	25 Fri
Air Temperature (°C)	20.17	19.86	20.38	20.66	20.91	21.13	21.14
Radiant Temperature (°C)	19.63	19.42	19.77	20.10	20.35	20.65	20.65
Operative Temperature (°C)	19.90	19.64	20.08	20.38	20.63	20.89	20.90
Outside Dry-Bulb Temperature (°C)	4.59	6.45	6.16	6.05	6.95	6.03	6.45
External Infiltration (kWh)	-23.88	-20.48	-21.69	-22.26	-21.30	-23.14	-22.58
General Lighting (kWh)	7.96	6.41	8.01	7.84	7.81	7.79	7.82
Computer + Equip (kWh)	9.37	5.49	9.37	9.37	9.37	9.37	9.37
Occupancy (kWh)	80.87	62.52	80.50	76.22	72.69	70.93	71.06
Solar Gains Exterior Windows (kWh)	0.86	3.03	1.66	2.14	3.22	2.57	3.23
Mech Vent + Nat Vent + Infiltration (ac/h)	0.21	0.21	0.21	0.21	0.21	0.21	0.21

Fig. 5 The accompanying table illustrate temperature distribution, surface heat transfer, and environmental factors (basic case)

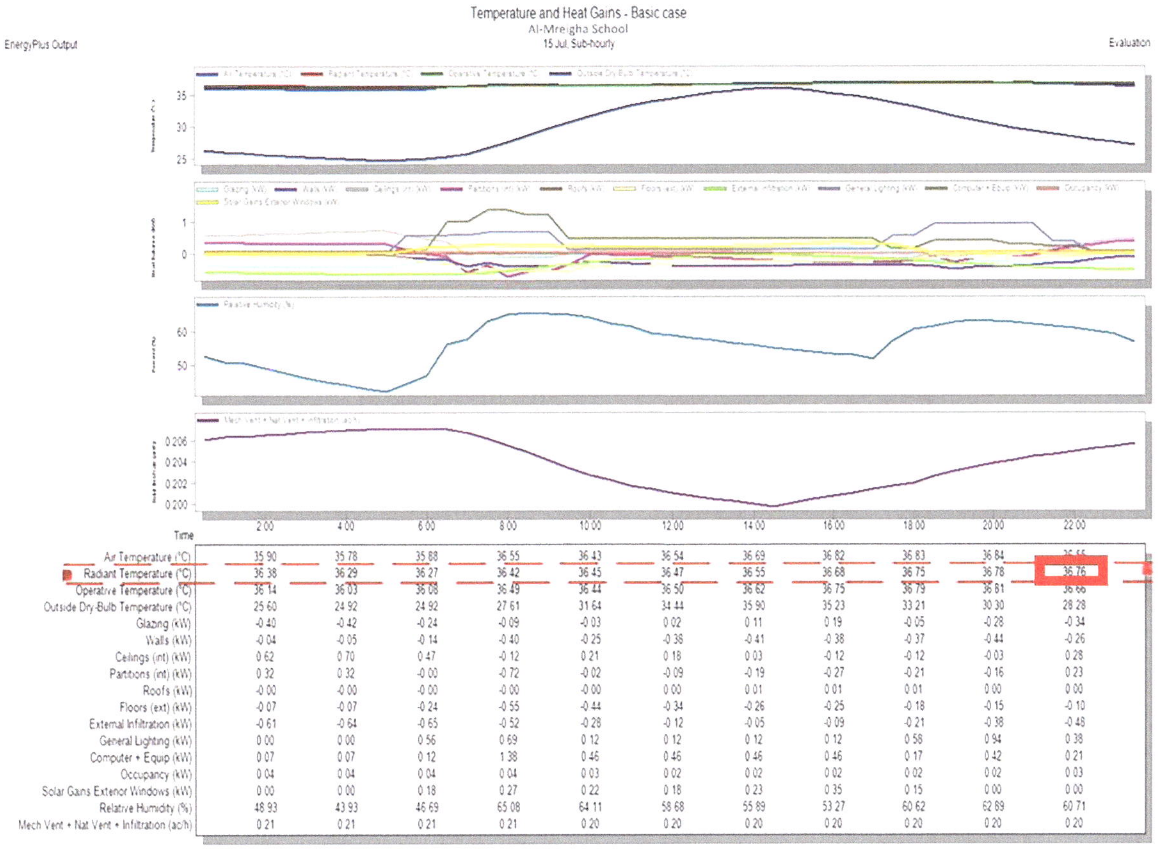

Fig. 6 Cooling design (temperature, heat gain/loss, ventilation)—basic case

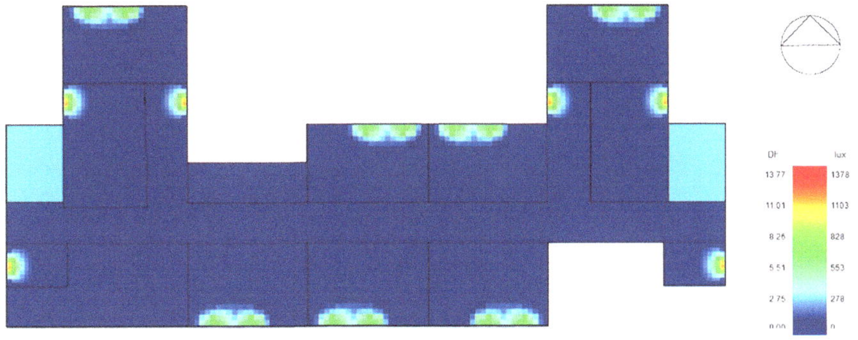

Fig. 7 Daylighting, the school building

30.77 °C on 9 August. It increases the heat gain from the outside and the high summer temperature inside the school.

Temperature, Heat Gains, and Energy consumption re-design school building.

During the Winter Period, Spanning from January 19th to January 25th

The findings are presented in Fig. 9 and are compared with the results of the baseline case calculations depicted in Fig. 5. It was observed that internal temperature fluctuations vary based on the wall insulation type. The temperature peaked at 21.89 °C on January 25th and reached 21.11 °C on January 19th, indicating reduced heat transfer between the interior and exterior.

Cooling Design

The result is in Fig. 10 and is compared to the results of the basic case calculations in Fig. 6. It has been observed that the internal temperature varies with the type and thickness of insulation in the wall. The temperature rises to 37.21 and increases the heat gain from outside, which reduces the cooling rate in the school.

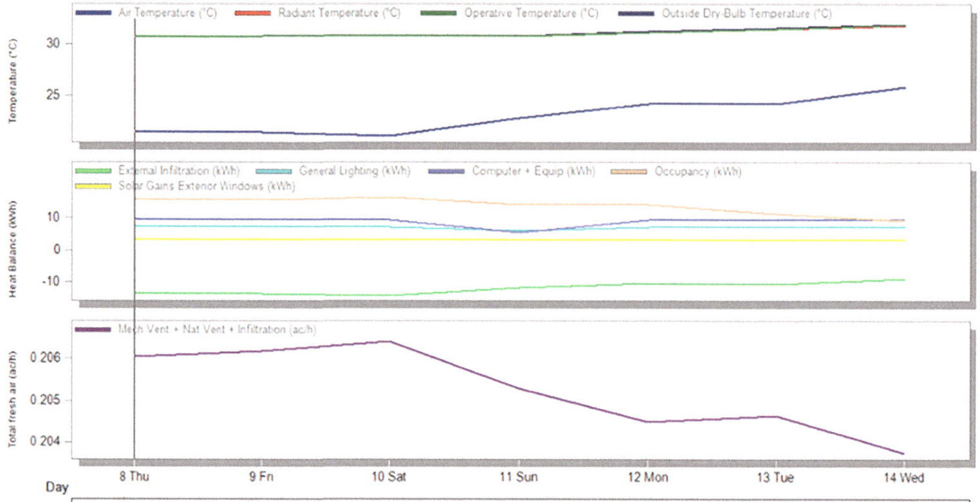

Day	8 Thu	9 Fri	10 Sat	11 Sun	12 Mon	13 Tue	14 Wed
Air Temperature (°C)	30.72	30.78	30.75	30.76	31.18	31.51	[illegible]
Radiant Temperature (°C)	30.70	30.77	30.75	30.75	31.07	31.45	31.83
Operative Temperature (°C)	30.71	30.77	30.75	30.75	31.13	31.48	31.88
Outside Dry-Bulb Temperature (°C)	21.56	21.46	21.08	22.85	24.21	24.24	25.92
External Infiltration (kWh)	-13.43	-13.65	-14.14	-11.59	-10.19	-10.62	-8.75
General Lighting (kWh)	7.12	7.13	7.13	5.94	7.14	7.14	7.14
Computer + Equip (kWh)	9.34	9.34	9.34	5.45	9.34	9.34	9.34
Occupancy (kWh)	15.59	15.71	16.43	14.13	14.10	11.38	9.14
Solar Gains Exterior Windows (kWh)	3.27	3.23	3.13	3.30	3.31	3.34	3.29
Mech Vent + Nat Vent + Infiltration (ac/h)	0.21	0.21	0.21	0.21	0.20	0.20	0.20

Fig. 8 The corresponding table depict temperature distribution, surface heat transfer, and environmental factors in the redesigned building

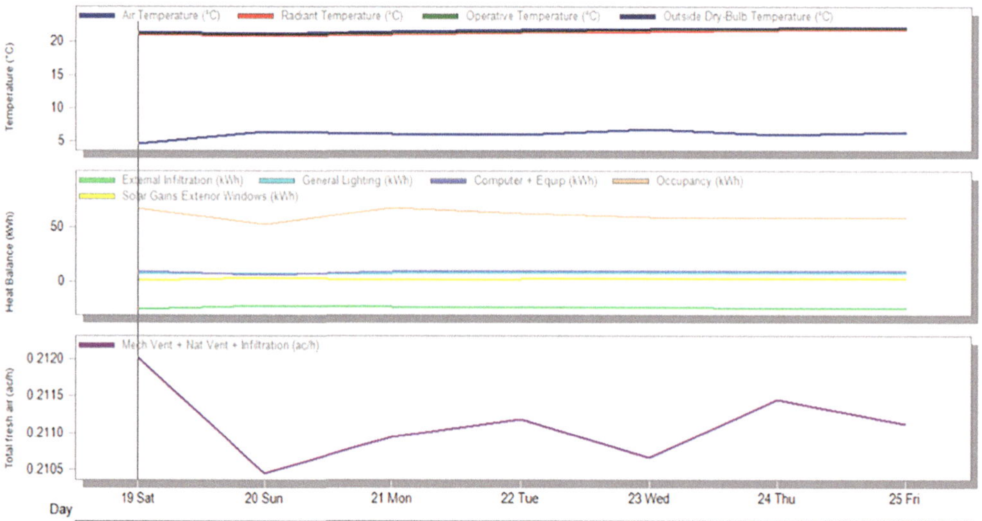

Day	19 Sat	20 Sun	21 Mon	22 Tue	23 Wed	24 Thu	25 Fri
Air Temperature (°C)	[illegible]	21.17	21.56	21.82	22.03	22.20	[illegible]
Radiant Temperature (°C)	21.11	20.91	21.14	21.46	21.67	21.91	21.89
Operative Temperature (°C)	21.28	21.04	21.35	21.64	21.85	22.06	22.04
Outside Dry-Bulb Temperature (°C)	4.59	6.45	6.16	6.05	6.95	6.03	6.45
External Infiltration (kWh)	-25.97	-22.65	-23.63	-24.19	-23.18	-24.96	-24.37
General Lighting (kWh)	7.85	6.31	7.90	7.74	7.70	7.69	7.72
Computer + Equip (kWh)	9.34	5.45	9.34	9.34	9.34	9.34	9.34
Occupancy (kWh)	66.98	52.10	67.72	62.71	59.33	58.10	58.29
Solar Gains Exterior Windows (kWh)	0.86	3.03	1.66	2.14	3.22	2.57	3.23
Mech Vent + Nat Vent + Infiltration (ac/h)	0.21	0.21	0.21	0.21	0.21	0.21	0.21

Fig. 9 The accompanying table display temperature distribution, surface heat transfer, and environmental factors in the redesigned building

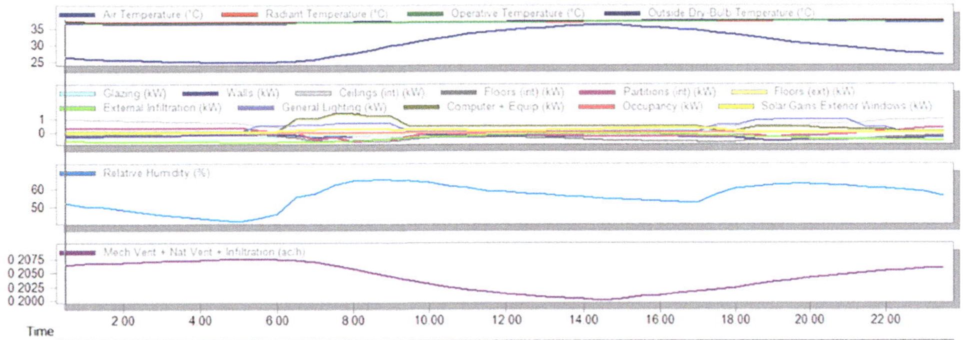

Time	2:00	4:00	6:00	8:00	10:00	12:00	14:00	16:00	18:00	20:00	22:00
Air Temperature (°C)	36.34	36.22	36.29	36.92	36.78	36.89	37.04	37.17	37.18	37.21	[illegible]
Radiant Temperature (°C)	36.84	36.74	36.68	36.77	36.81	36.82	36.90	37.02	37.12	37.19	37.21
Operative Temperature (°C)	36.59	36.48	36.48	36.84	36.79	36.86	36.97	37.09	37.15	37.20	[illegible]
Outside Dry-Bulb Temperature (°C)	25.60	24.92	24.92	27.61	31.64	34.44	35.90	35.23	33.21	30.30	28.28
Glazing (kW)	-0.42	-0.43	-0.26	-0.10	-0.04	0.01	0.10	0.18	-0.06	-0.29	-0.35
Walls (kW)	-0.23	-0.17	-0.16	-0.28	-0.11	-0.21	-0.25	-0.28	-0.37	-0.53	-0.44
Ceilings (int) (kW)	0.89	0.76	0.39	-0.24	0.02	0.09	0.16	0.27	0.45	0.62	0.86
Floors (int) (kW)	-0.10	-0.01	-0.23	-0.64	-0.34	-0.35	-0.46	-0.65	-0.63	-0.58	-0.36
Partitions (int) (kW)	0.33	0.36	0.10	-0.57	0.00	-0.11	-0.23	-0.30	-0.23	-0.20	0.17
Floors (ext) (kW)	-0.00	-0.00	-0.00	-0.00	-0.00	-0.00	-0.00	-0.00	-0.00	-0.00	-0.00
External Infiltration (kW)	-0.64	-0.67	-0.68	-0.55	-0.30	-0.14	-0.07	-0.11	-0.23	-0.41	-0.51
General Lighting (kW)	0.00	0.00	0.55	0.68	0.12	0.12	0.12	0.12	0.56	0.93	0.38
Computer + Equip (kW)	0.07	0.07	0.12	1.38	0.46	0.46	0.46	0.46	0.17	0.42	0.21
Occupancy (kW)	0.03	0.03	0.04	0.03	0.03	0.02	0.02	0.01	0.01	0.02	0.02
Solar Gains Exterior Windows (kW)	0.00	0.00	0.18	0.27	0.22	0.18	0.23	0.35	0.15	0.00	0.00
Relative Humidity (%)	48.52	43.54	45.95	64.93	64.04	58.55	55.78	53.19	60.56	62.67	60.40
Mech Vent + Nat Vent + Infiltration (ac/h)	0.21	0.21	0.21	0.21	0.20	0.20	0.20	0.20	0.20	0.20	0.21

Fig. 10 Cooling design (temperature 'heat gain/loss' ventilation)—re-design building

4.2.6 The Second Alternative: Adding a Type of Insulation in the Ceiling, Type of Waterproof Membrane, Thickness (10) mm

In Summer Week from 8 to 14 Aug

The findings are presented in Fig. 11 and compared with the results of the baseline case calculations shown in Fig. 4. It was observed that internal temperature variations correlate with wall thickness. The temperature peaked at 31.57 °C on August 14th and reached 30.41 °C on August 9th, indicating an increase in heat gain between the interior and exterior.

During the Winter Period, Spanning from January 19th to January 25th

The result is in Fig. 12 and is compared to the results of the basic case calculations in Fig. 5. It has been observed that the internal temperature varies by adding the insulation layer to the ceiling. The temperature rises to 21.71 on 25 January and reaches 20.78 ° C on 19 January. It reduces heat gain from the outside and creates a comfortable atmosphere to live without feeling cold during the winter days.

Cooling Design

The findings are presented in Fig. 13 and compared with the results of the baseline case calculations shown in Fig. 6. It was observed that internal temperature variations correlate with the roof insulation layer. The temperature decreased to 36.62 °C, leading to a reduction in heat gain from outside and an increase in the cooling rate within the school.

4.2.7 The Third Alternative: Change (the Opening) Glazing Type (Double Glazing LoE (e2-.1) Tint (6 mm/13 mm Air)

In Summer week from 8 to 14 Aug

The findings are shown in Fig. 14 and compared with the results of the baseline case calculations shown in Fig. 4. It was noted that internal temperature variations are influenced by the type of glass and emission characteristics.

Daylighting

The result shown Fig. 15 on the type of glazing, and compared to the findings of the base case calculations in Fig. 7 that the daylight distribution and the reference temperature and this type of glass allow light to enter more than the heat transferred through it.

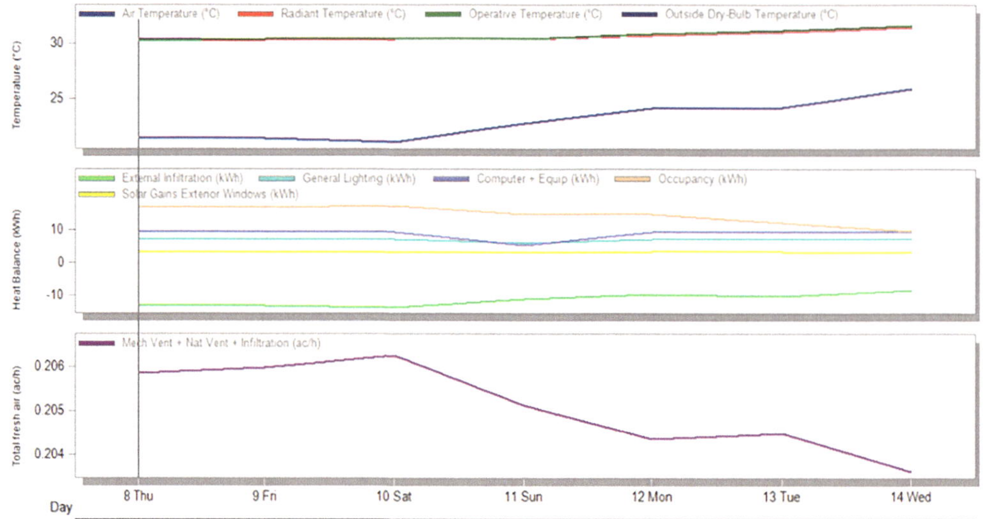

Day	8 Thu	9 Fri	10 Sat	11 Sun	12 Mon	13 Tue	14 Wed
Air Temperature (°C)	30.34	[illegible]	30.45	30.48	30.91	31.25	[illegible]
Radiant Temperature (°C)	30.29	30.41	30.42	30.44	30.79	31.17	31.57
Operative Temperature (°C)	30.31	30.42	30.44	30.46	30.85	31.21	[illegible]
Outside Dry-Bulb Temperature (°C)	21.56	21.46	21.08	22.85	24.21	24.24	25.92
External Infiltration (kWh)	-12.89	-13.17	-13.71	-11.19	-9.81	-10.25	-8.40
General Lighting (kWh)	7.12	7.13	7.13	5.94	7.14	7.14	7.14
Computer + Equip (kWh)	9.34	9.34	9.34	5.45	9.34	9.34	9.34
Occupancy (kWh)	16.91	16.79	17.46	14.90	15.01	12.26	9.79
Solar Gains Exterior Windows (kWh)	3.27	3.23	3.13	3.30	3.31	3.34	3.29
Mech Vent + Nat Vent + Infiltration (ac/h)	0.21	0.21	0.21	0.21	0.20	0.20	0.20

Fig. 11 The accompanying table illustrate temperature distribution, surface heat transfer, and environmental factors in the redesigned building

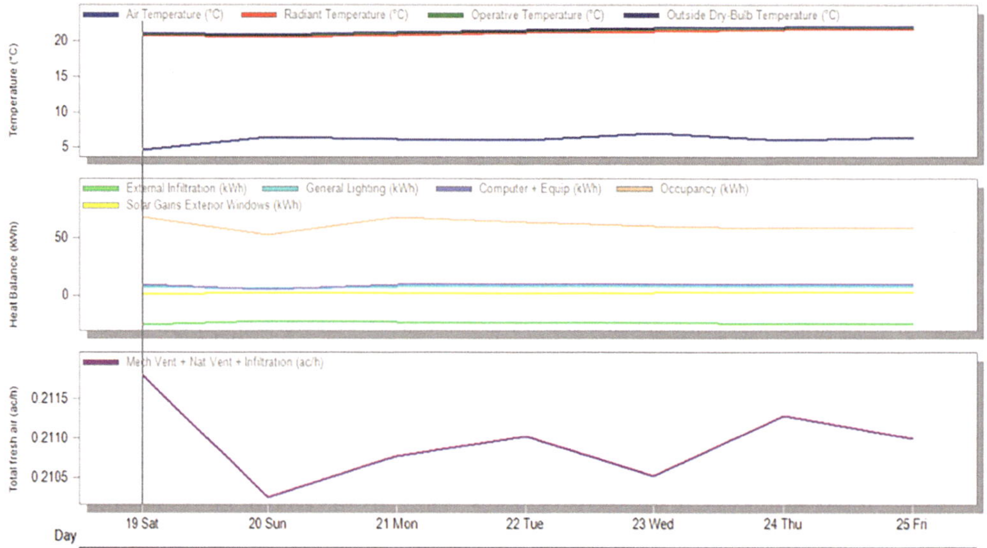

Day	19 Sat	20 Sun	21 Mon	22 Tue	23 Wed	24 Thu	25 Fri
Air Temperature (°C)	[illegible]	20.86	21.29	21.53	21.79	21.97	[illegible]
Radiant Temperature (°C)	20.78	20.61	20.90	21.20	21.44	21.69	21.71
Operative Temperature (°C)	[illegible]	20.74	21.10	21.36	21.61	21.83	[illegible]
Outside Dry-Bulb Temperature (°C)	4.59	6.45	6.16	6.05	6.95	6.03	6.45
External Infiltration (kWh)	-25.44	-22.18	-23.23	-23.75	-22.81	-24.59	-24.07
General Lighting (kWh)	7.85	6.31	7.90	7.74	7.70	7.69	7.72
Computer + Equip (kWh)	9.34	5.45	9.34	9.34	9.34	9.34	9.34
Occupancy (kWh)	68.01	52.93	67.59	63.68	60.00	58.61	58.58
Solar Gains Exterior Windows (kWh)	0.86	3.03	1.66	2.14	3.22	2.57	3.23
Mech Vent + Nat Vent + Infiltration (ac/h)	0.21	0.21	0.21	0.21	0.21	0.21	0.21

Fig. 12 The corresponding table depict temperature distribution, surface heat transfer, and environmental factors in the redesigned building

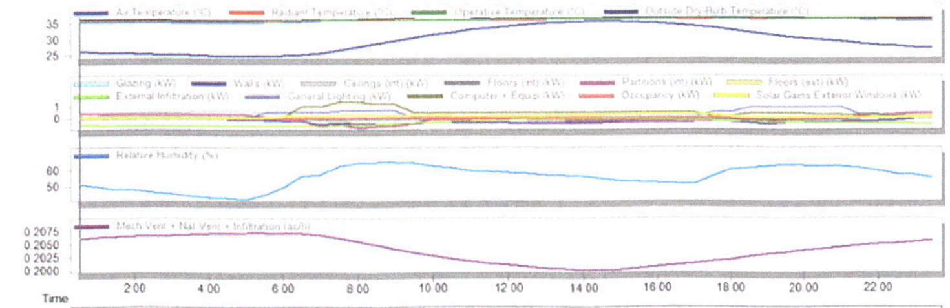

Time	2:00	4:00	6:00	8:00	10:00	12:00	14:00	16:00	18:00	20:00	22:00
Air Temperature (°C)	35.80	35.68	35.84	36.53	36.39	36.52	36.66	36.77	36.74	36.72	36.38
Radiant Temperature (°C)	36.26	36.17	36.19	36.40	36.44	36.44	36.50	36.61	36.65	36.65	36.62
Operative Temperature (°C)	36.03	35.92	36.01	36.46	36.41	36.48	36.58	36.69	36.70	36.68	[illegible]
Outside Dry-Bulb Temperature (°C)	25.60	24.92	24.92	27.61	31.64	34.44	35.90	35.23	33.21	30.30	28.28
Glazing (kW)	-0.40	-0.42	-0.24	-0.09	-0.03	0.02	0.11	0.19	-0.05	-0.28	-0.34
Walls (kW)	-0.01	-0.04	-0.17	-0.45	-0.28	-0.39	-0.40	-0.37	-0.32	-0.38	-0.22
Ceilings (int) (kW)	0.52	0.62	0.35	-0.35	-0.06	-0.04	-0.14	-0.29	-0.29	-0.20	0.12
Floors (int) (kW)	-0.04	-0.03	-0.07	-0.11	-0.07	-0.07	-0.08	-0.10	-0.08	-0.08	-0.05
Partitions (int) (kW)	0.37	0.34	-0.04	-0.80	-0.07	-0.11	-0.20	0.26	0.16	-0.10	0.30
Floors (ext) (kW)	-0.00	-0.00	-0.00	-0.00	-0.00	-0.00	0.00	0.00	-0.00	-0.00	-0.00
External Infiltration (kW)	-0.60	-0.64	-0.65	-0.53	-0.28	-0.12	-0.05	-0.09	-0.21	-0.38	-0.48
General Lighting (kW)	0.00	0.00	0.55	0.68	0.12	0.12	0.12	0.12	0.56	0.93	0.38
Computer + Equip (kW)	0.07	0.07	0.12	1.38	0.46	0.46	0.46	0.46	0.17	0.42	0.21
Occupancy (kW)	0.04	0.04	0.04	0.04	0.03	0.03	0.02	0.02	0.02	0.03	0.03
Solar Gains Exterior Windows (kW)	0.00	0.00	0.18	0.27	0.22	0.18	0.23	0.35	0.15	0.00	0.00
Relative Humidity (%)	47.87	43.28	49.36	65.09	63.09	58.66	55.95	52.45	60.49	63.09	59.73
Mech Vent + Nat Vent + Infiltration (ac/h)	0.21	0.21	0.21	0.21	0.20	0.20	0.20	0.20	0.20	0.20	0.20

Fig. 13 Cooling design (temperature, heat gain/loss, ventilation)—redesign building

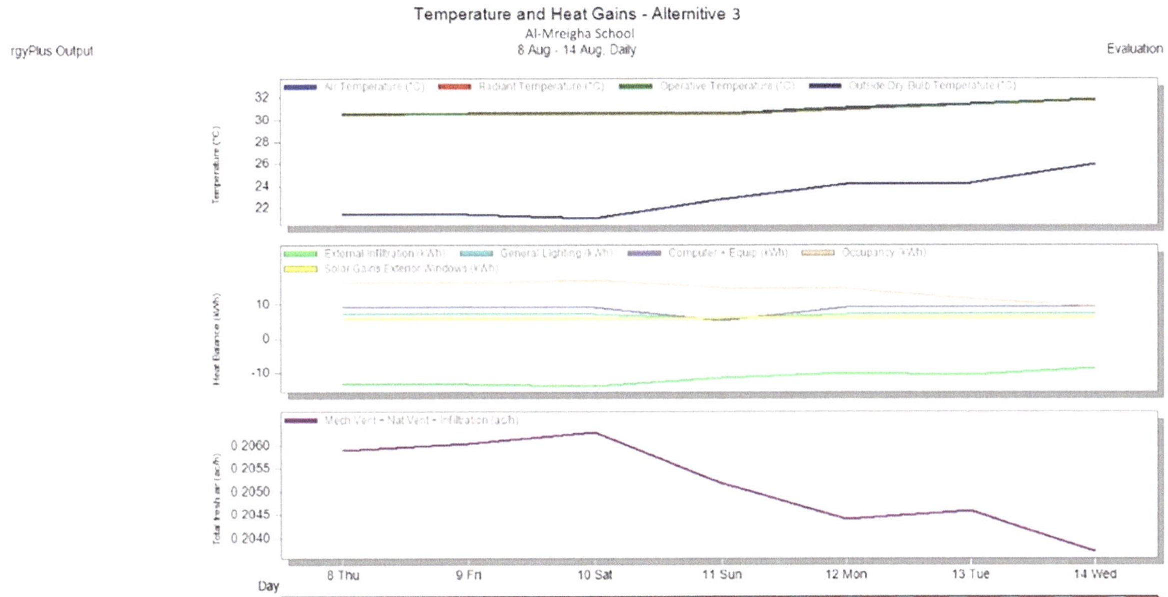

Day	8 Thu	9 Fri	10 Sat	11 Sun	12 Mon	13 Tue	14 Wed
Air Temperature (°C)	30.48	30.57	30.55	30.56	31.02	31.41	31.85
Radiant Temperature (°C)	30.42	30.52	30.51	30.51	30.89	31.32	31.74
Operative Temperature (°C)	30.45	30.54	30.53	30.53	30.96	31.36	31.79
Outside Dry-Bulb Temperature (°C)	21.56	21.46	21.08	22.85	24.21	24.24	25.92
External Infiltration (kWh)	-13.09	-13.35	-13.86	-11.30	-9.97	-10.47	-8.65
General Lighting (kWh)	7.17	7.18	7.18	5.98	7.19	7.20	7.20
Computer + Equip (kWh)	9.34	9.34	9.34	5.45	9.34	9.34	9.34
Occupancy (kWh)	16.67	16.56	17.26	14.99	14.84	11.80	9.44
Solar Gains Exterior Windows (kWh)	5.99	5.92	5.74	6.05	6.06	6.12	6.03
Mech Vent + Nat Vent + Infiltration (ac/h)	0.21	0.21	0.21	0.21	0.20	0.20	0.20

Fig. 14 The accompanying table present temperature distribution, surface heat transfer, and environmental factors in the redesigned building

During the Winter Period, Spanning from January 19th to January 25th

The result is in Fig. 16 and compared to the results of the basic case calculations in Fig. 5. It has been observed that the internal temperature varies with the type of glass (opening) [LoE double glass (e2-.1), light color (6 mm/13 mm air)]. The temperature rises to 20.64 °C on 19 January and reaches 21.63 °C on 25 January and reduces the gain/loss of heat between inside and outside.

The solar heat gain through exterior windows refers to the portion of solar radiation that enters a window, either by direct transmission or absorption, and is subsequently released as heat inside a school. Higher solar heat gain is beneficial for capturing solar heat during winter, which impacts energy

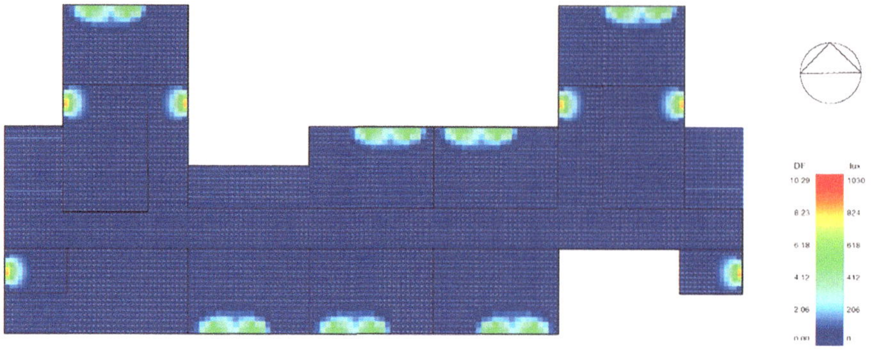

Fig. 15 Explains the temperature distribution and daylighting of the re-design building

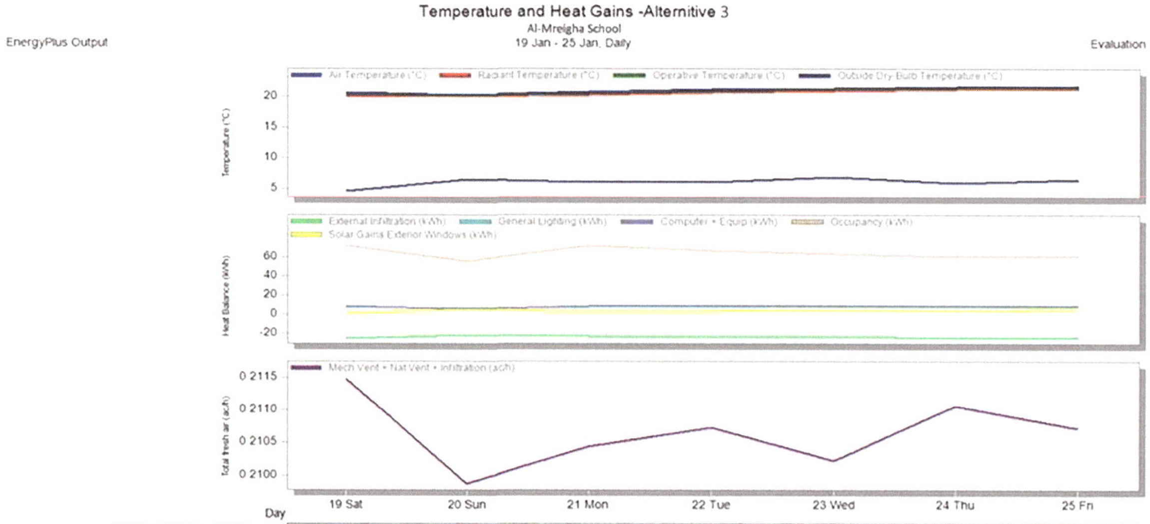

Day	19 Sat	20 Sun	21 Mon	22 Tue	23 Wed	24 Thu	25 Fri
Air Temperature (°C)	20.64	20.33	20.83	21.15	21.39	21.64	21.63
Radiant Temperature (°C)	20.17	19.96	20.30	20.69	20.92	21.25	21.22
Operative Temperature (°C)	20.40	20.15	20.57	20.92	21.16	21.45	21.42
Outside Dry-Bulb Temperature (°C)	4.59	6.45	6.16	6.05	6.95	6.03	6.45
External Infiltration (kWh)	-24.75	-21.38	-22.53	-23.18	-22.21	-24.10	-23.50
General Lighting (kWh)	7.90	6.34	7.96	7.79	7.75	7.74	7.77
Computer + Equip (kWh)	9.34	5.45	9.34	9.34	9.34	9.34	9.34
Occupancy (kWh)	71.90	55.99	72.23	66.97	63.39	61.43	61.97
Solar Gains Exterior Windows (kWh)	1.58	5.60	3.06	3.95	5.95	4.77	5.98
Mech Vent + Nat Vent + Infiltration (ac/h)	0.21	0.21	0.21	0.21	0.21	0.21	0.21

Fig. 16 The corresponding table display temperature distribution, surface heat transfer, and environmental data for the redesigned building

consumption for heating. Conversely, a lower U-factor indicates greater energy efficiency in windows, thereby affecting energy consumption.

Cooling Design

- **In summer week from 15 July-sub hourly**

The result is in Fig. 17 and is compared to the results of the basic case calculations in Fig. 6. It has been observed that the transmission of heat loss through glazing at 2 °C is (0.21) and the air temperature at 2 °C is 37.05°. Thus, the heat gain from the outside decreases, which increases the cooling rate in the school as shown in Fig. 18.

4.2.8 The Fourth Alternative: Adding Shading Device

In Summer Week from 8 to 14 Aug

The findings are shown in Fig. 19 and compared with the results of the baseline case calculations shown in Fig. 4. It was observed that internal temperature variations are influenced by the type of glass and emission characteristics, as shown in Fig. 20.

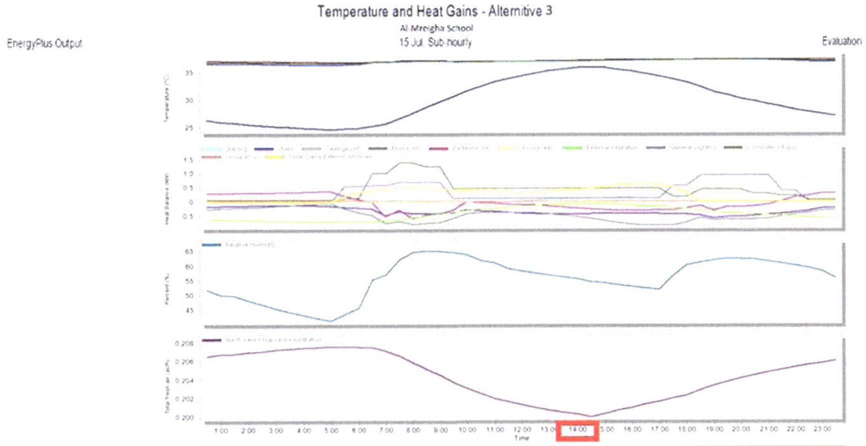

Air Temperature (°C)	36.54	36.42	36.49	37.12	36.97	37.05	37.19	37.32	37.34	37.38	37.14
Radiant Temperature (°C)	37.04	36.93	36.88	36.98	36.99	36.98	37.05	37.18	37.28	37.35	37.39
Operative Temperature (°C)	36.79	36.68	36.69	37.05	36.98	37.02	37.12	37.25	37.31	37.37	37.26
Outside Dry-Bulb Temperature (°C)	25.60	24.92	24.92	27.61	31.64	34.44	35.90	35.23	33.21	30.30	28.28
Glazing (kW)	-0.24	-0.25	-0.04	0.11	0.11	0.12	0.21	0.31	0.04	-0.17	-0.21
Walls (kW)	-0.14	-0.13	-0.21	-0.42	-0.28	-0.40	-0.43	-0.42	-0.43	-0.52	-0.37
Ceilings (int) (kW)	0.77	0.67	0.27	-0.38	-0.07	0.04	0.11	0.18	0.37	0.54	0.76
Floors (int) (kW)	-0.21	-0.10	-0.37	-0.79	-0.41	-0.38	-0.52	-0.80	-0.72	-0.63	-0.45
Partitions (int) (kW)	0.31	0.35	0.08	-0.59	0.01	-0.09	-0.22	-0.31	-0.21	-0.19	0.16
Floors (ext) (kW)	-0.00	-0.00	-0.00	-0.00	-0.00	-0.00	-0.00	-0.00	-0.00	-0.00	-0.00
External Infiltration (kW)	-0.65	-0.68	-0.69	-0.56	-0.31	-0.15	-0.08	-0.12	-0.24	-0.42	-0.52
General Lighting (kW)	0.00	0.00	0.55	0.70	0.12	0.12	0.12	0.12	0.56	0.93	0.38
Computer + Equip (kW)	0.07	0.07	0.12	1.38	0.46	0.46	0.46	0.46	0.17	0.42	0.21
Occupancy (kW)	0.02	0.03	0.03	0.02	0.02	0.02	0.01	0.01	0.01	0.01	0.01
Solar Gains Exterior Windows (kW)	0.00	0.00	0.32	0.50	0.39	0.33	0.43	0.63	0.28	0.00	0.00
Relative Humidity (%)	48.29	43.32	45.73	64.75	63.89	58.45	55.70	53.09	60.44	62.55	60.24
Mech Vent + Nat Vent + Infiltration (ac/h)	0.21	0.21	0.21	0.21	0.20	0.20	0.20	0.20	0.20	0.20	0.21

Fig. 17 Cooling design (temperature, heat gain/loss, ventilation) (redesign building)

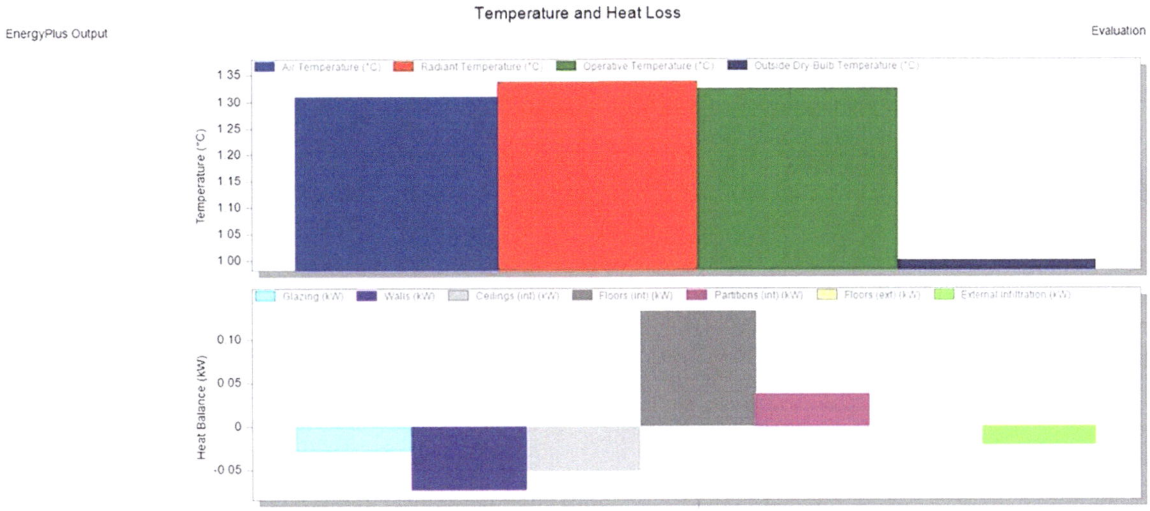

Air Temperature (°C)	1.31
Radiant Temperature (°C)	1.34
Operative Temperature (°C)	1.32
Outside Dry-Bulb Temperature (°C)	1.00
Glazing (kW)	-0.03
Walls (kW)	-0.07
Ceilings (int) (kW)	-0.05
Floors (int) (kW)	0.13
Partitions (int) (kW)	0.04
Floors (ext) (kW)	0.00
External Infiltration (kW)	-0.02

Fig. 18 Daily (temperature, heat gain/loss, ventilation) (redesign building)

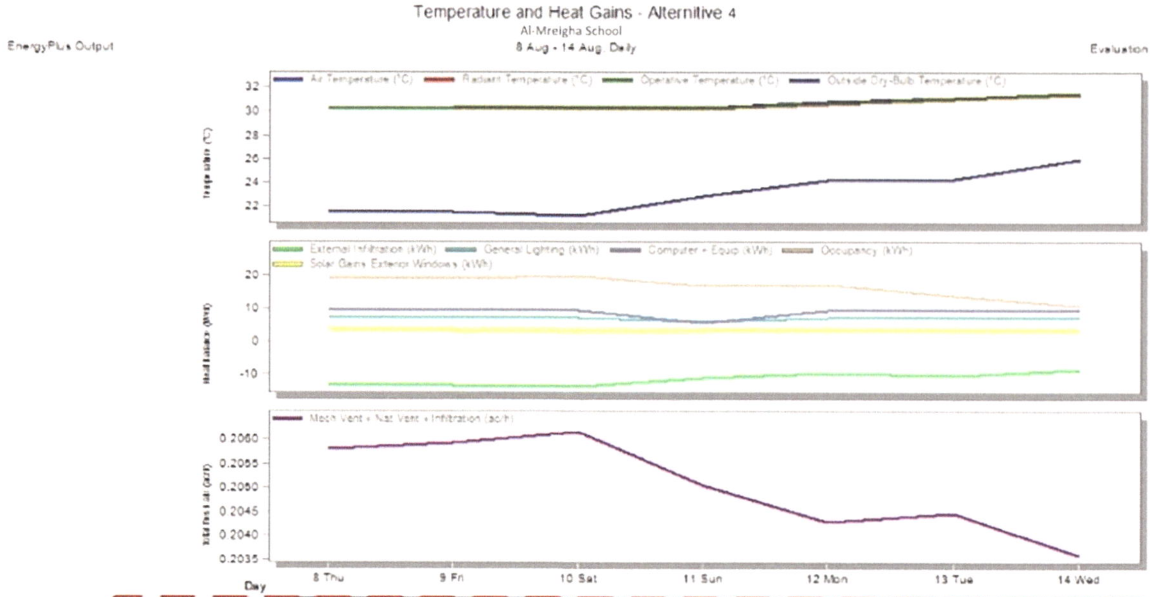

Day	8 Thu	9 Fri	10 Sat	11 Sun	12 Mon	13 Tue	14 Wed
Air Temperature (°C)	30.31	30.39	30.38	30.37	30.85	31.23	31.67
Radiant Temperature (°C)	30.23	30.33	30.31	30.30	30.70	31.13	31.54
Operative Temperature (°C)	30.27	30.36	30.34	30.34	30.77	31.18	31.61
Outside Dry-Bulb Temperature (°C)	21.56	21.46	21.08	22.85	24.21	24.24	25.92
External Infiltration (kWh)	-12.84	-13.09	-13.61	-11.03	-9.73	-10.22	-8.40
General Lighting (kWh)	7.12	7.13	7.13	5.94	7.14	7.14	7.14
Computer + Equip (kWh)	9.34	9.34	9.34	5.45	9.34	9.34	9.34
Occupancy (kWh)	18.89	18.84	19.63	16.56	16.92	13.65	10.95
Solar Gains Exterior Windows (kWh)	3.27	3.23	3.13	3.30	3.31	3.34	3.29
Mech Vent + Nat Vent + Infiltration (ac/h)	0.21	0.21	0.21	0.21	0.20	0.20	0.20

Fig. 19 The accompanying table illustrate temperature distribution, surface heat transfer, and environmental factors in the redesigned building

5 Cooling Design

5.1 In Summer Week from 15 July-Sub Hourly

The result is in Fig. 21 and is compared to the results of the basic case calculations in Fig. 6. It was observed that the transmission of heat loss through glazing at 2 °C is (0.10) and the air temperature at 2 °C is 36.75°.

6 Discussion

6.1 Discussion of the Simulation Software

In the simulation program, the researcher explored four different alternatives: enhancing wall insulation, changing insulation in the ceiling, upgrading window types, and adding shading devices. Table 4 presents a comprehensive overview illustrating the impact of these alternatives on air and radiant temperatures, as well as heat loss and gain through windows and walls, alongside the operative temperature.

In the simulation software, the results were analyzed over the entire year, from January to December, with peak dry temperatures ranging from 8 to 33 °C observed in August. The air temperature also reached 24 °C, aligning with the users' (students, teachers, and other staff) perception of comfortable temperatures as indicated in the questionnaire findings. The total capacity recorded was 130.05 kW, with a total flow rate of 6.76 m^3/s. The total cooling load amounted to 106.03 kW, with sensible power at 92.09 kW and latent power at 16.63 kW over the entire year. The air comfort temperature was found to be 21.18 °C, consistent with the students' perceived comfortable temperatures according to the questionnaire findings. These results are shown in Fig. 21.

The researcher found better findings of air temperature on (8 Aug to 14 Aug) in alternative (3) when we changed the type of glass in windows, in which the air temperature is decreased to 30.63° from the base case. The best findings of radiant temperature on (19 Jan–25 Jan) in Alternative (1)

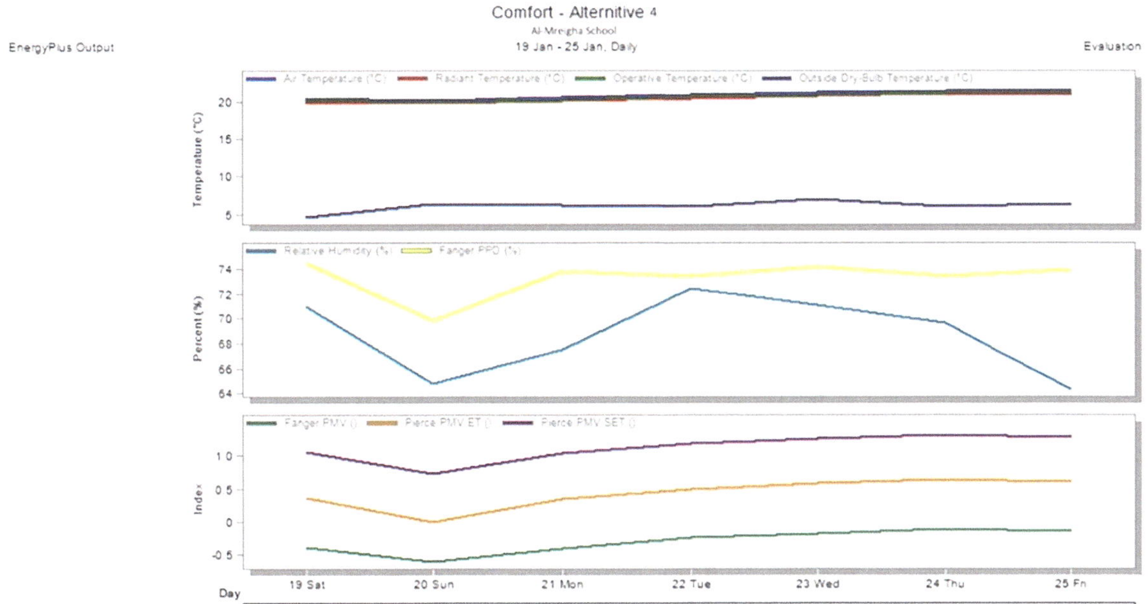

Day	19 Sat	20 Sun	21 Mon	22 Tue	23 Wed	24 Thu	25 Fri
Air Temperature (°C)	20.45	20.13	20.64	20.96	21.20	21.44	21.42
Radiant Temperature (°C)	19.98	19.75	20.10	20.48	20.72	21.04	21.00
Operative Temperature (°C)	20.21	19.94	20.37	20.72	20.96	21.24	21.21
Outside Dry-Bulb Temperature (°C)	4.59	6.45	6.16	6.05	6.95	6.03	6.45
Relative Humidity (%)	70.96	64.84	67.56	72.44	71.09	69.62	64.31
Fanger PPD (%)	74.49	69.87	73.83	73.44	74.18	73.38	73.90
Fanger PMV ()	-0.40	-0.62	-0.41	-0.26	-0.19	-0.12	-0.15
Pierce PMV ET ()	0.36	-0.00	0.33	0.49	0.58	0.64	0.60
Pierce PMV SET ()	1.07	0.72	1.04	1.19	1.26	1.33	1.29

Fig. 20 The corresponding table present temperature distribution, surface heat transfer, and comfort metrics based on the Fanger scale in the redesigned building

when we changed the type of wall insulation to rock wool with a thickness of (5) cm, the researcher noticed that it reached 25.06° from the base case. The best result also of operative temperature on (15 July, at 2 PM) in Alternative (2) when we added a type of insulation in the ceiling, the researcher noticed that it reached 37.02° from the base case.

6.2 Discussion Questionnaire

According to our survey, which involved users including students, teachers, and other staff, a total of 28 questions were included in the questionnaire. The questions varied in format, requiring short responses or selecting from multiple options, and respondents typically completed the questionnaire within 20 min. The survey was conducted on October 2, 2023, with 63 participants. Approximately 84% of respondents answered most questions related to thermal, visual, and acoustic comfort. Through the questionnaire, students and teachers provided suggestions, the most prominent of which included improving insulation in walls, floors, and roofs to enhance energy efficiency, mitigate high temperatures in summer, and utilize more effective insulation for windows. Additionally, there were suggestions to increase sunlight and airflow into interior spaces, particularly classrooms. These suggestions were aligned with the alternatives implemented in the Design-Builders simulation software to optimize user comfort.

6.3 Discussion of the Observation

The researcher's on-site observations at the school played a significant role in the findings and subsequent discussions presented in this study, particularly in correlation with the questionnaire results and simulation software outcomes. For instance, it was noted that in the northern classrooms, sunlight penetration occurs throughout the day due to the absence of shading elements. This observation aligns with the findings of the questionnaire. A potential solution to address this issue is to implement effective shading materials on the classroom windows.

The researcher observed that each classroom in the school is equipped with two glass windows, contributing to noise transmission between spaces. This observation corresponds with the questionnaire results, where students reported experiencing elevated noise levels from adjacent buildings. To mitigate this issue, upgrading these windows to double-glazed types could effectively reduce noise transfer.

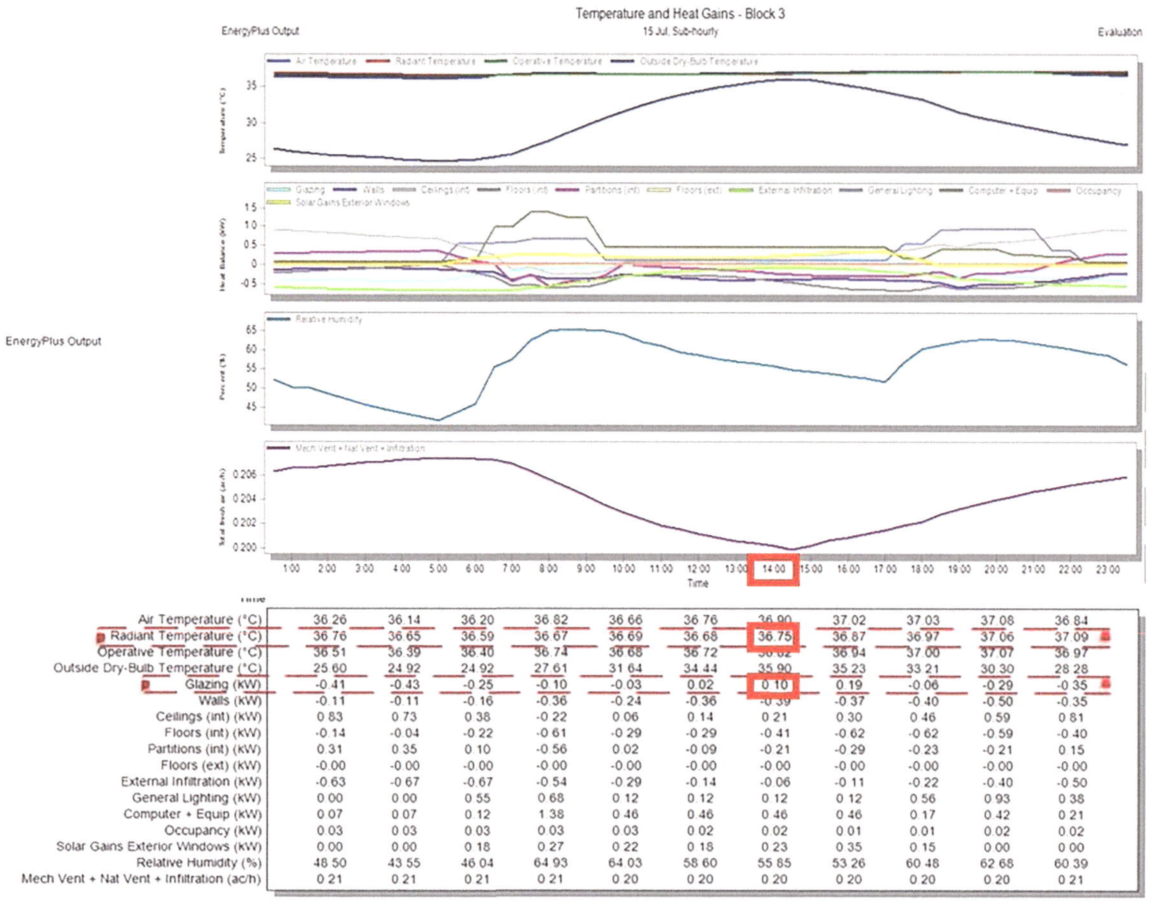

Fig. 21 Cooling design (temperature, heat gain/loss, ventilation) (re-design building)

Table 4 Air temperature, radiant temperature and operative temperature (comparisons between alternatives)

Alternative	Air temperature (°C) (8 Aug–14 Aug, daily)	Radiant temperature (°C) (19 Jan–25 Jan, daily)	Operative temperature (°C) (15 of July, at 2 PM)
Base case	30.5	20.08	36.48
Alternative no. (1)	31.1	21.44	36.97
Alternative no. (2)	30.79	21.19	36.58
Alternative no. (3)	30.92	21.08	37.12
Alternative no. (4)	30.74	20.43	36.9

Additionally, the researcher remarked that the lighting distribution in the school appears adequate for users, though utilizing energy-efficient lamps would further decrease energy consumption within the school premises.

The researcher noticed that the ceiling material significantly influences the cooling load, particularly the waterproof membrane, attributed to its color, texture, design, and energy-saving properties. Simulation results indicate that temperatures in the cooling mode range from 20 to 23 °C, aligning with the ideal temperature reported in the questionnaire responses from users. However, these findings could be further optimized by implementing triple or double-glazed windows.

7 Conclusions

This research employed a diverse set of assessment tools to aid architects, designers, and planners in decision-making processes aimed at enhancing sustainability in educational institutions, particularly schools. The primary objective was to cultivate a healthy environment characterized by ample natural light and ventilation, which are crucial for ensuring user comfort (students, teachers, and other staff). School designs that integrate computational architecture have been shown to foster a sense of belonging among users and positively influence their behavior. The analysis conducted in this research identified sustainability factors contributing to environmental comfort—both physically and psychologically—while proposing alternative strategies to reduce energy consumption in schools. Data collection and analysis involved utilizing questionnaires, interviews, and simulation software (Design Builder), a tool of computational architecture, focusing on evaluating the strengths and weaknesses of existing school designs concerning the implementation of sustainability standards.

Findings from both the questionnaire and interviews revealed dissatisfaction among students and teachers regarding the thermal, visual, and acoustic comfort within the Al-Mreigha school building. Key issues highlighted included high temperatures during summer, cold conditions in winter, inadequate airflow, elevated humidity levels, unacceptable lighting intensity, and disruptive external noises from adjacent buildings. Simulation software results indicated that classrooms facing south require enhanced cooling and sun protection, while those facing north would benefit from additional insulation layers to address discomfort arising from sounds emitted by neighboring buildings, particularly apartments.

Proposed solutions include implementing double or triple-glazed windows in the school building to improve thermal and acoustic comfort, demonstrating how computational architecture can effectively inform sustainable design strategies.

Competing Interests The authors have no conflicts of interest to declare that are relevant to the content of this chapter.

References

Adamkiewicz, G., Zota, A.R., Fabian, M.P., Chahine, T., Julien, R., Spengler, J.D., Levy, J.I.: Moving environmental justice indoors: understanding structural influences on residential exposure patterns in low-income communities. Am. J. Public Health **101**(S1), S238–S245 (2011)

Allesina, G., Mussatti, E., Ferrari, F., Muscio, A.: A calibration methodology for building dynamic models based on data collected through survey and billings. Energy Build. **158**, 406–416 (2018)

Asfour, O.S., Al Shurafa, E.A.Y.: A Study of Daylighting Performance in Apartment Buildings with Reference to the Gaza Strip (2016)

Atzeri, A.M.: Energy Efficiency, Thermal and Visual Comfor Integrated Building Performance Modelling and Measurement. Bozen-Bolzano University Press (2017)

Ayeni, D., Olalusi, C., Tukur, R.: Sustainable site planning and the architects' role. Acad. J. Sci. **7**, 47–55 (2017)

Bataineh, A., Ali, H.: A study of thermal performance analysis of low-income housing in Jordan: case of SOS buildings. Department of Architecture, College of Architecture and Design, Jordan University of Science and Technology, Irbid, Jordan (2010)

Creswell, J.W.: Planning, Conducting and Evaluating Quantitative Ad Quantitative Research. Educational Research. Pearson Education Inc., Upper Saddle River, NJ (2008)

Gough, A.: Sustainable schools in the UN decade of education for sustainable development: Meeting the challenge? Southern Afr. J. Environ. Educ. **23**, 48–63 (2006)

Labia, L., Shtrepi, L., Astolfi, A.: Improved room acoustics quality in meeting rooms: investigation on the optimal configurations of sound-absorptive and sound-diffusive panels. Acoustics 2(3) (2020)

Maachi, I.N., Mokhtari, A., Slimani, M.E.A.: The natural lighting for energy saving and visual comfort in collective housing: a case study in the Algerian building context. J. Build. Eng. **24**, 100760 (2019)

Omar, M.: Al-Mreigha Mixed Basic School in Ma'an suffers from a serious problem. Roya News Channel. https://royanews.tv/news/305585 (2023)

Ramli, N.H., Masri, M.H., Zafrullah, M., Taib, H.M., Abd Hamid, N.: A comparative study of green school guidelines. Proc. Soc. Behav. Sci. **50**, 462–471 (2012)

Rodriguez, C.M., Medina, J.M., Pinzón, A.: Thermal comfort and satisfaction in the context of social housing: case study in Bogotá, Colombia. J. Constr. Dev. Countr. **24**(1), 101–124 (2019)

Saidu, A.I., Yeom, C.: Success criteria evaluation for a sustainable and affordable housing model: a case for improving household welfare in Nigeria cities. Sustainability **12**(2), 656 (2020)

Sulaiman, S.J., Beithou, N.: Residential building walls and environment in Amman, Jordan. Int. J. Therm. Environ. Eng. **3**(2), 101–107 (2011)

Zhang, D., Tenpierik, M., Bluyssen, P.M.: Individual control as a new way to improve classroom acoustics: a simulation-based study. Appl. Acoust. **179**, 108066 (2021)

Mapping Temperature Evolution: A Four-Dimensional Perspective with Smooth Kernel Distribution

Arrigo Bertacchini, Pierpaolo A. Fusaro, and Pietro S. Pantano

Abstract

The rising average temperatures of Earth's surface, seas, and oceans constitute the primary drivers behind the increase in global temperatures, intricately intertwined with the issue of climate change. Global industrialization is among the main causes of the rise in global temperatures, which are constantly increasing. When compared to the beginning of the Industrial Revolution, this increase is approximately 0.98 °C. Projections derived from the analysis of trends spanning from 2000 to the present suggest that a temperature rise of 1.5 °C could be reached by 2030. Within the context of the current global situation, this paper aims to analyze temperature changes through an innovative technique: the smooth kernel distribution (SKD). This spatiotemporal framework provides a method for assessing the probability density function of a continuous stochastic variable, applied across both temporal and spatial dimensions, resulting in a four-dimensional function. This approach is based on constructing a comprehensive dataset that offers insights into the sequential temporal progression and geographical distribution of temperature metrics. The key strength of this method lies in its ability to seamlessly integrate temperature data, geographical coordinates, and the advanced application of the smooth kernel distribution. These elements synergistically produce results that offer greater flexibility for analyzing data across different scales, ranging from local to regional and national levels. This combination makes it possible to capture significant trends and to distinguish between occasional fluctuations and meaningful changes. The resulting maps offer a dynamic visual representation of temperature evolution over time and space.

Keywords

Smooth kernel distribution · Trends · Temperature maps · Climate change · Spatio-temporal analysis

The analysis of climate data can suggest sustainable solutions to mitigate the impacts of climate change. Specifically, the findings of the research can raise public awareness about the urgency of addressing climate change and the importance of adopting measures to protect the environment. This could influence public opinion and promote international cooperation, urging governments to take more decisive actions to tackle the climate crisis.

1 Introduction

Climate change holds a pivotal position in the contemporary world, with human actions and activities inevitably accelerating the rise in temperatures, commonly referred to as global warming. The concept of 'climate change' is not limited to global warming; it encompasses a wider range of transformations occurring on our planet. These include rising sea levels, glacier melt, and the increasing prevalence of mega-fires, (GISTEMP 2023; Lenssen et al. 2019; Moriondo et al. 2006; Lindsey 2021; Gray and Brady 2016; Demarco et al. 2020; Bertacchini et al. 2011; Lee et al. 2021; Marziliano et al. 2024). This issue not only exerts

A. Bertacchini
Department of Architecture and Industrial Design, University of Campania "Luigi Vanvitelli", Aversa, Italy
e-mail: arrigo.bertacchini@unicampania.it

P. A. Fusaro (✉) · P. S. Pantano
Department of Physics, University of Calabria, Rende, Italy
e-mail: pierpaolo.fusaro@unical.it

P. S. Pantano
e-mail: pietro.pantano@unical.it

P. Mohammad et al. (eds.), *Geographic Approaches to Climate Change and Mitigation: Urban and Rural Perspectives (Volume 2)*, Advances in Science, Technology & Innovation, https://doi.org/10.1007/978-3-031-92115-5_2

immediate effects on the environment but also puts the long-term sustainability of Earth at risk, emphasizing a collective responsibility to safeguard our delicate ecosystem. The Intergovernmental Panel on Climate Change (IPCC), the primary international body for assessing climate change, provides a precise definition of climate change. It refers to a change in the state of the climate that can be identified by changes in its average and/or variability of properties, persisting for an extended period typically decades or longer. It encompasses any climate change over time, whether due to natural variability or human activity (Bernstein et al. 2008). Historically, the onset of the First Industrial Revolution marked the beginning of large-scale human activities that have released millions of tons of carbon dioxide and other gases into the atmosphere. The planet's average temperature has increased by 0.98 °C compared to pre-industrial levels, with projections suggesting that this could rise by up to 1.5 °C between 2030 and 2050. NOAA's 2022 global climate report reveals that each month within this timeframe ranked among the warmest recorded for that period (NOAA 2023). Furthermore, according to the National Centers for Environmental Information (NCEI), 2022 ranked as the sixth warmest year since 1880, the beginning of global temperature records (NOOA 2023). Global warming does not follow a uniform pattern across the Earth; rather, the trend shows a general rise in average temperatures, with certain regions experiencing more warming than others. The extent of future warming will predominantly depend on the volume of carbon dioxide and other greenhouse gases emitted into the atmosphere in the coming decades (Lindsey 2021). Addressing this challenge requires a collaborative effort to seek sustainable solutions and mitigate the impacts of climate change. Studies such as those by Lewis et al. (2023) and Perčec Tadić et al. (2023) tend to display temperature maps in two dimensions, without incorporating the temporal dimension. This research, however, aims to analyze temperature changes within a spatiotemporal framework, identifying significant trends and distinguishing between occasional fluctuations and meaningful changes. Visual representations of temperature evolution over time and space will be provided.

2 Study Area

Italy occupies a central position in the Mediterranean Sea, located between 36° N and 47° N latitude and 6° E and 18° E longitude. Due to its distinctive geography, characterized by complex terrain and diverse landscapes, the country experiences highly variable climatic conditions. The Alps act as a protective barrier for the Po Valley and the Venetian Plain, shielding them from cold northern winds. The Apennines, which extend along the entire peninsula, limit the influence of humid air from the west to the Tyrrhenian side, which is also protected from cold eastern winds (Pinna 1997; Mennella 1967; Fratianni and Acquaotta 2017). Utilizing one of the most recognized climate classifications, as introduced by W. Köppen (Pinna 1978), Italy is categorized into various regions based on the distribution of average annual temperatures and precipitation. As shown in Fig. 1, the primary climatic regions in Italy include:

- Ligurian-Tyrrhenian, Middle Adriatic, Ionian, and Mediterranean coastal regions
- Inland sub-coastal region
- Po and Veneto plains, Upper Adriatic, and inland peninsular regions
- Pre-Alpine and central Apennine region
- Alpine and upper Apennine region
- Alpine region.

3 Smooth Kernel Distribution Technique

The smooth kernel distribution (SKD) is a non-parametric estimation technique within spatial geostatistics, enabling the creation of a continuous surface along with its corresponding density function (Waller and Gotway 2004). This methodology has proven successful in diverse applications, ranging from assessing fire hazard to mapping the spread of Covid and estimating the peak frequency of neural pulses (Bertacchini et al. 2020, 2023; Shimazaki and Shinomoto, 2010).

The kernel density function is defined as (Shuo et al. 2021) (1):

$$\hat{f}(x) = \frac{1}{nh^2} \sum_{i=1}^{n} K\left\{\frac{s - s_i}{h}\right\} \tag{1}$$

where n is the total number of observed events, K is the kernel function, h is the bandwidth that produces a density estimate, s is the position at which the density estimate is calculated, and s_i is the position of each observed event. The kernel function is used to smooth the distribution of data, offering a continuous estimation of the underlying distribution. The density estimate varies with the bandwidth, which determines the extent of smoothing and, consequently, the accuracy of the fit. A narrower bandwidth yields a more detailed density estimate, while a wider bandwidth produces a less detailed estimate (Jones and Kappenman 1992). The SKD allows you to construct a 4D distribution function (Bertacchini et al. 2021) (2), transitioning from discrete to continuous data as follows:

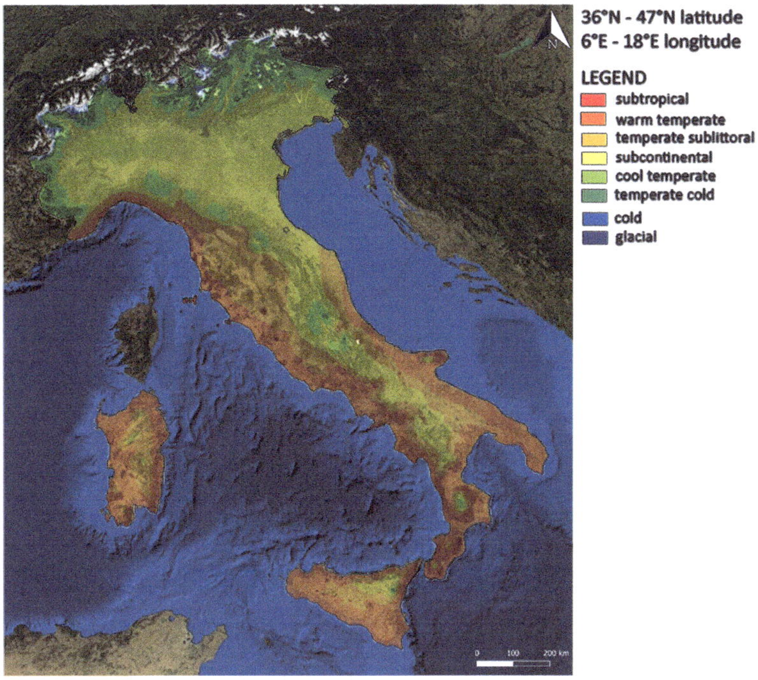

Fig. 1 Climate classifications by W. Köppen

$$w = f(x, y, T, t) \tag{2}$$

where x represents the longitude, y the latitude, T the daily mean temperature, and t the time. From this distribution, we construct the probability density function PDF as follows (3):

$$(\text{PDF})w = f(x, y, T, t) \tag{3}$$

The probability density function (PDF) operates as a standard four-variable function. To visualize this probability distribution at a specific point and time, having data on longitude, latitude, temperature, and time is crucial. These parameters enable the construction of a hypersurface that represents the temporal variation on temperature distribution at that specific location. To obtain a broader and more simplified version of the phenomenon, a specific value can be set for the time variable (t_{365}), and by visualizing the relationship between the variables x, y, and w in a two-dimensional space, the change in temperature can be displayed in 2D. This is expressed by the formula (4):

$$w = f_{t_{365}}(x, y) \tag{4}$$

This formulation then allows us to explore the temperature variation on a two-dimensional map in (x, y) space while keeping time constant at (t_{365}).

4 Methods

The methodological steps taken to collect the data, analyze the results, and visualize them are outlined below, as shown in Fig. 2.

The scientific computing system Wolfram Mathematica, which contains a number of useful methods and tools, was used to process and analyze the data (Wolfram 2015). Below is an overview of the steps and functions used in Mathematica:

Step 1. *Data collection and organization*

The process begins with the collection and organization of temperature data for the specified periods. Specifically, the temperature data used can be found within the Mathematica software. The resource is called the Wolfram Data Repository and is a public resource that hosts an expanding collection of computable datasets, curated and structured to be suitable for immediate use in computation, visualization, and analysis. Utilizing Wolfram Mathematica software provided direct access to a continuous stream of real-time meteorological data, complemented by an extensive archive of historical data. Weather data were made readily computable through symbolic representations, encompassing measured quantities, geographical locations, dates, and time series. The analysis focused on five specific years: 1982, 1992, 2002, 2012, and 2022. For each year, all available values were considered, incorporating the most recent measurement for the specific meteorological property at the corresponding location.

Fig. 2 Workflow (by authors)

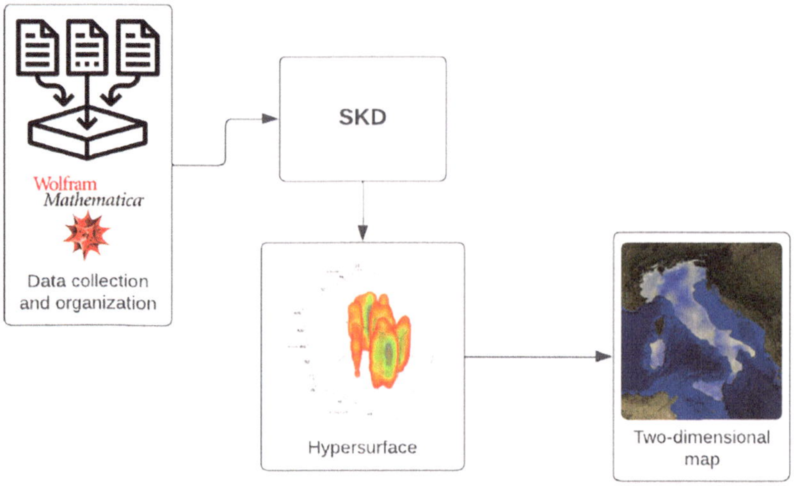

Step 2. *Global scale representation as a hypersurface*

The smooth kernel distribution technique, accessed in Mathematica via the *SmoothKernelDistribution*, allows the creation of a distribution function:

$$w = f(x, y, T, t) \tag{5}$$

where x represents the longitude, y the latitude, T the average daily temperature, and t the time, together with the related probability density function PDF, i.e., the probability density function given by the following formula (PDF)$w = f(x, y, T, t)$. The function used for visualization is Mathematica's DensityPlot3D function. This function evaluates a given function in $\mathbb{R}^3$ within its domain, assigning a color and opacity value to each point in space, evaluating the function f and mapping the results independently. The generated three-dimensional graph displays a color map, as shown in Fig. 3.

The temporal succession of 3D plots generates 4D patterns, represented as 3D iso-surfaces. Section (a) shows a spatiotemporal 3D density plot describing the potential temperature distribution. Orthographic projections (b), (c), and (d) provide a top view with latitude and longitude axes, a side projection with longitude and time axes, and another side projection with latitude and time axes, respectively. Observing from different angles provides insight into the temporal and spatial progression of the phenomenon on a global scale. These representations make it possible to visualize temperature variation over the periods considered, offering an analysis of the dynamics of the phenomenon over time. The resulting hypersurface spans space and time, providing a comprehensive visualization of the complexity of the phenomenon and enabling an analysis of temperature variation behavior.

Step 3. *Hypersurface reduction for distribution representation in two dimensions*

The three-dimensional representation of temperature changes over time offers a comprehensive visualization of the phenomenon's complexity but can be challenging to interpret. This challenge is addressed by employing the formula $w = f_{t365}(x, y)$, establishing a value for the temperature variable. For a two-dimensional visualization in Mathematica, the GeoDensityPlot function proves invaluable, serving as a robust tool for visualizing geographic data. This function enables the effective representation of values linked to geographic locations, creating a dynamic and intuitive visual depiction through color variations on the map. With GeoDensityPlot, as shown in Fig. 4, distinct colors can be assigned to different data intensities, ensuring that the map vividly illustrates spatial variations in geographic values. The use of color shading facilitates a rapid understanding of data distributions and concentrations, simplifying complex information into an easily interpretable visual format.

5 Results and Conclusion

The research has revealed that analyzing climate data using geostatistical techniques like the smooth kernel distribution enables the examination of temperature fluctuations over time, offering a visual depiction of the spatial distribution of this phenomenon. The adaptability of the smooth kernel distribution method allows for representation at various scales. Italy, with its distinctive geographical location and diverse landscapes, presents an intriguing setting for investigating these phenomena. This approach not only enhances the understanding of climate variations but also provides visual aids to clarify complex concepts through graphical representations (Fig. 3). To overcome this challenge, a 2D representation, illustrated in Fig. 4, was used to generate two-dimensional maps. This type of representation facilitates the delineation of spatial variations, enabling a detailed analysis of potential

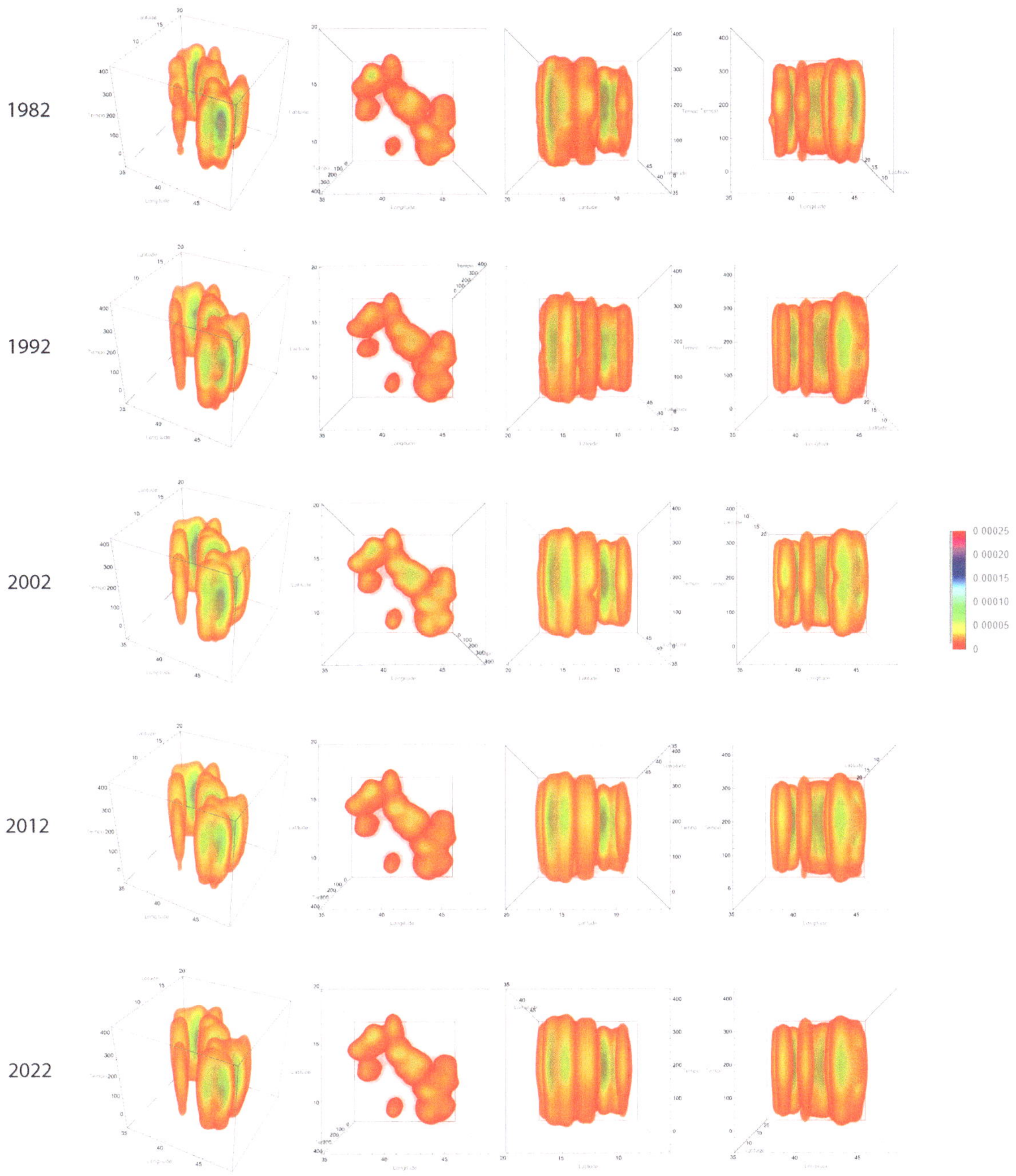

Fig. 3 3D density plot of the 4D space–time hypersurface (by authors)

changes, highlighting the spatial distribution, and simplifying the interpretation of dynamics. The process of gathering and organizing meteorological data, facilitated by Wolfram Mathematica software, enabled a thorough examination of five specific years (1982, 1992, 2002, 2012, and 2022), providing a comprehensive temporal perspective. In summary, the current climate conditions, dominated by climate change and global warming, stand out as one of the most crucial challenges of our time. The impact of these transformations is evident in weather observations, revealing a consistent rise in global temperatures. In conclusion, this exploration underscores the complexity of climate change, its manifestation in Italy, and the importance of advanced scientific approaches to fully comprehend its impacts. Addressing this collective challenge requires a shared commitment to adopting sustainable solutions and preserving our environment.

Fig. 4 GeoDensityPlot

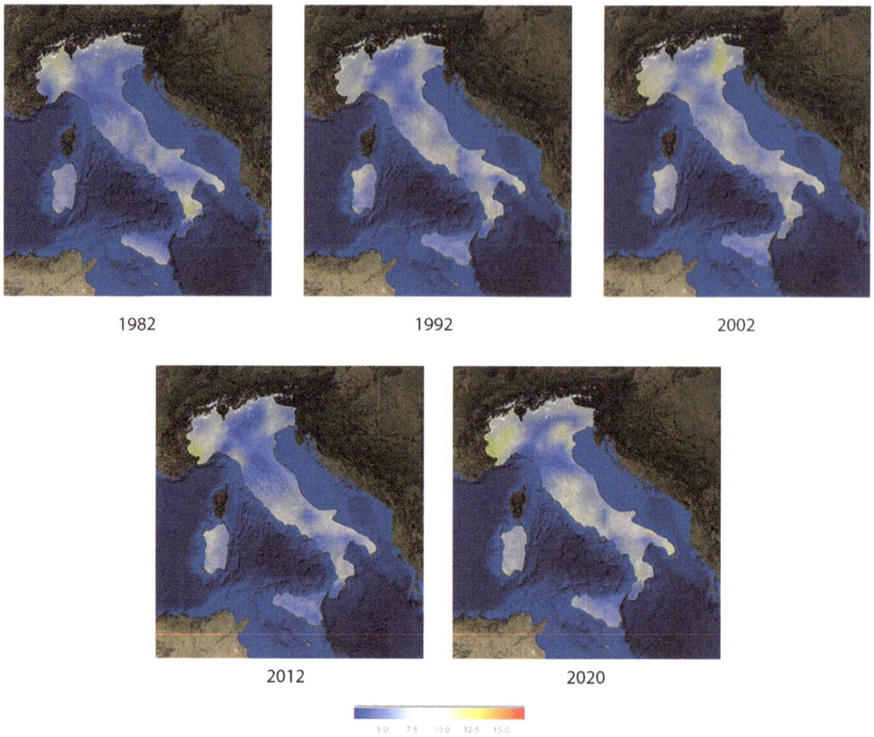

References

Bernstein, L., Bosch, P., Canziani, O., Chen, Z., Christ, R., Riahi, K.: IPCC, 2007: Climate Change 2007: Synthesis Report. IPCC (2008)

Bertacchini, F., Gabriele, L., Tavernise, A.: Bridging educational technologies and school environment: implementations and findings from research studies. Educ. Theory 63–82 (2011)

Bertacchini, F., Bilotta, E., Pantano, P.: On the temporal spreading of the SARS-CoV-2. PLoS ONE **15** (2020)

Bertacchini, F., Pantano, P., Bilotta, E.: SARS-CoV-2 emerging complexity and global dynamics. Chaos Interdisc. J. Nonlinear Sci. (2021)

Bertacchini, A., Fusaro, P.A., Zupi, M.: New probabilistic methods for generating risk maps. In: Numerical Computations: Theory and Algorithms NUMTA 2023, p. 56 (2023)

Demarco, F., Bertacchini, F., Scuro, C., Bilotta, E., Pantano, P.: The development and application of an optimization tool in industrial design. Int. J. Interact. des. Manufact. (IJIDeM) **955–970**(3), 14 (2020)

Fratianni, S., Acquaotta, F.: The Climate of Italy. Landscapes and Landforms of Italy, pp. 29–38 (2017)

GISTEMP: GISS Surface Temperature Analysis. Global Temperature Trends (2023)

Gray, S.B., Brady, S.M.: Plant developmental responses to climate change. Dev. Biol. **64–77**(1), 419 (2016)

Jones, M.C., Kappenman, R.F.: On a class of kernel density estimate bandwidth selectors. Scand. J. Stat. 337–349 (1992)

Lee, J.Y., Marotzke, J., Bala, G., Cao, L., Corti, S., Dunne, J.P., Zhou, T.: Future global climate: scenario-based projections and near-term information. In: Climate Change 2021: The Physical Science Basis. Contribution of Working Group I to the Sixth Assessment Report of the Intergovernmental Panel on Climate Change, pp. 553–672. Cambridge University Press (2021)

Lenssen, N.J., Schmidt, G.A., Hansen, J.E., Menne, M.J., Persin, A., Ruedy, R., Zyss, D.: Improvements in the GISTEMP uncertainty model. J. Geophys. Res. Atmosp. **6307–6326**(12), 124 (2019)

Lewis, P.G., Chiu, W.A., Nasser, E., Proville, J., Barone, A., Danforth, C., Craft, E.: Characterizing vulnerabilities to climate change across the United States. Environ. Int. **172**, 107772 (2023)

Lindsey, R.: Climate Change: Global Sea Level. Available online: Climate.gov. Accessed on 14 Aug 2020

Marziliano, P.A., Lombardi, F., Cataldo, M.F., Mercuri, M., Papandrea, S.F., Manti, L.M., Scuro, C.: Forest fires: silvicultural prevention and mathematical models for predicting fire propagation in Southern Italy. Fire **7**(8), 278 (2024)

Mennella, C.: Il clima d'Italia nelle sue caratteristiche e varietà e quale fattore dinamico del paesaggio. Conte (1967)

Moriondo, M., Good, P., Durao, R., Bindi, M., Giannakopoulos, C., Corte-Real, J.: Potential impact of climate change on fire risk in the Mediterranean area. Clim. Res. **85–95**(1), 31 (2006)

NOAA: NOAA National Centers for Environmental Information, Monthly Global Climate Report for Annual 2022 (2023)

NOOA: NOAA National Centers for Environmental Information, Climate at a Glance: Global Time Series (2023)

Perčec Tadić, M., Pasarić, Z., Guijarro, J.A.: Croatian high-resolution monthly gridded dataset of homogenised surface air temperature. Theoret. Appl. Climatol. **151**(1), 227–251 (2023)

Pinna, M.: L'atmosfera e il clima. Utet (1978)

Pinna, M.: Climatologia. Unione tipografico-editrice torinese, Torino (1997)

Shimazaki, H., Shinomoto, S.: Kernel bandwidth optimization in spike rate estimation. J. Comput. Neurosci. **171–182**, 29 (2010)

Shuo, Z., Jingyu, Z., Zhengxiang, Z., Jianjun, Z.: Identifying the density of grassland fire points with kernel density estimation based on spatial distribution characteristics. Open Geosci. **796–806**(1), 13 (2021)

Waller, L.A., Gotway, C.A.: Applied Spatial Statistics for Public Health Data. Wiley (2004)

Wolfram, S.: An Elementary Introduction to the Wolfram Langauge (2015)

The Nexus Between Tourism, Economic Growth and Environment: Empirical Evidence from European Countries

Alejandro Alcalá-Ordóñez, Juan Gabriel Brida, Pablo Juan Cárdenas-García, and Verónica Segarra

Abstract

Since the beginning of the century, following the work published by Balaguer and Cantavella-Jordá (Appl Econ 34:877–884, 2002), a significant number of studies have emerged analyzing the tourism-led growth hypothesis (TLGH). However, in the recent years, studies have begun to incorporate the effects of tourism on the environment into this analysis, given that this activity often entails significant environmental degradation. In this context, the aim of this study is to analyze the relationship between tourism, economic growth, and pollutant emissions for the 27 countries of the European Union during the period 1995–2022. A cointegration test and an econometric estimation model will be applied to analyze the relationships between the described variables, utilizing the Granger causality test. The results obtained determine that the variables are significant, indicating that tourism has a causal relationship with economic growth. However, this expansion of tourism also has significant negative effects on environmental pollution.

Keywords

Tourism · Economic growth · CO_2 emissions · Causality analysis · TLGH

A. Alcalá-Ordóñez · P. J. Cárdenas-García (✉)
Department of Economics, University of Jaén, Jaén, Spain
e-mail: pcgarcia@ujaen.es

A. Alcalá-Ordóñez
e-mail: aaordone@ujaen.es

J. G. Brida · V. Segarra
GIDE, Faculty of Economics, Universidad de la República, Montevideo, Uruguay
e-mail: gabriel.brida@fcea.edu.uy

V. Segarra
e-mail: veronica.segarra@fcea.edu.uy

1 Introduction

Tourism is one of the sectors with the greatest weight in the global economy (Ady et al. 2022; Badimo and Yuhuan 2024), making the study of its impact from an economic standpoint a topic of great interest in scientific literature. In fact, both developed and developing countries are betting on this sector due to the significant economic benefits it generates in the destination area (Nadeem et al. 2020).

In this regard, tourism has made a significant contribution to global production (Fahimi et al. 2018; Habibi and Amani 2022), and employment is also affected both directly and indirectly (Albaladejo et al. 2023; Dogru and Bolut 2018), as it impacts not only the agents within the sector but also other sectors such as construction, transportation, food, and hospitality which are highly benefited (Lenzen et al. 2018). Thanks to the creation of these jobs, households see an increase in their income, which also results in higher government revenue (Shahbaz et al. 2018).

These positive impacts lead countries worldwide to bet on tourism as an activity that can enable higher levels of economic growth. In fact, this improvement in economic growth can lead to better socioeconomic conditions for the resident population, increasing their quality of life or reducing inequality and poverty, thereby enabling economic development (Thao and Bakucz 2022).

Therefore, scientific literature opens a new line of research analyzing the relationship between tourism and economic growth, which originated from the work published by Balaguer and Cantavella-Jordá (2002), where the tourism-led growth hypothesis (TLGH) was proposed for the first time. Since then, numerous studies have been published worldwide on this line of research, predominantly determining that tourism enables economic growth in receiving destinations (Alcalá-Ordóñez et al. 2023).

P. Mohammad et al. (eds.), *Geographic Approaches to Climate Change and Mitigation: Urban and Rural Perspectives (Volume 2)*, Advances in Science, Technology & Innovation, https://doi.org/10.1007/978-3-031-92115-5_3

However, tourism is not immune to certain negative consequences associated with its expansion in receiving destinations. In this context, the negative impacts generated in destination areas, especially on the environment, due to tourism activity, are of particular relevance (Kumail et al. 2022). Indeed, this sector generates significant negative externalities such as coastal degradation or the increase of pollutant gases (Lee and Syah 2018). The consideration of these negative impacts has led current empirical literature to address the analysis of the relationship between tourism and economic growth by including environmental variables (Segarra et al. 2024) and the importance of sustainable tourism (Li 2022).

As a result of these negative aspects, countries have undertaken the mission to mitigate or compensate for the negative environmental impact through, e.g., the practice of sustainable tourism (Ben Jebli et al. 2019), as well as conscious tourism, a type of tourism based on principles of ethics, social responsibility, and preservation of the natural environment (Montesdeoca et al. 2015).

Therefore, considering, on the one hand, that tourism can contribute to economic growth and, on the other hand, that the expansion of tourism activity can lead to environmental degradation in receiving destinations, the objective of this study is to empirically analyze the relationship between tourism, economic growth, and pollutant emissions in Europe. This region is the most visited worldwide, supported by intraregional demand and travel from the United States and China (UN Tourism 2024).

In this regard, there is currently an important gap in the scientific literature in two ways: on the one hand, although the European region is the most visited in the world, there are few studies that use this region as a case study and, on the other hand, none of them include, in addition to the analysis of the relationship between tourism and economic growth, the analysis of the effects on the environment.

To achieve this objective, this research empirically analyzes whether there is a causal relationship between tourism, economic growth, and pollutant emissions, using panel data for the 27 countries of the European Union during the period 1995–2022. A Granger causality test, through the adaptation of Dumitrescu and Hurlin, is applied, and some estimations are made to establish the magnitude of the relationships found.

The contribution of this research lies in providing a knowledge base regarding the analysis between tourism, economic growth, and environmental pollution, determining the behavior of these variables in the European Union, as it has generally not been the subject of study by scientific literature. The results of this study will allow determining whether the expansion of tourism activity precedes economic growth, but at the cost of greater environmental degradation in receiving countries.

The rest of the paper is organized as follows: section two presents a literature review on the relationship between tourism, economic growth, and pollutant emissions; section three presents the data used for this study; section four details the methodology employed in this study; section five provides the results obtained after applying various econometric tests; and section six presents the final conclusions, along with a series of policy recommendations, limitations, and future lines of research.

2 Tourism, Economic Growth, and CO_2 Emissions

The expansion of tourism activity leads to the improvement of economic growth in receiving countries due to its positive impact on GDP, employment, increased tax revenues, economies of scale, and production growth (Brida et al. 2016). However, as Khan et al. (2022) point out, environmental degradation resulting from the increase in tourism activity means that these negative impacts must be considered as a threat that the governments of the receiving destinations, which have bet on the development of this activity, must face.

On the one hand, within the scientific literature, there is a large number of empirical studies published on the relationship between tourism and economic growth, with important literature reviews summarizing the main findings. These reviews identify three fundamental types of relationships between tourism and economic growth (Alcalá-Ordóñez et al. 2023; Brida et al. 2016).

First, tourism drives economic growth by generating foreign exchange, creating employment, fostering new investments, and stimulating other economic sectors (Tang et al. 2019). This is the most commonly found relationship in studies published so far.

Second, economic growth promotes the expansion of tourism, as higher economic growth facilitates higher income levels, better infrastructure at the destination, and greater productive specialization (Kumail et al. 2022). However, this relationship is observed in fewer studies within the scientific literature.

Finally, there is a bidirectional relationship between tourism and economic growth, which has also been demonstrated in a significant number of empirical studies.

On the other hand, within the current empirical literature, there are various studies that analyze the impact of tourism on the environment, as tourism activity is considered to contribute to carbon dioxide emissions (UN Tourism 2019). Indeed, due to the close relationship between tourism and the transportation sector, an increase in tourism activity increases transportation usage, ultimately raising the level of pollutant emissions (Boonyasana and Chinnakum 2020).

In this context, some authors have found that tourism negatively affects the environment (Balsalobre-Lorente et al. 2020), while others have concluded that there is a Environmental Kuznets Curve (EKC), where beyond a certain point, tourism activity does not influence pollutant emissions (Anser et al. 2020). Furthermore, in line with the aforementioned discussion on sustainable tourism, some studies highlight the existence of a positive relationship between the production of renewable energy and the expansion of tourism activity (Liu et al. 2022; Moslehpour et al. 2022).

As can be seen, the analysis between tourism, economic growth, and the environment has captured the interest of an increasing number of research studies within the scientific literature. However, there is no unanimity regarding the effects produced. While a logical relationship has been established between the variables indicating that the tourism sector has a positive impact on economic growth and, in turn, this economic growth, which is expanded as a result of tourism activity, justifies the increase in pollution due to the greater use of the transportation sector (Deng et al. 2022) and energy such as electricity or gas at destination sites (Xiangyu et al. 2021).

The variety of research in different regions, to a greater or lesser extent, highlights the importance placed on the environmental dimension and how countries are focusing on controlling negative impacts on the environment. Some empirical studies stand out, primarily conducted in Turkey within the European region or China, India, or Pakistan within the Asian region.

For example, in Turkey, tourist arrivals cause short-term CO_2 emissions, while they decrease in the long term (Uzuner et al. 2020). Eyuboglu and Uzar (2020) find a feedback relationship between tourism and CO_2, as well as that increased energy consumption causes long-term tourism and pollutant emissions. Additionally, De Vita et al. (2015) indicate that tourism, economic growth, and energy consumption increase carbon dioxide emissions, while Akadiri et al. (2019) determine a bidirectional causality between economic growth, tourism, and pollutant emissions.

Other studies analyzed groups of European countries, as in the case of Khan et al. (2021), where they find that renewable energy use impacts tourism. Likewise, Jambor and Leitão (2017) demonstrate that tourism impacts economic growth, and economic growth impacts CO_2 emissions, which are reduced in the long term.

In the Asian region, numerous studies examine environmental variables along with tourism and economic growth. Deng et al. (2022) analyzed 20 regions in China where tourism impacts both growth and CO_2 emissions. In Malaysia, tourism impacts economic growth and CO_2 emissions, while a bidirectional relationship exists between economic growth and carbon dioxide emissions (Sun et al. 2022). In Bangladesh, electricity consumption causes pollutant emissions, and both contribute to economic growth (Rahaman et al. 2022). In India, tourist arrivals and economic growth measured by GDP per capita cause CO_2 emissions (Villanthenkodath et al. 2022). In Pakistan, both tourism and economic growth, along with energy consumption, increase CO_2 emissions (Oad et al. 2022), while Bano et al. (2021) identified a bidirectional relationship between tourism and growth and tourism and CO_2, as well as the causality of renewable energy consumption on economic growth. Niu (2022) indicates a bidirectional causality between economic growth and CO_2 emissions, while the latter is caused by international tourist arrivals in China.

On the other hand, the analysis of eight Asian countries by Kongbuamai et al. (2020) identified that tourism and long-term natural resource use reduce CO_2 emissions. Conversely, CO_2 emissions and energy consumption impact tourism in twelve Asian countries (Shakouri et al. 2017).

However, research on tourism, economic growth, and the environment in Africa and the Americas has attracted less interest. Nevertheless, the development of tourism has generally been identified as a causal factor for CO_2 emissions in 27 African countries (Agyeman et al. 2022), in 41 countries (Bekun et al. 2022) including Morocco, Algeria, Tunisia, and Egypt (El Menyari 2021), in Argentina, where economic growth is also a cause of CO_2 emissions (Raihan et al. 2022), or in the United States (Xiangyu et al. 2021).

Therefore, ultimately, it can be observed that, after reviewing the scientific literature, it is common to find that tourism activity and economic growth cause CO_2 emissions, but it cannot be conclusively stated as there is no unanimity in empirical literature. Additionally, it has also been noted from this review process that, despite the importance of tourism in Europe, this region has been less analyzed in terms of the effects of tourism on economic growth and pollutant emissions. Hence, this work, capitalizing on this gap in scientific literature, aims to address the effects between these variables at the European level, given the increasing importance in recent years, apart from the economic effects generated by tourism, in the need for tourism to preserve the environment.

3 Data

To conduct this empirical study at the country level in the European region, the analysis has focused on the 27 countries of the European Union. The details of these countries, along with descriptive statistics of the relevant variables for this research work, are shown in Table 1.

Table 1 Descriptive statistics: annual averages of each variable (period 1995–2022)

Country	Tourism	CO_2	GDP
Austria	28,142,494	69,903	34,218
Belgium	12,589,426	111,252	32,346
Bulgaria	4,582,482	50,761	4807
Croatia	9,717,847	19,636	10,278
Cyprus	2,414,677	7262	22,142
Czech Republic	13,636,344	119,011	14,404
Denmark	6,108,095	47,027	44,370
Estonia	2,244,147	18,697	11,261
Finland	9,647,082	55,970	33,416
France	130,390,467	367,846	30,285
Germany	129,124,958	810,149	31,602
Greece	18,194,250	83,406	18,211
Hungary	8,036,506	54,981	10,102
Ireland	9,684,896	40,177	42,022
Italy	93,480,806	409,289	26,861
Latvia	1,366,483	7969	8678
Lithuania	1,734,171	13,900	9228
Luxembourg	956,733	9803	78,475
Malta	1,588,294	2247	16,936
Netherlands	30,318,608	171,617	37,416
Poland	20,745,207	323,305	9067
Portugal	15,151,406	54,938	16,593
Romania	7,545,608	94,686	6253
Slovakia	3,607,874	39,158	11,625
Slovenia	3,078,252	15,942	16,920
Spain	90,009,780	292,271	22,515
Sweden	22,557,414	51,341	38,489
European Union	**25,061,271**	**123,798**	**23,649**

Source Own elaboration

The indicators used in this analysis of the relationship between tourism, economic growth, and environmental pollution, along with the databases from which these indicators originate, are detailed below:

- Tourism: To measure tourism, the total number of tourists (both domestic and international tourists) visiting the analyzed countries has been used, measured in the number of tourists. This indicator has been obtained from the database provided by Eurostat, specifically called the Tourism Database.
- Economic growth: To measure economic growth, Gross Domestic Product (GDP) at current prices, measured in euros, has been used. This indicator has been obtained from the database provided by Eurostat, specifically called National Accounts and GDP.
- Environmental pollution: To measure environmental pollution, carbon dioxide emissions have been used, measured in kilotonnes of CO_2 equivalents. This indicator has been obtained from the database provided by Eurostat, specifically called the Air emissions accounts dataset.

The time horizon considered in this study spans from 1995 to 2022, with the aim of conducting as comprehensive a time analysis as possible, covering a period of 28 years. In this regard, there is an initial temporal restriction, as the earliest available data for the set of these three variables is for the year 1995, with the most recent complete data available for the year 2022.

4 Methodology

To analyze the potential relationship between tourism activity, economic growth, and CO_2 emissions, the Granger causality test is employed. The objective of this test is to determine if one variable predicts the other, without implying a cause-and-effect type of causality. However, when working with panel data structure, one of the main drawbacks of this test is that it assumes that the coefficients of all components are equal. In this regard, the Dumitrescu and Hurlin test (2012) is an adaptation of Granger causality test, which considers different coefficients among the units composing the panel.

Given two stationary variables $x_{i,t}$ and $y_{i,t}$ for country i in period t, the Dumitrescu and Hurlin (2012) test proposes the following equation:

$$y_{i,t} = \alpha_i + \sum_{k=1}^{K} \gamma_{ik} y_{i,t-k} + \sum_{k=1}^{K} \beta_{ik} x_{i,t-k} + \varepsilon_{i,t}$$
$$\text{with } i = 1, \ldots, N;\ t = 1, \ldots, T \quad (1)$$

where

- α_i is the intercept term.
- γ_{ik} are the coefficients of lagged $y_{i,t-k}$ terms.
- β_{ik} are the coefficients of lagged $x_{i,t-k}$ terms.
- $\varepsilon_{i,t}$ is the error term.

Just like in Granger (1969), the procedure to determine the existence, or absence, of causality between the variables is to test whether the coefficients of past values of x on the current value of y are significant. The null hypothesis, corresponding to the absence of causality for all individuals in the panel, is defined as follows:

$$H_0\colon \beta_{i1} = \cdots = \beta_{iK} = 0, \quad \forall i = 1, \ldots, N \quad (2)$$

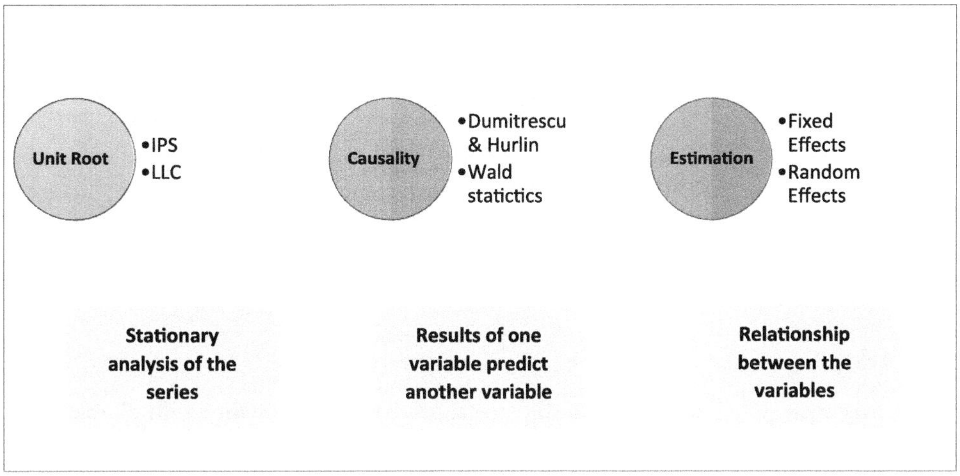

Fig. 1 Methodology description. *Source* Own elaboration

Outside of the null hypothesis, the Dumitrescu and Hurlin test allows for the possibility of causality for some individuals in the panel, but not necessarily for all. Therefore, the alternative hypothesis is defined as follows:

$$H_1: \beta_{i1} = \cdots = \beta_{iK} = 0, \quad \forall i = 1, \ldots, N_1 \tag{3}$$

$$\beta_{i1} \neq 0 \quad \text{or} \cdots \text{or} \quad \beta_{iK} \neq 0 \quad \forall i = N_1 + 1, \ldots, N$$

In the Dumitrescu and Hurlin (2012) test, N regressions of Eq. (1) are run, and tests of the K linear hypotheses $\beta_{i1} = \cdots = \beta_{iK} = 0$ are performed, providing the individual Wald statistic W_i. From these statistics, the average Wald statistic W is calculated. These statistics and p-values are obtained for each country, providing additional information, indicating which units in the panel verify the causal relationship and which do not.

The Dumitrescu and Hurlin (2012) test assume that all series are stationary. Therefore, in the first step of the analysis, unit root tests are conducted using the Im et al. (2003) tests and the Levin et al. (2002) test. Once the stationarity of the series is verified, causality is analyzed using the Dumitrescu and Hurlin test, also examining the Wald statistics. Finally, in the last step of this analysis, some estimations are made to quantify the magnitude of the relationship found. The methodological framework used can be seen in Fig. 1.

5 Results

In Table 2, the statistics of the unit root tests for panel data proposed by Levin, Lin, and Chu (LLC) and Im, Pesaran, and Shin (IPS) are detailed. In all cases, the considered series are stationary.

Next, we proceed to analyze the causality between tourism, economic growth and CO_2 emissions. Through this test, we analyze the causal relationship between these variables in both directions. Specifically, Table 3 gives the results of the Dumitrescu and Hurlin Causality test, considering as null hypothesis that there is no causal relationship between each of the pairs of variables indicated. This technique has been used in the empirical literature in different researches (Badulescu et al. 2021; Khan et al. 2021; Roudi et al. 2019; Tugcu 2014).

Table 2 Panel root tests of variables (all variables are in first differences)

Variables	LLC	IPS
Tourism	−16.0041***	−23.0640***
GDP	−17.8316***	−17.1687***
CO_2	−19.5829***	−19.8918***

Source Own elaboration
Note *Indicate the significance level: *10%; **5%; ***1%. *Lags* AIC

Table 3 Dumitrescu and Hurlin causality test

	Statistic	
Tourism → CO_2	Z-bar	4.9081***
	Z-bar tilde	3.8677***
Tourism → GDP	Z-bar	21.3610***
	Z-bar tilde	2.8233***
GDP → CO_2	Z-bar	2.0727**
	Z-bar tilde	14.618
GDP → Tourism	Z-bar	−11.562
	Z-bar tilde	−14.490
CO_2 → GDP	Z-bar	14.193
	Z-bar tilde	0.9073
CO_2 → Tourism	Z-bar	3.6444***
	Z-bar tilde	−5.382

Source Own elaboration
Note *Indicate the significance level: *10%; **5%; ***1%

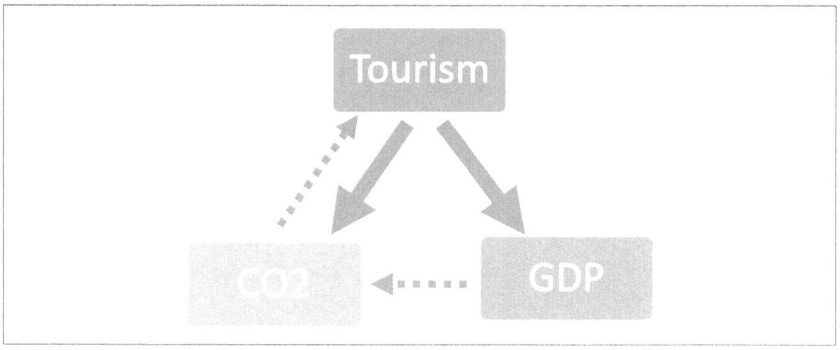

Fig. 2 Causality relationship between tourism, GDP, and CO_2 emissions. *Source* Own elaboration

The existence of a causal relationship from tourism to GDP and to CO_2 emissions is observed. This finding is consistent with the results obtained by Sun et al. (2022) in Malaysia, where they indicate that GDP and CO_2 emissions are caused by tourism activity, as well as with the work conducted by Ghosh (2022), who analyzed a set of one hundred countries worldwide and the investigation carried out by Wang and Wu (2022) on 5 countries worldwide considering GDP per capita, international tourist arrivals, and carbon dioxide emissions. For these two causal relationships identified, both from tourism to economic growth and to CO_2 emissions, in both cases, the statistic is significant at the 1% level, indicating strong evidence supporting such relationships. It is worth noting that for small sample period, the most appropriate statistic for this test is the *Z*-bar-tilde. The results of the Wald statistics and corresponding *p*-values can be seen in Table 5 of the Appendix. The Wald statistic indicates that the causal relationship found from tourism to economic growth applies to the following countries: Cyprus, Estonia, Greece, Ireland, Italy, and Slovakia. Meanwhile, the causal relationship from tourism to CO_2 emissions would be verified for the following countries: Czech Republic, Cyprus, France, Germany, Greece, Ireland, Italy, Luxembourg, Poland, and Slovakia.

Additionally, it is also noted that one of the test statistics (Z-bar only) is significant for the unidirectional causality relationship from GDP to CO_2 emissions. Other studies also validated this result, such as the research on 20 countries worldwide by Sofi et al. (2022), an analysis using Argentina as a case study (Raihan et al. 2022), in 41 countries of Africa (Bekun et al. 2022), in the European Union (Musa et al. 2021) or an investigation focusing on India (Villanthenkodath et al. 2022). It is also noted that one of the test statistics (Z-bar only) is significant for the unidirectional causality from CO_2 emissions to tourism. This result aligns with the work of Chien et al. (2022) on the United States, where they indicate a causality from CO_2 emissions to tourism. Similarly, in an analysis using Saudi Arabia as a case study, the same result is obtained (Ozturk et al. 2022) as in the analysis of Pakistan (Amir et al. 2021). However, it should be noted that for small *T*, the most appropriate statistic is the *Z*-bar tilde.

A graphical representation of the relationship between the variables analyzed (tourism, economic growth and CO_2 emissions) is shown in Fig. 2.

Finally, after testing the causality between the variables using fixed effects and random effects regression models, the relationship between the variables is estimated. Both in the fixed effects and random effects models, the equation is:

$$y_{it} = x'_{it}\beta + \mu_i + \varepsilon_{it} \quad (4)$$

where μ_i is a time-invariant fixed value.

In the fixed effects model, the μ_i is parameters to be estimated, while in the random effects model, the μ_i is treated as omitted random variables, which are included in the error term of the model. Thus, the fixed effects estimator is basically the Ordinary Least Squares estimator with panel individual dummy variables, while the random effects estimator is a feasible Generalized Least Squares estimator. Additionally, the Hausman test is performed. This test compares the parameter estimators obtained in both models, fixed effects and random effects, and assesses whether the differences between the estimators are systematic or random. If evidence is found that these differences are systematic, the fixed effects model is suggested, given that the random effects model is inconsistent. On the other hand, if evidence is found that the differences are random, this implies that both models are consistent, but the random effects model might be more efficient. Thus, once the test is conducted, if the *p*-value obtained is less than a significance level, the null hypothesis that random effects are consistent is rejected, and the fixed effects model is chosen. If the *p*-value is greater than the established significance level, the null hypothesis is accepted, indicating that both models are consistent, and the choice between them depends on additional considerations (the chosen model is marked with * in the Table 4).

On the one hand, as can be observed in panel (b) of Table 4, the coefficient of tourism is positive, meaning that an improvement in the tourism activity of countries trans-

Table 4 Random-effect estimation model and fixed-effect estimation model

(a) Random-effect estimation model					Fixed-effect estimation model*				
d_Tourism	Coefficient	Std. error	*z*	*P* > IzI	d_Tourism	Coefficient	Std. error	*t*	*P* > ItI
d_GDP	2107.819	296.3734	7.11	0.000	d_GDP	2225.346	315.4768	7.05	0.000
d_CO_2	88.33314	18.9934	4.65	0.000	d_CO_2	210.7432	28.1611	7.48	0.000
Constant	−366,594.2	318,788.9	−1.15	0.250	Constant	−563,188.6	315,815.8	−1.78	0.075
(b) Random-effect estimation model*					Fixed-effect estimation model				
d_GDP	Coefficient	Std. error	*z*	*P* > IzI	d_GDP	Coefficient	Std. error	*t*	*P* > ItI
d_Tourism	0.0000305	4.18e−06	7.31	0.000	d_Tourism	0.0000298	4.23e−06	7.05	0.000
d_CO_2	0.0091056	0.0029508	3.09	0.002	d_CO_2	0.0115429	0.0033597	3.44	0.001
Constant	405.55	71.07288	5.71	0.000	Constant	403.1166	33.32503	12.10	0.000
(c) Random-effect estimation model*					Fixed-effect estimation model				
d_CO_2	Coefficient	Std. Error	*z*	*P* > IzI	d_CO_2	Coefficient	Std. error	*t*	*P* > ItI
d_GDP	1.434.868	0.4169297	3.44	0.001	d_GDP	1.436.637	0.4181542	3.44	0.001
d_Tourism	0.0003509	0.0000469	7.48	0.000	d_Tourism	0.0003515	0.000047	7.48	0.000
Constant	329.7845	2324.067	0.14	0.887	Constant	3286.137	408.6075	0.80	0.422

Source Own elaboration

lates into an improvement in GDP. This is in line with many previous works within the scientific literature that analyze this relationship, among others, this relationship has been demonstrated in the study on the island of Mauritius in Africa (Opuala-Charles et al. 2022), in Malaysia (Tang et al. 2019), in the analysis of Greece (Lolos et al. 2021), for the case of Portugal (Bento 2016), and in the case of China (Su et al. 2021), showing that it is a well-established relationship.

On the other hand, as observed in panel (c) of Table 4, the coefficient of tourism is positive. This would indicate that with an increase in Tourism, CO_2 emissions also increase, which is consistent with various studies examining the impact of tourism on the environment, as, e.g., in the case of the seven island countries analyzed by Adedoyin and Bakun (2020), in 27 countries of Africa (Agyeman et al. 2022), the research carried out in 7 countries worldwide (Gyamfi et al. 2022), the OECD countries analyzed by Gedikli et al. (2022) or in the research on Pakistan by Bano et al. (2021). This effect of tourism activity on the increase in CO_2 emissions is explained by the relationship between tourism and the transport sector. In this regard, several authors have found evidence supporting the idea that tourism negatively affects the environment (e.g., Balsalobre-Lorente et al., 2020).

6 Conclusions

The present study has aimed to analyze the relationship between tourism, economic growth, and the environment, given that the inclusion of the environmental variable in TLGH analysis is crucial due to the increasing importance placed in recent years on ensuring that tourism is environmentally friendly. Indeed, the expansion of tourism often entails certain negative impacts, such as environmental degradation, increased congestion, or pollution, making it important to include this environmental dimension when analyzing the economic impacts generated by tourism, that is, to analyze whether the economic growth generated by tourism is balanced with environmental preservation.

This approach has been applied to the countries that are part of the European Union, representing a significant contribution to scientific literature, as although the European region receives the most tourists globally, few studies have addressed this region from this perspective. Specifically, the causal relationship between tourism, economic growth, and pollutant emissions has been analyzed for the 27 countries that are part of the European Union, and an estimation of these relationships has been made.

The results obtained reflect the existence of a unidirectional relationship from tourism to economic growth and from tourism to carbon dioxide emissions. Additionally, economic growth also presents a unidirectional causal relationship to pollutant emissions, while these emissions showed a unidirectional causal relationship to tourism (although these latter two causal relationships are not very robust).

Thus, the results obtained in this study reflect the importance of incorporating the relationship with the environment that tourism has, avoiding analyzing TLGH in isolation.

Furthermore, the estimation model confirms that the expansion of tourism implies a positive increase in economic growth, thus confirming the potential of tourism activity, from an economic point of view, in destination sites. However, it has also been confirmed, in line with much of the existing literature, that an increase in tourism has negative effects on the environment.

In this context, it should be noted that the increase in pollutant emissions may be due to an increase in the transportation sector at the destination, since this sector is one of the most closely related to the expansion of tourism. Thus, the European region, being the most visited worldwide, translates into tourism generating a significant carbon footprint, not only due to transportation but also due to its relationship with other productive sectors such as construction or services (Gedikli et al. 2022).

Therefore, the commitment to tourism made by many European Union countries has led to increased levels of economic growth, but at the cost of significant environmental degradation in the form of higher pollutant emissions. This imbalance between economic prosperity and environmental degradation must be taken into account by the administrations responsible for developing policies related to this activity, in that tourist flows should not be reduced, as they are crucial for economic growth in the region, but specific plans should be developed in tourist destinations to raise awareness among tourists about environmental care and, furthermore, to develop new, more sustainable transportation systems.

In this regard, the actions being taken by many local administrations in this region to limit access to cities by combustion vehicles are in line with this approach, and it is also necessary to promote sustainable tourism through the use of clean energy sources and the practice of tourism under the premise of environmental care, in line with the thinking of Bano et al. (2021), who suggest investment in renewable energy sources that improve economic growth and, at the same time, reduce pollutant emissions, leading to increased tourist demand for destination countries as they become environmentally sustainable tourist destinations, increasingly sought after by tourists.

It is also necessary to highlight that this study has some limitations: (i) the analyzed time horizon, which, due to data availability, configures it as a short analysis period, (ii) the use of indicators to measure the environment has been limited to pollutant emissions, as it has not been possible to create a composite index to explain the environmental variable considering other indicators related to the environment, (iii) the use of linear models is not necessarily the most appropriate way to explain causality, although most studies use this system to analyze causality and, (iv) the use of panel data should be interpreted with some caution, since, although it allows to increase the database for the study, such data need not always be homogeneous and, even so, results can be given in a general way for the panel or set of countries.

Finally, as future lines of research, based on the empirical results obtained, there is the possibility of replicating this analysis in other territories, even at the level of individual countries, or using a regional-level analysis approach, in which the characteristics of the places can be analyzed to identify which destination-associated factors condition the relationships between the variables, in order to better guide policy recommendations to the managers of these destinations.

Appendix

See Table 5.

Table 5 Wald test results and *p*-values (tourism to GDP and tourism to CO_2)

Country	Tourism → GDP		Tourism → CO_2	
	Wald	*p*-value	Wald	*p*-value
Austria	9.0447034	0.40228265	0.34126507	0.56478625
Belgium	1.0112397	0.35427714	0.03026816	0.86340539
Bulgaria	2.0758601	0.12473247	1.7438298	0.19964471
Croatia	1.7387149	0.16687046	0.00199514	0.9647583
Cyprus	6.8039935	**0.01175746**	6.9422068	**0.01480697**
Czech	1.7929057	0.94799638	4.6809393	**0.04113178**
Denmark	2.1246741	0.11988422	0.99664588	0.32851141
Estonia	2.6718857	**0.07973188**	0.07080146	0.79254419
Finland	3.0532355	0.06205671	0.34232379	0.56418755
France	2.7877932	0.86907271	6.6816578	**0.01656007**
Germany	3.8266779	0.77420438	2.9390272	**0.09990906**
Greece	3.991916	**0.03655664**	4.5180759	**0.04449487**
Hungary	4.3838286	0.72355777	0.7036045	0.4102088
Ireland	3.5941421	**0.045145**	3.9493433	**0.05892521**
Italy	2.0189013	**0.00094454**	5.566444	**0.02717319**
Latvia	1.3859026	0.23478501	0.04157418	0.84022696
Lithuania	2.4753699	0.89602425	0.00601925	0.93883009
Luxembourg	6.0675571	0.58434803	4.1631601	**0.05295489**
Malta	1.6305049	0.95833272	0.575248	0.45588331
Netherlands	1.2693002	0.2653382	0.52459805	0.47619053
Poland	1.5422406	0.20079156	7.996141	**0.009541**
Portugal	6.8891515	0.52607182	0.01548059	0.90206295
Romania	6.8063024	0.53165316	0.77910904	0.38654645
Slovakia	3.3734292	**0.05116054**	8.147571	**0.0089717**
Slovenia	3.56713	0.79813162	0.66118376	0.42448012
Spain	4.4957263	0.71358456	0.56826011	0.45859954
Sweden	2.2820532	0.91188955	0.08047229	0.77919632

Source Own elaboration
In bold: statistically significant at 10%

References

Adedoyin, F.F., Bekun, F.V.: Modelling the interaction between tourism, energy consumption, pollutant emissions and urbanization: renewed evidence from panel VAR. Environ. Sci. Pollut. Res. **27**(31), 38881–38900 (2020)

Ady, S.U., Moslehpour, M., Van, D.N., Johari, S.M., Hieu, V.M.: The impact of sustainable tourism growth on the economic development: evidence from a developing economy. Cuad. Econ. **45**(127), 130–139 (2022)

Agyeman, F.O., Zhiqiang, M., Li, M., Sampene, A.K., Dapaah, M.F., Kedjanyi, E.A.G., Buabeng, P., Li, Y., Hakro, S., Heydari, M.: Probing the effect of governance of tourism development, economic growth, and foreign direct investment on carbon dioxide emissions in Africa: the African experience. Energies **15**(13), 4530 (2022)

Akadiri, S.S., Alola, A.A., Akadiri, A.C.: The role of globalization, real income, tourism in environmental sustainability target. Evidence from Turkey. Sci. Total. Environ. **687**, 423–432 (2019)

Albaladejo, I.P., Brida, J.G., González-Martínez, M.I., Segarra, V.: A new look to the tourism and economic growth nexus: a clustering and panel causality analysis. World Econ. **46**(9), 2835–2856 (2023)

Alcalá-Ordóñez, A., Brida, J.G., Cárdenas-García, P.J.: Has the tourism-led growth hypothesis been confirmed? Evidence from an updated literature review. Curr. Iss. Tour. 1–37 (2023)

Amir, M., Siddique, M., Ali, K., Bukhari, A.A.A., Kausar, N.: Asymmetric relationship of environmental degradation and economic growth with tourism demand in Pakistan: evidence from non-linear ARDL and causality estimation. Environ. Sci. Pollut. Res. **29**(4), 5891–5901 (2022)

Anser, M.K., Yousaf, Z., Nassani, A.A., Abro, M.M.Q., Zaman, K.: International tourism, social distribution, and environmental Kuznets curve: evidence from a panel of G-7 countries. Environ. Sci. Pollut. Res. **27**(3), 2707–2720 (2020)

Badimo, D., Yuhuan, Z.: The relationship between inbound tourism and economic growth: case of Botswana. Nat. Resourc. For. **48**(1), 16–34 (2024)

Badulescu, D., Simut, R., Mester, I., Dzitac, S., Sehleanu, M., Bac, D.P., Badulescu, A.: Do economic growth and environment quality contribute to tourism development in EU countries? A panel data analysis. Technol. Econ. Dev. Econ. **27**(6), 1509–1538 (2021)

Balaguer, J., Cantavella-Jordá, M.: Tourism as a long-run economic growth factor: the Spanish case. Appl. Econ. **34**(7), 877–884 (2002)

Balsalobre-Lorente, D., Driha, O.M., Shahbaz, M., Sinha, A.: The effects of tourism and globalization over environmental degradation in developed countries. Environ. Sci. Pollut. Res. **27**(7), 7130–7144 (2020)
Bano, S., Alam, M., Khan, A., Liu, L.: The nexus of tourism, renewable energy, income, and environmental quality: an empirical analysis of Pakistan. Environ. Dev. Sustain. **23**(10), 14854–14877 (2021)
Bekun, F.V., Gyamfi, B.A., Bamidele, R.O., Udemba, E.N.: Tourism-induced emission in Sub-Saharan Africa: a panel study for oil-producing and non-oil-producing countries. Environ. Sci. Pollut. Res. **29**(27), 41725–41741 (2022)
Ben Jebli, M., Ben Youssef, S., Apergis, N.: The dynamic linkage between renewable energy, tourism, CO_2 emissions, economic growth, foreign direct investment, and trade. Latin Am. Econ. Rev. **28**(2), 1–19 (2019)
Bento, J.P.: Tourism and economic growth in Portugal: an empirical investigation of causal links. Tour. Manag. Stud. **12**(1), 164–171 (2016)
Boonyasana, P., Chinnakum, W.: Linkages among tourism demand, human development, and CO_2 emissions in Thailand. Abac J. **40**(3), 78–98 (2020)
Brida, J.G., Cortes-Jimenez, I., Pulina, M.: Has the tourism-led growth hypothesis been validated? A literature review. Curr. Issue Tour. **19**(5), 394–430 (2016)
Chien, F., Zhang, Y., Sharif, A., Sadiq, M., Hieu, M.V.: Does air pollution affect the tourism industry in the USA? Evidence from the quantile autoregressive distributed lagged approach. Tour. Econ. **0**(0) (2022)
De Vita, G., Katircioglu, S., Altinay, L., Fethi, S., Mercan, M.: Revisiting the environmental Kuznets curve hypothesis in a tourism development context. Environ. Sci. Pollut. Res. **22**(21), 16652–16663 (2015)
Deng, Z., Zhou, M., Xu, Q.: How to decouple tourism growth from carbon emissions? A spatial correlation network analysis in China. Sustainability **14**(19), 11961 (2022)
Dogru, T., Bulut, U.: Is tourism an engine for economic recovery? Theory and empirical evidence. Tour. Manage. **67**, 425–434 (2018)
Dumitrescu, E.I., Hurlin, C.: Testing for Granger non-causality in heterogeneous panels. Econ. Model. **29**(4), 1450–1460 (2012)
El Menyari, Y.: The effects of international tourism, electricity consumption, and economic growth on CO_2 emissions in North Africa. Environ. Sci. Pollut. Res. **28**(32), 44028–44038 (2021)
Eyuboglu, K., Uzar, U.: The impact of tourism on CO_2 emission in Turkey. Curr. Issue Tour. **23**(13), 1631–1645 (2020)
Fahimi, A., Saint Akadiri, S., Seraj, M., Akadiri, A.C.: Testing the role of tourism and human capital development in economic growth. A panel causality study of microstates. Tour. Manag. Perspect. **28**, 62–70 (2018)
Gedikli, A., Erdoğan, S., Çevik, E.I., Çevik, E., Castanho, R.A., Couto, G.: Dynamic relationship between international tourism, economic growth and environmental pollution in the OECD countries: evidence from panel VAR model. Econ. Res. Ekon. Istraž. **35**(1), 5907–5923 (2022)
Ghosh, S.: Effects of tourism on carbon dioxide emissions, a panel causality analysis with new data sets. Environ. Dev. Sustain. **24**(3), 3884–3906 (2022)
Granger, C.W.J.: Investigating causal relations by econometric models and cross-spectral methods. Econometrica **37**(3), 424–438 (1969)
Gyamfi, B.A., Bein, M.A., Adedoyin, F.F., Bekun, F.V.: How does energy investment affect the energy utilization-growth-tourism nexus? Evidence from E7 countries. Energy Environ. **33**(2), 354–376 (2022)
Habibi, F., Amani, R.: The impact of geopolitical risk, world economic policy uncertainty on tourism demand: evidence from Malaysia. Iran. Econ. Rev. **26**(2), 477–488 (2022)
Im, K., Pesaran, M.H., Shin, Y.: Testing for unit roots in heterogeneous panels. J. Econ. **115**(1), 53–74 (2003)
Jambor, A., Leitão, N.C.: Economic growth and sustainable development: evidence from central and eastern Europe. Int. J. Energy Econ. Policy **7**(5), 171–177 (2017)
Khan, I., Han, L., Khan, H.: Renewable energy consumption and local environmental effects for economic growth and carbon emission: evidence from global income countries. Environ. Sci. Pollut. Res. **29**(9), 13071–13088 (2022)
Khan, A.M., Naseem, S., Khan, U.: Tourism industry and its impact on economic growth of five GCC countries. Int. J. Entrepr. **25**, 1–9 (2021)
Kongbuamai, N., Bui, Q., Yousaf, H.M.A.U., Liu, Y.: The impact of tourism and natural resources on the ecological footprint: a case study of ASEAN countries. Environ. Sci. Pollut. Res. **27**(16), 19251–19264 (2020)
Kumail, T., Ali, W., Sadiq, F., Abbas, S.M.: A step toward tourism development: do economic growth, energy consumption and carbon emissions matter? Evidence from Pakistan. Environ. Dev. Sustainab. 1–21 (2022)
Lee, J.W., Syah, A.M.: Economic and environmental impacts of mass tourism on regional tourism destinations in Indonesia. J. Asian Financ. Econ. Bus. **5**(3), 31–41 (2018)
Lenzen, M., Sun, Y.Y., Faturay, F., Ting, Y.P., Geschke, A., Malik, A.: The carbón footprint of global tourism. Nat. Clim. Chang. **8**(6), 522–528 (2018)
Levin, A., Lin, C.F., Chu, C.S.J.: Unit root tests in panel data: asymptotic and finite-sample properties. J. Econ. **108**(1), 1–24 (2002)
Li, X.: Green innovation behavior toward sustainable tourism development: a dual mediation model. Front. Psychol. **13**, 930973 (2022)
Liu, Z., Lan, J., Chien, F., Sadiq, M., Nawaz, M.A.: Role of tourism development in environmental degradation: a step towards emission reduction. J. Environ. Manag. **303**, 114078 (2022)
Lolos, S., Palaios, P., Papapetrou, E.: Tourism-led growth asymmetries in Greece: evidence from quantile regression analysis. Port. Econ. J. 1–24 (2021)
Montesdeoca, E.A.C., Roget, F.M., Rozas, E.V.: O turismo en Ecuador. Novas tendencias no turismo sostible e contribución ao crecemento económico. Rev. Galega Econ. Publ. Interdiscip. Fac. Cien. Econ. Empr. **24**(2), 69–84 (2015)
Moslehpour, M., Shalehah, A., Wong, W.-K., Ismail, T., Altantsetseg, P., Tsevegjav, M.: Economic and tourism growth impact on the renewable energy production in Vietnam. Environ. Sci. Pollut. Res. **29**(53), 81006–81020 (2022)
Musa, M.S., Jelilov, G., Iorember, P.T., Usman, O.: Effects of tourism, financial development, and renewable energy on environmental performance in EU-28: does institutional quality matter? Environ. Sci. Pollut. Res. **28**(38), 53328–53339 (2021)
Nadeem, M.A., Liu, Z., Xu, Y., Nawaz, K., Malik, M.Y., Younis, A.: Impacts of terrorism, governance structure, military expenditures and infrastructures upon tourism: Empirical evidence from an emerging economy. Eurasian Bus. Rev. **10**, 185–206 (2020)
Niu, J.: Novel research methods on evaluating the nexus between environment and energy use: evaluating the role of tourism in the pre-COVID period. Econ. Res. Ekon. Istraz. **36**(1), 1490–1509 (2022)
Oad, S., Jinliang, Q., Shah, S.B.H., Memon, S.U.R.: Tourism: economic development without increasing CO_2 emissions in Pakistan. Environ. Dev. Sustain. **24**(3), 4000–4023 (2022)
Opuala-Charles, S., Omoke, P.C., Uche, E.: The dynamic role of financial development in analyzing the N-shaped tourism-led growth hypothesis in mauritius. J. Knowl. Econ. 1–26 (2022)
Ozturk, I., Aslan, A., Altinoz, B.: Investigating the nexus between CO_2 emissions, economic growth, energy consumption and pilgrimage tourism in Saudi Arabia. Econ. Res. Ekon. Istraz. **35**(1), 3083–3098 (2022)

Rahaman, M.A., Hossain, M.A., Chen, S.: The impact of foreign direct investment, tourism, electricity consumption, and economic development on CO_2 emissions in Bangladesh. Environ. Sci. Pollut. Res. **29**(25), 37344–37358 (2022)

Raihan, A., Muhtasim, D.A., Pavel, M.I., Faruk, O., Rahman, M.: Dynamic impacts of economic growth, renewable energy use, urbanization, and tourism on carbon dioxide emissions in Argentina. Environ. Process. **9**(2), 38 (2022)

Roudi, S., Arasli, H., Akadiri, S.S.: New insights into an old issue–examining the influence of tourism on economic growth: evidence from selected small island developing states. Curr. Issue Tour. **22**(11), 1280–1300 (2019)

Segarra, V., Brida, J.G., Cárdenas-García, P.J.: On the relationships between tourism demand, carbon dioxide emissions and economic growth: a literature review. J. Policy Res. Tour. Leisure Events, 1–42 (2024)

Shahbaz, M., Ferrer, R., Shahzad, S.J.H., Haouas, I.: Is the tourism–economic growth nexus time-varying? Bootstrap rolling-window causality analysis for the top 10 tourist destinations. Appl. Econ. **50**(24), 2677–2697 (2018)

Shakouri, B., Khoshnevis, S., Ghorchebigi, E.: Does tourism development promote CO_2 emissions? Anatolia **28**(3), 444–452 (2017)

Sofi, A.A., Bhat, M.Y., Ahmad, L., Aara, R.R., Aswani, R.S.: Renewable energy and transitioning towards sustainable tourism: Inferences from kernel density and nonparametric approach. Renew. Energy **193**, 963–975 (2022)

Su, Y., Cherian, J., Sial, M.S., Badulescu, A., Thu, P.A., Badulescu, D., Samad, S.: Does tourism affect economic growth of China? A panel granger causality approach. Sustainability **13**(3), 1349 (2021)

Sun, Y.Y., Gossling, S., Zhou, W.: Does tourism increase or decrease carbon emissions? A systematic review. Ann. Tour. Res. **97**, 1035–1102 (2022)

Tang, C.F., Cheah, Y.K., Chua, S.Y.: Does educational tourism significantly influence economic growth? Evidence from a macro-econometric modelling. Int. J. Bus. Soc. **20**(3), 924–935 (2019)

Thao, H.T.P., Bakucz, M.: The impact of growth in tourism on economic and human development-incorporating a systematic literature review. J. Environ. Manag. Tour. **13**(1), 287–299 (2022)

Tugcu, C.T.: Tourism and economic growth nexus revisited: a panel causality analysis for the case of the Mediterranean Region. Tour. Manage. **42**, 207–212 (2014)

U.N. Tourism: Transport-related CO_2 emissions of the tourism sector. UNWTO, Madrid, Spain (2019)

U.N. Tourism: UN Tourism Barometer 2023. UN Tourism, Madrid, Spain (2024)

Uzuner, G., Akadiri, S.S., Lasisi, T.T.: The asymmetric relationship between globalization, tourism, CO_2 emissions, and economic growth in Turkey: implications for environmental policy making. Environ. Sci. Pollut. Res. **27**(26), 32742–32753 (2020)

Villanthenkodath, M.A., Ansari, M.A., Shahbaz, M., Vo, X.V.: Do tourism development and structural change promote environmental quality? Evidence from India. Environ. Dev. Sustain. **24**(4), 5163–5194 (2022)

Xiangyu, S., Jammazi, R., Aloui, C., Ahmad, P., Sharif, A.: On the nonlinear effects of energy consumption, economic growth, and tourism on carbon footprints in the USA. Environ. Sci. Pollut. Res. **28**(16), 20128–20139 (2021)

Environmental Implications of the NEOM Project in Saudi Arabia: A Geographical Analysis

Giovanna Giulia Zavettieri

Abstract

This paper will analyze the environmental impacts of the NEOM project in Saudi Arabia, with a particular focus on desertification and its implications for human geography and culture in the region. The discussion will examine the environmental dynamics that characterize the Saudi desert and investigate the climate change mitigation strategies adopted in the context of the project. Furthermore, the role of human geography and culture in shaping and adapting to these environmental challenges will be explored, identifying opportunities and obstacles arising from this intersection. The objective of the presentation is to contribute to the understanding of the complex interaction between environment, society and culture in desert contexts such as Saudi Arabia, providing insights for future research and intervention strategies.

Keywords

Desertification · Human geography · Climate change mitigation · Saudi Arabia

1 Sustainable Adaptations in Arid Environments: Human–Nature Interactions and Technological Innovations

The techniques employed for the control of droughts are of ancient provenance and have been handed down from generation to generation. They are employed throughout the territory in a variety of contexts, including soil management, use and protection of natural areas, rural architecture, and organization of urban centers. These techniques represent the historical knowledge of humanity that has allowed the realization of architectural and landscape designs of universal value. Through the application of traditional knowledge, the appropriate use of natural resources, including water, soil, and solar energy, is realized. The concept of architectural and environmental harmony is defined by the integration of spatial organization techniques, traditions, social customs, and spiritual values, resulting in a harmonious fusion of functionality and beauty (Gagliardo 2005).

The desert, a challenging environment for humans, has been a site of trial and struggle for survival for numerous civilizations.

Desertification represents a significant environmental challenge that affects local communities in diverse ways, contingent upon factors such as climate, geology, and socio-cultural-political structure. The vulnerability of resources, adverse weather conditions, and recurring droughts are further exacerbated by global warming. While humans are affected by land degradation, they are also a significant contributing factor. Despite the extensive research and publications on degradation and desertification, it remains unclear to what extent these findings have reached government decision-makers and have been implemented (Farshad et al. 2024).

The United Nations Convention to Combat Desertification (UNCCD) defines desertification as land degradation in arid, semi-arid, and sub-humid dry areas, which is attributable to various causes, including climatic variations and human activities. The technologies of most interest to desert areas are those related to forecast support systems, decision and pattern finding, rain simulation (cloud seeding), and biofertility enhancement (Gagliardo 2005). In particular, the advent of rain simulation technology has attracted considerable media attention in the wake of the events that transpired in the United Arab Emirates and Oman in

G. G. Zavettieri (✉)
Tor Vergata University of Rome, Rome, Italy
e-mail: giovanna.zavettieri@uniroma2.it

P. Mohammad et al. (eds.), *Geographic Approaches to Climate Change and Mitigation: Urban and Rural Perspectives (Volume 2)*,
Advances in Science, Technology & Innovation, https://doi.org/10.1007/978-3-031-92115-5_4

April 2024. Through the clouds seeding with silver iodide crystals, this technology has been employed for years with the objective of overcoming drought in numerous countries. In addition to the aforementioned technique, it is fundamental to acknowledge the role of remote sensing systems that not only enhance our understanding of the environmental state and its processes but also facilitate the prediction of future scenarios. The current IT potential enables the production and visualization of large amounts of information through the use of models that integrate diverse data types, thereby providing support for decision-making and guiding interventions to counter desertification.

It is of utmost importance that policy-making processes take into account the technical and scientific recommendations developed by the UNCCD, the primary international instrument for mitigating desertification. In particular, the Convention to combat desertification, a global action plan to reverse the trend of desertification by 2030, should be given significant consideration. Furthermore, the Global Environment Facility (GEF) provides funding for projects aimed at reducing desertification and soil degradation on a global scale. Among its principal objectives is the implementation of sustainable soil and water resource management projects in arid areas.

It is also important to consider anthropic and cultural issues, as some contemporary urban realities represent examples of creations "from scratch", developed through public–private partnerships that have acquired considerable media prominence due to their uniqueness. Precisely because of their uniqueness, these cities cannot be considered replicable models for other territories, as they represent extremely utopian visions of development. Advertised as examples of technological innovation and urban intelligence, these places are exceptions and often appear detached from the reality of their home countries. In these contexts, territories face significant challenges related to social exclusion, pollution, traffic, and other common issues of daily urban life.

Saudi Arabia is facing significant challenges in water scarcity, particularly in view of the country's growing reliance on desalination, which gives rise to both sustainability and adaptability concerns. The resilience of infrastructure in key cities such as Riyadh and Jeddah is noteworthy, underscoring the need for innovative solutions to counter unforeseeable environmental disruptions. Furthermore, the country's approach to integrating its deeply rooted cultural heritage with emerging environmental imperatives is noteworthy. This represents a unique convergence of tradition and contemporaneity that resonates on a global scale. Geopolitical dynamics, accentuated by Saudi Arabia's dominant role in international arenas, bring climate change-induced socio-political and economic changes to the forefront. Furthermore, the advent of technological advances highlights the pivotal role of robust digital frameworks in addressing climate challenges. Saudi Arabia's governance policies and initiatives offer valuable insights into effective adaptation strategies, underscoring a triad of collaboration, vision, and ingenuity (Mani and Goniewicz 2023).

This study aims to analyze desertification and its implications for human geography and culture, with a specific focus on the Saudi desert and the ultra-technological NEOM project (see Fig. 1a, b).

The primary research questions are as follows: How do the environmental dynamics of the Saudi desert influence climate change mitigation strategies? What are the interactions between human geography, culture, and adaptation to environmental challenges in a desert context? What opportunities and obstacles arise from the integration of advanced technologies in traditionally arid regions? By addressing these research questions, the study seeks to elucidate how the intersection of environment, society, and culture contributes to sustainability and human adaptation in desert contexts, and to offer insights for future research and intervention strategies.

The study involved a comprehensive review of the existing literature on desertification, climate change, and sustainable

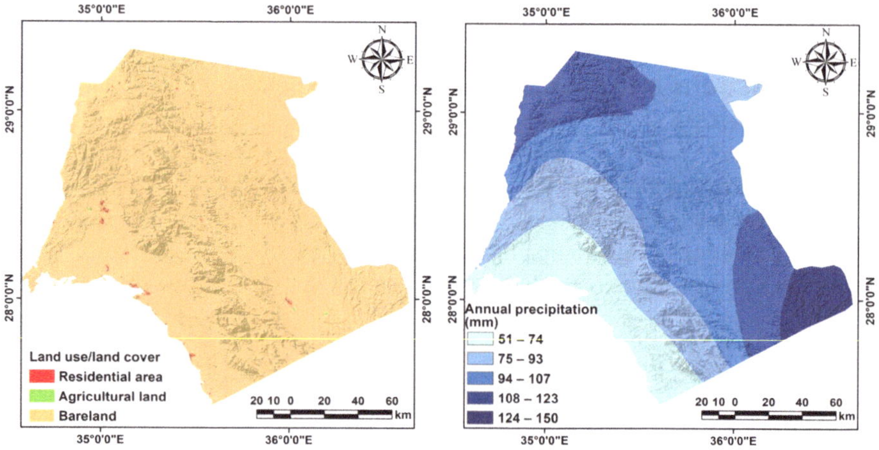

Fig. 1 **a** Land use/land cover map showing that the NEOM area is a typical desert area primarily made up of bare land. The residential and agricultural areas represent less than 1% of the total area (Alharbi 2023). **b** Map presenting the annual precipitation average for the study region. The northwest and southwest regions of NEOM receive the most precipitation (Alharbi 2023)

Fig. 2 The line project (https://www.neom.com/en-us/regions/theline)

urbanization, with a particular focus on arid environments. It encompassed academic articles, government reports, and case studies pertaining to the NEOM project and analogous initiatives. Additionally, data were gathered through field studies, which enabled the evaluation of environmental conditions, resource availability, and the influence of existing urbanization on local ecosystems. The involvement of local communities, stakeholders, and geoscience and environmental management experts facilitated the collection of qualitative data pertaining to perceptions, challenges, and opportunities associated with the NEOM project. The utilization of satellite imagery and geographic information systems (GIS) could facilitate the analysis of alterations in land utilization, water resource management, and the ecological footprint of the NEOM project.

2 Sustainable Urbanization in the Desert and Environmental Considerations: NEOM Project

Despite the proximity of the Arabian Peninsula and North Africa in terms of geography and culture, particularly in regard to religion and language, the formation of cities in the two regions is strikingly different, due to the history and geography of each of them. Cities in North Africa tend to be an amalgam of Arab, Berber, and colonial influences, reflecting the various dominations and cultural interactions that have occurred throughout history. Cities in the Arabian Peninsula, on the other hand, have developed distinctive characteristics, influenced primarily by Arab culture and the unique environmental conditions of the surrounding desert.

The presence of National Visions geared toward economic diversification emerges as a common denominator among countries in the Arabian Peninsula that are committed to reduce dependence on natural resources and hydrocarbons, a significant challenge in achieving sustainable economic development. Even the waterfront areas of the region have benefited from targeted investments towards the development of coastal infrastructure. In contrast, the internal desert and mountainous areas highlight the need for more balanced planning, considering environmental and social equilibrium, in order to ensure harmonious and sustainable development.

The NEOM project represents an ambitious initiative by the Kingdom of Saudi Arabia to construct a new, futuristic metropolis in Tabuk province, situated in proximity to the Red Sea. The project's objective is to set up a state-of-the-art region that is powered by 100% renewable energy, and which employs a range of innovative technologies and a diversified economy. The area of the proposed development is 26,500 km^2, while its location is in northwestern Saudi Arabia, in proximity to the Red Sea.

The primary urban project planned for the NEOM area is The Line, a linear megacity (see Fig. 2). This ambitious undertaking will encompass a global commercial center, floating industrial complexes, and tourist resorts. Currently, it is still under development, with an estimated completion date of 2025.

In essence, NEOM is a new regional development part of Saudi Vision 2030, born from the idea of sustainability, conservation of nature and development.

Its landscape includes geomorphological features such as canyons, mesa, butte, and mushroom rocks and geological features such as sandstone cross bed, faults, dyke and soft sedimentary deformation structures.[1] The esthetic beauty of nature is an added advantage for ecotourism development in

[1] Given that NEOM espouses environmental sustainability and urbanization, it is imperative that natural heritage be made accessible to the geoscientific community and explained in an open and transparent manner to the general public. Consequently, geotourism can be a valuable tool for promoting geoscientific education. It is similarly evident that geoconservation will be of significant consequence in this context. However, a series of strategic measures will be required to safeguard the

Fig. 3 Architectural rendering of the reservoir and surrounding structures in Trojena (Patella 2024)

NEOM, for example, the Sindalah island and Trojena initiative in NEOM mountain. In general, there are several opportunities for nature-based tourism development (Sen et al. 2024).

The aim of the Trojena project is to create a state-of-the-art tourist resort, including an artificially snowed ski area. The website uses the word 'alpine' to refer to the types of activities and sports that will be available. A controversial aspect of the project is its location in the desert, which raises doubts about its environmental sustainability, due to the need to use large amounts of water and energy to produce artificial snow. In addition, the project has had a significant impact on local communities who had to move to make way for the construction. Amongst the affected communities are the tribes of Al-Huwaitat, whose inhabitants were forced to relocate. The proximity of Tayma, an ancient city in northwestern Saudi Arabia near Tabuk, also presents a challenge to the preservation of its ruins and archaeological artifacts. The well of Tayma, a renowned water well, is among the oldest ever discovered in Arabia.

To serve Trojena, three dams will be constructed, which will facilitate the creation of a freshwater lake (see Fig. 3). *The Bow*, an architectural structure that will have the shape of the bow of a ship and extend the surface of the lake beyond the front of the main dam, will be constructed in the basin. The lake will also contain a luxury hotel, a residential area, and a central atrium with accommodation and recreational facilities. The artificial lake will cover an area of 1.5 km^2 in total and will feature an island dedicated to diving and botanical walks. The basin is a central element in the Trojena project, serving as a basic support for the natural environment of the area and offering residents and customers an open-air mountain lifestyle. As well, the body of water will become a global hub for water sports and all-round entertainment.

3 Bridging Sustainable Urban Futures: NEOM Project in the Context of Saudi Arabia's Climate Policies

The objective of *The Line* project is to redefine the concept of urban development with the intention of establishing the design standards of the new cities of the future. The concept of *The Line* entails a predominantly altimetric trend, whereby the city develops in elevation and no longer on the level ground. This development in elevation is then repeated in plan but following the specific path of a straight line. The project will alter the manner in which communities typically reside, and it will also affect the manner in which the surrounding environment interacts with the city in a punctual manner (one might consider, for instance, the distribution of sunlight in the project area).

The proposed structure will be 500 m tall and will alternate between urban and "natural" areas along its height. The objective is to preserve an important green area, which will not be concentrated in specific points but will be immersed in the urban context. This fact will also allow the climate in the city to be managed to ensure the right environmental comfort. From the lowest to the highest areas, it should be possible to have access to the services a person needs on a daily basis, thereby minimizing travel times and allowing for the gain of free time, which can then be used for other activities.

A significant aspect of the design is the geotechnical engineering of the foundation systems for an extension of 170 km. While it is conceivable that technological disruptions could occur during the development of the structures along the line, the geotechnical characterization of all the foundation soils being crossed remains relevant, as does the potential for a unified design solution to be implemented throughout the city's development.

natural heritage and facilitate geotourism, which in turn will contribute to local economic development.

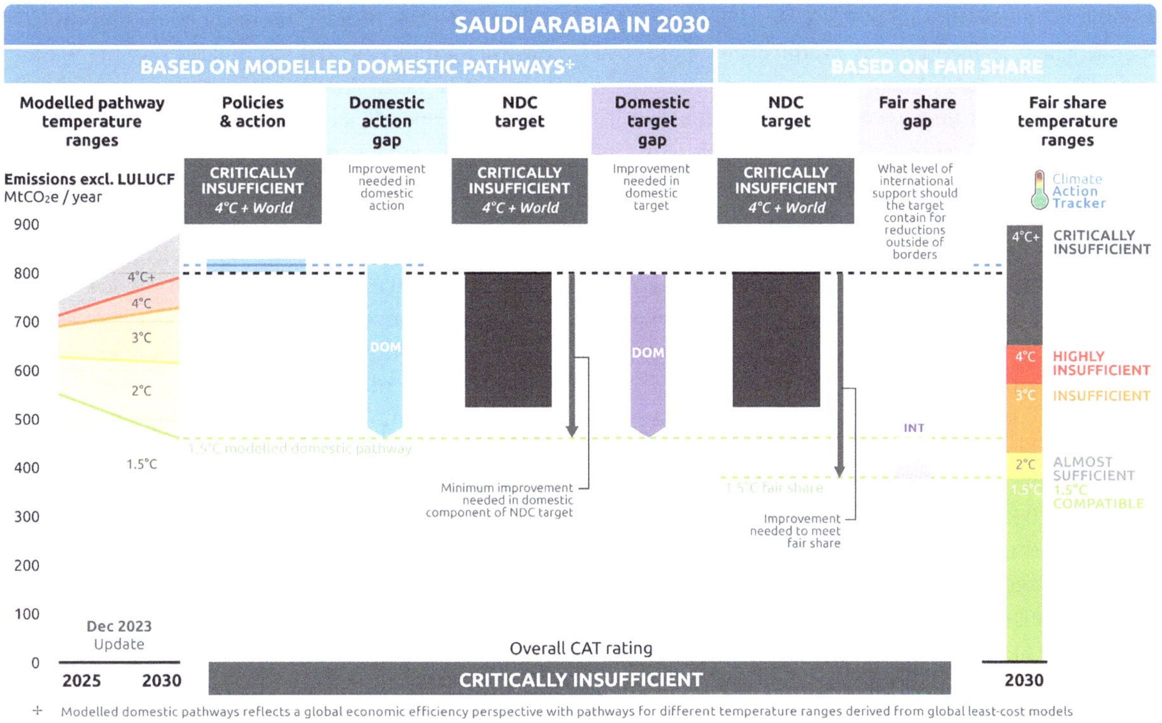

Fig. 4 Saudi Arabia CAT rating (https://climateactiontracker.org/countries/saudi-arabia/policies-action/)

Climate change has a profound impact on Saudi Arabia, resulting in arid climate, elevated temperatures, flash floods, reduction in water resources, and alterations in rainfall patterns (Saghir 2022; Ahmad and Rizvi 2023). Saudi Arabia has demonstrated a profound commitment to mitigating the effects of climate change through the reduction of greenhouse gas emissions and has pledged to achieve net-zero greenhouse gas emissions by 2060 (The Associated Press 2021; Ahmad and Rizvi 2023). The country has set a goal of planting 10 billion trees in various landscapes to improve air quality, decrease sandstorms, combat desertification, and mitigate the microclimate impact of lower temperatures (Saudi Green Initiative 2021).

Nevertheless, the Climate Action Tracker (CAT) has evaluated Saudi Arabia's policies and actions in modeled domestic pathways and determined that they are "critically insufficient"[2] (see Fig. 4). This assessment shows that Saudi Arabia's policies and actions for 2030 show minimal action and are inconsistent with the 1.5 °C warming limit. If all countries were to follow Saudi Arabia's approach, warming would exceed 4 °C.

Also, the report shows compelling evidence that Saudi Arabia has not implemented any policies capable of substantially decreasing its emissions and reducing its dependence on fossil fuels. Emissions are projected to increase to 800–830 MtCO2e by 2030, representing a 15–19% increase compared to 2021 levels. Even if Saudi Arabia were to achieve its target of 50% renewable energy in the energy mix, emissions are still expected to rise to 730–770 MtCO2e by 2030.

Saudi Arabia is relying on carbon capture and storage (CCS) technologies as a means of maintaining its oil and gas production levels, rather than implementing measures to reduce emissions from fossil fuels. Furthermore, sectoral policies are being examined, including initiatives related to methane, coal phase-outs, electric vehicles, forests, and beyond oil and gas.

In addition to the aforementioned factors, other aspects are also being analyzed, including voluntary carbon markets, energy production, fossil fuel subsidies, the transportation sector, and building management.

An article published on Al Arabiya (2023) reports that the construction of the $8.4 billion green hydrogen production plant in the heart of Saudi Arabia's NEOM will establish the Kingdom as a global leader in the transition to clean energy (Bell 2023). In an exclusive interview, David Edmondson, CEO of the NEOM Green Hydrogen Company (NGHC), revealed that construction of the site, located in NEOM's Oxagon, is progressing according to plan and is expected to be completed by the end of 2026.

On completion, the joint venture between Acwa Power, Air Products, and NEOM will become the world's largest green hydrogen production facility, spanning over 300 km^2 of land. It will be equipped with approximately 5.6 million solar

[2] https://climateactiontracker.org/countries/saudi-arabia/policies-action/.

panels, capable of generating up to 2.2 GW of solar energy. When operational, the facility will produce an impressive quantity of carbon-free hydrogen.

The CEO highlighted that upon completion, the facility will be capable of producing up to 600 tons of carbon-free hydrogen per day for global transportation. This will be achieved through the integration of up to 4 GW of solar and wind energy, with the potential to export up to 1.2 million tons of green ammonia annually. Edmondson emphasized that Saudi Arabia is pursuing a transition towards renewable energy in order to fulfill its ambitious climate commitments and to diversify its oil exports.

Furthermore, he asserted that Saudi Arabia possesses considerable potential for hydrogen production, with plans to generate 2.9 million tons per year by 2030 and 4 million tons per year by 2035. The plant is designed to support the Kingdom's climate mandate by facilitating the large-scale adoption of green hydrogen.

The company has concluded agreements with financial institutions valued at $8.4 billion for the construction of the plant. It is anticipated that the initial investment will result in the creation of up to 300 jobs. Furthermore, there is potential for further hiring in the future.

Another noteworthy aspect to consider is the response of Shady Khalil, Campaigns Director for Greenpeace MENA, to the letter sent by OPEC to its members (Flores 2023). Khalil expresses profound dismay at OPEC's desperate and self-serving request to its members to reject any action against fossil fuels during COP28. This deliberate attempt by the oil and gas industry to obstruct crucial climate action demonstrates that they are feeling the pressure as negotiations move towards a gradual reduction of fossil fuels. Khalil strongly urges oil-producing countries, including Saudi Arabia, to embrace the goal set by the COP presidency of the United Arab Emirates and the International Energy Agency (IEA) to reduce fossil fuels in this decade to achieve the 1.5 °C target, and to work together to make this conference a historic milestone. Khalil underscores the significance of Arab countries at COP28 prioritizing the safety and well-being of their citizens over the short-term interests of the industry.

4 Conclusions

Upon consideration of the data presented on Saudi Arabia's policies and actions, including its continued commitment to expanding oil and gas production and its limited implementation of renewable energies, it becomes evident how this intersects with the NEOM project.

The NEOM project represents an ambitious effort to construct a futuristic and sustainable city in the Saudi desert. However, if Saudi Arabia persists on its current trajectory of fossil fuel dependence and limited commitment to renewable energies, there is a possibility that NEOM's sustainability vision may diverge from the country's actual policies and actions.

To ensure the sustainability of NEOM, Saudi Arabia should undertake a comprehensive review of its energy and environmental policies. This may necessitate a more extensive adoption of renewable energies throughout the country, not merely in NEOM, and a more robust commitment to reducing greenhouse gas emissions and mitigating environmental impacts. Furthermore, NEOM's approach to water resource management, urban planning, and environmental conservation should be integrated with Saudi Arabia's national policies to ensure coherence and a more sustainable environmental footprint for the country as a whole.

References

Ahmad, F., Rizvi, J.: GIS modeling of environmental parameters for identification of landscape for expansion of biodiversity conservation, and greenery projects (a case study for Neom region, Saudi Arabia). Model. Earth Syst. Environ. **9**(2), 2987–2992 (2023)

Alharbi, T.: Mapping of groundwater, flood, and drought potential zones in neom, Saudi Arabia, using GIS and remote sensing techniques. Water **15**(5), 966 (2023)

Bell, J.: NEOM's green hydrogen plant will secure Saudi Arabia's clean energy transition: CEO. Al Arabiya. https://english.alarabiya.net/News/2023/05/25/Saudi-s-green-hydrogen-plant-will-put-Kingdom-on-global-map-for-clean-energy-CEO (2023)

Climate Action Trackers. https://climateactiontracker.org/countries/saudi-arabia/policies-action/

Farshad, A., Shrestha, D.P., Abdelfattah, M.A.: Desertification seen through the eyes of a group of international students. Labyrinth Fayoum J. Sci. Interdiscip. Stud. **2**, 1–7 (2024)

Flores, G.: Greenpeace response to OPEC letter at COP28. Greenpeace. https://www.greenpeace.org/international/press-release/64308/greenpeace-response-to-opec-letter-at-cop28/?_gl=1*blwirz*_up*MQ..*_ga*MTYyMjU5MTc4Ni4xNzEzNTM5MzMz*_ga_94MRTN8HG4*MTcxMzUzOTMzMy4xLjAuMTcxMzUzOTcyNy4wLjAuMTc2MTQxNzkyMQ..*_ga_0CCB1GTVV6*MTcxMzUzOTMzMy4xLjEuMTcxMzUzOTcyNy4wLjAuMA..*_ga_99CSX66YC1*MTcxMzUzOTMzMy4xLjEuMTcxMzUzOTcyNy4wLjAuMA (2023)

Gagliardo, P.: Verso una integrazione delle conoscenze tradizionali e nuove tecnologie per la lotta alla siccità e alla desertificazione. Geotema **25**, 27–30 (2005)

Mani, Z.A., Goniewicz, K.: Adapting disaster preparedness strategies to changing climate patterns in Saudi Arabia: a rapid review. Sustainability **15**(19), 1–19 (2023)

Patella, A.: L'Arabia saudita costruirà 3 colossali dighe per aprire una stazione di sci. Wired. https://www.wired.it/gallery/arabia-saudita-dighe-stazione-sci-neom-webuild/ (2024)

Saghir, B.: Climate change in Saudi Arabia: rethinking the role of local government. In: King Salman Center for Local Governance. http://www.ksclg.org/en/publication-project/climate-change-in-saudi-arabia-rethinking-the-role-of-local-government/ (2022)

Saudi Green Initiative. https://www.greeninitiatives.gov.sa/

Sen, S., Abouelresh, M.O., Al-Musabeh, A.H., Al-Ismail, F.S.: Potential geoheritage resources in Saudi Arabia for geotourism development: In the context of IUCN theme. Int. J. Geoherit. Parks **12**(1), 98–112 (2024)

The Associated Press: Saudi Arabia pledges net-zero greenhouse gas emissions by 2060, NPR. https://www.npr.org/2021/10/23/1048655294/saudi-arabia-zero-emissions-climate-change-2060 (2021)

Variability and Long-Term Tendencies in CO Content and Their Links with Urban Climate Changes in the Atmosphere of the Moscow Megapolis

Vadim S. Rakitin, Andrey I. Skorokhod, Natalia S. Kirillova, Arseny V. Shilkin, and Eugenia I. Fedorova

Abstract

We present the results of a comprehensive analysis of long-term spectroscopic data sets on the CO total column (CO TC) at stations of the A.M. Obukhov Institute of Atmospheric Physics, Russian Academy of Sciences (OIAP RAS) in Moscow and Moscow province. The aim of the work was to continue our research on long-term trends in CO TC in the Moscow megalopolis and its surroundings and their interrelations with meteorological conditions. One of our objectives was to study the contribution of meteorological factors to the improvement of air quality in Moscow. The long-term variability of CO content and meteorological parameters in the atmospheric boundary layer (ABL) was studied. A decrease of CO TC in the period 2000–2023 was established for Moscow, (− 2.2 ± 0.4% yr^{-1}) and for Zvenigorod Scientific Station ZSS, − 1.1 ± 0.3% yr^{-1}. The rate of decrease in both sites is different in different seasons and time periods. After about 2007, the rate of CO TC reduction decreased at both sites. In the period 2009–2023 at the ZSS, the statistically insignificant increase of CO TC was recorded in the summer and autumn months. A decrease in the wind speed in the ABL of Moscow in different periods of 2003–2020 was found at the rate of 0.7–1.3% yr^{-1}. At the same time, no statistically significant changes in wind speed were found in the Kaluga region. Characteristics of CO trends in both sites in different seasons and periods are presented and discussed. The decrease in urban CO TC and concentration is primarily due to the reduction in anthropogenic emissions; however, the additional possible reason for the improvement in urban air quality could be the impact of some climatic factors, such as spatial changes in the urban "heat island" and its intensity and related factors leading to the intensification of turbulent processes in the ABL.

Keywords

Carbon monoxide · Atmospheric composition · Atmospheric spectroscopy · Remote sensing · Megapolis · Trends

1 Introduction

Carbon monoxide (CO) is considered one of the most important minor gaseous atmospheric pollutants, despite its low global warming potential. CO contributes to the removal of hydroxyl radicals from the atmosphere, thereby preventing the methane oxidation, and participates in the formation of tropospheric ozone and carbon dioxide. Taking indirect effects into account, the radiative forcing of CO is estimated to be greater than, for example, that of nitrous oxide and halocarbon hydrocarbons. Globally, the concentration of CO determines the level of hydroxyl, the main oxidant in photochemical reactions, a reduction which leads to the depletion of stratospheric ozone (Novelli et al. 1998; Jacob 1999). All of the above confirms the important role of carbon monoxide in the

V. S. Rakitin (✉) · A. I. Skorokhod · N. S. Kirillova · E. I. Fedorova
Obukhov Institute of Atmospheric Physics, Russian Academy of Sciences, Moscow, Russia
e-mail: vadim@ifaran.ru

A. I. Skorokhod
e-mail: skorokhod@ifaran.ru

N. S. Kirillova
e-mail: nkirillova@ifaran.ru

E. I. Fedorova
e-mail: fedorova@ifaran.ru

A. I. Skorokhod
Department of Meteorology and Geophysics, University of Vienna, Vienna, Austria

A. V. Shilkin
Laboratory of Research and Production Association Taifun, Obninsk, Russia
e-mail: dogget@list.ru

P. Mohammad et al. (eds.), *Geographic Approaches to Climate Change and Mitigation: Urban and Rural Perspectives (Volume 2)*, Advances in Science, Technology & Innovation, https://doi.org/10.1007/978-3-031-92115-5_5

processes of change in atmospheric composition and climate (IPCC 2021). The monitoring of atmospheric CO is also important in the study of urban air pollution as the concentrations of CO is one of the main parameters for assessing the air quality index.

Numerous studies report the decrease in concentrations and content of CO and some other atmospheric pollutants in the air of large cities (IPCC 2021; Timofeev et al. 2020; Elansky et al. 2022). This decrease is usually explained by the reduction of anthropogenic emissions (IPCC 2021; Elansky et al. 2018). However, we found that in 2005–2014 the CO concentrations in Moscow decreased at the rate of $-$ 3.6 ± 0.7%/year, while the rate of emission reduction was lower, $-$ 1.9 ± 0.3%/year (Elansky et al. 2018). Similar results for Moscow (the rate of emission reduction is lower than the rate of concentration reduction) were obtained in the work (Elansky et al. 2022) for several air pollutants.

In our recent study (Rakitin et al. 2021), we tried to explain the discrepancy between the rate of change of CO TC and the rate of emission reduction by the meteorological factor, i.e. the changes in meteorological conditions in Moscow during recent years. It was found that the trends of wind speed for different periods of 2000–2018 in the city and in rural areas are different (an increase in wind speed in Moscow and absence of changes in this parameter in rural areas. The average annual number of calm days in Moscow from 2005 to 2017 have decreased at a rate $-$ 7.1 ± 4.0% yr^{-1}. During the same period, the anthropogenic part of the CO content decreased at almost the same rate, $-$ 6.7 ± 3.5% yr^{-1} (Rakitin 2021). The purpose of this work was to continue our research (Rakitin et al. 2021) using new experimental information of recent years. In particular, the analysis of the relationship of CO content trends with changes in wind speed in the atmospheric boundary layer (ABL) of urban and nearby rural areas was continued (Fig. 1).

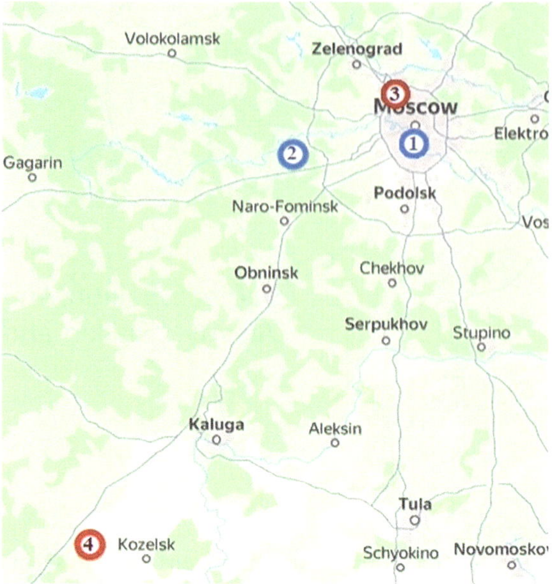

Fig. 1 Layout of the observation sites: **1** OIAP RAS, Moscow center (CO and H_2O TC); **2** Zvenigorod Scientific Station (ZSS) of the OIAP RAS, Moscow province (CO, CH_4, and H_2O TC); **3**, **4** Aerological sounding stations № 27,713 (Dolgoprudny, Moscow suburb, 3) and № 27,707 (Sukhinichi, rural Kaluga region, 4)

2 Research Methods

Long-term spectroscopic data sets have been used to analyze the variations of the total CO content (CO TC) in the atmosphere of the Moscow megacity and its surroundings. The results were obtained from two different locations: from the city center (OIAP RAS site) and the Zvenigorod Scientific Station (ZSS site or Zvenogorod), located 53 km west of the center of Moscow. Both sites are equipped with identical medium-resolution spectrometers with a resolution of 0.2 cm^{-1}, which analyze the absorption of solar radiation in the spectral region around the main CO absorption band (4.67 μm) (Yurganov et al. 2002; Rakitin et al. 2021). Measurements were taken on sunny days (70–80 days per year), and usually synchronously at both sites. The difference between measured CO TC in Moscow and ZSS allows to estimate an anthropogenic part of CO, i.e. the contribution of Moscow sources to the CO content.

To study the long-term trends of ABL parameters, the data from aerological soundings at Dolgoprudny, station № 27,713, Moscow outskirts, and Sukhinichi, station № 27,707, 235 km southwest of Moscow, were used.

Interannual variations and long-term trends of the studied parameters were analyzed for different observation periods and different seasons of 2000–2023. All of trend estimates were calculated on the basis of annual (or seasonal) mean values, followed by linear regression to calculate trends and their 95% confidence intervals. This method is often used to assess long-term changes in various parameters, including atmospheric parameters.

3 Results

The analysis shows a decrease of the total CO content in Moscow in the period 2000–2023 at the rate of $-$ 2.2 ± 0.4% yr^{-1} (estimated by annual averages), see Fig. 2a and Table 1. The annual decrease in CO TC for Moscow is more pronounced than for Zvenigorod. This could be explained by the location of the ZSS and the virtual absence of Moscow emissions in the ZSS CO TC measurements (Rakitin et al. 2021). While the CO trends in the Moscow observations are mainly characterized by changes in emissions and urban meteorology, the trends in Zvenidorog mainly reflect global background tendencies. The rate of decrease at both sites varies

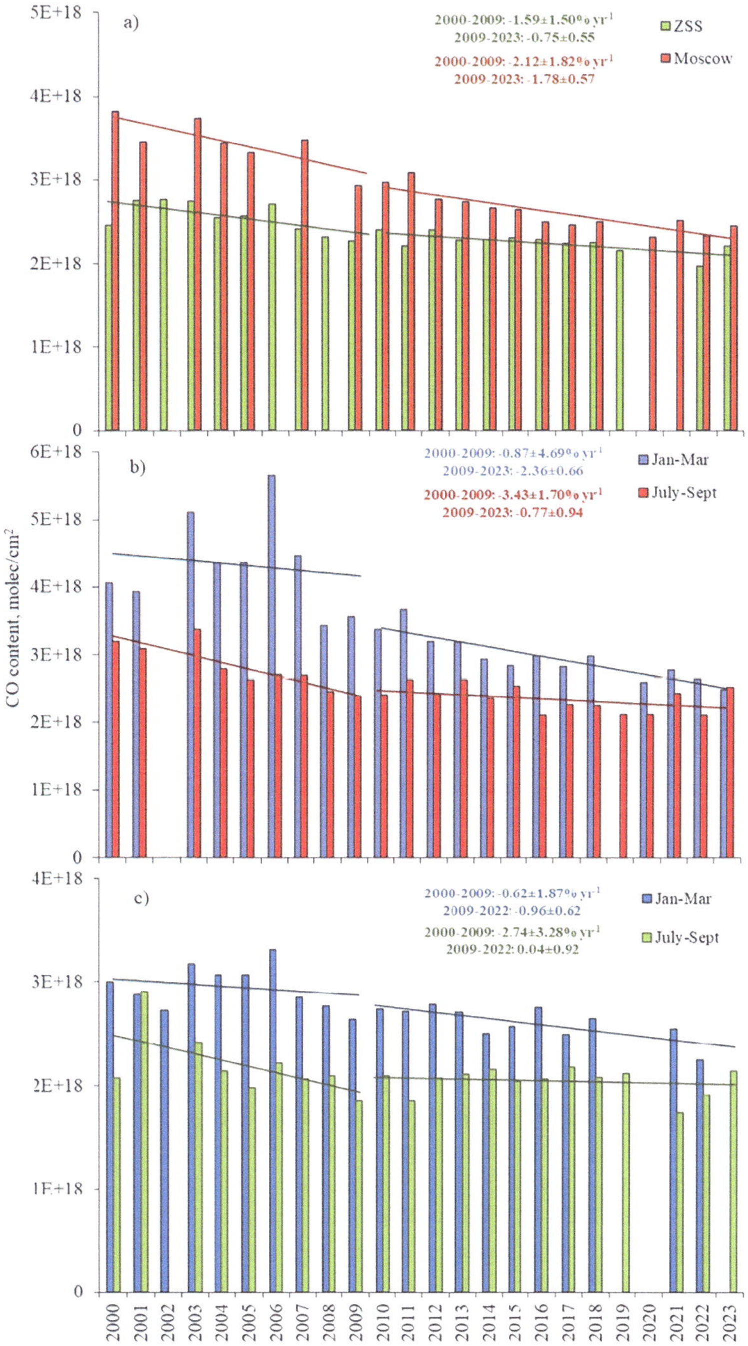

Fig. 2 Interannual variations and trends of the CO TC in different periods and seasons 2000–2023.: **a** in Moscow and Zvenigorod (ZSS), annual means; **b** in Moscow during cold (January–March) and warm (July–September) seasons; **c** in Zvenigorod during cold (January–March) and warm (July–September) seasons

depending on the season and time period. Figure 2b and c show the interannual variation of CO TC in Moscow and Zvenigorod, respectively, for the cold (January–March) and warm (July–September) seasons. It can be seen that there is no change in CO TC in ZSS in the warm season from 2009 to 2023. For Moscow the minimum rate of decrease is typical for the warm season (July–September) of 2009–2023, $-$ 0.8 $\pm$ 0.9% yr^{-1}, see Fig. 2b and Table 1. The annual and warm season trends for the whole period 2000–2023 were calculated without considering 2002 (the year of prolonged summer and autumn wildfires in the ETR, lasting from July to October). The period of fires in 2010 was significantly shorter (22.07–20.08), therefore the annual and seasonal averages of CO columns were calculated without considering a small amount of data from this period.

The characteristic trends of CO TC in Moscow and ZSS during different periods and seasons are also shown in Table 1.

The decrease in the content and concentration of CO in Moscow atmosphere in recent years and the improvement of air quality are confirmed by the latest results (Wang et al.

Table 1 Trends in the CO total column in Moscow and Zvenigorod during different seasons and periods. Some data were calculated without considering 2002 (the year of prolonged summer and autumn wildfires in the European Territory of Russia)

Point	Period	Season	Trend ±, % yr^{-1}
Moscow	2000–2009	Average year	– 2.1 ± 1.8
		January-March	– 0.9 ± 4.7
		July–September	– 3.4 ± 1.7
	2009–2023	Average year	– 1.8 ± 0.6
		January-March	– 2.4 ± 0.7
		July–September	– 0.8 ± 0.9
Zvenigorod	2000–2009	Average year	– 1.6 ± 1.5
		January-March	– 0.6 ± 1.9
		July–September	– 2.7 ± 3.3
	2009–2023	Average year	– 0.8 ± 0.6
		January-March	– 1.0 ± 0.7
		July–September	0.0 ± 0.9

Table 2 Wind speed trends in the 100–500 m layers above the Earth's surface according to the average annual data from aerological soundings in Dolgoprudny (suburb of Moscow, station № 27,612), period 2003–2020

Dolgoprudny	Annual	Cold season	Warm season
Day (% yr^{-1})	0.9 ± 0.4	1.3 ± 0.7	0.7 ± 0.4
Night (% yr^{-1})	1.0 ± 0.4	1.1 ± 0.6	1.0 ± 0.5

Table 3 Wind speed trends in layers 100–500 m above the Earth's surface according to the average annual data from aerological soundings in Sukhinichi (Kaluga province, station № 27,707), the period 2003–2020

Sukhinichi	Annual	Cold season	Warm season
Day (% yr^{-1})	0.1 ± 0.4	0.1 ± 0.5	– 0.1 ± 0.6
Night (% yr^{-1})	0.6 ± 0.6	0.5 ± 0.8	0.3 ± 0.9

2018; Elansky et al. 2022; Rakitin et al. 2021) and indicate primarily a reduction in urban emissions that due to the closure, re-profiling, and modernization city industrial enterprises, as well as their relocation outside the city borders and changes in the composition of the car fleet, improvements in the quality of car engines and fuel, and restrictive measures by the Moscow government on heavy-duty vehicles.

The influence of the "climatic factor" associated with improved conditions for the removal of impurities from the urban air basin, for example, a reduction in the duration of temperature inversions or the frequency of calm conditions (Rakitin et al. 2021), is also possible. Meanwhile, the CO TC trends in Zvenigorod differ from those in Moscow in terms of magnitude and direction (depending on the period and season). Negative trends are observed during the winter and spring months, – 1.1 ± 0.4% yr^{-1} and – 1.0 ± 0.7% yr^{-1}, for the periods 2000–2023 and 2009–2023, respectively; the background CO value stopped decreasing in the summer months of 2009–2023, – 0.04 ± 0.92% yr^{-1}, see Table 1 for more details.

Using data from aerological soundings in the suburbs of Moscow (Dolgoprudny) and the Kaluga region (Sukhinichi), the interannual variations of wind speed in ABL from 2003 to 2020 were studied. It was found that changes in wind speed in Moscow and the Kaluga regions differ in different layers and periods, but overall is a positive wind speed dynamic, which is most evident at the Dolgoprudny site (see Table 2). It was found that in the 100–500 m layer, the average annual trends in daily wind speed at Dolgoprudny are positive (0.9 ± 0.4% yr^{-1}), while at the Sukhinichi site, close to zero and statistically insignificant estimates of 0.1 ± 0.4% yr^{-1} were obtained. During the nighttime, wind speed increases at both sites, but the trend value at Dolgoprudny is significantly higher, 1.0 ± 0.4% yr^{-1} versus 0.6 ± 0.6% yr^{-1} at Sukhinichi. The parameters for the wind speed trends during different seasons in 2003–2020 are shown in Tables 2 and 3.

To obtain seasonal trend estimates, the year was divided into cold (October–March) and warm (April–September) seasons. During the winter period, the daily wind speed trend value at altitudes of 100–500 m in the Moscow region is positive, 1.3 ± 0.7% yr^{-1} and is significantly higher than that in the Kaluga region (0.1 ± 0.5% yr^{-1}). For the warm season, a similar situation occurred—in Dolgoprudny, the trend was positive and statistically significant (0.7 ± 0.4% yr^{-1}), while in Sukhinichi it was close to zero, – 0.1 ± 0.6% yr^{-1}. The nocturnal values of the trend at both stations were similar to those during the day in terms of orientation, but were higher in magnitude.

In (Rakitin et al. 2021) was provided the analysis of data from the VDNKh meteorological station (2005–2017) on the wind speed at a level of 2 m above the earth's surface. Days were selected on which the sum of the wind speeds over eight 3-h observation periods did not exceed 3 m/s per day. These days were considered as calm.

The average annual number of calm days from 2005 to 2017 have decreased at a rate – 7.1 ± 4.0% yr^{-1}. During the same period, the anthropogenic part of the CO content decreased at almost the same rate, – 6.7 ± 3.5% yr^{-1} (Rakitin et al. 2021). These results provide additional confirmation of the influence of the meteorological factor on the long-term changes in air quality in Moscow.

Thus, a possible explanation for the differences in the characteristics of changes in meteorological parameters in the ABL of Moscow and its surroundings may be an increase in the influence of the urban "heat island" (due to the expansion of metropolises and an increase in height of urban development), which is expressed in increased wind speed in the ABL.

4 Conclusions

1. The rate of decrease of CO TC in Moscow is maximum during the cold months, $-2.9 \pm 0.9\%\ yr^{-1}$ for 2000–2023. In the warm season, on the other hand, the decrease is less pronounced, $-1.6 \pm 0.5\%\ yr^{-1}$ for the same period.
2. In rural Zvenigorod, a decrease of the annual mean values of CO TC was observed, $-1.1 \pm 0.3\%\ yr^{-1}$ for 2000–2023. After 2008, the decline of the background CO TC slowed down; moreover, in the summer and autumn months of 2009–2023, a trend of CO TC close to zero was observed. This change may be due to the intensification of economic activities and automobile traffic in rural areas. Our estimates for Moscow and its outskirts are consistent with similar results of other studies (Elansky et al. 2018; 2022). The relatively small decrease in CO TC in Zvenigorod reflects the typical global tendencies for background European areas.
3. Analysis of meteorological conditions in the ABL revealed an increase in wind speed in Moscow and no significant changes in rural Kaluga province. Thus, in the layer of 100–500 m above the ground level, the annual and seasonal trend values, $0.7 - 1.3\%\ yr^{-1}$ for Moscow differs from the similar estimates for the rural region, $-0.1 - 0.6\%\ yr^{-1}$.
4. The CO TC decrease in Moscow indicates not only a reduction in anthropogenic emissions, but also the possible influence of a meteorological factor manifested in an increase in wind speed in the city.
5. Obtained trend estimates on CO and ABL parameters are in good agreement with the results of our recent studies and the findings of other authors concerning other urban areas.

Funding This work was supported by the Russian Science Foundation, project № 20-17-00200.

References

Elansky, N.F., Ponomarev, N.A., YaM, V.: Air quality and pollutant emissions in the Moscow megacity in 2005–2014. Atmos. Environ. 175(2), 54–64 (2018). https://doi.org/10.1016/j.atmosenv.2017.11.057

Elansky, N.F., Shilkin, A.V., Ponomarev, N.A., et al.: Spatiotemporal Variations in the content of pollutants in the Moscow air basin and their emissions. Izv. Atmos. Ocean. Phys. 58, 80–94 (2022). https://doi.org/10.1134/S0001433822010029

IPCC Climate Change.: The physical science basis. Contribution of Working Group I to the Sixth Assessment Report of the Intergovernmental Panel on Climate Change. Cambridge University Press, Cambridge (2021)

Jacob, D.J.: Introduction to Atmospheric Chemistry. Princeton University Press, New Jersey, Princeton (1999). http://acmg.seas.harvard.edu/people/faculty/djj/book/

Novelli, P., Masarie, K., Lang, P.: Distributions and recent changes in carbon monoxide in the lower troposphere. J. Geophys. Res. 103(19), 015–033 (1998). https://doi.org/10.1029/98JD01366

Rakitin, V.S., Elansky, N.F., Skorokhod, A.I., et al.: Long-term tendencies of carbon monoxide in the atmosphere of the Moscow megapolis. Izv. Atmos. Ocean. Phys. 57(1), 116–125 (2021). https://doi.org/10.1134/s0001433821010102

Timofeev, Y.M., Polyakov, A.V., Virolainen, Y.A., et al.: Estimates of trends of climatically important atmospheric gases near St. Petersburg. Izv. Atmos. Ocean. Phys. 56, 79–84 (2020). https://doi.org/10.1134/S0001433820010119

Wang, P., Elansky, N.F., Timofeev, Yu.M., et al.: Long-term trends of carbon monoxide total columnar amount in urban areas and background regions: ground- and satellite-based spectroscopic measurements. Adv. Atmos. Sci. 35(7), 785–795 (2018). https://doi.org/10.1007/s00376-017-6327-8

Yurganov, L.N., Grechko, E.I., Dzhola, A.V.: Long-term measurements of carbon monoxide over Russia using a spectrometer of medium resolution. Recent Res. Dev. Geophys. **4**, 249–265 (2002)

Why Education of Undergraduate Tourism Students Is a Key Factor for Climate Change Mitigation

Manuel Ramón González Herrera

Abstract

Education is a unique opportunity to confront the climate crisis, and it is necessary to ask how we can prepare undergraduate tourism students for mitigating and adapting to climate change. This study's aim was to determine how the education of undergraduate tourism students can play a key role in mitigating climate change through tourism entrepreneurship at the local level. The research had a qualitative and applied design, using theoretical and empirical didactic methods. The methodology was supported by developing a learning experience in the subject of tourism and the environment with 66 students enrolled in the Bachelor's degree in tourism at the Autonomous University of Ciudad Juarez in Mexico between 2021 and 2023. Eighteen nature hotels were proposed by students, incorporating bio-architecture solutions and environmental technologies to promote sustainable tourism development and contribute to minimizing the adverse impacts of climate change. In this activity, the project-based learning approach was implemented. Through their work, the students demonstrated that education is a key factor in climate change mitigation by making innovative proposals with viable solutions for climate change adaptation. These solutions were presented through investigative reports using posters and the manual creation of sustainable models for nature hotels that respond to the challenges induced by changes in temperature and climate patterns.

Keywords

Climate change · Tourism · Sustainable · Education · Mitigation and adaptation · University

M. R. G. Herrera (✉)
Autonomous University of Ciudad Juarez, Ciudad Juarez, Mexico
e-mail: manuel.gonzalez@uacj.mx

1 Introduction

Climate change is a global concern with growing adverse impacts on the economy, the environment, and society (Leung and You 2023). The United Nations (UN) has stated that climate change will affect all countries worldwide due to various extreme and changing weather phenomena and rising sea levels (ONU 2023). It warns that a climate catastrophe is approaching and that we are not prepared for its possible consequences; meanwhile, important segments of the international community are avoiding the commitments necessary to reverse it (ONU n.d.).

Within this context, it is key to recognize that among the 17 Sustainable Development Goals (SDGs) proposed by the UN in 2015, SDG 13 corresponds to climate action. This goal aims to promote the adoption of urgent measures to combat climate change and its effects—even though we are far from achieving this (ONU n.d.). A recent SDG report (ONU 2023) was also far from optimistic, reporting the need to "urgently reduce global greenhouse gas emissions" and suggesting "[that] the unprecedented rise in sea level poses a serious threat to hundreds of million people … [and] the climate financing objective of developed countries has not been met" (pp. 38–39).

For this reason, climate change constitutes a challenging geographic-environmental threat to tourism. Accordingly, professionals in the sector must understand climate change, assume a critical attitude toward the actions that cause it, and take measures that support innovation by mitigating climate solutions that promote sustainable travel. This affirmation led to the formulation of the following research question: How can we promote the education of undergraduate tourism students to help them understand, mitigate, and adapt to climate change?

The research problem formulation was based on the fact that education is an unexploited opportunity to confront

P. Mohammad et al. (eds.), *Geographic Approaches to Climate Change and Mitigation: Urban and Rural Perspectives (Volume 2)*,
Advances in Science, Technology & Innovation, https://doi.org/10.1007/978-3-031-92115-5_6

climate change; accordingly, educational policies and practices must be investigated to address it (Anderson 2012). It is striking that despite the potential of educational programs aimed at raising awareness and climate action, there is little evidence of their effectiveness (Keller et al. 2024; Tang 2023). Additionally, there has not been a favorable response to the demand of educating young people on climate change (ONU 2023). Since these educational opportunities are not being taken advantage of, it is impossible to guarantee the knowledge, skills, and behavioral changes necessary for successful adaptation to and mitigation of climate change (Apollo and Mbah 2021; Mochizuki and Bryan 2015). The situation outlined above involves a contradiction between the insufficiency of effective systematic educational practices and the urgent action demanded by the climate crisis.

In response to the established problem, it is necessary to improve the understanding of the causes and consequences of climate change among university students and commit them to taking action based on the development of responsible behaviors (Mochizuki and Bryan 2015). Accordingly, there is a need to plan and implement educational programs that improve learning outcomes on climate change (Kumar et al. 2023) by applying a holistic approach—as studies addressing the multidimensional impact of climate change education in universities are lacking (Tolppanen et al. 2022). At the same time, climate change must be addressed as a social issue involving student participation, and there is a lack of educational programs focused on a narrow view of the problem (Karsgaard and Davidson 2023).

For this purpose, it will be convenient to develop communication and educational interventions to mobilize and involve university students in conscious actions that respond to global climate change (Baldwin et al. 2023), especially since education can enable individuals and communities to make informed decisions that favor climate resilience and sustainability (Apollo and Mbah 2021). This effort implies reducing vulnerability and generating resilience by proposing new tourism development projects that address climate change-related challenges. Therefore, this study's general aim is to investigate how the education of undergraduate tourism students can become a key factor in mitigating climate change through tourism entrepreneurship at the local level.

Following this approach, the object of this research was the education of undergraduate tourism students as a key factor for climate change mitigation. The context of the developed experience was the tourism program of the Autonomous University of Ciudad Juarez in Chihuahua, Mexico. This report includes an analysis of education as a key factor in mitigating climate change and the development of innovative proposals by students with viable solutions for adaptation. These solutions were enhanced through research reports using posters and the manual creation of models of sustainable and responsible nature hotels that respond to the challenges posed by changes in temperature and weather patterns.

2 Literature Review

2.1 Education for Climate Change Mitigation

The construction of the theoretical framework of the research took into consideration that the problem of education and climate change is multidimensional and has a marked spatial–temporal scope. In the 1990s, education about climate change was examined among international bodies as an indispensable and global strategy (Borde et al. 2022), while the population was mainly skeptical about the issue. Given this apparent contradiction, it may be that no one was prepared for the coming paradigm shift. Over the years, the issue has remained without due attention. Climate change is more than simply a change; it is an emergency situation: "It is happening, and its adverse impacts could reverse progress towards meeting the Sustainable Development Goals." (Apollo and Mbah 2021, n.d.).

According to Garbarino (2023), the data contained in the reports of the Intergovernmental Group of Experts on Climate Change highlight the serious nature of the climate crisis facing the planet and society, making it vital to undertake mitigation measures and stop the problem. In this regard, it is accepted as support for this research that formal, non-formal, and informal education, when conveniently planned and implemented, can be powerful tools for promoting the understanding of and adaptation to climate change (Feinstein and Mach 2020). This can in turn contribute to reducing vulnerability and building resilience as a strategy to manage the multiple challenges posed by the climate crisis.

Indeed, there has been an increase in research on education and climate change at the international level, as reflected in the number of articles available on the Internet. However, recent literature systematization research carried out by Perwitasari et al. (2023) recognizes that it is still advisable to increase the number of publications on education for sustainable development and climate change education, strengthening formal and non-formal pathways by incorporating less conventional solutions that encourage learning. Innovative proposals, methods, and appropriate designs for climate change are also necessary, encouraging researchers, professionals, governments, and society to dynamically and permanently involve themselves in this commitment.

Morote and Olcina (2023) clarify that education about climate change in major official international documents reflects the inclusion of interesting measures but often without specificity regarding the regional and local environment, including the student population and society. However, the authors do recognize that the teaching–learning process has progressively gained prominence in this context, with greater scope given to concrete measures and government actions, especially in Spain, where the study was developed.

As a theoretical frame for this research, it is assumed that education for climate change must be addressed in the context of education for sustainable development (Mochizuki and Bryan 2015). In this respect, education for climate change is considered by some authors as a branch of education for sustainability, whose purpose is timely preparation for climate change (Tang and Hadibarata 2022). In line with this approach, we agree with Perwitasari et al. (2023), who point out that education for sustainable development should focus on the quality of learning results, emphasizing the educational content and its contribution to future sustainability.

This research accepts that education for climate change must be interpreted as a complex socio-cultural and scientific dimension of the teaching–learning process (Mochizuki and Bryan 2015), considering the current scenario and the uncertainty of each context in different future scenarios. This requires educators to actively participate in methodological research and engage in co-learning with students (Stevenson et al. 2017). These efforts must be a comprehensive and multidisciplinary process including not only content on climate change, socio-environmental situations, disaster risk reduction, consumption, and sustainable lifestyles but also the institutional and community environment to ensure that educational institutions are resistant to climate change (Anderson 2012).

In terms of the transition strategy towards climate action, it is accepted that scientific knowledge is a key factor for achieving the desired change. It can help in adopting pro-climatic commitments that promote a clear and explicit understanding of climate dynamics and their causal relationships (Kolenatý et al. 2022) as well as in developing positive attitudes and the willingness to commit to measures to mitigate climate change (Sinatra et al. 2012) at the local level. Therefore, the operationalization of the main concepts of the theoretical framework indicates that education for climate change is one of the mitigation measures with the most significant potential for change (Garbarino 2023). This makes it essential to urgently and systematically evaluate its implementation possibilities.

In the case of Mexico, numerous studies have been carried out to contribute to climate change education and the integration of adaptation and mitigation measures. These efforts, as suggested in research carried out by Valle et al. (2020) and Gutiérrez et al. (2023), are a collaborative endeavor to reduce vulnerability and prepare civil society, taking into account gender differences (García et al. 2020). These contributions cover various sectors of society and the economy (Maldonado et al. 2024), helping to develop a climate culture that provides stakeholders with various tools to take action.

2.2 Empowering Undergraduate Tourism Students to Take Climate Action in a Scenario of Change

The theoretical foundation of this research is based on the close relationship between climate change and tourism (Aygün-Oğur and Baycan 2023; Gössling et al. 2023) and includes elements of both climate and tourism theories. There is great concern about how to rethink tourism management in the face of these challenges and raise tourists' awareness about where, when, and how often they should travel. Companies and destinations are preparing for operational actions given the current scenario. However, in the medium and long term, new professionals with a different worldview contextualized according to the new realities are required, making undergraduate tourism students' education a key factor in climate change mitigation. In this context, empowerment, and social transformation begin with education (Boyadjieva and Ilieva-Trichkova 2023; Kyle 2020).

In response to this social demand, university and institutional support for educational interventions on climate change must become a source of new opportunities (Ssekamatte 2023) to train tourism students to deal with the uncertain situation of climate change. Improving both students' and professors' understanding of climate change is required as a transition strategy, and professors must encourage students to think critically and creatively about mitigation and adaptation approaches and develop their capacity to respond with meaningful actions (Stevenson et al. 2017). At the same time, academic and non-academic spaces must be created to involve young students in education for climate change.

It is necessary to consider that young people entering university will live most of their adult lives exposed to a warmer climate caused by climate change (Baldwin et al. 2023). In recent years, some youths have been mobilizing due to the injustices associated with the causes, consequences, and policies of climate change (Karsgaard and Davidson 2023). Despite this, they are not often aware of the reasons that should drive them to participate in climate mitigation and protection activities (Zeigermann et al. 2023). This awareness must be raised within a wide range of disciplines (Filho et al. 2021) since many students see these topics as limited to natural, geographical, biological, and related science programs.

A recent study carried out in Australia found that young people showed a good level of understanding of the impacts of climate change but a lack of knowledge about measures to mitigate it and little hope that society would take mitigating and adaptive actions (Baldwin et al. 2023). These findings serve as a guiding basis for this research to the extent that they incorporate the cognitive-instrumental and affective-motivational factors as well as the behaviors of the students who participated. When put to use, these can promote the understanding of problems related to responding to climate change.

Faced with widespread skepticism, a profound debate has emerged in the field of climate change education, with particular attention paid to predictors that influence the behavior of young people with respect to the climate and what educational strategies and practices can stimulate their behavior (Kolenatý et al. 2022). For this reason, a pedagogical understanding of the factors that drive climate change and its mitigation is crucial for the training of future tourism professionals (Kumar et al. 2023). In this regard, it is essential to identify what and how professors should teach and how students can engage in learning (Stevenson et al. 2017) to ensure their responsible and sustainable performance in the face of an uncertain future arising from the risks and impacts of climate change.

In response to UN (2023) recommendations, the implementation of climate action by tourism students must be aligned with SDG 13, focusing on limiting and adapting to climate change issues and their impacts (Leung and You 2023). This necessitates establishing congruence between climate change mitigation and adaptation actions in the strategic and operational policies and objectives of each tourism entrepreneurship effort. In this way, issues related to climate change can be included as part of teaching and research practices in universities from an educational perspective, with a focus on the training needs of teaching staff (Filho et al. 2021).

This research incorporates the principle that education for climate change adaptation should be supported by research and experiential learning based on practical projects to activate social and political change. In this way, meaningful learning can be achieved before and during adaptive decision-making (Feinstein and Mach 2020). The abovementioned authors emphasize that the transformative contribution of education consists in supporting learning for adaptation, according to which new curricular, pedagogical, and technological resources must be developed that prepare all students to make complex adaptive decisions and solidify their knowledge about the topic (Feinstein and Mach 2020).

Therefore, changes in study plans/programs and pedagogical practices must be implemented in the teaching–learning process for tourism to promote the understanding of climate change, ensuring the emergence of initiatives that can inspire future sustainable tourism developments (Bar-Yam et al. 2002). This corresponds to the use of empirical scientific education in university tourism training to reduce the adverse effects of climate change and improve quality of life and human well-being (Kumar et al. 2023). This approach will facilitate social climate change acceptance and the materialization of adaptation strategies by exponentially raising awareness about the impacts of climate change among the population, especially among young people as future decision-makers (Feja et al. 2019).

According to the literature, students must cognitively and affectively confront climate change during their training (Feja et al. 2019). If this process occurs effectively, tourism students will be better trained and motivated to understand, evaluate, and communicate climate adaptation strategies. The relationship between cognitive and motivational variables can increase students' willingness to take mitigation measures that reduce the effects of socially induced climate change (Sinatra et al. 2012). This will require the guidance of motivating teaching tasks that lead to the desired change in attitude and the expression of willingness to take action against climate change among university tourism students (Baldwin et al. 2023; Sinatra et al. 2012).

As another essential contribution to this research, Tang and Hadibarata (2022) identified barriers to changes in the attitudes and behavior towards climate actions among students of higher education institutions. These barriers include the complexity and multidimensionality of the topic, prejudices and misconceptions, the perception that some of these changes are difficult, and the fact that habitual behaviors can unconsciously lead students to contradict climate actions. A critical reflective analysis of these barriers reveals their potential consideration in tourism education for climate change in the context of sustainability.

In correspondence with the theoretical background, it is considered convenient to explore the methodological assumptions of the theory of change for the analysis and interpretation of the research findings. This theory "explains how it is understood that activities produce a series of results that contribute to achieve the intended final impacts" (Rogers 2014, p. 1). Thus, it is a "visual representation of how a program or intervention works… [allowing one] to identify the implicit mechanisms through which change is expected to be achieved" (Cassetti and Paredes-Carbonell 2020, pp. 305–306).

3 Methodology Framework

This research used a qualitative and applied design based on theoretical and empirical didactic methods. The pedagogical methodology was supported by developing a learning experience in the subject of tourism and the environment with undergraduate tourism students at the Autonomous University of Ciudad Juarez in Mexico. As indicated in Fig. 1, the

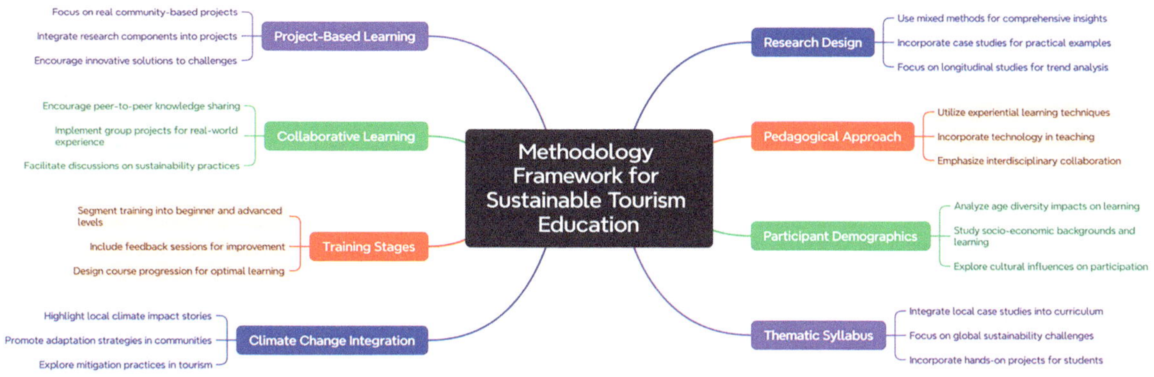

Fig. 1 Methodology flowchart. *Source* Author's elaboration using Mapify.AI https://mapify.so/es/app/cf359c69-dcbc-4c51-82cf-061b732a2843

methodological design was based on the integration and operationalization of the main conceptual variables. The pedagogical experience was developed between August, 2021, and June, 2023, with the participation of 66 students. There were no notable differences in age, while the majority of the students were female. There were 22 students in the February–June 2023 semester, 15 in the August–December 2022 semester, 17 in the February–June 2022 semester, and 12 in the August–December 2021 semester. In total, 18 tourism entrepreneurship projects were developed, incorporating bio-architecture principles and environmental technologies to promote sustainable tourism development and minimize the adverse impacts of climate change.

The syllabus on tourism and the environment was selected given its high potential for introducing the topic. This syllabus is structured in four thematic units: (1) the environment as a setting in which tourism takes place, focusing on the environmental impacts of tourism, global environmental problems, the carbon footprint and best environmental practices; (2) tourism environmental management; (3) visitors' environmental education and interpretation; and (4) the formulation of sustainable tourism projects. These units focus on the analysis of both traditional and sustainable tourism development models, the methodology for the formulation of environmental tourism investment projects, environmental evaluation, and the design of sustainable products and services. The topics of sustainability and climate change were incorporated as transversal axes.

The introduction to the study problem was based on the diagnosis of prior knowledge through brainstorming and the group construction of learning using the following questions: Have you ever heard about climate change? What factors affect the climate crisis? Are we responsible for global warming? Would you like to take climate action but do not know how to start? Before carrying out this activity, the following information was also provided: The planetary climate is approximately 1.2 °C warmer than in the nineteenth century, and the amount of CO_2 in the atmosphere has increased by about 50%.

Based on the results of the diagnosis, the pedagogical experience was planned in two stages: (1) curricular training of tourism students for climate action in the subject of tourism and the environment and (2) planning and designing a natural tourism project with a low environmental impact based on bio-architecture principles and environmental technologies. The theoretical–practical training for climate action aimed to (i) promote the understanding of climate change, (ii) assume a critical attitude towards the actions that provoke it and (iii) demonstrate mitigation and adaptation measures that support innovation through climate mitigation solutions that promote sustainable travel.

Theoretical learning was developed through collaborative workshops in which students in the classroom, guided by the professor, contributed to constructing group knowledge. The developed content was organized using the following questions: What are climate and climate change? What are the human causes of climate change? What irreversible changes have occurred in the global climate? What is the impact of climate change? What should be the mitigation and adaptation strategies for climate change? Five professors of undergraduate tourism courses validated these questions. As a result, the following impact chain was built to understand climate change: action–change–impact–responses (Fig. 2).

For practical training, project-based learning (PBL) was selected since it offers opportunities for students to develop knowledge and skills by carrying out projects that respond to specific problems and challenges. It is related to collaborative learning and improves student participation and commitment, making it a recommended option for educational use in universities (Almulla 2020). PBL was implemented as a team task focused on the students' work to involve

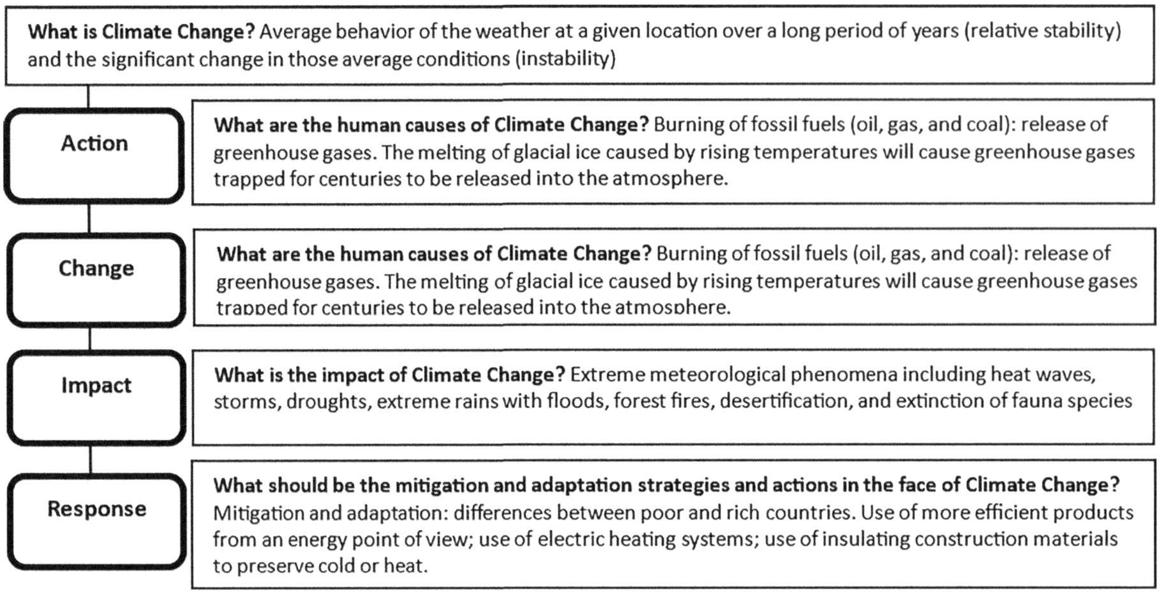

Fig. 2 Example of an impact chain for understanding climate change. (*Source* By author)

them in their education through the practical application of bio-architecture solutions and environmental technologies that could be incorporated as sustainability strategies to minimize the impacts of climate change. As a result, two academic projects were developed from the local action level as an education strategy for sustainable development, taking proactive measures to align with SDG 13.

These two academic projects were posed as follows:

- Bio-architecture—Based on the projected hotel for your tourist destination and the design of its outside areas using the conservation of natural vegetation and low-impact gardening as criteria, develop and model the following: construction materials; design and construction techniques; assembly and construction of the project; and solutions for project functionality (comfort, aesthetics, quality, and the fulfillment of tourist expectations).
- Environmental technologies project—Based on the projected hotel for your tourist destination, develop and model the following using environmental technologies for your entrepreneurship project: water consumption; energy consumption; waste treatment; the use of chemicals, fuels, and lubricants; hygiene/sanitary conditions; vector control; and management of flora and fauna.

4 Results and Discussion

4.1 Curricular Training of Undergraduate Students in Tourism for Climate Action

The initial survey revealed that all of the participants had heard about climate change at some time. However, they had not paid enough attention to the topic to contribute concrete ideas. Regarding the factors that cause climate change, 45% thought it was due to natural causes, while 45% did not know its origin. This confirms that the students were not convinced humans are responsible for the climate crisis. Furthermore, 75% wanted to take action to reduce climate change but did not know what or how to do it. It was thus determined that the previous level of knowledge the students possessed was insufficient for taking proactive mitigation and adaptation measures against climate change. This validated the presented theoretical premise that it is valuable to create spaces to involve young people in climate change education, helping them develop knowledge, skills, and behavioral changes for successful climate action.

Through the organization of collaborative workshops, the students consciously participated in various teaching tasks, including an introductory talk for motivation, video obser-

Table 1 Results of the tests on learning related to climate change

Questions	High (%)	Medium (%)	Low (%)
What are climate and climate change?	85	15	–
What are the human causes of climate change?	70	25	5
What irreversible changes have occurred in the climate?	60	25	15
What is the impact of climate change on society?	56	20	24
What should be the mitigation and adaptation strategies?	40	20	40

vation, interpretation of photos, reading and debate, and knowledge tests through a drawing. After carrying out these teaching activities, they answered an individual test and developed an impact chain about climate change in teams. The individual test results on the formative efficiency of education for sustainable development in learning about climate change (Table 1) reflected a high level of understanding of climate change and its causes, a trend from high to medium in terms of changes and impacts and predominantly medium and low levels with respect to mitigation and adaptation strategies.

These findings are consistent with the empirical results of Tolppanen et al. (2022), who found a significant increase in knowledge but minimal changes in the willingness to implement mitigation measures among their students. These results were based on the application of questionnaires that examined how knowledge, values, worldview, and willingness to undertake climate mitigation actions change. Other authors, such as Garbarino (2023), observed changes in attitudes and mitigation habits before and after the application of surveys, highlighting their effectiveness in light of the theory of change. Kolenatý et al. (2022) verified that a greater willingness to act translates into real climate action. This agrees with the theoretical approach of this research, which assumes that tourism students must learn to guarantee sustainability in their entrepreneurship to prevent and reduce the adverse effects of climate change.

The cognitive strengthening of students to promote a better understanding of climate change occurred through elaborating the chain of impacts (see Fig. 1). This assignment was supported by the contributions of Sinatra et al. (2012), who consider that openness to change and the willingness to think about problems precede changes in attitude and will. Of the total of 18 teams in which the students were grouped, 86% of the tasks were evaluated at the high-quality level and the rest at the medium level. The problems detected in the medium-level works corresponded to weak relationships among the different links in the chain of impacts, the imprecise identification of the impacts generated, and the proposal of non-innovative mitigation and adaptation strategies. Referring back to Sinatra et al. (2012), it is necessary to generate more favorable conditions for openness to change as well as a willingness to deeply understand climate change problems related to tourism.

4.2 Planning and Design of Low-Impact Nature Tourism Ventures Based on Bio-Architecture and Environmental Technologies

As a result of the planning and design of hotel entrepreneurship projects, 18 semi-annual investigative reports and final exhibitions were carried out using posters and manual models (mockups) for different tourist destinations in the Chihuahua State. These incorporated proactive measures aligned with SDG 13 to mitigate and adapt to climate change (Figs. 3 and 4). Incorporating bio-architecture led to solutions for the conservation of natural vegetation, the proposal of local construction materials (Adobe) integrated into the environment, and assembly and construction involving minimal impacts. Environmental technologies solutions included measures for minimal water and energy consumption, minimal waste generation, the use of alternative energy sources (without greenhouse gases), the substitution of polluting chemical products, and environmentally friendly hygienic and sanitary products.

Fig. 3 Exhibitions of two nature hotels with minimal environmental impact. *Source* Author's photos

Fig. 4 Student presenting her project. *Source* Author's photos

The most valuable contribution of the presentations was the proposal of environmental education actions to engage visitors and hosts with sustainability and the minimization of climate change (100%), the promotion of sustainable and lifestyles (93%), and the integration of community members into climate management and action to strengthen the resilience of locals as well as the institutional and community environment (88%). The value of these contributions lies in the fact that they are socially transformative, as recognized by Stevenson et al. (2017), who verified that educational results for climate change are more efficient if they are based on active social learning that develops the capacity for transformative practices on a personal and social level.

As a result of the qualitative assessment of the 18 semiannual investigative reports and final presentations, the students' instrumental cognitive training (knowledge and skills about climate change) was highly favorable; their affective-motivational formation was favorable (pro-climatic commitments and critical attitude towards actions that cause climate change); and their behavioral training was moderately favorable (taking mitigation and adaptation measures). However, in general terms, a paradigm shift was achieved in the students' views and behavior towards sustainability. This was observed based on three predictors that influence the behavior of young people concerning the climate: holistic conceptualization of the environment; preventive impact management; and integration of the environmental, social, and economic dimensions to local tourism development.

Among the students who participated in this experience, the quality of the learning results regarding climate change education and its contribution to sustainability were differentiated at both the individual and group level. In particular, they depended on the level of motivation, active participation in the search for mitigation and adaptation solutions in each of the teaching tasks, the state of previous knowledge regarding the subject, familiarization with the innovative nature of tourism projects, a lack of previous practical experiences concerning work methods as a basis for transformative intervention proposals, and a lack of interest in engaging in teamwork based on PBL.

The results of this study were also compared with the results of other similar investigations. In research by Tang (2023), significant changes in belief were achieved in relation to the anthropogenic causes of climate change and vulnerability to its impacts; however, no significant attitudinal changes were recorded. On the contrary, Garbarino (2023) concluded that their students' attitudes towards climate change improved; they began to develop mitigation habits and greater awareness and interest, indicating the need to deepen research on the topic using various methodologies.

The qualitative approach to the efficiency of sustainable development education in learning about climate change and its impacts allowed us to evaluate students' preparation for adopting mitigation solutions and adapting at the local level in tourism entrepreneurship. This result is consistent with the findings of Anderson (2012), who suggests that educational interventions are more successful when they focus on local, tangible, and viable aspects of sustainable development, climate change, and environmental education. Working at a local scale allowed the participants to feel empowered as they identified with familiar places, making their learning more conscious and committed.

Among the factors that most limited sustainable education for climate action in this study, it is essential to mention the multidisciplinary and multidimensional nature of climate change; uncertainty that the proposals could be put into practice; a systematic and transversal approach to the topic; and contrasting beliefs and vision. These limitations are similar to those reported by Tang and Hadibarata (2022) in their study of changes in attitudes and behavior among higher education students toward climate action. They also agree with Leung and You (2023), who add that fostering resilience to climate change requires constant joint efforts by policymakers, market regulators, commercial entities, and the community.

Despite this, numerous challenges can still be recognized related to planning and designing low-impact nature tourism ventures based on bio-architecture and environmental technologies. Among these, there is a lack of previous studies on best practices to guarantee desirable attitudes and effective commitments in students concerning climate change, which would help achieve more proactive and collaborative participation. At the same time, it is necessary to continue studying the climate change in all its complexity using a multidisciplinary approach to guarantee viable and contextualized project proposals. This is especially true in the current scenario where contrary beliefs and opinions continue to prevail concerning the causes of climate change and the actions required by all stakeholders.

5 Conclusion and Implications

It was demonstrated that formal university education is a valuable opportunity to promote the preparation and empowerment of undergraduate students in tourism for climate action in response to SDG 13 and train them to act on the problems that arise in everyday life. This finding responds to the detected contradiction between the insufficiency of effective systematic educational practices to promote the urgent action demanded by the climate crisis and the adoption of proactive climate actions in a scenario of dynamic change.

The students were able to both understand climate change and propose practical and local mitigation and adaptation measures based on PBL. These actions favored awareness and climate action with acceptable levels of training efficiency. The students developed knowledge and skills and demonstrated behavioral changes for successful adaptation to and mitigation of climate change through their proposals for innovative nature hotel entrepreneurship committed to sustainability and climate solutions. This commitment to the climate at the local level allowed us to affirm that tourism students can be a key factor in climate change mitigation by contributing to social transformation as tourism professionals.

The practical value of this research is the creation of an academic space to prepare students to face the challenging consequences of the climate crisis, which affect tourism as well as other social and economic activities at a global level. It responds to curricular education for sustainable development and learning about climate change to promote understanding, mitigation, and adaptation to the climate crisis among students, helping them transfer their knowledge to the rest of the community and society with which they coexist. This will allow them to replicate, through their responsible behavior, the knowledge, skills, and behavioral changes necessary for successful adaptation to and mitigation of climate change.

The main limitations of this research include the fact that awareness of climate change must cover a wide range of disciplines. In this case, the pilot experience is limited to the subject of tourism and the environment, which suggests a rethinking of the study plan to address the issue systematically and transversally with multiple efforts among various disciplines. The experience presented was also limited to formal education about climate change, and non-formal and informal routes can be incorporated for such purposes.

In future research, it will also be necessary to deepen the study of evidence on the factors that most influence changes in students' skills, attitudes, and behaviors. Future research should incorporate all subjects in the curriculum and involve all students in this activity to strengthen climate responsibility with a focus on sustainability. This local contribution to the understanding and management of climate change can be developed with other types of students and made relevant to other regions of the world where similar locally based actions can be undertaken to improve the Earth's climate.

References

Almulla, M.A.: The effectiveness of the project-based learning (PBL) approach as a way to engage students in learning. SAGE Open **10**(3) (2020). https://doi.org/10.1177/2158244020938702

Anderson, A.: Climate change education for mitigation and adaptation. J. Educ. Sustain. Dev. **6**(2), 191–206 (2012). https://doi.org/10.1177/0973408212475199

Apollo, A., Mbah, M.F.: Challenges and opportunities for climate change education (CCE) in East Africa: a critical review. Climate **9**(6), 93 (2021). https://doi.org/10.3390/cli9060093

Aygün-Oğur, A., Baycan, T.: Assessing climate change impacts on tourism demand in Turkey. Environ. Dev. Sustain. **25**(3), 2905–2935 (2023). https://doi.org/10.1080/13683500.2022.2049711

Baldwin, C., Pickering, G., Dale, G.: Knowledge and self-efficacy of youth to take action on climate change. Environ. Educ. Res. **29**(11), 1597–1616 (2023). https://doi.org/10.1080/13504622.2022.2121381

Bar-Yam, M., Rhoades, K., Booth Sweeney, L., Kaput, J., Bar-Yam, Y.: Complex systems perspectives on education and the education system. New England Complex Systems Institute (2002). https://necsi.edu/changes-in-the-teaching-andlearning-process-in-a-complex-education-system

Borde, B., Léna, P., Lescarmontier, L.: Education as a strategy for climate change mitigation and adaptation. In: Lackner, M., Sajjadi, B., Chen, W.Y. (eds.) Handbook of Climate Change Mitigation and Adaptation. Springer, Cham (2022). https://doi.org/10.1007/978-3-030-72579-2_149

Boyadjieva, P., Ilieva-Trichkova, P.: Adult education as a pathway to empowerment: challenges and possibilities. In: Lifelong Learning, Young Adults and the Challenges of Disadvantage in Europe, pp. 169–191 (2023). https://doi.org/10.1007/978-3-031-14109-6

Cassetti, V., y Paredes-Carbonell, J.: La teoría del cambio: una herramienta para la planificación y la evaluación participativa en salud comunitaria. Gac. Sanit. **34**(3), 305–307 (2020). https://doi.org/10.1016/j.gaceta.2019.06.002

Feinstein, N.W., Mach, K.J.: Three roles for education in climate change adaptation. Clim. Policy **20**(3), 317–322 (2020). https://doi.org/10.1080/14693062.2019.1701975

Feja, K., Lütje, S., Neumann, L., Mönter, L., Otto, KH., Siegmund, A. (2019). Climate ChangeS Cities—A Project to Enhance Students' Evaluation and Action Competencies Concerning Climate Change Impacts on Cities. In: Leal Filho, W., Lackner, B., McGhie, H. (eds.) Addressing the Challenges in Communicating Climate Change Across Various Audiences. Climate Change Management. Springer, Cham. https://doi.org/10.1007/978-3-319-98294-6_11

Filho, W., Sima, M., Sharifi, A., Luetz, J., Lange Salvia, J., Mifsud, M., Motunrayo Olooto, F., Djekic, I., Anholon, R., Rampasso, I., Kwabena, F., Pimenta Dinis, M., Klavins, M., Finnveden, G., Munashe Chari, M., Molthan-Hill, P., Mifsud, A., Sen, S., Lokupitiya, E.: Handling climate change education at universities: an overview. Environ. Sci. Eur. **33**, 109 (2021). https://doi.org/10.1186/s12302-021-00552-5

Garbarino, P.: La educación para el cambio climático como herramienta de mitigación. MLS Educ. Res. (MLSER) **7**(1) (2023). https://doi.org/10.29314/mlser.v7i1.1015

García, A., Bello, L., Iglesias, M.: Desigualdades de género en la educación para el cambio climático. Estudio de caso: México y España. Revista mexicana de investigación educativa **25**(87), 1013–1041 (2020). https://www.scielo.org.mx/scielo.php?pid=S1405-66662020000401013&script=sci_arttext

Gössling, S., Balas, M., Mayer, M., Sun, Y.Y.: A review of tourism and climate change mitigation: the scales, scopes, stakeholders and strategies of carbon management. Tour. Manage. **95**, 104681 (2023). https://doi.org/10.1016/j.tourman.2022.104681

Gutiérrez, M., Vargas, M., Campillo, J.: Adaptación al cambio climático con enfoque de economía circular para reducir la vulnerabilidad del sector ganadero extensivo en México: estado del arte. Regiones y Desarrollo Sustentable **23**(44) (2023). http://coltlax.edu.mx/openj/index.php/ReyDS/article/view/252

Karsgaard, C., Davidson, D.: Must we wait for youth to speak out before we listen? International youth perspectives and climate change education. Educ. Rev. **75**(1), 74–92 (2023). https://doi.org/10.1080/00131911.2021.1905611

Keller, J., Eichinger, M., Bechtoldt, M., Liu, S., Neuber, M., Peter, F., Pohle, C., Reese, G., Schäfer, F., Heinzel, S.: Evaluating the Public Climate School, a multi-component school-based program to promote climate awareness and action in students: a cluster-controlled pilot study. J. Clim. Chang. Health **15**, 100286 (2024). https://doi.org/10.1016/j.joclim.2023.100286

Kolenatý, M., Kroufek, R., Činčera, J.: What triggers climate action: the impact of a climate change education program on students' climate literacy and their willingness to act. Sustainability **14**, 10365 (2022). https://doi.org/10.3390/su141610365

Kumar, P., Sahani, J., Rawat, N., Debele, S., Tiwari, A., Mendes Emygdio, A., Abhijith, V., Kukadia, V., Holmes, K., Pfautsch, S.: Using empirical science education in schools to improve climate change literacy. Renew. Sustain. Energy Rev. **178**, 113232 (2023). https://doi.org/10.1016/j.rser.2023.113232

Kyle, W.C.: Expanding our views of science education to address sustainable development, empowerment, and social transformation. Discipl. Interdiscip. Sci. Educ. Res. **2**(2) (2020). https://doi.org/10.1186/s43031-019-0018-5

Leung, T.C.-H., You, C.S.-X.: Climate change in hong kong with the integration of UN SDG 13 climate action. In: Leal Filho, W., Ng, T.F., Iyer-Raniga, U., Ng, A., Sharifi, A. (eds.) SDGs in the Asia and Pacific Region. Implementing the UN Sustainable Development Goals—Regional Perspectives. Springer, Cham (2023). https://doi.org/10.1007/978-3-030-91262-8_18-1

Maldonado, M., Monterroso, A., Escárraga, L., Bustos, E., Sibelet, N.: Barreras y potencialidades de los productores agrícolas que inciden en su capacidad de adaptación al cambio climático en la Meseta Comiteca Tojolabal, Chiapas, México. Sociedad y Ambiente (27), 1–26 (2024). https://doi.org/10.31840/sya.v2024i27.2897

Mochizuki, Y., Bryan, A.: Climate change education in the context of education for sustainable development: rationale and principles. J. Educ. Sustain. Dev. **9**(1), 4–26 (2015). https://doi.org/10.1177/0973408215569109

Morote, F., y Olcina, J.: Cambio climático y educación: una revisión de la documentación oficial. Documents d'anàlisi geogràfica **69**(1), 107–134 (2023). https://dag.revista.uab.cat/article/view/v69-n1-morote-olcina

Organización de las Naciones Unidas-ONU.: Informe de los Objetivos de Desarrollo Sostenible. Edición Especial (2023). https://unstats.un.org/sdgs/report/2023/The-Sustainable-Development-Goals-Report-2023_Spanish.pdf?_gl=1*knc84y*_ga*NjA2MjUxMjc0LjE3MDY5MDQ3MTQ.*_ga_TK9BQL5X7Z*MTcwNjkwNDcxMy4xLjEuMTcwNjkwNDgyMi4wLjAuMA

Organización de las Naciones Unidas -ONU. Objetivo 13: Adoptar medidas urgentes para combatir el cambio climático y sus efectos (n.d.). https://www.un.org/sustainabledevelopment/es/climate-change-2/

Organización de las Naciones Unidas-ONU. Objetivos y metas de desarrollo sostenible (n/d/b). https://www.un.org/sustainabledevelopment/es/sustainable-development-goals/

Perwitasari, S.I., Hariyono, E., Susantini, E.: Implementation of ESD (Education for Sustainable Development) in climate change learning: a literature review. IJORER: Int. J. Recent. Educ. Res. **4**(4), 399–415 (2023). https://doi.org/10.46245/ijorer.v4i4.317

Rogers, P.: Síntesis metodológica n.º 1: La teoría del cambio. Fondo de las Naciones Unidas para la Infancia (UNICEF) (2014). https://www.unicef-irc.org/publications/pdf/Brief%202%20Theory%20of%20Change_ES.pdf

Sinatra, G.M., Kardash, C.M., Taasoobshirazi, G., Lombardi, D.: Promoting attitude change and expressed willingness to take action toward climate change in college students. Instr. Sci. **40**, 1–17 (2012). https://doi.org/10.1007/s11251-011-9166-5

Ssekamatte, D.: The role of the university and institutional support for climate change education interventions at two African universities. High. Educ. **85**, 187–201 (2023). https://doi.org/10.1007/s10734-022-00828-6

Stevenson, R.B., Nicholls, J., Whitehouse, H.: What is climate change education? Curric Perspect **37**, 67–71 (2017). https://doi.org/10.1007/s41297-017-0015-9

Tang, K.H.D.: Climate change education in China: a pioneering case of its implementation in tertiary education and its effects on students' beliefs and attitudes. Int. J. Sustain. High. Educ. **24**(5), 1058–1081 (2023). https://doi.org/10.1108/IJSHE-05-2022-0151/full/html

Tang, K.H.D., Hadibarata, T.: What are stopping university students from acting against climate change? Community Engag. High. Educ. **1**(1), 1–13 (2022). https://tecnoscientifica.com/journal/cehe/article/view/88

Tolppanen, S., Kang, J., Riuttanen, L.: Changes in students' knowledge, values, worldview, and willingness to take mitigative climate action after attending a course on holistic climate change education.

J. Clean. Prod. **373**, 133865 (2022). https://doi.org/10.1016/j.jclepro.2022.133865

United Nations (UN).: Annual Report 2023 (2023). https://www.un.org/en/annualreport2023

Valle, B., Valdés, O., Conde, C., Zavaleta, L.: Las organizaciones de la sociedad civil y su papel en la adaptación al cambio climático en México. Revista mexicana de investigación educativa **25**(87), 1149–1182 (2020). https://www.scielo.org.mx/scielo.php?pid=S1405-66662020000401149&script=sci_arttext

Zeigermann, U., Kammerer, M., Böcher, M.: What drives local communities to engage in climate change mitigation activities? Examining the rural–urban divide. Rev. Policy Res. **40**(6), 894–919 (2023). https://doi.org/10.1111/ropr.12528

Meat Consumption and Intention to Reduce Meat Intake by the Generation Z Cohort

Irene Kamenidou, Spyridon Mamalis, Stergios Gkitsas, Aikaterini Stavrianea, and Ifigeneia Mylona

Abstract

The production and consumption of meat are considered among the most harmful dietary patterns due to their high carbon, water, and ecological footprint. Therefore, nonprofit organizations and policymakers worldwide are seeking strategies to persuade citizens to reduce their meat consumption. However, these strategies will not yield the desired outcomes if citizens' behaviors, attitudes, and perceptions toward reducing meat consumption are negative, or if their interest in environmental issues is low. This study examines the meat consumption patterns, attitudes toward reducing meat consumption, and purchasing behaviors of adult Generation Z cohort members. It also investigates their environmental concern, general pro-environmental beliefs, and their willingness to both reduce meat consumption and recommend it to others. Finally, Gen Zers are segmented based on these variables. An online questionnaire targeting adult Gen Z participants ($N = 920$) was utilized. Mean Factor Scores (MFS) were calculated per factor to use in segmentation analysis. Results revealed that nearly half of the participants (48.2%) reported regularly eating meat at least 1–3 times a week. On a 7-point Likert and Likert-type scale, participants somewhat agreed that reducing meat consumption benefits both humans and the environment (MFS = 4.55). They also somewhat agreed that environmental issues are important (MFS = 5.42). However, Gen Zers tend to neither agree nor disagree that they are "green consumers" (MFS = 3.97) and that they intend to decrease meat consumption (MFS = 3.69). Additionally, they were unlikely to recommend reducing meat consumption to others (MV = 2.97). K-means cluster analysis identified three segments within the Gen Z group: those who were indifferent toward the studied variables and unwilling to reduce or recommend meat reduction ($n = 421$; 45.7%), those who were favorably disposed ($n = 296$; 32.2%), and those who were unfavorably disposed ($n = 203$; 22.1%). These findings underscore the need for environmental education and marketing strategies to foster long-term changes in attitudes, perceptions, and behavior.

Keywords

Sustainable food consumption · Segmentation · Generational cohorts · Attitudes · Perceptions · Marketing communications · Environmental education

1 Introduction

Nutritional behavior is closely linked to sustainability. Consumers food choices and eating habits, which have a direct impact on the livestock sector, are linked to problems such as overuse of water, land and oil, climate change and loss of biodiversity (Foley et al. 2011; Landrigan et al. 2018; Mekonnen and Hoekstra 2010, 2011; Newbold et al. 2016; Tilman and Clark 2014). In particular, animal-based food consumption, especially meat, is widely acknowledged as harmful to the environment (Godfray et al. 2018). With regard to meat consumption, there is a strong scientific agreement

I. Kamenidou (✉) · S. Mamalis · I. Mylona
Democritus University of Thrace, Kavala, Greece
e-mail: eikamenidou@mst.duth.gr

S. Mamalis
e-mail: mamalis@mst.duth.gr

I. Mylona
e-mail: imylona@mst.duth.gr

S. Gkitsas
Aristotle University of Thessaloniki, Thessaloniki, Greece
e-mail: sggkitsas@eled.auth.gr

A. Stavrianea
National and Kapodistrian University of Athens, Athens, Greece
e-mail: aikstavria@media.uoa.gr

P. Mohammad et al. (eds.), *Geographic Approaches to Climate Change and Mitigation: Urban and Rural Perspectives (Volume 2)*, Advances in Science, Technology & Innovation, https://doi.org/10.1007/978-3-031-92115-5_7

that plant-based foods are significantly better than animal-based foods from both an environmental and energy efficiency perspective (Bryant 2022; Marlow et al. 2009; Pimentel and Pimentel 2003). In addition, eating meat also increases the risk of chronic diseases and raises animal welfare issues (Bonnet et al. 2020). Therefore, reducing meat consumption is crucial to make individual diets more sustainable and healthier and also to reduce the environmental footprint of food systems (Trewern et al. 2022).

Around the world, non-profit organizations and policy makers are looking for strategies to convince citizens to reduce their meat consumption, adopt a vegetarian or vegan diet or replace meat protein with other types of protein (e.g. Scalabre et al. 2024; Kamenidou et al. 2023; Trent Grassian 2020). These strategies will come to nothing if citizens' behavior, attitudes, and perceptions towards reducing meat consumption are negative and their environmental awareness is low. Given the importance of the issue, all people, but especially the younger generations, need to understand the problem and learn to make appropriate, environmentally conscious decisions regarding meat consumption. Age is one of the most important factors in product selection (e.g. Drolet et al. 2010). Today, researchers believe that generational cohorts are more important than age, as they also include behavioral aspects (e.g. Lissitsa and Kol 2021). Generation Z (Gen Z) is the youngest adult generational cohort and comprises of individuals born between 1995 and 2009 (Williams and Page 2011). Understanding the sustainable or unsustainable consumption behavior of the Gen Z cohort is crucial as they will be future parents and influential leaders in the educational, political, social, and financial spheres, ultimately shaping sustainable consumption and development policies in future societies (e.g. Kamenidou et al. 2019). Therefore, academics argue in favor of paying much attention to this cohort as they are the future consumers and policy makers.

There is ample literature on the environmental and public health consequences of meat consumption, as well as on animal welfare issues. Furthermore, many studies advocate the importance of reducing meat consumption and try to seek ways to communicate this to a wide audience, including younger generations.

An extensive search revealed a small number of studies on meat reduction from a generational cohort perspective and Gen Z in particular (e.g. Juhasz et al. 2023). Most of the studies identified focused on sustainable food consumption (not specifically meat reduction) and the Gen Z cohort. Previous studies claim that attitudes directly or indirectly impact behaviors (Ajzen et al. 2018; Bamberg 2003). Additionally, extant research has found that individuals' concern for the environment has an impact on their pro-environmental choices (Albayrak et al. 2013). Dunlap and Jones (2002) define environmental concern as "the degree to which people are aware of problems regarding the environment and support efforts to solve them and or indicate the willingness to contribute personally to their solution" and refers to a person's (positive) attitude towards protecting the environment (Crosby et al. 1981). Regarding concern for the environment and attitudes towards reducing meat consumption, Cheah et al. (2020) found that "environmental concern is linked to consumer attitudes toward reducing meat consumption". Carfora et al. (2020) describe pro-environmental beliefs as recognizing the interconnectedness of nature preservation and human progress. These beliefs influence individuals' awareness of the consequences, i.e. the extent to which people acknowledge the negative impact of their actions on the environment. Extant research reveals that people with a high pro-environmental orientation reduce their meat consumption (Lai et al. 2020). With regard to the Gen Z cohort and their intention to reduce meat consumption, Neff et al. (2018) found that the 18–29 age group (which falls into the Gen Z category) is more reluctant to reduce meat consumption compared to other age groups. Also, Balińska et al. (2023) found in their study that Gen Z consider that their least environmental behavior is reducing meat consumption. The lack of studies on generational cohorts and meat consumption reduction provides fertile ground for further research. In addition, previous research has shown that attitudes can influence behavior and that environmentally concerned people tend to make more environmentally friendly choices (Lee and Holden 1999), among which is consuming less meat (Lentz et al. 2018).

For this reason, and taking into account the above points, this study focuses on the adult Gen Z cohort and has the following objectives (OB): to examine the adult Gen Z cohorts' meat consumption patterns (OB1); to explore their attitudes towards reducing meat consumption (OB2); to investigate their meat purchasing behavior (OB3); to examine Gen Zers' environmental concerns (OB4); to study their general pro-environmental beliefs (OB5); to explore Generation Z cohorts' intention to reduce meat consumption (OB6); to examine their intention to recommend a reduction in meat consumption (OB7); and finally, to group Gen Zers based on the above variables (OB8).

This study complements previous research as (a) meat reduction from the perspective of generational cohorts is still under-researched, in combination with the studies that simultaneously examine the abovementioned issues (OB1-7) focusing on adult Gen Zers, (b) few studies (to the best of our knowledge) conduct a segmentation analysis on meat consumption reduction and generational cohorts, and, (c) the above issues are under-researched in the Greek context.

2 Methodology

An online questionnaire aimed at adult members of the Generation Z cohort was used (i.e., born between 1995 and 2005). Multi-item questions were adopted and/or modified from previous research, thus ensuring the validity of the scale. Specifically, Table 1 shows the issues analyzed, the sources from which the items were adapted, the number of variables, the scale used, the reliability (Cronbach's alpha) of each multi-item question, and its mean factor score (MFS). In all cases, the Cronbach's alpha coefficient was greater than 0.700, indicating at least acceptable internal consistency. Multi-item questions were provided with answers on a 7-point Likert scale. The 7-point Likert scale was rated as 1 = completely disagree, 2 = mostly disagree, 3 = slightly/somewhat disagree, 4 = neither agree/neither disagree (neutral), 5 = slightly/somewhat agree, 6 = mostly agree, 7 = completely agree. The 7-point Likert-type scale was used for intention to reduce and intention to recommend to others to reduce meat consumption (1 = very unlikely, 2 = most unlikely, 3 = quite/somewhat unlikely, 4 = neither likely/neither unlikely (neutral), 5 = quite/somewhat likely, 6 = most likely, 7 = very likely).

A pilot test ($N = 111$) was carried out to ensure face validity (Tanner 2018) and to correct any comprehension problems in the wording of the questions. After finalizing the questionnaire, a non-probability sample (criteria and snowball sampling) was used for data collection. For a questionnaire to be eligible for data analysis, the respondent had to be born between 1995 and 2005 and give consent for their data to be used for analysis, which was the first question of the questionnaire (Government Gazette of the Hellenic Republic GDPR Law 4624 2019). In this way, and using the internet as a data collection channel, a final sample of 920 adult members of the Greek Generation Z cohort was collected during a five-month period (May–September 2023). The data analysis included descriptive statistics, i.e., frequencies, percentages, and mean values (MV). It also included factor analysis (and their mean factor scores; MFS), and k-means cluster analysis. Factor analysis in all cases produced one dimension, except for the question regarding attitudes towards meat purchasing which produced three dimensions ("Environmentally conscious meat purchasing", "Price-dependent meat purchasing", and "Locally oriented meat purchasing"). In all cases the MFS and the MV (for intention to recommend) were used in the

Table 1 Issues examined, sources of questions, scale used, reliability and MFS

Questions-issues	No. of variables	Adopted and/or modified from	Scale used	Reliability of scale (α)	MFS/MV
Meat consumption pattern	5	Kamenidou et al. (2019)	Multiple choice	–	–
Benefits of low meat consumption	4	Siegrist et al. (2015)	7-point Likert scale	0.873	4.55
Environmentally conscious meat purchasing	6	Roberts (1996)		0.910	3.96
Price-dependent meat purchasing	3	Roberts (1996)		0.790	3.26
Locally oriented meat purchasing	3	Roberts (1996)		0.715	4.19
Environmental concern	4	Untaru et al. (2016)		0.960	5.42
General pro-environmental beliefs	3	Carfora et al. (2021)		0.918	5.02
Self-identity as "green consumer"	3	Carfora et al. (2021)		0.922	3.96
Intention to reduce meat consumption (IRMC)	7	Armitage and Conner (2001); Untaru et al. (2016)	7-point Likert -type scale	0.980	3.69
Intention to recommend the consumption of less meat	1			–	2.97

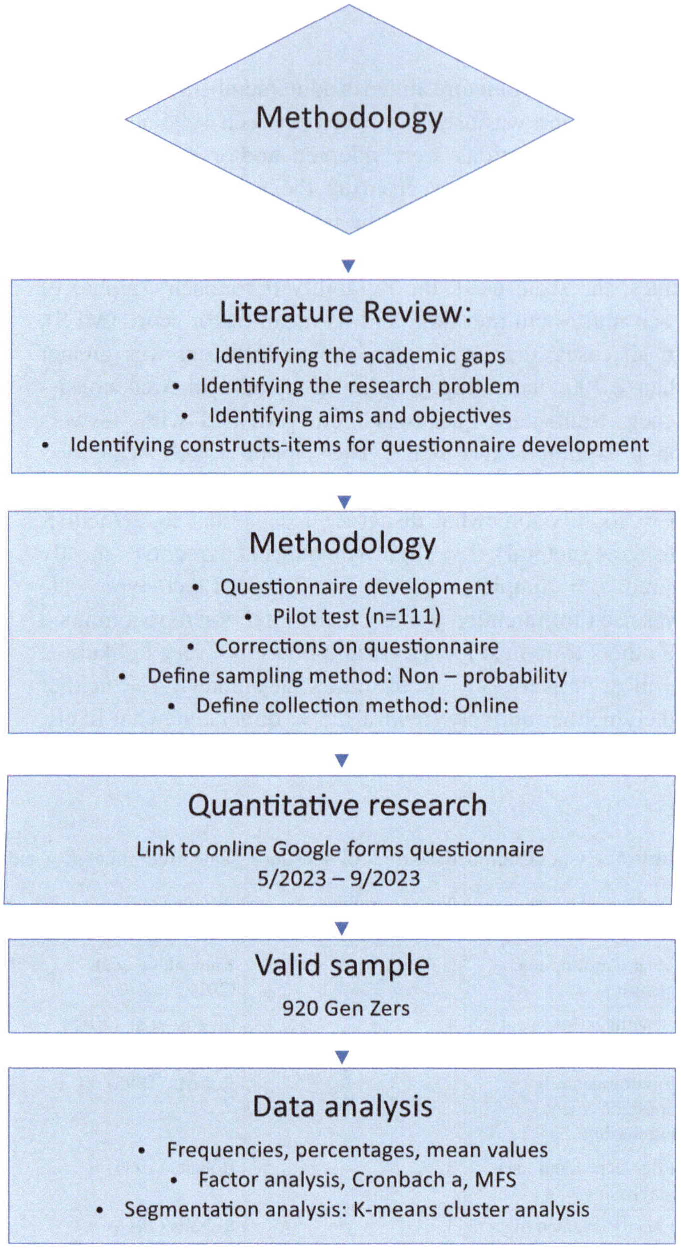

Fig. 1 Methodology flowchart (by authors)

segmentation analysis. Figure 1 presents the flowchart with an overview of the methodology.

3 Results

3.1 Sample Profile

Males (50.9%) and females (49.1%) were equally represented. The Gen Z participants were divided into two age groups, those born between 1995 and 1999 (41.7%) and those born between 2000 and 2005 (58.3%). The vast majority were single (91.7%), had completed secondary school (61.4%), and were studying at a university (63.9%). They also lived in cities (80.9%) and had a personal net income of up to 600€ (61.4%). The remaining demographic characteristics of Gen Zers (education, area of residence, and net family income per month) are presented in Figs. 2, 3 and 4.

3.2 Attitudes, Behavior, and Intention

The results reveal that half of the participants (49.8%) reported being "meat-eaters", i.e., eating meat every day or several times a week. Additionally, 41.5% stated that they follow a Mediterranean diet, i.e. they eat all kinds of food, emphasizing vegetables, legumes, fruit, and fish, while they

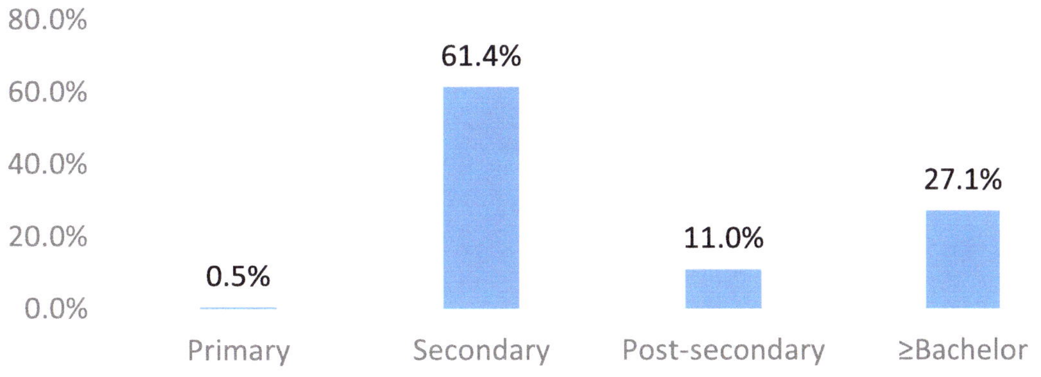

Fig. 2 Gen Zers education level (by authors)

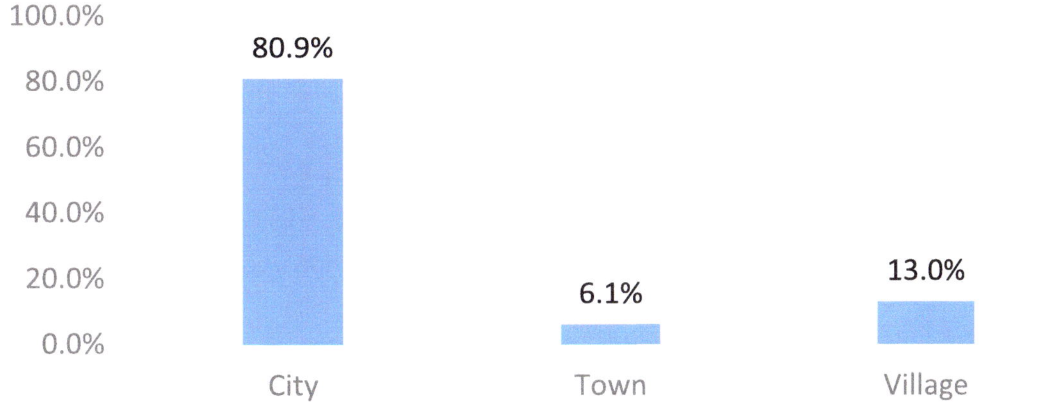

Fig. 3 Gen Zers area of residence (by authors)

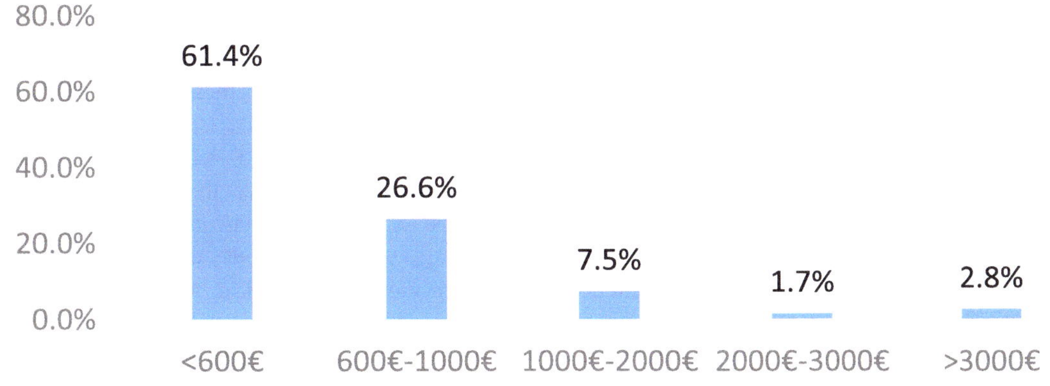

Fig. 4 Gen Zers net monthly family income (by authors)

do not eat meat very often. Pescatarians, flexitarians, vegetarians, and vegans together consist of a very small proportion of the sample (8.6%).

In terms of frequency of consumption, the majority of participants consume meat 1–3 times per week (48.2%), while 17.8% of respondents do not eat meat, either at all, on special occasions or 1–3 times per month. There are also 34.0% "heavy consumers", who consume meat more than four times a week up to several times a day.

Participants tend to somewhat agree (on a 7-point Likert scale) that consuming less meat has benefits for their health and the environment (MFS = 4.55). They also somewhat agree that they have general pro-environmental beliefs (MFS = 5.02) and have environmental concern (MFS = 5.42), while

Table 2 Segmentation analysis

Dimensions/ Factors	Segments and FCC			ANOVA tests	
	1st «Indifferent but unwilling» $N = 421$	2nd «Environmentally conscious» $N = 296$	3rd «Non-environmentally sensitive» $N = 203$	F	Sign
Benefits of low meat consumption	4.44	5.84	2.88	414.744	< 0.001
Environmentally conscious meat purchasing	3.72	5.39	2.41	531.924	< 0.001
Price-dependent meat purchasing	3.24	3.65	2.75	23.962	< 0.001
Locally oriented meat purchasing	4.26	4.83	3.09	141.241	< 0.001
Environmental concern	5.72	6.51	3.25	546.663	< 0.001
General pro-environmental beliefs	5.17	6.23	2.96	505.002	< 0.001
Self-identity as "green consumer"	3.68	5.46	2.38	471.884	< 0.001
Intention to reduce meat consumption (IRMC)	3.24	5.32	2.30	390.358	< 0.001
Intention to recommend the consumption of less meat	2.54	3.92	2.51	160.324	< 0.001

they neither agree nor disagree that they self-identify as a "green consumer" (MFS = 3.96).

Regarding their purchasing behavior, they neither agree nor disagree that they are environmentally conscious (MFS = 3.97) or purchase locally produced meat (MS = 4.19), and they somewhat disagree that they focus on price and nonlocal meat (MFS = 3.26).

Finally, they tend to neither agree nor disagree (on a 7-point Likert-type scale) that they will reduce their meat consumption (MFS = 3.69), while they will not recommend others to do so (MV = 2.97).

These MFS and MV were used in the segmentation analysis.

3.3 Segmentation

The K-Means cluster analysis revealed three groups of Gen Zers, namely the first group refers to those who were indifferent to the above variables but are unwilling to reduce meat consumption nor recommend others to do so. The second group incorporates those who were favorably disposed and the third, those who were unfavorably disposed. Table 2 presents the three segments and their Final Cluster Centers (FCC-in the column under each segment) as well as the ANOVA statistics which revealed that the three segments differ in all cases.

Segment I is labeled as "Indifferent but unwilling" and is formed by 45.7% of the participants ($N = 421$). They generally tend to neither agree nor disagree with the statements posed to them. Their FCC range from 2.54 to 5.72, but mostly is around the neither agree nor disagree point (3.51–4.49). As compared to the other segments, they have the most moderate responses in almost all factors that emerged from the analysis (excluding price-dependent meat purchasing, intention to reduce meat consumption, and intention to recommend the consumption of less meat). They consider themselves to be people with environmental concerns and environmental beliefs, but do not intend to reduce their quantities of meat intake, nor are they going to recommend anyone to do so.

Segment II: "Environmentally conscious", comprises about one-third (32.2%) of the participants ($N = 296$), which tend to somewhat agree to completely agree ($4.83 < \text{FCC} < 6.51$) with almost all dimensions. Exceptions are "Intention to recommend the consumption of less meat" and "Price-dependent meat purchasing", in which cases they neither agree nor disagree (FCC = 3.92 and FCC = 3.65 respectively). Compared to the other segments, they have the highest FCCs for all dimensions and therefore labeled as the "Environmentally conscious" segment.

Segment III: "Non-environmentally sensitive" comprises 22.1% of participants (N = 203). In all cases, they have the lowest FCC value (2.30 < FCC < 3.25) compared to the other segments, and they have no intention of reducing meat consumption or recommending others to do so.

4 Discussion

Adopting a vegetarian lifestyle has increased rapidly in the Western world, but acceptance varies greatly between age groups. For example, while over 25% of Gen Zers in the UK are meat-free, only around 5% of Baby Boomers are (Ge et al. 2022). The present study found a different situation in Greece, as less than 10% of the Gen Zers abstained from meat.

The results also revealed that 34% of the sample consume meat more than four times a week to several times a day, while another 48.2% consume meat 1–3 times a week. These results are in line with the results of Bogueva and Marinova (2020) who in their study found that Australian Gen Zers also are (as regards frequency of meat consumption), "heavy meat consumers" with 44% consuming meat daily and 38% consuming meat a few times per week.

Gen Zers tend to somewhat agree (MFS = 4.55) that consuming less meat has benefits for their health and the environment. This result is partly consistent with the findings of Chen et al. (2022), who found that urban Gen Z tourists tend to agree that health benefits motivated them towards vegetarian consumption.

Our study found that Gen Zers somewhat agree that they have general pro-environmental beliefs (MFS = 5.02) and environmental concern (MFS = 5.42), and neither agree nor disagree that they are a "green consumer" (MFS = 3.96), nor are they environmentally conscious (MFS = 3.97) as regards their attitudes towards meat purchasing behavior.

Saut and Saing (2021) concluded in their study on Cambodian Gen Zers that environmental consciousness has an important effect on both attitude and purchase intentions, while Gen Zers who are environmentally conscious are more likely to purchase green products. Whilst previous research on Gen Zers point out that when they are environmentally aware and concerned, they tend to consume less meat (Food Standards Agency Research Report 2020 for UK Gen Zers), the results of the present study do not appear to be similar, as Gen Zers tend to neither agree nor disagree that they will reduce their meat consumption (MFS = 3.69), nor are they willing to recommend to other people to do so (MV = 2.97). This behavior is confirmed by previous research. Teather and Etterson (2023) came to the conclusion that there is a discrepancy between values and actions among US Generation Z, while their awareness and attitudes towards sustainability are high, their willingness to behave sustainably is low.

Segmentation analysis provided with three distinct groups, namely the «Indifferent but unwilling», the «Environmentally conscious», and the «Non-environmentally sensitive» who have distinct perceptions, attitudes, and behavior. The study by Götze and Brunner (2021) produced six segments which had similarities with the segments of the present study. Their six identified segments encompassed distinct dietary patterns: (1) "the environmentally and health-oriented meat-eaters," (2) "the uncompromising meat-eaters," (3) "the moderate meat-eaters who are willing to substitute meat," (4) "the indifferent but moderate meat-eaters," (5) "the environmentally-conscious regular meat-eaters," and (6) "the environmentally and health-conscious meat avoiders." These findings mirror those of the current study, revealing both environmentally conscious and indifferent groups, and partially with the "Non-environmentally sensitive" group.

Of the three segments, only Segment II (32.2%) shows a stronger inclination towards environmental issues. The fact that only a third of participants were categorized as environmentally conscious is in line with the findings of another study of Gen Zers, which found that despite their awareness of environmental issues and ethical considerations and their belief in the benefits of sustainable food consumption for their personal and societal wellbeing, their sustainable consumption is limited (Orea-Giner and Fusté-Forné 2023). The study by Slotnick et al. (2023) found that those who prioritized environmentally friendly and climate-conscious food choices and believed in the effectiveness of reduced meat consumption to combat climate change reported consuming 10–25% less red meat, which is also consistent with the results of the present study, as people who reported environmentally conscious attitudes also exhibited relatively corresponding behavior.

Although the first segment of Gen Zers self-identifies as environmentally conscious, they are not prepared to reduce their meat consumption and will certainly not recommend this to others. This also corresponds to the value-action gap of Gen Zers identified by Teather and Etterson (2023).

Finally, the third segment of Gen Zers do not identify as environmentally conscious and do not consider themselves to be "green consumers". The price and origin of the meat also play no role for them. This segment will not reduce its meat consumption and is not willing to recommend others to do so either. This segment is similar to the second segment "the uncompromising meat-eaters," from the study by Götze and Brunner (2021), i.e., Gen Zers that show strong meat consumption pattern.

The segmentation analysis shows that only 32.2% of the sample intend to reduce meat consumption to some extent and 67.8% of the sample do not. This means that these groups will not be taking any action towards sustainable food consumption in the near future and therefore strategic marketing measures are required.

4.1 Implications

Addressing the challenge of changing meat consumption habits, specifically aiming for meat reduction, requires an understanding of the social factors involved and the development of effective, focused interventions. Therefore, this study highlights the importance of environmental education and marketing strategies to shift attitudes and behaviors over time, especially towards reducing meat consumption. The findings offer insights for marketers and authorities to tailor their communication efforts to Gen Z by suggesting targeted strategies and the use of innovative tools such as virtual reality and augmented reality. Social marketing campaigns incorporating interactive platforms, digital applications and influencer support can complement traditional media in spreading environmentally friendly messages. Also, integrating communication initiatives into national education policies to promote environmental awareness and action among younger generations, including the promotion of meat reduction could be utilised.

5 Conclusions, Limitations, and Directions for Future Research

To summarize, this study of adult Gen Zers reveals different perspectives within this cohort: While some are indifferent to meat reduction and environmental issues, others are environmentally conscious and support a reduction in meat consumption, while another segment is not environmentally conscious. These findings highlight the value of targeted environmental education and marketing strategies to promote long-term changes in attitudes, perceptions and behaviors towards meat consumption among the Greek Gen Zers. This study is subject to certain limitations. It focused exclusively on a single country, Greece, a specific generational cohort (Generation Z) and was realized with a non-probability sampling method. These restrictions may limit the generalizability and the applicability of the results in general. Finally, although the sample size was relatively small, it was deemed appropriate for the specific aims of this study. These limitations provide valuable insights for guiding future research efforts in this area.

Acknowledgements "There are no ethical issues involved in the processing of the questionnaire data used in the study. The necessary consents have been obtained by the persons involved, and the anonymity of the participants has been secured. All procedures performed in studies involving human participants were in accordance with the ethical standards of the Democritus University of Thrace research committee and with the 1964 Helsinki declaration and its later amendments or comparable ethical standards." Law 4751/20.08.2024, Issue B', article 205, §1.

References

Ajzen, I., Fishbein, M., Lohmann, S., Albarracín, D.: The influence of attitudes on behavior. In: The Handbook of Attitudes, Volume 1: Basic Principles, pp. 197–255 (2018)

Albayrak, T., Aksoy, Ş, Caber, M.: The effect of environmental concern and scepticism on green purchase behaviour. Mark. Intell. Plan. **31**(1), 27–39 (2013). https://doi.org/10.1108/02634501311292902

Armitage, C.J., Conner, M.: Efficacy of the theory of planned behaviour: a meta-analytic review. Br. J. Soc. Psychol. **40**(4), 471–499 (2001). https://doi.org/10.1348/014466601164939

Balińska, A., Jaska, E., Werenowska, A.: Environmentally and socially sustainable behavior of Generation Z in the context of tourist trips. Eur. Res. Stud. J. **26**(4), 945–959 (2023)

Bamberg, S.: How does environmental concern influence specific environmentally related behaviors? A new answer to an old question. J. Environ. Psychol. **23**(1), 21–32 (2003). https://doi.org/10.1016/S0272-4944(02)00078-6

Bogueva, D., Marinova, D.: Cultured meat and Australia's generation Z. Front. Nutr. **7**, 148 (2020)

Bonnet, C., Bouamra-Mechemache, Z., Réquillart, V., Treich, N.: Viewpoint: regulating meat consumption to improve health, the environment and animal welfare. Food Policy **97**, 101847 (2020). https://doi.org/10.1016/j.foodpol.2020.101847

Bryant, C.J.: Plant-based animal product alternatives are healthier and more environmentally sustainable than animal products. Future Foods **6**, 100174 (2022). https://doi.org/10.1016/j.fufo.2022.100174

Carfora, V., Cavallo, C., Catellani, P., Del Giudice, T., Cicia, G.: Why do consumers intend to purchase natural food? Integrating theory of planned behavior, value-belief-norm theory, and trust. Nutrients **13**(6), 1904 (2021). https://doi.org/10.3390/nu13061904

Carfora, V., Conner, M., Caso, D., Catellani, P.: Rational and moral motives to reduce red and processed meat consumption. J. Appl. Soc. Psychol. **50**(12), 744–755 (2020). https://doi.org/10.1111/jasp.12710

Cheah, I., Sadat Shimul, A., Liang, J., Phau, I.: Drivers and barriers toward reducing meat consumption. Appetite **149**, 104636 (2020). https://doi.org/10.1016/j.appet.2020.104636

Chen, G., Tan, W., Ran, N., Zhang, J., Yan, B.: Do urban tourists prefer vegetarianism? An urban-rural comparison of vegetarian consumption in China. Front. Nutr. **9**, 996158 (2022)

Crosby, L.A., Gill, J.D., Taylor, J.R.: Consumer/voter behavior in the passage of the Michigan container law. J. Mark. **45**(2), 19–32 (1981)

Drolet, A.L., Schwarz, N., Yoon, C.: The Aging Consumer: Perspectives From Psychology and Economics. Routledge Academic (2010)

Dunlap, R.E., Jones, R.E.: Environmental concern: conceptual and measurement issues. Handb. Environ. Sociol. **3**(6), 482–524 (2002)

Foley, J.A., Ramankutty, N., Brauman, K.A., Cassidy, E.S., Gerber, J.S., Johnston, M., Mueller, N.D., O'Connell, C., Ray, D.K., West, P.C., Balzer, C., Bennett, E.M., Carpenter, S.R., Hill, J., Monfreda, C., Polasky, S., Rockström, J., Sheehan, J., Siebert, S., Tilman, D., Zaks,

D.P.M.: Solutions for a cultivated planet. Nature **478**(7369), 337–342 (2011). https://doi.org/10.1038/nature10452

Food Standards Agency.: The Future consumers—Food and Generation Z. FSA Research Report (2020). Retrieved 10 Oct 2024, from https://www.food.gov.uk/sites/default/files/media/document/generation-z-full-report-final.pdf

Ge, J., Scalco, A., Craig, T.: Social influence and meat-eating behaviour. Sustainability **14**(13), 7935 (2022). https://doi.org/10.3390/su14137935

Godfray, H.C.J., Aveyard, P., Garnett, T., Hall, J.W., Key, T.J., Lorimer, J., Pierrehumbert, R.T., Scarborough, P., Springmann, M., Jebb, S.A.: Meat consumption, health, and the environment. Science **361**(6399), eaam5324 (2018). https://doi.org/10.1126/science.aam5324

Götze, F., Brunner, T.A.: A consumer segmentation study for meat and meat alternatives in Switzerland. Foods **10**(6), 1273 (2021). https://doi.org/10.3390/foods10061273

Government Gazette of the Hellenic Republic. (2019). Hellenic Data Protection Authority (HDPA), measures for implementing Regulation (EU) 2016/679 of the European Parliament and of the Council of 27 April 2016 on the Protection of Natural Persons With Regard to the Processing of Personal Data, and Transposition of Directive (EU) 2016/680 of the European Parliament and of the Council of 27 April 2016, and Other Provisions (Law 4624/29-9-2019). Retrieved from https://www.dpa.gr/sites/default/files/2020-08/LAW%204624_2019_EN_TRANSLATED%20BY%20THE%20HDPA.PDF

Juhasz, M., Charlebois, S., Music, J.: Perceived value of cultured proteins as novel food in Canada: Generation Z consumers in a cross generational perspective. J. Int. Food Agribus. Mark. 1–28 (2023). https://doi.org/10.1080/08974438.2023.2190739

Kamenidou, I.C., Mamalis, S.A., Pavlidis, S., Bara, E.Z.: Segmenting the Generation Z Cohort University students based on sustainable food consumption behavior: a preliminary study. Sustainability **11**(3), 837 (2019). https://doi.org/10.3390/su11030837

Kamenidou, I. (Eirini), Mamalis, S., Gkitsas, S., Mylona, I., Stavrianea, A. (2023). Is Generation Z ready to engage in entomophagy? A segmentation analysis study. Nutrients **15**(3), 525. https://doi.org/10.3390/nu15030525

Lai, A.E., Tirotto, F.A., Pagliaro, S., Fornara, F.: Two sides of the same coin: environmental and health concern pathways toward meat consumption. Front. Psychol. **11**, 578582 (2020). https://doi.org/10.3389/fpsyg.2020.578582

Landrigan, P.J., Fuller, R., Acosta, N.J.R., Adeyi, O., Arnold, R., Basu, N. (Nil), Baldé, A.B., Bertollini, R., Bose-O'Reilly, S., Boufford, J.I., Breysse, P.N., Chiles, T., Mahidol, C., Coll-Seck, A.M., Cropper, M.L., Fobil, J., Fuster, V., Greenstone, M., Haines, A., … Zhong, M. (2018) The lancet commission on pollution and health. Lancet **391**(10119), 462–512. https://doi.org/10.1016/S0140-6736(17)32345-0

Lee, J.A., Holden, S.J.S.: Understanding the determinants of environmentally conscious behavior. Psychol. Mark. **16**(5), 373–392 (1999). https://doi.org/10.1002/(SICI)1520-6793(199908)16:5%3c373::AID-MAR1%3e3.0.CO;2-S

Lentz, G., Connelly, S., Mirosa, M., Jowett, T.: Gauging attitudes and behaviours: meat consumption and potential reduction. Appetite **127**, 230–241 (2018). https://doi.org/10.1016/j.appet.2018.04.015

Lissitsa, S., Kol, O.: Four generational cohorts and hedonic m-shopping: association between personality traits and purchase intention. Electron. Commer. Res. **21**(2), 545–570 (2021). https://doi.org/10.1007/s10660-019-09381-4

Marlow, H.J., Hayes, W.K., Soret, S., Carter, R.L., Schwab, E.R., Sabaté, J.: Diet and the environment: does what you eat matter? Am. J. Clin. Nutr. **89**(5), 1699S-1703S (2009). https://doi.org/10.3945/ajcn.2009.26736Z

Mekonnen, M.M., Hoekstra, A.Y.: The green, blue and grey water footprint of farm animals and animal products. Volume 2: Appendices. Value of Water: Research Report Series No. 48, UNESCO-IHE, Delft, the Netherlands (2010)

Mekonnen, M.M., Hoekstra, A.Y.: The green, blue and grey water footprint of crops and derived crop products. Hydrol. Earth Syst. Sci. **15**(5), 1577–1600 (2011). https://doi.org/10.5194/hess-15-1577-2011

Neff, R.A., Edwards, D., Palmer, A., Ramsing, R., Righter, A., Wolfson, J.: Reducing meat consumption in the USA: a nationally representative survey of attitudes and behaviours. Public Health Nutr. **21**(10), 1835–1844 (2018)

Newbold, T., Hudson, L.N., Arnell, A.P., Contu, S., De Palma, A., Ferrier, S., Hill, S.L.L., Hoskins, A.J., Lysenko, I., Phillips, H.R.P., Burton, V.J., Chng, C.W.T., Emerson, S., Gao, D., Pask-Hale, G., Hutton, J., Jung, M., Sanchez-Ortiz, K., Simmons, B.I., Whitmee, S., Zhang, H., Scharlemann, J.P.W., Purvis, A.: Has land use pushed terrestrial biodiversity beyond the planetary boundary? A global assessment. Science **353**(6296), 288–291 (2016). https://doi.org/10.1126/science.aaf2201

Orea-Giner, A., Fusté-Forné, F.: The way we live, the way we travel: generation Z and sustainable consumption in food tourism experiences. Br. Food J. **125**(13), 330–351 (2023). https://doi.org/10.1108/BFJ-11-2022-0962

Pimentel, D., Pimentel, M.: Sustainability of meat-based and plant-based diets and the environment. Am. J. Clin. Nutr. **78**(3), 660S-663S (2003). https://doi.org/10.1093/ajcn/78.3.660S

Roberts, J.A.: Green consumers in the 1990s: profile and implications for advertising. J. Bus. Res. **36**(3), 217–231 (1996)

Saut, M., Saing, T.: Factors affecting consumer purchase intention towards environmentally friendly products: a case of generation Z studying at universities in Phnom Penh. SN Bus. Econ. **1**(6), 83 (2021)

Scalabre, C., Boiral, O., Yuriev, A., Guillaumie, L., Selihi, Z., Gaudreau, C.: Promotional campaigns in the agri-food sector aimed at reducing meat consumption: a promising leverage for action. J. Rural. Stud. **105**, 103187 (2024). https://doi.org/10.1016/j.jrurstud.2023.103187

Siegrist, M., Visschers, V.H., Hartmann, C.: Factors influencing changes in sustainability perception of various food behaviors: results of a longitudinal study. Food Qual. Prefer. **46**, 33–39 (2015)

Slotnick, M.J., Falbe, J., Cohen, J.F.W., Gearhardt, A.N., Wolfson, J.A., Leung, C.W.: Environmental and climate impact perceptions in university students: sustainability motivations and perceptions correspond with lower red meat intake. J. Acad. Nutr. Diet. **123**(5), 740–750 (2023). https://doi.org/10.1016/j.jand.2022.09.015

Tanner, K.: Survey designs. In: Williamson, K., Johanson, G. (eds.) Research Methods. Chandos Publishing (2018)

Teather, A., Etterson, J.: Value-action gaps between sustainability behaviors, knowledge, attitudes and engagement in campus and curricular activities within a cohort of Gen Z university students. J. Sustain. Educ. **28** (2023)

Tilman, D., Clark, M.: Global diets link environmental sustainability and human health. Nature **515**(7528), 518–522 (2014). https://doi.org/10.1038/nature13959

Trent Grassian, D.: The dietary behaviors of participants in UK-based meat reduction and vegan campaigns—a longitudinal, mixed-methods study. Appetite **154**, 104788 (2020). https://doi.org/10.1016/j.appet.2020.104788

Trewern, J., Chenoweth, J., Christie, I.: Sparking change: evaluating the effectiveness of a multi-component intervention at encouraging more sustainable food behaviors. Appetite **171**, 105933 (2022). https://doi.org/10.1016/j.appet.2022.105933

Untaru, E.-N., Ispas, A., Candrea, A.N., Luca, M., Epuran, G.: Predictors of individuals' intention to conserve water in a lodging context: the application of an extended theory of reasoned action. Int. J. Hosp. Manage. **59**, 50–59 (2016). https://doi.org/10.1016/j.ijhm.2016.09.001

Williams, K., Page, R.: Marketing to the generations. J. Behav. Stud. Bus. **3**, 1–17 (2011)

Rural Adaptations and Nature-Based Solutions

Turning Climate Peril into Economic Catalyst: Adaptive Strategies for Enhancing Ecosystem Services and Local Industry in Grimsby, UK

Hao Huang, Guangda Li, Haochen Shi, and Tianyi Zhao

Abstract

This study probes nature-based adaptation strategies to address climate-induced flooding risks in Grimsby, UK, a historic fishing town facing economic decline. Acknowledging the town's historical economic downturn post-Brexit and the ongoing environmental degradation, we propose an innovative dual strategy. By constructing an interconnected lagoon wetland system utilising dredged sediments and advancing multi-trophic aquaculture, we aim to safeguard against storm surges while reviving the local fishing industry sustainably. Our spatial analysis pinpoints optimal locations for wetland restoration and aquaculture sites, promising to bolster biodiversity, protect coastal habitats, and foster economic growth. This research underscores the transformative potential of aligning ecological preservation with economic development to address global climate challenges, offering a replicable model for other coastal communities with ecologically regenerative local industry and an active response to achieving the Sustainable Development Goals of the United Nations.

Keywords

Climate change · Grimsby · Flooding risk · Nature-based adaptation · Aquaculture · Sustainable development goals

1 Introduction

Anthropogenic climate change has developed into a planetary existential threat, generating extreme meteorological events such as heatwaves, drought, rain-driven floods, and intense storms, which have undermined infrastructure and both natural and human habitats worldwide (IPCC 2023). The increasingly accelerated melting of glaciers and polar ice sheets is driving sea-level rise and exacerbating coastal erosion, permanent inundation, and the displacement of residents in many coastal and island regions globally (Church et al. 2013). The urgency of these threats demands rapid and effective climate change mitigation and adaptation policies. This study proposes a nature-based adaptation strategy to address these challenges in Grimsby, UK.

As one of the UK's few remaining coastal fishing towns, Grimsby faces a decisive moment regarding its future. Once the epicenter of the fishing industry, it has seen its fishing capacity decline due to rapidly changing climate conditions, Brexit-related restructuring of the fishing industry, and the downsizing of coastal infrastructure, including the relocation of fishing harbours. Rising sea levels and the other resulting climate impacts compund flood risks stemming from industrial alterations to the coastline, further reducing fish populations. Simultaneously, economic factors such as the declining incomes and a shrinking population have placed significant stress on the town in recent years. These challenges create a complex situation requiring inventive and sustainable solutions (Young et al. 2019). Adaptation strategies thus play a vital role in mitigating economic crises for

H. Huang (✉) · G. Li
Department of Architecture and Civil Engineering, City University of Hong Kong, Hong Kong (SAR), China
e-mail: hhuang95-c@my.cityu.edu.hk

G. Li
e-mail: guangdali3-c@my.cityu.edu.hk

H. Shi
School of Architecture and Urban Planning, Guangzhou University, Guangzhou, China
e-mail: shc_scut@gzhu.edu.cn

T. Zhao
School of Architecture, Southeast University, Nanjing, China
e-mail: zhaotianyi@seu.edu.cn

P. Mohammad et al. (eds.), *Geographic Approaches to Climate Change and Mitigation: Urban and Rural Perspectives (Volume 2)*, Advances in Science, Technology & Innovation, https://doi.org/10.1007/978-3-031-92115-5_8

coastal communities heavily reliant on fishing, particularly amid dwindling fisheries and intensifying climate impacts (Cisneros-Montemayor et al. 2021).

Nature-based solutions (NbS hereafter) are 'actions to protect, sustainably manage, restore, or harvest ecosystems to enhance human well-being and biodiversity.' They are increasingly recognized as valuable approaches for enhancing resilience to climate change (Cohen-Shacham et al. 2016). The core premise of this design-based study is that climate change threats can be transformed into opportunities for economic revitalization and ecological restoration. Green infrastructure investments in natural capital can enhance ecosystem services while supporting local industry development. By implementing nature-based adaptation strategies, specifically the creation of interconnected lagoon wetland systems and the integrating multi-trophic aquaculture through managed realignment, Grimsby can revitalize its fishing legacy and secure a sustainable future (Barbier et al. 2011; Roberts et al. 2017).

2 Related Work

2.1 Climate Change Impacts on Coastal Communities

Coastal communities are disproprtionately affected by climate change due to their proximity to the ocean. Key impacts include sea-level rise, increased storm frequency and intensity, ocean acidification, shoreline erosion, and altered precipitation patterns (Smith et al. 2001; Nicholls and Cazenave 2010). Low-lying coastal areas are particularly vulnerable, with flooding projected to affect hundreds of millions more people by 2050 (Knutson et al. 2010; Neumann et al. 2015). Coastal wetlands, such as mangroves, provide critical storm protection, but their degradatin compromises natural coastal defenses (Gedan et al. 2011). Simultanesously, climated-inducted fisheries collapse and heightened storm damage pose socioeconomic threats to fishing-dependent coastal communities (Cisneros-Montemayor et al. 2021).

Research on climate change has surged recently due to its profound effects on ecosystems and societies. Coastal flooding, driven by rising sea levels, is pressing issue worldwide, spurring extensive research. Nicholls and Cazenave (2010) examined the consequences of sea-level rise, warning of widespread displacement, property loss, and ecosystem damage. Industrial activities, such as commercial dredging and port development, alter coastlines, diminishing habitats and increasing flood risks (Pye and Blott 2014; Zhao et al. 2019). Extreme weather events, including intensified storms, have become more frequent due to climate change. Knutson et al. (2010) demonstrated that tropical cyclones are likely to grow more intense, posing catastrophic risks to infrastructure, economies, and ecosystems. Hallegatte et al. (2013) advocate for improved early warning systems and resilient infrastructure planning, particularly for low-lying areas. Rising sea levels and stronger storms also accelerate coastal erosion. Cooper and Pilkey (2004) critique hard engineering solutions (e.g., seawalls), arguing that managed retreat and sustainable approaches are more viable long-term strategies.

Socioeconomic risks and planning challenges are exacerbated by climate change, especially without robust adaptation strategies. Hinkel et al. (2014) modeled the economic and social impacts, emphasizing the need for adaptation efforts to avert massive damage, particularly in Asia and Africa. Adger et al. (2005) framed community vulnerability as a combination of climate exposure and their socioeconomic resilience, calling for policy-driven adaptation (Smit and Wandel 2006). Moser and Ekstrom (2010) focused on coastal governance, highlighting the need for integrated land, water, and resource management.

2.2 Nature-Based Adaptation Strategies for Coastal Resilience

Nature-based adaptation (NbA), or ecosystem-based adaptation (EbA), is now regarded as critical for coastal resilience (Colls et al. 2009). NbS leverage living organisms, soil, sediment, and topography to mitigate climate hazards. Compared to conventional methods, NbS are more adaptive, multifunctional, and dynamic (Hobbie and Grimm 2020). Seddon et al. (2020) define NbS as actions that protect, restore, or manage ecosystems to address societal challenges (e.g., food security, biodiversity loss) while benefiting land, water, and the air systems. The authors discuss the potential for co-benefits these systems provide, including biodiversity conservation, carbon storage, and improved opportunities for human well-being. The use of dredged sediments to develop lagoon wetland systems along coasts has been discussed and demonstrated in several situations (Temmerman et al. 2013). Wetland restoration (and other NbS) is increasingly seen as a potent means of improving climate resilience through ecosystem benefits (Sutton-Grier et al. 2015; Sutton-Grier and Sandifer 2019). Managed retreat/realignment and tidal wetland restoration are viable and cost-effective alternatives to grey infrastructure for coastal protection (Cheong et al. 2013). Controlled realignment, defined as converting farmland to the floodplain, has been proposed to increase wildlife habitat and biodiversity and improve ecosystem services (Ledoux et al. 2005; Esteves 2013).

There is a growing consensus on the value of NbA strategies for enhancing coastal resilience. Despite their great potential, implentation aligned with Target 14 of the Sustainable Development Goals (SDGs) faces socioeconomic challenges. The primary challenges in establishing NbA strategies include integrating governance across scales, fostering stakeholder engagement, and prioritising equity. Bennet and Ruef (2016) provide a case study of the socioeconomic dimensions of NbS, showing that the design and success of community-based ecosystem conservation projects require practitioners and policymakers to incorporate social science insights. Particular focus is placed on how nature-based adaptation strategies affect people's daily lives. Furthermore, Jones et al. (2012) examine the economic viability of NbS, demonstrating that, in many cases, the cost-effectiveness of these solutions outperforms that of grey infrastructure, especially when considering the full range of ecosystem service benefits. At the policy level, the Intergovernmental Panel on Climate Change (IPCC 2015) highlights the emerging role of NbS in climate adaptation strategies and recommends the inclusion of ecosystem services in integrated approaches to coastal management. Cohen-Shacham et al. (2016) explore ecosystem-based disaster risk reduction (Eco-DRR), noting its alignment with EbA and stressing its significance within sustainable development policies. Therefore, it is essential to integrate natural infrastructure into urban coastal adaptation, which requires an interdisciplinary methodology combining ecology, geomorphology, hydrology, and community engagement. Spatial analysis techniques are highly beneficial in identifying ideal sites for wetland restoration and aquaculture growth. This approach enables better integration of urban, agricultural, and natural systems, thereby promoting the establishment of robust ecotones along coasts (Naylor et al. 2000; Gittman et al. 2016) and improving the integration between grey and green infrastructure (Sutton-Grier et al. 2018).

2.3 Transitioning Coastal Economies to Sustainable Models

Most of the world's coastal economies are under pressure to transition toward sustainable development while ensuring environmental sustainability. This involves a complex multidimensional approach that balances the economic, social, and environmental aspects of sustainability. Barbier (2017) focuses on the 'blue economy', arguing that sustainable use of ocean resources can generate economic growth, create good jobs and improve quality of life while protecting marine ecosystems. The author provides a framework for sustainable coastal economic development that balances economic benefits with ecological and social outcomes. The UN established the SDGs in 2015, with a focus on protecting and conserving oceans, seas, and marine resources. These goals outline the global agenda for transitioning coastal economies to sustainability while respecting traditional ocean-dependent communities and supporting their transition to a more sustainable future. Fernandino et al. (2018) examine the role of ecosystem-based management (EBM) in coastal sustainability transitions, highlighting its 'dual function': enhancing ecological integrity ('conserving nature') and improving the resilience of communities ('conserving livelihoods and well-being'). Nicholls and Cazenave (2010) emphasize the impacts of sea-level rise, storm surge, and coastal floods caused by rising sea-surface temperatures, stressing the need for economic planning framework to address these risks. Mahon et al. (2008) discuss governance challenges in sustainable coastal economics, advocating for multi-level governance approaches that facilitate participatory decision-making and bridge science-policy gaps to advance sustainability transitions. Bennett et al. (2014) analyze marine spatial planning (MSP) as a framework for balancing economic and environmental objectives in ocean resource use. They position MSP as a tool for developing sustainable coastal economies that support fishing and aquaculture livelihoods, deliver ecosystem services, ensure food security, and guarantee equitable ocean access.

Several coastal regions have begun transitioning toward sustainability. Gibson-Graham et al. (2013) document challenges and opportunities in the Pacific Northwest, highlighting community engagement and local knowledge as key drivers. For fishing-dependent communities facing declining fisheries and climate change impacts, adaptation pathways are essential to avert economic crisis (Cisneros-Montemayor et al. 2021). Van Hoof et al. (2020) study Dutch fisheries' sustainability transition, examining challenges like lost fishing areas (due to wind farms, conservation, and Brexit) and the shift toward sustainable methods. Their work shows how stakeholder involvement and policy support can transform coastal economies. This transition from exploitative to sustainable models requires a multidisciplinary approach that integrates economic, social, and environmental considerations across local, national, and international governance levels. Growing interest in community-based adaptation models combines local livelihoods with ecosystem-based approaches (Forsyth 2013). The challenges facing fisheries-dependent communities underscore the need for sustainable transformation. Multitrophic aquaculture systems offers a promising approach to enhance productivity and sustainability (Chopin 2008; Stévant et al. 2017), creating an ecological-economic symbiosis that reinforces sustainability.

2.4 Natural Capital and Ecosystem Services

Natural capital refers to the world's stocks of natural assets, which comprise geology, soil, air, all forms of water, and all living organisms (Jansson et al. 1994). It is the underlying basis for environmental health and ecosystem services that are indispensable to human survival and economic activities. The concept of natural capital has sparked heightened interest as a framework for economically understanding and valuing the natural environment. Costanza et al. (1997) were among the first to provide global estimates of the economic valuation of ecosystem services and natural capital. They estimated the economic value of 17 ecosystem services for 16 biomes and calculated that these accounted for around half of the world's ecological services, valued at $33 trillion (US). They highlighted the significance of acknowledging and quantifying nature's inherent worth in economic terms to demonstrate its impact on human well-being. This influential work established the foundation for discussion on nature's value and emphasized the urgent need for sustainable management of natural resources. Daily (1997) further expanded on the concept, emphasizing ecosystem services' importance to human well-being. She argued for integrating ecological and economic disciplines to develop sustainable management practices for natural capital and outlined societies' dependence on these natural systems and services, highlighting the need for greater understanding and respect for natural capital. Natural capital accounting represents a nascent analytical instrument designed to gauge the reserves and movements of natural assets within the interconnected environmental and economic systems (Edens and Hein 2013). This methodology encourages prudent stewardship of natural resources by quantifying the benefits of preservation versus resource depletion (Costanza et al. 2014). In the businesses sector, natural capital approaches expand environmental accounting beyond emissions to assess dependencies and impacts on external ecosystems (Houdet et al. 2016). However, biodiversity measurement has proven challenging (van der Esch 2017). Frameworks like the Natural Capital Protocol provide guidance, but empirical applications remain limited, highlighting the need for more integrated valuation.

At national scales, natural capital assessments like the UK's Natural Capital Committee inform policy to manage risks, realize opportunities, and ensure resilient growth (Helm 2015). Turner and Daily (2008) introduced the concept of ecosystem service flows and highlighted the necessity of maintaining natural capital's integrity to ensure service sustainability. They stressed incorporating natural capital considerations into land-use planning and policy-making. In the context of sustainable development, the Millennium Ecosystem Assessment (MEA 2005) provides a comprehensive analysis of ecosystem change's consequences for human well-being and the scientific basis for enhanced conservation and sustainable use. The MEA framework has been widely used to evaluate ecosystem services' status and economic valuation. De Groot et al. (2012) built upon this work by grouping and quantifying ecosystem services to better support decision-making. Their valuation framework has supported many assessments of biodiversity and ecosystem services' economic value, underlining healthy ecosystems' role in sustainable economic development. Barbier (2012) examines the economic and policy challenges in ecosystem service valuation and governance. Similar to Costanza and Daly's work, Barbier demonstrates natural capital's role in developing a green economy but focuses more on the policy and managerial implications. The relatively recent 'green infrastructure' concept reflects the importance of natural capital in urban planning. According to Benedict and McMahon (2012), green infrastructure is an interconnected network of natural areas and other environmental features capable of providing diverse ecosystem services. As awareness of green infrastructure grows, it increasingly guides innovative land-use planning and urban design approaches.

3 Global Practice

It is widely acknowledged that NbS can enhance resiliency to climate change in coastal regions, leveraging natural processes and ecosystems to provide sustainable and cost-effective climate impact mitigation. Here, we summarize some global practices demonstrating NbS implementation in coastal settings (Table 1).

While NbS show promise, they face challenges such as uncertain long-term viability, ongoing maintenance requirements, and difficulties in integration into existing policy frameworks. Studies consistently recommend inclusive governance models that integrates scientific, local, and indigenous knowledge (Bennett et al. 2016). Additionally, policies should promote adaptive management approaches to enhance both the resilience and sustainability of NbS (Reid 2016).

4 Design Context

Grimsby is a dock town located on the south bank of the Humber Estuary in northeastern England (Fig. 1). Since the medieval period, Grimsby has grown from a small village into one of England's largest fishing port by the mid-twentieth century. The thriving fishing industry in Grimsby spawned associated industries in fish processing, ice manufacturing, and shipping. At its height, Grimsby's docks were bustling

Table 1 Overview of NbS approaches for climate change adaptation in coastal communities worldwide

Reference	Region/Country	Description	Main approach
(Veettil et al. 2019)	Vietnam	Vietnam's extensive mangrove reforestation programme, designed to reduce vulnerability to typhoons and sea-level rise, involved community engagement and the replanting of mangrove forests along the coast, which has successfully decreased storm damage and improved the livelihoods of local fishers by enhancing fish stocks and biodiversity.	Mangrove reforestation
(Schulte et al. 2009)	Maryland, USA	Living shorelines incorporating natural elements like native plants and oysters are used in Maryland's Chesapeake Bay to stabilize the coast, which has proven effective in reducing erosion, improving water quality, and providing habitat for aquatic species.	Native plants and oysters
(Stive et al. 2013)	Netherlands	The Netherlands' innovative 'Sand Engine' project involves strategically placing sand to reinforce dunes and beaches. Powered by natural tidal currents, the sand spreads along the coast over time, bolstering the natural defence against sea-level rise and storm surges.	Dune reinforcement
(Ba et al. 2001; Foster et al. 2013)	UK	The UK has initiated salt marsh conservation projects to act as natural flood barriers. These marshes absorb wave energy and reduce flood risk while sequestering carbon and providing habitat for wildlife.	Salt marshes
(Van Katwijk et al. 2016; Tan et al. 2020)	Australia	Australia has embarked on large-scale seagrass planting initiatives in response to the loss of seagrass meadows. These underwater meadows protect against coastal erosion and serve as significant carbon sinks.	Seagrass planting
(Bakti et al. 2012)	Indonesia	Indonesia's use of bio-rock technology to accelerate coral growth and reinforce natural reef defences against tidal surges and erosion could be detailed in this study.	Bio-rock structures
(Madison et al. 2013; Raynie 2015)	Louisiana, USA	To combat land loss and buffer against hurricanes, Louisiana has embarked on efforts to restore coastal marshes and wetlands through sediment diversions, hydrologic restoration, and planting vegetation. These wetlands provide critical storm protection and habitat for wildlife and fisheries.	Wetland restoration
(Reguero et al. 2019; Randazzo-Eisemann et al. 2021)	Mesoamerican Reef, Mexico	Restoring coral reefs, which act as natural breakwaters, can reduce wave energy reaching shores and protect coastal communities. In the Mesoamerican Reef, efforts include rearing and transplanting corals, controlling invasive species, and establishing marine protected areas to enhance reef resilience.	Coral reef restoration and protection
(Boyer et al. 2017; Moosavi 2017)	San Francisco Bay, USA	Living shorelines use various structural and organic materials, such as native plants, sand, and rock, to provide shoreline protection and maintain coastal processes. In San Francisco Bay, living shorelines accommodate sea-level rise while enhancing biodiversity.	Native plants, sand, and rock
(Stokes et al. 2012)	Gulf of Mexico, USA	Oyster reefs can protect shorelines by attenuating wave energy and trapping sediments. Restoration projects in the Gulf of Mexico involve constructing reefs with oyster shells or other materials to provide a substrate for oyster larvae to attach and grow, forming natural barriers.	Oyster reefs

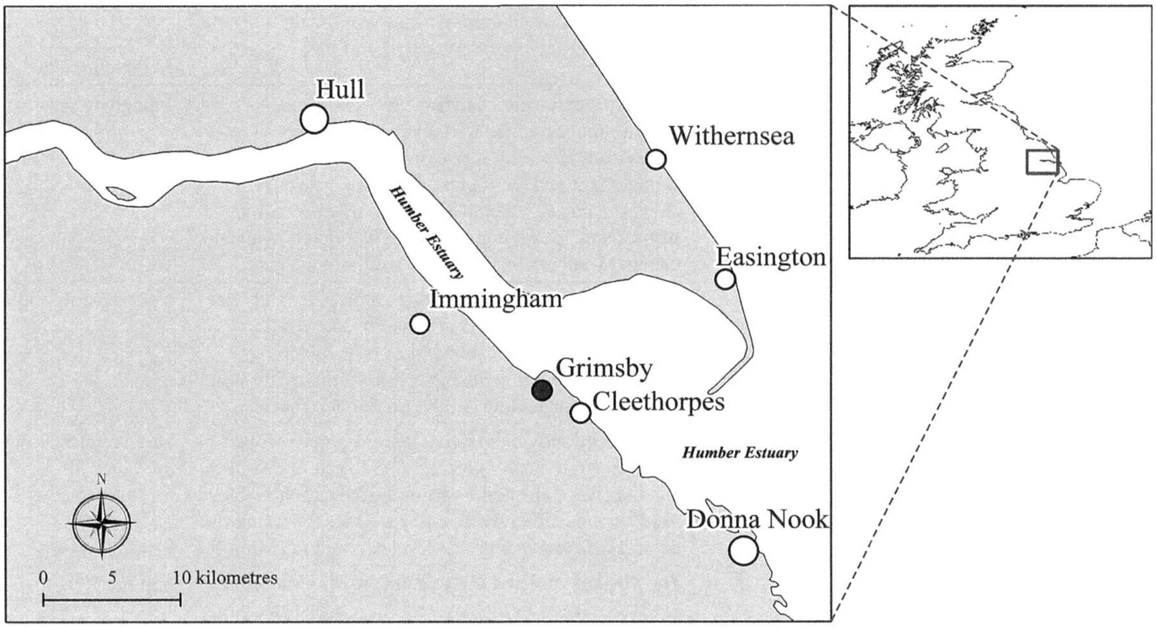

Fig. 1 Proposed design area: the location of Humber Estuary and Grimsby. (*Source* By author)

with Klondike trawlers that harvested vast quantities of cod from fishing grounds including the North Sea, while an extensive network of cold stores, ice factories, and fish processing plants processed and packaged fish for distribution across the UK and Europe (Poulsen 2008). At its peak, Grimsby's fishing industry employed approximately 35,000 workers both at sea and onshore (Chapman 2002; Whitfield 2009). However, Grimsby's fishing industry underwent a substantial decline, resulting in rising unemployment and deprivation. This was caused by multiple factors including overfishing, loss of access to Icelandic waters, increased competition from imported fish (particularly following Brexit), changing EU regulations, and growing competition from foreign fleets (Stewart et al. 2022). These developments created a crucial turning point in Grimsby's economic history, requiring strategic response to mitigate the impact of the industry's decline.

4.1 Socioeconomic Context

The cod fisheries of Hull and Grimsby in Northeast England's Humberside region faced a significant downturn following the Cod Wars and the 1960s–1970s fishing restrictions. Despite the geopolitical importance and Royal Navy's involvement during these conflicts, the industry succumbed to the loss of fishing grounds and increased competition after Britain joined the European Economic Community. While the fish processing sectors remained resilient, becoming Europe's largest by 2020, the demise of the local fishing fleets led to long-term economic and social challenges. Despite government measures like retraining programs and compensation schemes, Hull and Grimsby still face economic deprivation, persistent unemployment, and eroded cultural identity (McDowall 2022).

Brexit represents a pivotal event for Grimsby's fisheries. The UK's EU withdrawal raised hopes for sovereignty over British waters while creating uncertainty about trade tariffs and market access (Symes and Phillipson 2019). Potential tax hikes could disproportionately affect already-struggling coastal communities like Grimsby, now confronting Brexit-induced trade complexities alongside ongoing industry decline (Kierzenkowski et al. 2016). Grimsby's response to the decline of fishing was to diversify its industrial base, including developing renewable energy sectors, particularly offshore wind farms, and transforming the docks into a logistics and food processing hub. Regeneration efforts seek to transition Grimsby towards the renewable energy sector while revitalising the urban center areas (Invest North East Lincolnshire 2021). Crucially, integrating an ecosystem perspective requires deliberately providing public goods and ensuring equitable welfare distribution to achieve environmental, agricultural, and fisheries sustainability. Policymakers must address the unique needs and adopt innovative approaches from UK regions while developing new policy frameworks. Achieving these goals demands best practices and integrated policies that connect food production, environmental stewardship, and coastal/rural community vitality. Our research highlights grassroots, participatory methods as essential for developing both ecologically sustainable and community-supportive fisheries policies (Stewart et al. 2019).

4.2 Environmental Context

Within the UK, the Humber Estuary ranks second only to the Thames in flood risk, with sea levels rising by an average of 2 mm annually from 1993 to 2015. This poses a significant flood threat to the Humber region due to its geographic profile, endangering around 400,000 residents and 90,000 hectares of land, particularly near Hull and Grimsby. The economic impact of flooding is severe; the 2007 floods in Yorkshire and the Humber cost £2.7 billion, 4% of the region's Gross Value Added, with a 14-month recovery period. The 2013 Tidal Surge also heavily disrupted Hull's economy, damaging 115 businesses and 149 homes. These events highlight the region's vulnerability to both investment losses and economic instability (Humber Leadership Board 2021).

Conventional single-purpose flood defenses (levees, dikes, and floodwalls) frequently fail and harm marine environments (Ezban 2019). Modern landscape architecture increasingly favors transforming these structures into multi-functional systems that support aquatic life, agriculture, and species migration. This approach employs dynamic, 'safe-to-fail' methodologies to enhance coastal resilience, which is defined as the capacity to withstand and recover from disturbances like floods. Ecologist Nina-Marie Lister argues that effective resilience requires embracing adaptive change rather than resisting it through unrealistic notions of permanence (Lucas et al. 2019). This philosophy aligns with 'living infrastructure' principles. Given the prohibitive costs of constructing an invulnerable esturay-wide flood barrier, leveraging natural system offers economic advantages. Coastal ecosystems' natural wave attenuation capabilities, for instance, could enable more modest and cost-effective protective. Climate change adaptation thus benefits from environmentally synergistic approaches that minimize fiscal impacts.

The proposed strategy combines a defensive perimeter with strategic retreats to create new intertidal ecosystems, which enhances both flood protection and habitat conservation against coastal squeeze (Winn et al. 2003). Regional stakeholders have allocated over £150 million through the Humber Strategy to improve flood defences by 2021, protecting 70,000 properties via government subsidies, European Regional Development Fund, and private investment. The Environment Agency, twelve local authorities, and the Humber LEP are developing a new Humber Flood Risk Management Strategy to guide tidal flood risk management into the next century, with approval expected by late 2021. This strategy will incorporate environmental sustainability and natural flood management practices to increase the region's resilience to flooding (Humber Leadership Board 2021).

Critics note that initiatives like the Humber Enterprise Zones and Investing in Natural Capital intersect in their economic valuation of nature. These plans involve dredging to maintain shipping channels for offshore wind farms while supplying sediment for managed realignment projects. In concurrence with the Enterprise Zone's launch, the Humber Nature Partnership unveiled the 'Investing in Natural Capital' strategy, which explicitly links dredging operations to both economic development and ecological preservation, ensuring channel maintenance while supporting natural defenses (Humber Nature Partnership 2017).

5 Design Methods

5.1 Promoting a Dynamic Land Balance

The proposed framework introduces an innovative land exchange mechanism (land swap tool) designed to balance economic development with ecosystem preservation. This mechanism establishes a comprehensive compensation system for regional land reallocation, using a specialized visualization system—hereafter called 'the system'—to map land-use transitions within the estuary (Fig. 2). 'The system' requires collaborative engagement among government entities, development corporations, and local communities to create a land balancing and exchange protocol for the estuarine environment. The primary goals of this initiative include the restoration of natural habitats at the urban periphery, the integration of disparate ecological zones, the assimilation of natural resource bases, and the rehabilitation of estuarine ecosystems (Thom 2000; Temmerman et al. 2013).

When an urban developer earmarks a parcel of land for construction purposes within the estuary, 'the system' automatically generates compensatory natural habitats using an algorithmic process based on pre-established parameters. This approach maintains a dynamic balance of land distribution regionally, accounting for the total land mass. 'The system' evaluates metrics including land market value, ecological value, infrastructural accessibility, and the spatial footprint of the proposed development project. For instance, if a construction firm selects a high-value land for development, 'the system' determines the required compensatory habitat area using lower-cost land alternatives (Barbier et al. 2011). Similarly, when developers acquire large coastal areas, 'the system' calculates proportionally larger compensation areas, requiring greater financial investment (Daily et al. 2009).

As shown in Figure 3, when initiating construction on a designated parcel, developers must allocate equivalent land

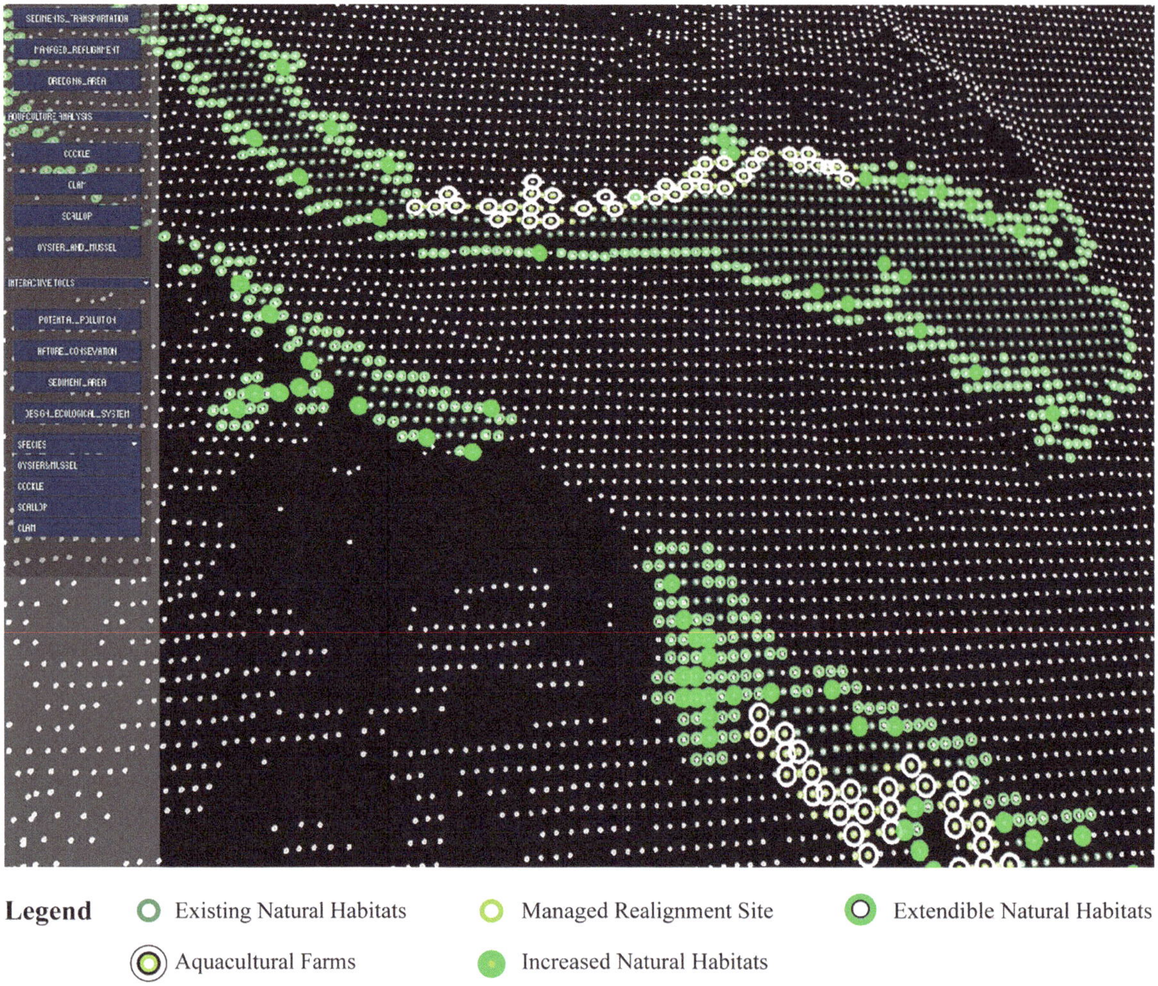

Fig. 2 Specialised visualisation system: land swap tool. (*Source* By author)

for aquaculture within managed realignment zones adjacent to the estuary. The system's algorithm requires that compensatory natural habitats be located adjacent to existing ecological preserves, thereby preserving landscape continuity. The extent of the newly established natural habitat is contingent upon the spatial separation from the original habitat. An inversely proportional relationship exists between the proximity of the original and new habitats and the requisite size of the latter; the further the new habitat is from the original, the larger it must be. Moreover, new compensatory habitats are constrained to feasible plots with lower elevation and suboptimal soil fertility. Upon the conversion of a parcel for aquaculture purposes, this development engenders the creation of communal land. This newly formed public domain, constituting 30% of the total land area, is designated for agricultural activities and the establishment of public amenities or verdant spaces. This land exchange tool limits economic expansion to protect natural habitats while effectively reducing their degradation. By mitigating land-use conflicts between development and conservation, the system clarifies land tenure and strengthens ecological network resilience (Bennet et al. 2013).

5.2 Implementation of the Coastal Managed Realignment

Traditional coastal defence engineering has often negatively affected environmental integrity, compromising ecosystem continuity and health. The use of impermeable artificial structures has been shown to disrupt marine species habitats (including fish and shellfish), causing degradation (Gray 1997). Offshore development significantly modifies benthic environments, leading to marine biodiversity decline that adversely affects fisheries and aquaculture (Halpern et al. 2008). Modifying coastal wetlands for construction projects implicates the loss of critical wildlife habitats. Such anthropogenic interventions can result in substantial changes to coastal geomorphology and hydrodynamic regimes, altering sediment budgets

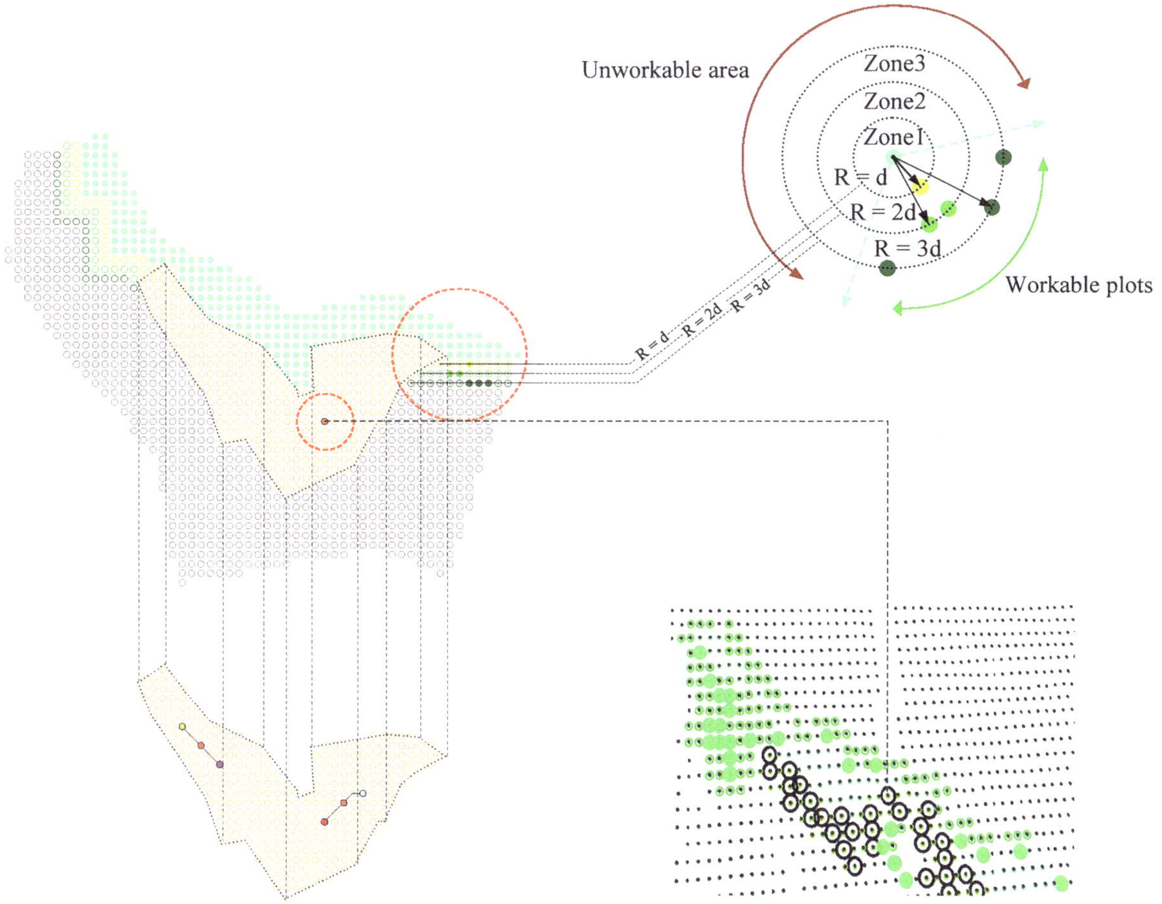

Fig. 3 Land swap and ecological compensation mechanism. (*Source* By author). (From the circle's centre to zone 1, zone 2, and zone 3, the plots to be compensated are the same area, two times the area, and three times the area, in that order.)

and transforming the morphology of navigational channels, potentially impacting maritime transport operations (Lonsdale et al. 2022). For instance, enhanced erosion may facilitate deeper channels for larger vessels, while sediment accumulation may require dredging to maintain navigability (Syvitski et al. 2009).

The Humber Estuary's intertidal zone has experienced physical changes due to commercial dredging activities and the global rise in sea levels. Drawing on insights from the Selsey Regional Coastal Control Re-engineering Project (Gregory and Goudie 2011), the proposed plan advocates for a coastal protection approach rooted in eco-engineering and ecosystem management principles. Deliberate removal of specific engineering structures could reconstitute the estuary land into habitats such as salt marsh wetlands and river floodplains. Complementing this, constructed sandy beaches, coastal protection forests, and wetlands are envisioned to coalesce with existing infrastructure to form a defense against storm surges and coastal erosion (French 2006). The primary aim of this protection system is to mitigate storm and flood hazards. However, integrating constructed features has proven to promote ecological connectivity and integrity by reuniting fragmented land parcels. This strategy has significant implications for rejuvenating tidal flats and biodiversity, enhancing urban resilience buffers, and minimizing expenditure on additional hard coastal defences (Temmerman et al. 2013).

Computerized simulations have been employed to investigate the geomorphological changes along the coast before and after commercial dredging, as well as following the implementation of the Coastal Adjustment Plan. To simulate the scenarios, we acquired tidal data every 15 min from the websites of UK government organisations, such as the Environment Agency. We overlaid it with predicted estimates of sea level rise in the UK and neighbouring waters. The findings indicate that current dredging practices may elevate flood risks in the Humber Estuary, while the Coastal Adjustment Program is projected to reduce such risks (Fig. 4). A pivotal element of the program involves the transformation of unproductive farmland near the intertidal zone into a large floodplain, fostering a wetland ecosystem that functions as a pre-urban flood buffer (Moss and Monstadt 2015). Furthermore, the establishment of new stormwater retention ponds, lagoons, and aquaculture zones, coupled with the creation of terraced topographies, will not only enhance

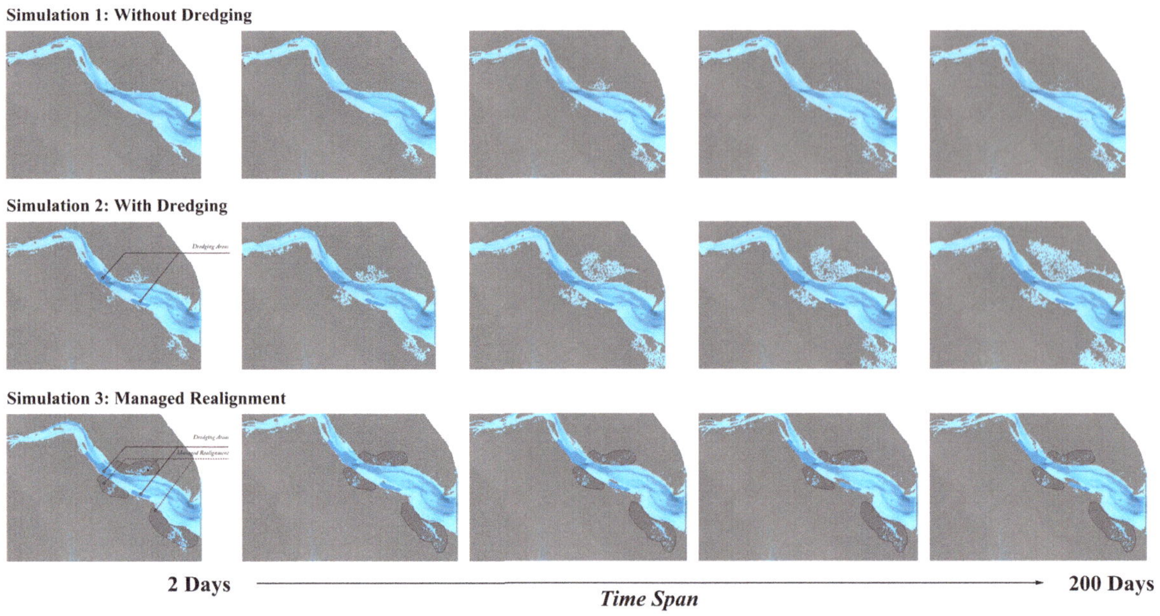

Fig. 4 Scenario simulation of dredging and managed realignment impact on Humber estuary. (*Source* By author). (The above figures were created using the CAESAR-Lisflood landscape evolution model. The first line is without any intervention like dredging; the second line is with dredging, and the third line is dredging overlapped with managed realignment. The second through fifth rows show a simulation of the scenario for the period from day 2 to day 200, with the simulation period being a multiple-speed virtual representation of the real event, not representing the actual number of days.)

agricultural productivity but also reinforce the protection of coastal communities and ecosystems from flooding events (Nguyen et al. 2019).

5.3 Creating Productive Landscapes

Urban planners increasingly harness productive spaces in metropolitan zones, orchestrating various civic activities—from food production and waste recycling to fostering social interactions. These spaces are adeptly woven into the urban ecological fabric, forming an integrated network of productive landscapes within the urban milieu (Pickett et al. 2004). Eco-urbanism principles are at the forefront of this movement, advocating for the creation of agricultural enclaves within urban settings. This dual-faceted approach strives to provide sustenance and jobs for city dwellers while repurposing underutilized urban lands (Beatley 2011). It facilitates reconnection with the natural environment through innovative, low-emission landscape development and departs from the passive aesthetics of conventional landscapes by emphasizing active participation and practical learning. Productive landscapes transcend mere leisure, sports, and food services; they are catalysts for urban agriculture, fisheries, and the preservation of local heritage (Tornaghi 2014). Furthermore, these landscapes are pioneering an emerging tourism paradigm that leverages authentic natural experiences, thereby substantially increasing the economic returns from landscape cultivation. The Humber Estuary presents a unique challenge in reimagining the relationship among urban areas, productive land, and ecological networks. The formation of coastal natural capital consortia is envisaged to transform these landscapes, providing material stability and delivering essential technical support for the strategy's realization (Benedict and McMahon 2012).

This comprehensive strategy entails integrating fisheries, aquaculture, crop cultivation, and livestock farming into an innovative ecological construct designed to navigate the sustainable evolution of the city through the maintenance of a productive landscape (Fig. 5). Concurrently with the strategic development, an exhaustive analysis of the region's existing agricultural and industrial structure is proposed. The insights from this analysis will shape an integrated network encompassing the entire value chain from production to distribution to fortify local businesses. The governance of public domains demands a tripartite partnership between governmental bodies, civil society actors, and the citizenry. The government extends policy support to civil society organizations, which in turn channel capital towards fostering urban advancement and developing novel, productive spaces, ultimately benefiting the local populace (Roseland 2012). Moreover, residents generate income by

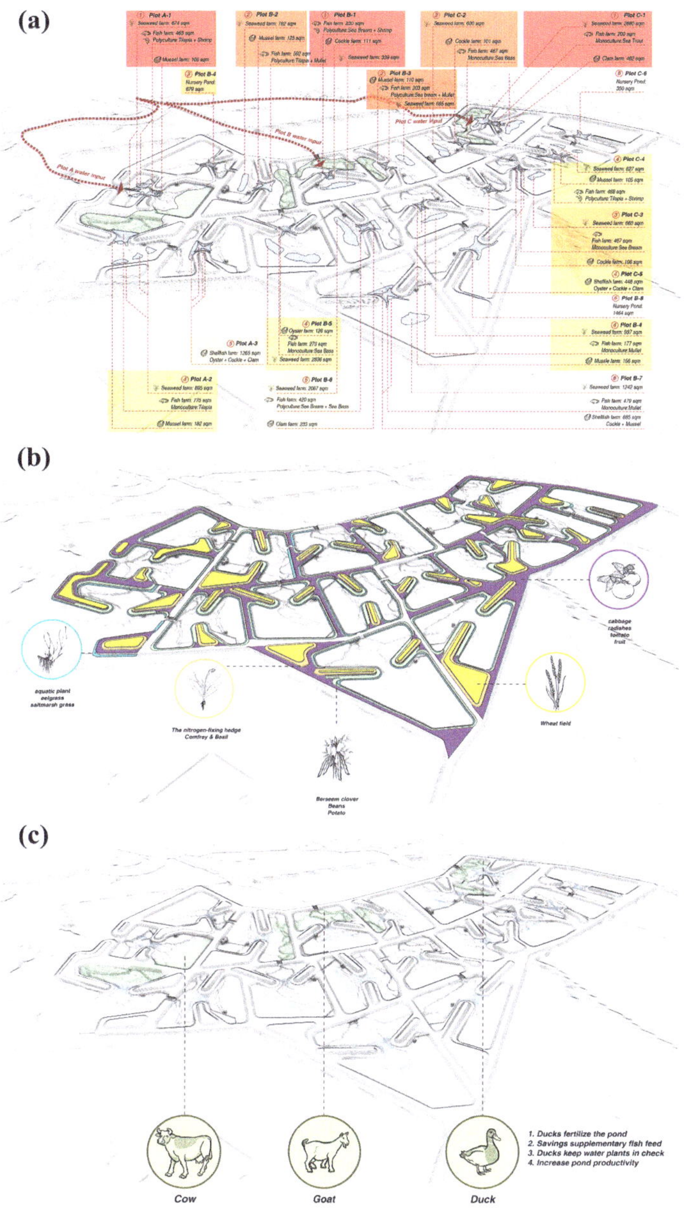

Fig. 5 Comprehensive strategy and productive landscapes system: **a** fisheries and aquaculture farm; **b** crop cultivation; **c** livestock farming. (*Source* By author)

engaging in agricultural ventures on their properties, thus reducing the government's need to directly create employment opportunities (Deelstra and Girardet 2000).

5.4 Building Ecologically Sustainable Systems

The goal of the proposed initiative is to implement hybrid agricultural practices, establishing an aquaculture enterprise that is both ecologically sound and sustainable. The strategy integrates the principles of Integrated Multi-Trophic Aquaculture (IMTA) and Integrated Agriculture–Aquaculture (IAA) systems to foster productive aquatic environment that accommodate the varied bathymetric preferences of diverse aquaculture species. The IMTA model serves as a cornerstone for marine pasture development, balancing aquatic resource use with increased biomass yield and economic value per unit area while maintaining ecological equilibrium (Chopin et al. 2012).

In this model, the photic zone supports macroalgae cultivation, which sequesters atmospheric carbon dioxide and releases oxygen through photosynthesis, thereby contributing to carbon cycling and climate change ditigation (Troell et al. 1997). The resulting detritus supports faunal communities in deeper marine strata, initiating a complex trophic cascade. Bivalves and other filter-feeders in the mesopelagic zone perform dual roles: water filtration (imporving water quality) and providing metabolic by-products for benthic organisms. The benthos then

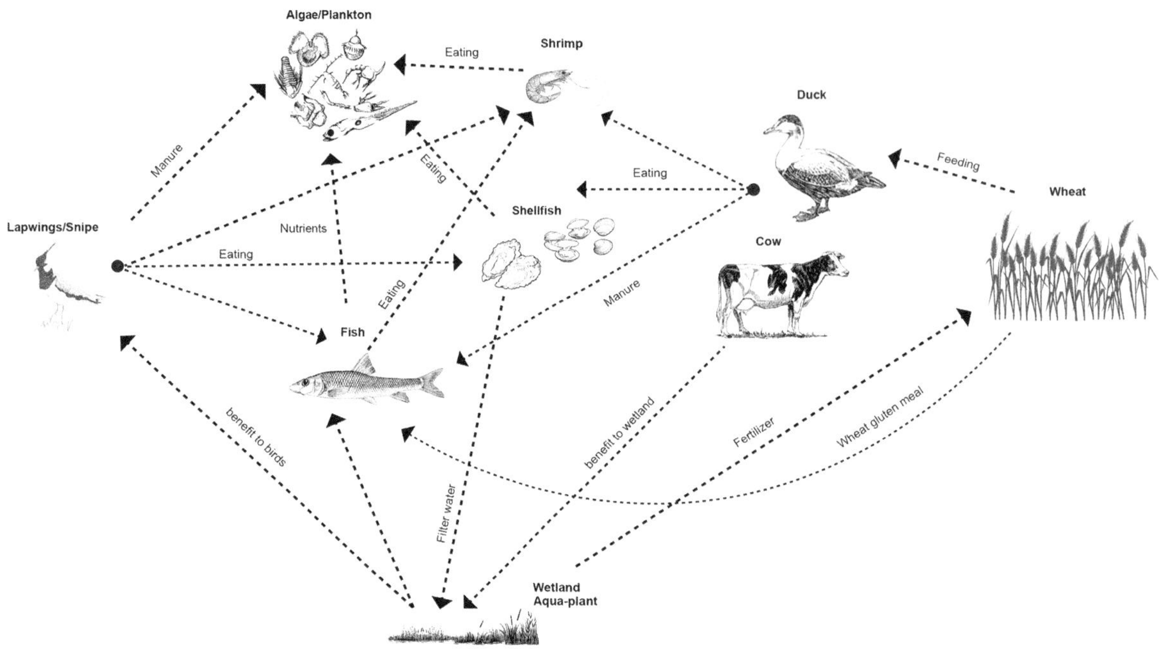

Fig. 6 Ecological cycle of agriculture and aquaculture production. (*Source* By author)

recycles nutrients to support algae growth, completing a self-sustaining nutrient loop (Soto 2009) (Fig. 6).

The proposed system features a dynamic and adaptive framework that accounts for tidal regimes oscillations, ensuring agricultural matrix robustness amid hydrological variability (Fig. 7). The aquaculture farm layout (including shellfish and fishponds) was developed using simulated bathymetry data. Sediment samples were gathered from potential dredging areas within the estuary and assessed for suitability. These sediments were digitally placed in various combinations within the managed realignment sites. Tidal hydrodynamics were then numerically simulated over 720 days to determine the resulting bathymetry and water depth conditions created by erosion across the sites. Based on shellfish and fish species preferences for different water depths, multiple farm layouts were iteratively modelled to optimise productivity. This computational approach integrates elevation data, tidal dynamics, sediment erosion and accretion patterns, and aquaculture species requirements to engineer shellfish and fish farm systems. The hydrodynamic simulations enabled creation of tailored bathymetric conditions for specific aquatic species cultivation. This system demonstrates ecological resilience by adapting to water level fluctuations while supporting a sustainable agricultural landscape (Barbara 2011).

6 Discussion

This study proposes several innovative methods tailored to Grimsby's context, including a land swap mechanism that uses a visualization system to balance development and ecosystem preservation through algorithmically generated compensatory habitats. Managed coastal realignment that builds on previous projects to restore wetlands and convert farmlands into floodplain buffers, supported by simulations showing reduced flood risks. Multifunctional productive landscapes that combine fisheries, aquaculture, agriculture, and livestock farming, applying eco-urbanism concepts to repurpose underutilised land. Sustainable aquaculture models like IMTA and IAA are adapted to foster an effective ecology-economy symbiosis. The proposed design provides NbS for Grimsby, UK, and demonstrates the multifaceted benefits of ecosystem restoration and sustainable aquaculture in mitigating climate-induced risks. The interconnected lagoon wetland system, created from dredged sediments, serves a dual function: enhancing coastal resilience against storm surges while creating habitats for aquaculture. This symbiosis between the environment and economy addresses both immediate need for flood protection and the long-term objective of reinvigorating the local fishing industry.

The managed realignment approach for expanding intertidal habitats aligns with global initiatives that use natural processes for shoreline stabilization. Our strategy is supported by Cooper and Pilkey (2004), who advocate sustainable, long-term approaches to coastal erosion. By

Fig. 7 Productive moundscape with different tidal heights: **a** low tide scenario at 1 m sea level; **b** middle tide scenario at 2 m sea level; **c** high tide scenario at 3 m sea level. The drawings above show the polyculture productive system: crops grown and sustainable farming, including grazing and aquaculture. (*Source* By author)

replacing hard engineering solutions, which often exacerbate coastal issues, and embracing a managed retreat, we align with their recommendation for softer NbS. Studies have shown that salt marshes, mangroves, and other coastal ecosystems can effectively attenuate waves and offer more ecologically sound alternatives to hard structures like seawalls (Temmerman et al. 2013; Sutton-Grier et al. 2015). However, uncertainty remains around the long-term robustness of nature-based defences (Narayan et al. 2016), requiring ongoing monitoring. Stakeholder perceptions also pose challenges, as some view managed retreat as land relinquishment (Myatt-Bell et al. 2002). Engaging local communities is, therefore, vital. The success of managed realignment and nature-based strategies hinges on robust community engagement and adaptive governance structures. Socioeconomic factors such as property rights, local industry buy-in, and the re-skilling of the workforce for aquaculture must be navigated with care to ensure equitable and sustainable development (Moser and Ekstrom 2010).

Regarding industry transition, integrating multi-trophic aquaculture presents opportunities to increase productivity while minimising environmental impacts (Chopin et al. 2001; Stévant et al. 2017). Our proposed interconnected lagoon wetland systems align with the findings of Barbier et al. (2011), who demonstrated that such ecosystems function as effective buffers against storms and provide valuable habitat for marine life, thereby enhancing biodiversity and ecosystem services. The integration of multi-trophic aquaculture within these systems is supported by Roberts et al. (2017), whose research underscores how such integrative approaches can preserve marine environments while bolstering food security and local economies. The proposed

IMTA system aims to foster symbiotic relationship between economic activity and ecological health. However, achieving this model's potential requires overcoming technical hurdles and ensuring profitability (Troell et al. 2009). Market development and policy incentives could facilitate wider adoption. Additionally, aquaculture faces sustainability challenges related to feed sources, nutrient pollution, and genetic impacts, which must be addressed through comprehensive planning and regulation (FAO 2023).

Overall, the study provides a framework for reimagining human-nature relationships in coastal zones. Climate change continues to test the resilience of coastal settlements globally, from rising seas flooding cities to declining fisheries weakening local economies. Adaptation requires a profound rethinking of how communities can not only cope with climate change impacts but thrive alongside them. Beyond physical infrastructure, this approach represents a shift in mindsets and economic models towards greater symbiosis with nature. Through strategic design, communities can transform climate risks into opportunities for regeneration.

7 Conclusion

The adaptive strategies detailed herein present a compelling case for transforming Grimsby's climate peril into a catalyst for economic and ecological revitalization. Our investigation concludes that coastal communities can achieve increased resilience to climate change by implementing NbS. The interconnected lagoon wetland system is projected to significantly mitigate flooding risks, while the integrated multi-trophic aquaculture offers potential to rejuvenate the town's fishing legacy in an environmentally responsible manner. Our study's implications are twofold:

(1) Ecological Benefits: Restoring coastal habitats is likely to increase local biodiversity and provide enhanced natural protection against extreme weather. Such prevention and sustainable restoration of British coasts aligns global targets to protect, restore, and promote sustainable use of terrestrial ecosystems, as outlined in the United Nations SDGs.

(2) Economic Implications: By transitioning to sustainable aquaculture, Grimsby could establish a regenerative economy that leverages its historical fishing industry. This shift would generates new employment opportunities and establishes a model for environmentally-conscious economic development that adapts to changing climatic conditions.

In the face of climate change, Grimsby's story—from crisis to possibility—could serve as a model for communities seeking to work with nature rather than against it. Moving forward, we must evaluate such programs not merely by their flood prevention capabilities or economic returns, but also by their contributions to long-term resilience and sustainability. Adaptation policies must remain nimble and responsive to emerging data, as these challenges will persist for generations. We recommend that future research expand upon these findings, exploring the long-term social, economic, and environmental outcomes of the NbS implementation.

References

Adger, W.N., Hughes, T.P., Folke, C., Carpenter, S.R., Rockström, J.: Social-ecological resilience to coastal disasters. Science **309**(5737), 1036–1039 (2005). https://doi.org/10.1126/science.1112122

Ba, I.M., Spencer, T., French, J.R., Leggett, D.J., Dixon, M.: The sea-defence value of salt marshes: field evidence from North Norfolk. Water Environ. J. **15**(2), 109–116 (2001). https://doi.org/10.1111/j.1747-6593.2001.tb00315.x

Bakti, L.A.A., Virgota, A., Damayanti, L.P.A., Radiman, T.H.U., Retnowulan, A., Sabil, A., Robbe, D.: Biorock reef restoration in gili trawangan, North Lombok, Indonesia. Innov. Methods Mar. Ecosyst. Restor. **59** (2012)

Barbara, S. (ed.): Aquaculture and the Environment—A Shared Destiny. IntechOpen (2011)

Barbier, E. B.: Natural Capital, Ecological Scarcity and Rural Poverty. The World Bank (2012). https://doi.org/10.1596/1813-9450-6232

Barbier, E.B.: Marine ecosystem services. Curr. Biol. **27**(11), R507–R510 (2017). https://doi.org/10.1016/j.cub.2017.03.020

Barbier, E.B., Hacker, S.D., Kennedy, C., Koch, E.W., Stier, A.C., Silliman, B.R.: The value of estuarine and coastal ecosystem services. Ecol. Monogr. **81**(2), 169–193 (2011)

Beatley, T.: Biophilic Cities: Integrating Nature into Urban Design and Planning. Island Press (2011)

Benedict, M.A., McMahon, E.T.: Green Infrastructure: Linking Landscapes and Communities. Island press (2012)

Bennet, G., Carroll, N., Hamilton, K.: Charting New Waters: State of Watershed Payments 2012 (2013)

Bennet, G., Ruef, F.: State of Watershed Investment 2016: Alliances for Green Infrastructure. Forest Trends' Ecosystem Marketplace (2016). https://www.forest-trends.org/wp-content/uploads/2017/03/doc_5463.pdf

Bennett, E.M., Solan, M., Biggs, R., McPhearson, T., Norström, A.V., Olsson, P., Pereira, L., Peterson, G.D., Raudsepp-Hearne, C., Biermann, F., Carpenter, S.R., Ellis, E.C., Hichert, T., Galaz, V., Lahsen, M., Milkoreit, M., Martin López, B., Nicholas, K.A., Preiser, R., Vince, G., Vervoort, J.M., Xu, J.: Bright spots: seeds of a good Anthropocene. Front. Ecol. Environ. **14**(8), 441–448 (2016). https://doi.org/10.1002/fee.1309

Bennett, N.J., Dearden, P., Murray, G., Kadfak, A.: The capacity to adapt?: communities in a changing climate, environment, and economy on the northern Andaman coast of Thailand. Ecol. Soc. **19**(2), art5 (2014). https://doi.org/10.5751/ES-06315-190205

Boyer, K., Zabin, C., De La Cruz, S., Grosholz, E., Orr, M., Lowe, J., Latta, M., Miller, J., Kiriakopolos, S., Pinnell, C.: San Francisco Bay living shorelines: restoring eelgrass and Olympia oysters for habitat and shore protection. In: Living Shorelines, pp. 333–362. CRC Press (2017)

Chapman, P.: Grimsby: The Story of the World's Greatest Fishing Port. Breedon Books Pub (2002)

Cheong, S.-M., Silliman, B., Wong, P.P., Van Wesenbeeck, B., Kim, C.-K., Guannel, G.: Coastal adaptation with ecological engineering. Nat. Clim. Chang. **3**(9), 787–791 (2013). https://doi.org/10.1038/nclimate1854

Chopin, T.: Models for analysis and practical realities of marine Aquaculure siting. In: Aquaculture, Innovation and Social Transformation, pp. 271–277. Springer (2008)

Chopin, T., Buschmann, A.H., Halling, C., Troell, M., Kautsky, N., Neori, A., Kraemer, G.P., Zertuche-González, J.A., Yarish, C., Neefus, C.: Integrating seaweeds into marine aquaculture systems: a key toward sustainability. J. Phycol. **37**(6), 975–986 (2001)

Chopin, T., Cooper, J.A., Reid, G., Cross, S., Moore, C.: Open-water integrated multi-trophic aquaculture: environmental biomitigation and economic diversification of fed aquaculture by extractive aquaculture. Rev. Aquac. **4**(4), 209–220 (2012). https://doi.org/10.1111/j.1753-5131.2012.01074.x

Church, J.A., Clark, P.U., Cazenave, A., Gregory, J.M., Jevrejeva, S., Levermann, A., Merrifield, M.A., Milne, G.A., Nerem, R.S., Nunn, P.D.: Sea Level Change. In: Stocker, T.F.T.F., Qin, D., Plattner, G.-K., Tignor, M.M.B., Allen, S.K., Boschung, J., Nauels, A., Xia, Y., Bex, V., Midgley, P.M. (eds.) The Physical Science Basis: Contribution of Working Group I to the Fifth Assessment Report of the Intergovernmental Panel on Climate Change, pp. 1137–1216. Cambridge University Press (2013)

Cisneros-Montemayor, A.M., Moreno-Báez, M., Reygondeau, G., Cheung, W.W.L., Crosman, K.M., González-Espinosa, P.C., Lam, V.W.Y., Oyinlola, M.A., Singh, G.G., Swartz, W., Zheng, C., Ota, Y.: Enabling conditions for an equitable and sustainable blue economy. Nature **591**(7850), 396–401 (2021). https://doi.org/10.1038/s41586-021-03327-3

Cohen-Shacham, E., Walters, G., Janzen, C., Maginnis, S. (eds.): Nature-Based Solutions to Address Global Societal Challenges. IUCN International Union for Conservation of Nature (2016). https://doi.org/10.2305/IUCN.CH.2016.13.en

Colls, A., Ash, N., Ikkala, N.: Ecosystem-Based Adaptation: A Natural Response to Climate Change. IUCN (2009)

Cooper, J.A.G., Pilkey, O.H.: Sea-level rise and shoreline retreat: time to abandon the Bruun rule. Glob. Planet. Change **43**(3–4), 157–171 (2004). https://doi.org/10.1016/j.gloplacha.2004.07.001

Costanza, R., d'Arge, R., De Groot, R., Farber, S., Grasso, M., Hannon, B., Limburg, K., Naeem, S., O'neill, R. V., Paruelo, J.: The value of the world's ecosystem services and natural capital. Nature **387**(6630), 253–260 (1997)

Costanza, R., De Groot, R., Sutton, P., Van Der Ploeg, S., Anderson, S.J., Kubiszewski, I., Farber, S., Turner, R.K.: Changes in the global value of ecosystem services. Glob. Environ. Chang. **26**, 152–158 (2014). https://doi.org/10.1016/j.gloenvcha.2014.04.002

Daily, G.C.: Nature's Services: Societal Dependence on Natural Ecosystems. Island Press (1997)

Daily, G.C., Polasky, S., Goldstein, J., Kareiva, P.M., Mooney, H.A., Pejchar, L., Ricketts, T.H., Salzman, J., Shallenberger, R.: Ecosystem services in decision making: time to deliver. Front. Ecol. Environ. **7**(1), 21–28 (2009). https://doi.org/10.1890/080025

De Groot, R., Brander, L., Van Der Ploeg, S., Costanza, R., Bernard, F., Braat, L., Christie, M., Crossman, N., Ghermandi, A., Hein, L., Hussain, S., Kumar, P., McVittie, A., Portela, R., Rodriguez, L.C., Ten Brink, P., Van Beukering, P.: Global estimates of the value of ecosystems and their services in monetary units. Ecosyst. Serv. **1**(1), 50–61 (2012). https://doi.org/10.1016/j.ecoser.2012.07.005

Deelstra, T., Girardet, H.: Urban agriculture and sustainable cities. In: Bakker, N., Dubbeling, M., Gündel, S., Sabel-Koshella, U., de Zeeuw, H. (eds) Growing Cities, Growing Food. Urban Agriculture on the Policy Agenda, vol. 43, pp. 66. Zentralstelle Für Ernährung Und Landwirtschaft (ZEL), Feldafing, Germany (2000)

Edens, B., Hein, L.: Towards a consistent approach for ecosystem accounting. Ecol. Econ. **90**, 41–52 (2013). https://doi.org/10.1016/j.ecolecon.2013.03.003

Esteves, L.S.: Is managed realignment a sustainable long-term coastal management approach? J. Coastal Res. **65**, 933–938 (2013). https://doi.org/10.2112/SI65-158.1

Ezban, M.: Aquaculture Landscapes: Fish Farming and the Public Realm. Routledge (2019)

FAO.: Report of the Expert Workshop on Aquaculture Co-management (Fisheries and Aquaculture Report NFIAN/R1399 (En); Hybrid Meeting Kigali, Rwanda, 8–10 June 2022). Food and Agriculture Organization of the United Nations (2023). https://doi.org/10.4060/cc8315en

Fernandino, G., Elliff, C.I., Silva, I.R.: Ecosystem-based management of coastal zones in face of climate change impacts: challenges and inequalities. J. Environ. Manage. **215**, 32–39 (2018). https://doi.org/10.1016/j.jenvman.2018.03.034

Forsyth, T.: Community-based adaptation: a review of past and future challenges. Wires Clim. Change **4**(5), 439–446 (2013). https://doi.org/10.1002/wcc.231

Foster, N.M., Hudson, M.D., Bray, S., Nicholls, R.J.: Intertidal mudflat and saltmarsh conservation and sustainable use in the UK: a review. J. Environ. Manage. **126**, 96–104 (2013). https://doi.org/10.1016/j.jenvman.2013.04.015

French, P.W.: Managed realignment—the developing story of a comparatively new approach to soft engineering. Estuar. Coast. Shelf Sci. **67**(3), 409–423 (2006). https://doi.org/10.1016/j.ecss.2005.11.035

Gedan, K.B., Kirwan, M.L., Wolanski, E., Barbier, E.B., Silliman, B.R.: The present and future role of coastal wetland vegetation in protecting shorelines: answering recent challenges to the paradigm. Clim. Change **106**, 7–29 (2011)

Gibson-Graham, J.K., Cameron, J., Healy, S.: Take Back the Economy: An Ethical Guide for Transforming our Communities. U of Minnesota Press (2013)

Gittman, R.K., Peterson, C.H., Currin, C.A., Joel Fodrie, F., Piehler, M.F., Bruno, J.F.: Living shorelines can enhance the nursery role of threatened estuarine habitats. Ecol. Appl. **26**(1), 249–263 (2016). https://doi.org/10.1890/14-0716

Gray, J.S.: Marine biodiversity: patterns, threats and conservation needs. Biodivers. Conserv. **6**(1), 153–175 (1997)

Gregory, K.J., Goudie, A. (eds.): The SAGE HANDBOOK of Geomorphology (1. publ). Sage (2011)

Hallegatte, S., Green, C., Nicholls, R.J., Corfee-Morlot, J.: Future flood losses in major coastal cities. Nat. Clim. Chang. **3**(9), 802–806 (2013). https://doi.org/10.1038/nclimate1979

Halpern, B.S., Walbridge, S., Selkoe, K.A., Kappel, C.V., Micheli, F., D'Agrosa, C., Bruno, J.F., Casey, K.S., Ebert, C., Fox, H.E., Fujita, R., Heinemann, D., Lenihan, H.S., Madin, E.M.P., Perry, M.T., Selig, E.R., Spalding, M., Steneck, R., Watson, R.: A global map of human impact on marine ecosystems. Science **319**(5865), 948–952 (2008). https://doi.org/10.1126/science.1149345

Helm, D.: Natural Capital: Valuing Our Planet. Yale University Press (2015)

Hinkel, J., Lincke, D., Vafeidis, A.T., Perrette, M., Nicholls, R.J., Tol, R.S.J., Marzeion, B., Fettweis, X., Ionescu, C., Levermann, A.: Coastal flood damage and adaptation costs under 21st century sea-level rise. Proc. Natl. Acad. Sci. **111**(9), 3292–3297 (2014). https://doi.org/10.1073/pnas.1222469111

Hobbie, S.E., Grimm, N.B.: Nature-based approaches to managing climate change impacts in cities. Philos. Trans. R. Soc. B: Biol. Sci. **375**(1794), 20190124 (2020). https://doi.org/10.1098/rstb.2019.0124

Houdet, J.R., Finisdore, J., Martin-Ortega, J., Ding, H., Maleganos, J.K., Spurgeon, J., Hartmann, T., Steuerman, D.: Accounting for ecosystem services in business. In: Routledge Handbook of Ecosystem Services, pp. 220–227. Routledge (2016)

Humber Leadership Board.: Humber Estuary Plan 2021, p. 35 (2021). https://heylep.com/wp-content/uploads/2021/05/Humber-Estuary-Plan-final-draft-Jan-2021.pdf

Humber Nature Partnership.: Investing in Natural Capital: Creating the Right Environment for Economic Investment, p. 23. Humber Nature Partnership (2017). https://humbernature.co.uk/assets/investing-in-natural-capital.pdf

Invest North East Lincolnshire.: Economic Strategy 2021, p. 67. North East Lincolnshire Council (2021). https://www.nelincs.gov.uk/assets/uploads/2022/05/NEL-Economic-Strategy-2021.pdf

IPCC.: Climate Change 2014: Synthesis Report, p. 151. Intergovernmental Panel on Climate Change (2015)

IPCC.: Climate Change 2022—Impacts, Adaptation and Vulnerability: Working Group II Contribution to the Sixth Assessment Report of the Intergovernmental Panel on Climate Change, 1st edn. Cambridge University Press (2023). https://doi.org/10.1017/9781009325844

Jansson, A., Hammer, M., Folke, C., Costanza, R. (eds.).: Investing in Natural Capital: The Ecological Economics Approach to Sustainability. Island Press (1994)

Jones, H.P., Hole, D.G., Zavaleta, E.S.: Harnessing nature to help people adapt to climate change. Nat. Clim. Chang. **2**(7), 504–509 (2012). https://doi.org/10.1038/nclimate1463

Kierzenkowski, R., Pain, N., Rusticelli, E., Zwart, S.: The Economic Consequences of Brexit: A Taxing Decision (2016)

Knutson, T.R., McBride, J.L., Chan, J., Emanuel, K., Holland, G., Landsea, C., Held, I., Kossin, J.P., Srivastava, A.K., Sugi, M.: Tropical cyclones and climate change. Nat. Geosci. **3**(3), 157–163 (2010). https://doi.org/10.1038/ngeo779

Ledoux, L., Cornell, S., O'Riordan, T., Harvey, R., Banyard, L.: Towards sustainable flood and coastal management: identifying drivers of, and obstacles to, managed realignment. Land Use Policy **22**(2), 129–144 (2005). https://doi.org/10.1016/j.landusepol.2004.03.001

Lonsdale, J.-A., Leach, C., Parsons, D., Barkwith, A., Manson, S., Elliott, M.: Managing estuaries under a changing climate: a case study of the Humber Estuary, UK. Environ. Sci. Policy **134**, 75–84 (2022). https://doi.org/10.1016/j.envsci.2022.04.001

Lucas, J.S., Southgate, P.C., Tucker, C.S. (eds.): Aquaculture: Farming Aquatic Animals and Plants, 3rd edn. Wiley (2019)

Madison, M.J., Mack, S.K., Lane, R.R., Day, J.W.: Wetland restoration using mangroves in southern Louisiana. In: SPE Health, Safety, Security, Environment, & Social Responsibility Conference-North America, pp. SPE-163789-MS (2013). https://doi.org/10.2118/163789-MS

Mahon, R., McConney, P., Roy, R.N.: Governing fisheries as complex adaptive systems. Mar. Policy **32**(1), 104–112 (2008). https://doi.org/10.1016/j.marpol.2007.04.011

McDowall, W.: The political economy of actively phasing out harmful industries: lessons from resource-based sectors beyond fossil fuels. Energy Res. Soc. Sci. **90**, 102647 (2022). https://doi.org/10.1016/j.erss.2022.102647

Mea, M.: Ecosystem assessment. In: Ecosystems and Human Well-Being, vol. 5. Island press Washington, DC (2005)

Moosavi, S.: Ecological coastal protection: pathways to living shorelines. Procedia Eng. **196**, 930–938 (2017). https://doi.org/10.1016/j.proeng.2017.08.027

Moser, S.C., Ekstrom, J.A.: A framework to diagnose barriers to climate change adaptation. Proc. Natl. Acad. Sci. **107**(51), 22026–22031 (2010). https://doi.org/10.1073/pnas.1007887107

Moss, T., Monstadt, J.: Restoring floodplains in Europe: policy contexts and project experiences. Water Intell. Online **7**, 9781780401966–9781780401966 (2015). https://doi.org/10.2166/9781780401966

Myatt-Bell, L.B., Scrimshaw, M.D., Lester, J.N., Potts, J.S.: Public perception of managed realignment: Brancaster West Marsh, North Norfolk, UK. Mar. Policy **26**(1), 45–57 (2002). https://doi.org/10.1016/S0308-597X(01)00033-1

Narayan, S., Beck, M.W., Reguero, B.G., Losada, I.J., Van Wesenbeeck, B., Pontee, N., Sanchirico, J.N., Ingram, J.C., Lange, G.-M., Burks-Copes, K.A.: The effectiveness, costs and coastal protection benefits of natural and nature-based defences. PLoS ONE **11**(5), e0154735 (2016). https://doi.org/10.1371/journal.pone.0154735

Naylor, R.L., Goldburg, R.J., Primavera, J.H., Kautsky, N., Beveridge, M.C.M., Clay, J., Folke, C., Lubchenco, J., Mooney, H., Troell, M.: Effect of aquaculture on world fish supplies. Nature **405**(6790), 1017–1024 (2000). https://doi.org/10.1038/35016500

Neumann, B., Vafeidis, A.T., Zimmermann, J., Nicholls, R.J.: Future coastal population growth and exposure to sea-level rise and coastal flooding—a global assessment. PLoS ONE **10**(3), e0118571 (2015). https://doi.org/10.1371/journal.pone.0118571

Nguyen, T.T., Ngo, H.H., Guo, W., Wang, X.C., Ren, N., Li, G., Ding, J., Liang, H.: Implementation of a specific urban water management—Sponge city. Sci. Total Environ. **652**, 147–162 (2019). https://doi.org/10.1016/j.scitotenv.2018.10.168

Nicholls, R.J., Cazenave, A.: Sea-level rise and its impact on coastal zones. Science **328**(5985), 1517–1520 (2010). https://doi.org/10.1126/science.1185782

Pickett, S.T.A., Cadenasso, M.L., Grove, J.M.: Resilient cities: meaning, models, and metaphor for integrating the ecological, socio-economic, and planning realms. Landsc. Urban Plan. **69**(4), 369–384 (2004). https://doi.org/10.1016/j.landurbplan.2003.10.035

Poulsen, B.: Dutch Herring: An Environmental History, c. 1600–1860, vol. 3. Amsterdam University Press (2008)

Pye, K., Blott, S.J.: The geomorphology of UK estuaries: the role of geological controls, antecedent conditions and human activities. Estuar. Coast. Shelf Sci. **150**, 196–214 (2014). https://doi.org/10.1016/j.ecss.2014.05.014

Randazzo-Eisemann, Á., Arias-González, J.E., Velez, L., McField, M., Mouillot, D.: The last hotspots of structural complexity as conservation targets in the Mesoamerican Coral Reef. Biol. Cons. **256**, 109021 (2021). https://doi.org/10.1016/j.biocon.2021.109021

Raynie, S.M.K.R.C.: Coastal restoration in Louisiana: an update. Shore Beach **83**(4), 4 (2015)

Reguero, B.G., Secaira, F., Toimil, A., Escudero, M., Díaz-Simal, P., Beck, M.W., Silva, R., Storlazzi, C., Losada, I.J.: The risk reduction benefits of the mesoamerican reef in Mexico. Front. Earth Sci. **7**, 125 (2019). https://doi.org/10.3389/feart.2019.00125

Reid, H.: Ecosystem- and community-based adaptation: learning from community-based natural resource management. Clim. Dev. **8**(1), 4–9 (2016). https://doi.org/10.1080/17565529.2015.1034233

Roberts, C.M., O'Leary, B.C., McCauley, D.J., Cury, P.M., Duarte, C.M., Lubchenco, J., Pauly, D., Sáenz-Arroyo, A., Sumaila, U.R., Wilson, R.W., Worm, B., Castilla, J.C.: Marine reserves can mitigate and promote adaptation to climate change. Proc. Natl. Acad. Sci. **114**(24), 6167–6175 (2017). https://doi.org/10.1073/pnas.1701262114

Roseland, M.: Toward Sustainable Communities: Solutions for Citizens and their Governments, 4th edn. New Society Publication (2012)

Schulte, D.M., Burke, R.P., Lipcius, R.N.: Unprecedented restoration of a native oyster metapopulation. Science **325**(5944), 1124–1128 (2009). https://doi.org/10.1126/science.1176516

Seddon, N., Chausson, A., Berry, P., Girardin, C.A.J., Smith, A., Turner, B.: Understanding the value and limits of nature-based solutions to climate change and other global challenges. Philos. Trans. R. Soc. B: Biol. Sci. **375**(1794), 20190120 (2020). https://doi.org/10.1098/rstb.2019.0120

Smit, B., Wandel, J.: Adaptation, adaptive capacity and vulnerability. Glob. Environ. Chang. **16**(3), 282–292 (2006). https://doi.org/10.1016/j.gloenvcha.2006.03.008

Smith, J.B., Schellnhuber, H.-J., Mirza, M.M.Q., Fankhauser, S., Leemans, R., Erda, L., Ogallo, L., Pittock, B., Richels, R., Rosenzweig, C.: Vulnerability to climate change and reasons for concern: a synthesis. Clim Change 913–967 (2001)

Soto, D. (ed.): Integrated Mariculture: A Global Review. Food and Agriculture Organization of the United Nations (2009)

Stévant, P., Rebours, C., Chapman, A.: Seaweed aquaculture in Norway: recent industrial developments and future perspectives. Aquacult. Int. **25**(4), 1373–1390 (2017). https://doi.org/10.1007/s10499-017-0120-7

Stewart, B.D., Burns, C., Hejnowicz, A.P., Gravey, V., O'Leary, B.C., Hicks, K., Farstad, F.M., Hartley, S.E.: Making Brexit work for the environment and livelihoods: delivering a stakeholder informed vision for agriculture and fisheries. People Nature **1**(4), 442–456 (2019). https://doi.org/10.1002/pan3.10054

Stewart, B.D., Williams, C., Barnes, R., Walmsley, S.F., Carpenter, G.: The Brexit deal and UK fisheries—has reality matched the rhetoric? Maritime Stud. **21**(1), 1–17 (2022). https://doi.org/10.1007/s40152-022-00259-0

Stive, M.J.F., De Schipper, M.A., Luijendijk, A.P., Aarninkhof, S.G.J., Van Gelder-Maas, C., Van Thiel De Vries, J.S.M., De Vries, S., Henriquez, M., Marx, S., Ranasinghe, R.: A new alternative to saving our beaches from sea-level rise: the sand engine. J. Coast. Res. **290**, 1001–1008 (2013). https://doi.org/10.2112/JCOASTRES-D-13-00070.1

Stokes, S., Wunderink, S., Lowe, M., Gereffi, G.: Restoring Gulf Oyster Reefs, p. 60. Center on Globalization, Governance & Competitiveness, Duke University (2012). http://www.oyster-restoration.org/wp-content/uploads/2012/06/CGGC_Oyster-Reef-Restoration.pdf

Sutton-Grier, A.E., Gittman, R., Arkema, K., Bennett, R., Benoit, J., Blitch, S., Burks-Copes, K., Colden, A., Dausman, A., DeAngelis, B., Hughes, A., Scyphers, S., Grabowski, J.: Investing in natural and nature-based infrastructure: building better along our coasts. Sustainability **10**(2), 523 (2018). https://doi.org/10.3390/su10020523

Sutton-Grier, A.E., Sandifer, P.A.: Conservation of wetlands and other coastal ecosystems: a commentary on their value to protect biodiversity, reduce disaster impacts, and promote human health and well-being. Wetlands **39**(6), 1295–1302 (2019). https://doi.org/10.1007/s13157-018-1039-0

Sutton-Grier, A.E., Wowk, K., Bamford, H.: Future of our coasts: the potential for natural and hybrid infrastructure to enhance the resilience of our coastal communities, economies and ecosystems. Environ. Sci. Policy **51**, 137–148 (2015). https://doi.org/10.1016/j.envsci.2015.04.006

Symes, D., Phillipson, J.: "A sea of troubles" (2): Brexit and the UK seafood supply chain. Mar. Policy **102**, 5–9 (2019). https://doi.org/10.1016/j.marpol.2019.01.015

Syvitski, J.P.M., Kettner, A.J., Overeem, I., Hutton, E.W.H., Hannon, M.T., Brakenridge, G.R., Day, J., Vörösmarty, C., Saito, Y., Giosan, L., Nicholls, R.J.: Sinking deltas due to human activities. Nat. Geosci. **2**(10), 681–686 (2009). https://doi.org/10.1038/ngeo629

Tan, Y.M., Dalby, O., Kendrick, G.A., Statton, J., Sinclair, E.A., Fraser, M.W., Macreadie, P.I., Gillies, C.L., Coleman, R.A., Waycott, M., Van Dijk, K., Vergés, A., Ross, J.D., Campbell, M.L., Matheson, F.E., Jackson, E.L., Irving, A.D., Govers, L.L., Connolly, R.M., McLeod, I.M., Rasheed, M.A., Kirkman, H., Flindt, M.R., Lange, T., Miller, A.D., Sherman, C.D.H.: Seagrass restoration is possible: insights and lessons from Australia and New Zealand. Front. Mar. Sci. **7**, 617 (2020). https://doi.org/10.3389/fmars.2020.00617

Temmerman, S., Meire, P., Bouma, T.J., Herman, P.M.J., Ysebaert, T., De Vriend, H.J.: Ecosystem-based coastal defence in the face of global change. Nature **504**(7478), 79–83 (2013). https://doi.org/10.1038/nature12859

Thom, R.M.: Adaptive management of coastal ecosystem restoration projects. Ecol. Eng. **15**(3–4), 365–372 (2000). https://doi.org/10.1016/S0925-8574(00)00086-0

Tornaghi, C.: Critical geography of urban agriculture. Prog. Hum. Geogr. **38**(4), 551–567 (2014). https://doi.org/10.1177/0309132513512542

Troell, M., Joyce, A., Chopin, T., Neori, A., Buschmann, A.H., Fang, J.-G.: Ecological engineering in aquaculture—potential for integrated multi-trophic aquaculture (IMTA) in marine offshore systems. Aquaculture **297**(1–4), 1–9 (2009). https://doi.org/10.1016/j.aquaculture.2009.09.010

Troell, M., Halling, C., Nilsson, A., Buschmann, A.H., Kautsky, N., Kautsky, L.: Integrated marine cultivation of Gracilaria chilensis (Gracilariales, Rhodophyta) and salmon cages for reduced environmental impact and increased economic output. Aquaculture **156**(1–2), 45–61 (1997). https://doi.org/10.1016/S0044-8486(97)00080-X

Turner, R.K., Daily, G.C.: The ecosystem services framework and natural capital conservation. Environ. Resour. Econ. **39**(1), 25–35 (2008). https://doi.org/10.1007/s10640-007-9176-6

Van der Esch, S.: Exploring Future Changes in Land Use and Land Condition and the Impacts on Food, Water, Climate Change and Biodiversity: Scenarios for the UNCCD Global Land Outlook (2017)

Van Hoof, L., Steins, N.A., Smith, S., Kraan, M.: Change as a permanent condition: a history of transition processes in Dutch North Sea fisheries. Mar. Policy **122**, 104245 (2020). https://doi.org/10.1016/j.marpol.2020.104245

Van Katwijk, M.M., Thorhaug, A., Marbà, N., Orth, R. J., Duarte, C.M., Kendrick, G.A., Althuizen, I.H.J., Balestri, E., Bernard, G., Cambridge, M.L., Cunha, A., Durance, C., Giesen, W., Han, Q., Hosokawa, S., Kiswara, W., Komatsu, T., Lardicci, C., Lee, K.-S., Meinesz, A., Nakaokam, M., O'Brien, K.R., Paling, E.I., Pickerell, C., Ransijn, A.M.A., Verduin, J.J.: Global analysis of seagrass restoration: the importance of large-scale planting. J. Appl. Ecol. **53**(2), 567–578 (2016). https://doi.org/10.1111/1365-2664.12562

Veettil, B.K., Ward, R.D., Quang, N.X., Trang, N.T.T., Giang, T.H.: Mangroves of Vietnam: historical development, current state of research and future threats. Estuar. Coast. Shelf Sci. **218**, 212–236 (2019). https://doi.org/10.1016/j.ecss.2018.12.021

Whitfield, M.: 'Grimsby Fishdocks–An Assessment of Character and Significance. Historic England Kasbah Report (Great Grimsby Ice Factory Trust (2009)

Winn, P., Young, R., Edwards, A.: Planning for the rising tides: the humber estuary shoreline management plan. Sci. Total Environ. **314–316**, 13–30 (2003). https://doi.org/10.1016/S0048-9697(03)00092-5

Young, T., Fuller, E.C., Provost, M.M., Coleman, K.E., St. Martin, K., McCay, B.J., Pinsky, M.L.: Adaptation strategies of coastal fishing communities as species shift poleward. ICES J. Mar. Sci. **76**(1), 93–103 (2019). https://doi.org/10.1093/icesjms/fsy140

Zhao, T., Chen, Y., Palwatwichai, P., Huang, H.: Researching on the tidal flooding through the coastal simulation: developing potential managed realignment areas of humber estuary in England. J. Digit. Landsc. Arch. **4**, 68–75 (2019). https://doi.org/10.14627/537663007

Mainstreaming and Implementation of Nature-Based Solutions for Climate Change Mitigation, Adaptation, and Resilience in Zambia

Kalapula Marrien

Abstract

The study aimed to determine the influence of institutional factors on the mainstreaming of Nature-Based Solutions (NBS) into Local Development Plans and the actual implementation of NBS projects for climate change adaptation and mitigation in Zambia's municipalities, using Mumbwa Town Council as a case study. A comprehensive review of literature on climate change adaptation, mitigation, resilience building, NBS mainstreaming, implementation, and institutional factors was conducted. Qualitative methods were employed for data collection and analysis, utilizing semi-structured interview guides to gather primary data from key informants and document reviews to collect secondary data. Data analysis was performed using Atlas.ti in which co-occurrence analyses were conducted to determine the relationship between the various institutional factors and the NBS mainstreamed and implemented. The research findings indicate that Mumbwa has successfully integrated NBS into its local development plans across five main categories: sustainable forest management, water ecosystem conservation, conservation agriculture, wildlife ecosystem restoration, and urban greening, all of which have been implemented through various projects. The study identifies several institutional factors that have facilitated the mainstreaming and implementation of NBS, including institutionalization, supportive local and national policies, a holistic, flexible, and inclusive planning system, good governance, and international funding. The study concludes that to a high extent, these facilitating factors have had a positive influence on NBS mainstreaming and implementation in Zambia, outweighing the hindering factors.

Keywords

Nature-based solutions · Climate change mitigation · Adaptation and resilience · Institutional factors · Mainstreaming · Implementation

1 Introduction

This is the introductory chapter, which outlines the aim and objectives of the study and discusses its relevance to climate change mitigation and adaptation in Zambia. The study aims to assess the mainstreaming and implementation of Nature-Based Solutions (NBS) into Local Development Plans, using Mumbwa Town Council as a case study. The key objectives include evaluating the extent to which NBS have been mainstreamed into local planning, identifying implemented projects, and analyzing institutional factors that either facilitate or hinder their success. The chapter also highlights the importance of this research in informing sustainable urban development strategies and enhancing climate resilience by addressing critical institutional challenges.

1.1 Study Background

Climate change is one of the most profound urban challenges being faced globally and has adversely impacted the social, economic, and natural environments, costing billions of dollars in global damage and collectively affecting many people worldwide (Zhao et al. 2020, 2019). Zambia has not been spared from the adverse impacts of climate change and the country remains highly vulnerable (Government of The Republic of Zambia 2016). In Zambia, alternating floods and droughts have historically been the primary climatic

K. Marrien (✉)
University: Erasmus University Rotterdam, Rotterdam, The Netherlands
e-mail: kmarrien@gmail.com

P. Mohammad et al. (eds.), *Geographic Approaches to Climate Change and Mitigation: Urban and Rural Perspectives (Volume 2)*,
Advances in Science, Technology & Innovation, https://doi.org/10.1007/978-3-031-92115-5_9

hazards, significantly impacting the economy, infrastructure, natural environment, livelihoods, and overall human well-being (Government of The Republic of Zambia 2016; Government of The Republic of Zambia 2017). Specifically, floods and droughts in Zambia have mainly affected the agricultural sector and natural resources which are the main source of livelihood for the majority of the population in rural and peri-urban areas, impacting food security at the community level thereby increasing poverty levels (2018, 2017).

To combat climate change, reduce its effects, and strengthen adaptation and resilience, Nature-Based Solutions (NBS) have emerged with high recognition in urban planning and policy as a solution to climate change due to the growing evidence that they have the potential to help reduce vulnerability and enhance climate change adaptation and resilience of cities and communities (Cohen-Shacham et al. 2016; Kabisch et al. 2016; Kalantari et al. 2018; Seddon et al. 2020; Mendes et al. 2020). The International Union for Conservation of Nature (IUCN) (2020) defines NBS as, "actions to protect, sustainably manage, and restore natural or modified ecosystems, that address societal challenges effectively and adaptively, while simultaneously providing human well-being and biodiversity benefits" (Cohen-Shacham et al. 2016). While the European Commission defines NBS as "living solutions inspired by, continuously supported by and using nature designed to address various societal challenges in a resource-efficient and adaptable manner and to provide simultaneously economic, social and environmental benefits" (2015). NBS for climate change mitigation, adaptation, and resilience also offers numerous environmental, social, and economic co-benefits that provide societally, and ecosystem benefits simultaneously (Cohen-Shacham et al. 2016; Kabisch et al. 2016; Frantzeskaki 2019). Kabisch et al. (2016) define co-benefits as additional benefits that NBS offer to humans and the ecosystem when the initiatives were initially implemented for another purpose.

Zambia is one of the African countries that have embraced the concept of NBS by turning to nature to adapt to climate change impacts (2021). The country has integrated the NBS actions in its Nationally Determined Contributions (NDCs) plans submitted to the United Nations Framework Convention on Climate Change (UNFCCC) as part of the international fight against climate change following the Paris Agreement (2021). The country has also included some NBS initiatives in its national climate policy and the seventh national development plan with some NBS projects being implemented at the national level (Government of The Republic of Zambia 2016, United Nations Environment Programme (UNEP) 2021; Government of The Republic of Zambia 2017). At the local level, NBS have been mainstreamed in the Local Development Plans (LDPs) as part of the climate change mitigation, adaptation, and resilience strategies. Some of these initiatives have been successfully implemented as various NBS projects.

Various scholars have highlighted numerous institutional factors associated with the mainstreaming and implementation of NBS across the sectors at global, regional, national, and local levels (Mendes et al. 2020; Nelson et al. 2020; Sarabi et al. 2020, 2021). These factors may act as enablers or barriers to the mainstreaming and implementation of NBS. In Zambia, various institutional factors responsible for the mainstreaming and implementation of NBS have been identified. Kalantari et al., (Kalantari et al. 2018) highlight policy and legal frameworks, land-use planning, availability of space to accommodate new NBS projects, and land ownership complexities. According to Cooper (2020), institutionalized and decision-makers prioritization, Actors' technical knowledge on how to mainstream the NBS and integrate them with gray infrastructure, and well as financial capacity play a key role in influencing NBS mainstreaming and implementation. With all these institutional factors identified, there is still a knowledge gap relating to the questions; to what extent have these institutional factors facilitated or hindered the mainstreaming and implementation of NBS at the local level in Zambia? This study aims to clarify this gap using the case study of Mumbwa Town Council, one of the 116 municipalities in Zambia.

1.2 Objectives

This study explains the extent to which the institutional factors influence the mainstreaming of NBS into Local Development Plans and actual implementation of NBS projects for climate change mitigation, adaptation, and resilience building in Zambia based on a case study of Mumbwa Town Council. The following were the specific objectives of the study:

1. To assess how NBS has been mainstreamed in Local Development Plans for climate change mitigation, adaptation, and resilience building in Mumbwa.
2. To investigate the NBS projects that have been implemented in Mumbwa.
3. To assess the institutional factors that have facilitated and/or hindered the mainstreaming and actual implementation of NBS projects in Mumbwa.
4. To assess the climate mitigation, adaptation, and resilience co-benefits of the NBS projects implemented in Mumbwa.

1.3 Relevance of the Study

The relevance of this study in urban and environmental planning lies in its potential to inform more effective mainstreaming of Nature-Based Solutions (NBS) into Zambia's urban development strategies. By examining institutional factors such as policy frameworks, land-use planning, land ownership complexities, and financial capacity, the research provides critical insights for planners and decision-makers on how to navigate the institutional context that influences NBS mainstreaming. Understanding these factors enables actors to address key challenges in spatial planning, infrastructure integration, and stakeholder collaboration. Additionally, the study highlights the importance of enhancing decision-makers technical knowledge and capacity for integrating NBS with gray infrastructure, which is crucial for creating climate-responsive and sustainable cities. Ultimately, this research contributes to long-term strategic planning, helping planners and decision-makers design cities that prioritize environmental sustainability while addressing local institutional constraints. Additionally, understanding the institutional factors that influence NBS implementation can help local authorities tap into global carbon markets by developing and maintaining NBS projects that sequester carbon, such as urban forests, wetlands, and green spaces. With adequate policy frameworks, land-use planning, and technical knowledge, local authorities can leverage these investments to generate revenue, fund further NBS initiatives, and promote sustainable urban development.

2 Literature/Theory Review

This chapter provides a comprehensive review of the concept of Nature-Based Solutions (NBS) and their application in addressing environmental and societal challenges. It begins by defining NBS and categorizing the approaches, including ecosystem restoration, infrastructure-related approaches, and ecosystem protection. The chapter then discusses the three main types of NBS: conservation, restoration, and the creation of new ecosystems, with examples illustrating their roles in climate change mitigation and adaptation. A key section reviews NBS contributions to climate resilience, emphasizing their multifunctional benefits, such as carbon sequestration and disaster risk reduction. Additionally, the chapter explores the co-benefits of NBS, including social, economic, and environmental advantages, and presents a framework for assessing these benefits. The final sections examine the mainstreaming of NBS in urban plans and policies and institutional factors influencing their successful implementation, highlighting the importance of supportive governance, planning, and financial mechanisms. This literature review sets the foundation for understanding how NBS can be leveraged to build climate resilience in urban environments.

2.1 The Concept of Nature-Based Solutions (NBS)

In recent years, the concept of Nature-Based Solutions (NBS) has emerged as a pivotal approach to addressing a multitude of ecological and societal challenges. Cohen-Shacham et al. (Cohen-Shacham et al. 2016) describe NBS as an umbrella concept for a wide range of approaches aimed at addressing societal challenges effectively and adaptively, while simultaneously providing human well-being and biodiversity benefits. They cluster these approaches into five (2018) main categories as shown in Table 1 together with their examples.

According to Cohen-Shacham et al. (2016) approaches focusing on ecological restoration, assist in the recovery of degraded ecosystems; ecological engineering integrates human society with natural environments for mutual benefit; forest landscape restoration combines reforestation with sustainable land management. Additionally, ecosystem-based adaptation and mitigation use biodiversity and ecosystem services to help people adapt to climate change and reduce greenhouse gas emissions, respectively. Ecosystem-based disaster risk reduction focuses on managing ecosystems to mitigate natural hazards, while green infrastructure involves creating networks of natural and semi-natural areas to deliver ecosystem services. Ecosystem-based management takes a holistic approach to sustainably manage land, water, and living resources (Cohen-Shacham et al. 2016; Notte et al. 2022), and area-based conservation approaches, such as protected area management, aim to conserve biodiversity and ecosystem services within designated areas. Collectively, these NBS approaches contribute to sustainable development by

Table 1 Categories and examples of NBS approaches

Category of NBS approaches	Examples
Ecosystem restoration approaches	Ecological restoration Ecological engineering Forest landscape restoration
Issue-specific ecosystem-related approaches	Ecosystem-based adaptation Ecosystem-based mitigation Climate adaptation services Ecosystem-based disaster risk reduction
Infrastructure-related approaches	Natural infrastructure Green infrastructure
Ecosystem-based management approaches	Integrated coastal zone management Integrated water resources management
Ecosystem protection approaches	Area-based conservation approaches including protected area management

Source Cohen-Shacham (2016)

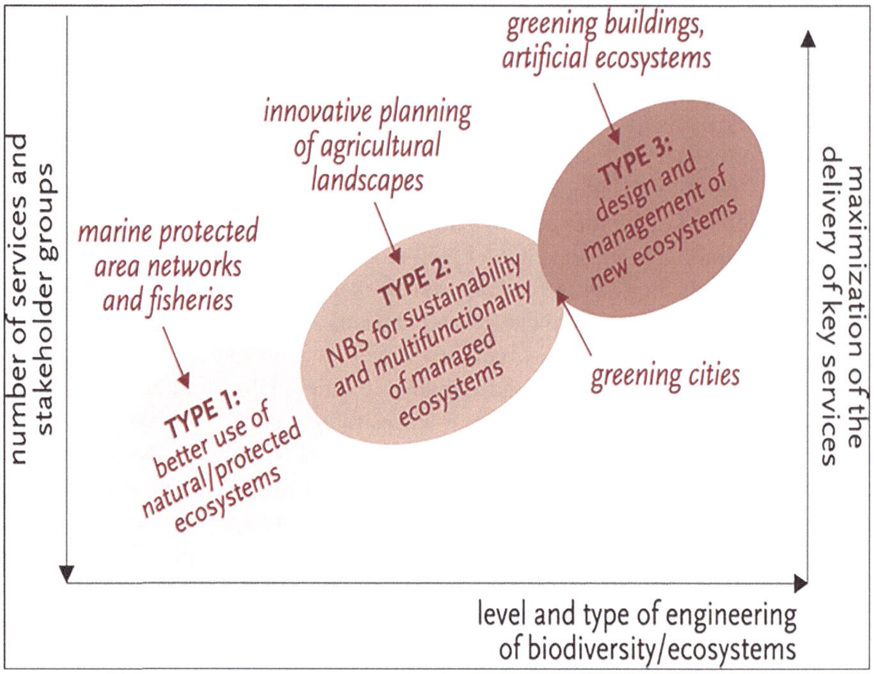

Fig. 1 Types of NBS. *Source* Eggermont et al. (2015)

enhancing ecosystem resilience, supporting biodiversity, and providing multiple benefits to human societies. They not only tackle issues such as climate change, food and water security, and natural disasters but also enhance human well-being and biodiversity (Cohen-Shacham et al. 2016).

The approaches presented in Table 1 can be categorized into three types based on the solutions they offer: protection, restoration, and creation of new ecosystems (Cohen-Shacham et al. 2016, 2019). Eggermont (Eggermont et al. 2015) presents a typology of these three types of NBS, in an X–Y axis diagram as shown in Fig. 1. Type 1 focuses on nature conservation through the protection, maintenance, or improvement of existing ecosystems. An example of this is wetland protection to enhance fish stocks. Type 2 involves establishing long-term management strategies aimed at managing and/or restoring ecosystems to enhance their multifunctionality. This includes practices such as innovative smart agriculture. Type 3 encompasses the creation of new ecosystems, which involves the establishment of blue and green infrastructure, such as green rooftops (Cohen-Shacham et al. 2016; Eggermont et al. 2015). This highlights the multifunctionality of NBS in emphasizing the critical roles of protecting, restoring, and creating ecosystems to enhance biodiversity, ecosystem services, and human well-being.

2.2 NBS for Climate Change Mitigation, Adaptation, and Resilience Building

There is a growing awareness that NBS have significant potential to mitigate climate change impacts and can help people adapt to the effects of climate change and disasters and build resilience (Kabisch et al. 2016; Dunlop et al. 2024). UNEP (2021) defines climate change mitigation as "efforts to reduce or prevent emission of greenhouse gases." Field (2015) defines climate adaptation as, "the process of adjusting to actual or expected climate and its effects in order to moderate or avoid harm or exploit beneficial opportunities in human and natural systems." Resilience on the other hand is defined as "the capacity of social, economic, and environmental systems to cope with a hazardous event or trend or disturbance, responding or reorganizing in ways that maintain their essential function, identity, and structure, while also maintaining the capacity for adaptation, learning, and transformation" (2020). NBS contribute to global climate mitigation through their role in carbon sequestration, surface flooding control through absorption of stormwater, coastal flooding protection by buffering coastal storms, water retention and soil erosion protection, air purification, pollution control, and temperature buffering (2015). This helps human beings adapt the to effects of climate change and build resilience to climatic hazards such as flooding, droughts, strong winds, and the heat island effects (Kabisch et al. 2016; Maes and Jacobs 2015).

Mosca et al. (2023) highlight the critical role of Nature-Based Solutions (NBS) in addressing climate change through both mitigation and adaptation strategies. Their study guides quantifying the effectiveness of NBS in climate change adaptation and mitigation through Key Performance Indicators (KPIs). By identifying common climate change drivers and their effects, the study establishes a set of KPIs to measure performance in relation to these specific effects and how and to what extent NBS can be used to improve CC adaptation and mitigation. Additionally, the study links NBS Sustainable Development Goals (SDGs) and Ecosystem Services (ES), demonstrating how NBS can support these frameworks. The study underscores the multifunctionality of NBS, noting their potential to enhance biodiversity, improve human well-being, and provide various ecosystem services (Mosca et al. 2023).

Bartlett and Mistry (2022) argue that NBS play a critical role in enhancing urban resilience by addressing multiple interconnected challenges, such as climate change, biodiversity loss, and socio-economic inequality. They highlight NBS as a versatile, cost-effective approach that can offer a wide range of benefits from flood control to biodiversity conservation and public health improvement by utilizing ecosystem processes in urban planning. Despite their growing recognition in global policy discussions, the authors note a significant gap between the widespread discourse on NBS and their actual implementation in cities. They emphasize that scaling NBS, accounting for climate impacts, and enhancing investment are crucial to unlocking their full potential in building urban resilience. The paper concludes with guiding principles for maximizing NBS efficacy, including scaling interventions, planning for climate impacts, improving quantification, and leveraging NBS for economic recovery.

2.3 NBS and Climate Mitigation, Adaptation, and Resilience Co-Benefits

Besides the climate mitigation, adaptation, and resilience benefits, NBS also offer multiple environmental, social-cultural, and economic co-benefits for biodiversity and human beings (Cohen-Shacham et al. 2016; Kabisch et al. 2016; Frantzeskaki 2019). Kabisch et al., (2016) define co-benefits as additional benefits that NBS offer to humans and the ecosystem when the initiatives were initially implemented for another purpose. For instance, trees increase green space in cities and foster human well-being by providing recreation and health benefits when they were initially planted for the purpose of carbon sequestration and temperature reduction. That's a co-benefit which is besides the main benefits and is one of the reasons for wide mainstreaming and implementation of NBS in policy and practice.

Raymond et al. (2017) developed a framework to assess the co-benefits of NBS. The framework depicts an interplay of four dimensions: (i) co-benefits for human health and well-being; (ii) integrated environmental performance; (iii) trade-offs and synergies to biodiversity, health, and the economy; (iv) potential for citizen participation in governance and monitoring. With this interplay, Raymond and colleagues concur with the views of other scholars such as Kabisch et al. (2016) and Cohen-Shacham et al. (2016) that NBS have environmental, social, and economic co-benefits besides the direct benefits and costs of NBS obtained from current ecosystem services. The framework also identifies 10 key societal areas of concern in which the NBS co-benefits occur. These include climate mitigation and adaptation, water management, coastal resilience, green space management, air quality, urban regeneration, participatory planning and governance, social justice and social cohesion, public health and well-being, economic opportunities, and green jobs (Raymond et al. 2017).

Seddon et al. (2020) however observe that, despite the growing evidence that NBS can strongly contribute to climate mitigation, adaptation, and resilience, they have been underutilized due to a bias toward the engineered approaches globally. Thus, to demonstrate the important role of NBS in supporting climate mitigation, human adaptation to climate change, and resilience building, Seddon and the co-authors used the IPCC vulnerability framework for social-ecological systems (Fig. 2).

This conceptual framework places NBS at the interface of the socio-economic system and the ecosystem to reduce socio-ecological vulnerability to climate change (Seddon et al. 2020). The framework recognizes three dimensions of vulnerability to climate change namely, exposure which entails the extent to which a region, ecosystem, resource, or community is impacted by climate change; sensitivity which is the degree to which a system is affected by, or responsive to, those effects; adaptive capacity which is the ability of the system to adjust or innovate in response to changing conditions. In this case, NBS is believed to bring all these elements together by enhancing the adaptive capacity of the system to buffer the potential impacts created by a combination of exposure and sensitivity. They, therefore, conclude that through the protection, restoration, and careful management of ecosystems, NBS can positively influence all three dimensions of socio-economic vulnerability and promote climate mitigation, adaptation, and resilience building (Seddon et al. 2020).

2.4 Mainstreaming of NBS in Urban Plans and Policies for Climate Mitigation, Adaptation, and Resilience Building

Mainstreaming of NBS in this context means the integration of NBS into local policies for climate mitigation, adaptation, and resilience building (2017). Literature shows that there is an increasing interest by urban policymakers and practitioners in making NBS part of the "mainstream" of urban development, international policy, and urban research agenda (Wamsler et al. 2017; Magni et al. 2020). This is because mainstreaming NBS into urban plans and policies provides

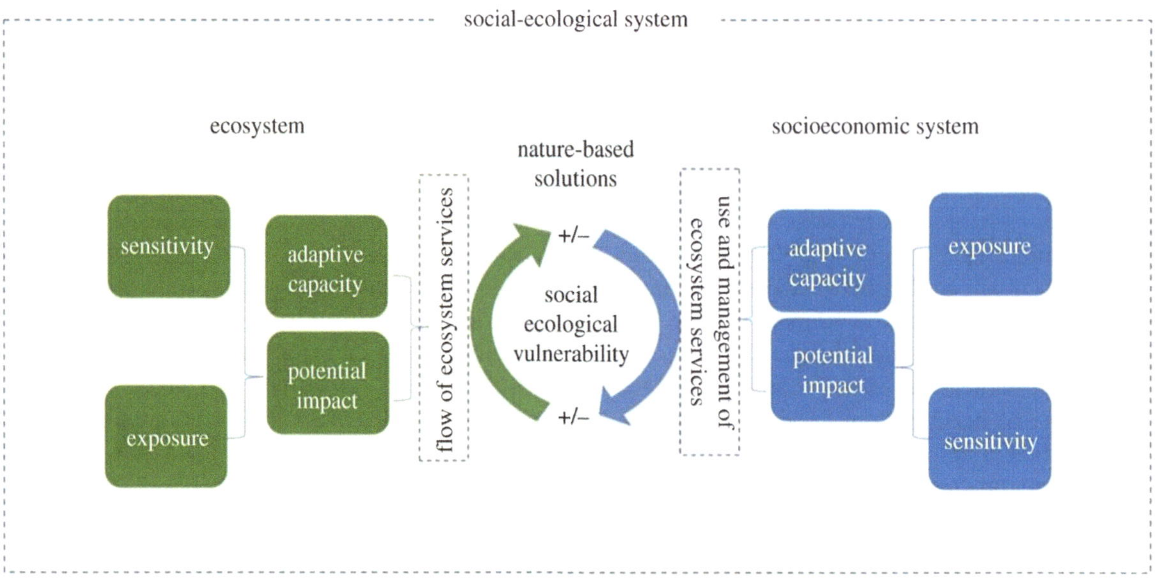

Fig. 2 Integrating NBS to climate change impacts into the social-ecological vulnerability framework. *Source* Seddon et al. (2020)

sustainable solutions to address climate change mitigation and adaptation challenges (Magni et al. 2020).

Various scholars note that mainstreaming NBS into urban plans at the local level is often challenging due to various institutional factors such as poor/rigid planning and governance (Magni et al. 2020) and lack of political will and a sense of urgency by the policymakers (Sarabi et al. 2020). Local authorities are also often not sure of the best approaches to use to systematically mainstream NBS into planning and urban governance (2017). Thus, they have attempted to find solutions to these challenges and barriers to NBS mainstreaming (Seddon et al. 2020; Magni et al. 2020).

Wamsler et al. (2017) for instance proposed an NBS mainstreaming framework involving four climate risk-reduction approaches within which NBS can be applied comprehensively. These are hazard exposure reduction, vulnerability reduction, effective response, and effective recovery. The framework presents two types of mainstreaming namely vertical and horizontal mainstreaming, with six mainstreaming strategies operating at three levels; local, institutional, and interinstitutional. Wamsler (2017) argues that in order to achieve a holistic and distributive governance system, climate mitigation and adaptation should be institutionalized at local, institutional, and interinstitutional levels by uniting vertical and horizontal mainstreaming. Here, the local level refers to the community or household levels and how the approaches are mainstreamed in the initiatives at the grassroot level (Strategy I & II). The institutional level implies internal organization (Strategy III) which is the implementing organization and its cooperation with other departments (Strategy IV) and the internal policies and regulations governing their operation and cooperation (Strategy V) while the interinstitutional level entails external cooperation (Strategy VI) or interconnections between the implementing organization and other players such as local, national, regional, international, and the citizens. Correspondingly, vertical mainstreaming in this case entails a top-down mainstreaming of mitigation and adaptation strategies from the interinstitutional level to the local level while horizontal entails a bottom-up approach in which strategies are mainstreamed from the local level to interinstitutional levels. The framework aims at promoting, comprehensiveness, inclusiveness, and flexibility rather than dependence on a single approach with emphasis being placed on effective urban planning and holistic governance (2017).

2.5 Institutional Factors Influencing Mainstreaming and Implementation of NBS

Institutional factors play a crucial role in both the mainstreaming and successful implementation of Nature-Based Solutions (NBS). Numerous scholars have examined these factors and classified them into several categories, which include planning, policy and legal frameworks, actors, financial capacity, institutionalization, and governance. Each of these factors has the potential to act as a driver or barrier to the incorporation of NBS into urban policies and practices (Kabisch et al. 2016; Seddon et al. 2020; Maes and Jacobs 2015; Wamsleret al. 2017; Mendes et al. 2020; Sarabi et al. 2020; Zingraff-Hamed et al. 2020; Mahmoud and Morello 2021). For instance, effective planning supported by flexible

Table 2 Common institutional factors that drive or hinder the mainstreaming and implementation of NBS

Institutional factors	Drivers	Barriers	Source
Planning, policy, and legal framework	• Holistic, flexible, and inclusive urban planning • Supportive policy and legal frameworks	• Poor and rigid land-use planning • Conflicting land uses • Lack of space for NBS projects • Conflicting policies and regulations	Mendes et al. (2020), Sarabi et al. (2020), Maes and Jacobs (2015)
Actors	• Professional knowledge of NBS • Teamwork • Clear guidelines and tools	• Lack of professional knowledge • Sectoral silos and communication challenges • Conflicting mandates • Lack of urgency among policymakers	Mendes et al. (2020), Sarabi et al. (2020), Kabisch et al. (2016)
Financial capacity	• Availability of financial resources • Co-financing mechanisms	• Lack of financial resources • Resistance from the private sector and citizens to invest • Limited international support	Seddon et al. (2020), Sarabi et al. (2020)
Governance	• Multi-stakeholder engagement • Co-creation, co-design, and co-implementation • Good governance	• Lack of established urban governance systems • Limited stakeholder collaboration • Misconceptions about NBS responsibilities	Mahmoud and Morello (2021), Wamsler et al. (2017), Zingraff-Hamed et al. (2020)
Institutionalization	• Clear institutional arrangements • Administrative structures	• Lack of institutionalization at the local level	Mendes et al. (2020), Maes and Jacobs (2015), Nesshöver et al. (2017)

Source Author 2021

policies is essential for NBS mainstreaming but rigid legal frameworks and conflicting land-use regulations often create barriers (Mendes et al. 2020; Sarabi et al. 2020). Additionally, actors like urban planners and policymakers must possess adequate knowledge about NBS, yet knowledge gaps and siloed communication hinder progress (Mendes et al. 2020; Sarabi et al. 2020). Financial constraints also pose significant challenges, with local governments often lacking the budget or incentives to fund NBS projects (Seddon et al. 2020). Governance models that promote multi-stakeholder collaboration, such as co-creation and co-design, are crucial, but fragmented decision-making limits NBS mainstreaming (Zingraff-Hamed et al. 2020; Mahmoud and Morello 2021). Finally, institutionalization is vital for NBS sustainability, but without formal rules and evolving institutional frameworks, long-term mainstreaming remains difficult (Mendes et al. 2020). Table 2 summarizes the common institutional factors that can either drive or hinder the mainstreaming and implementation of NBS.

3 Material and Methods

This chapter outlines the research design and methodology employed in the study, which focuses on exploring the institutional factors influencing the mainstreaming and implementation of Nature-Based Solutions (NBS) within Mumbwa Town Council, Zambia. A qualitative case study approach was selected to provide an in-depth understanding of the local context and challenges faced by the council in integrating NBS into its development plans. Data was collected from both primary and secondary sources, including semi-structured interviews with nine key informants and a review of local policy documents. Due to travel restrictions during the COVID-19 pandemic, a research assistant conducted physical interviews. The collected data was analyzed using Atlas.ti software, with a focus on identifying the relationships between institutional factors and NBS mainstreaming and implementation.

3.1 Research Design

This study employs a qualitative research design, specifically utilizing a case study approach to explore the institutional factors influencing the mainstreaming and implementation of Nature-Based Solutions (NBS) within Mumbwa Town Council, Zambia. The case study method was chosen to provide an in-depth understanding of the local context and the specific challenges and opportunities faced by the Mumbwa Town Council. The case study approach allows for a comprehensive examination of Mumbwa Town Council as a single unit of analysis. This method is particularly suitable for exploring complex phenomena within their real-life context, providing detailed insights into how institutional factors influence the adoption and implementation of NBS at the local level.

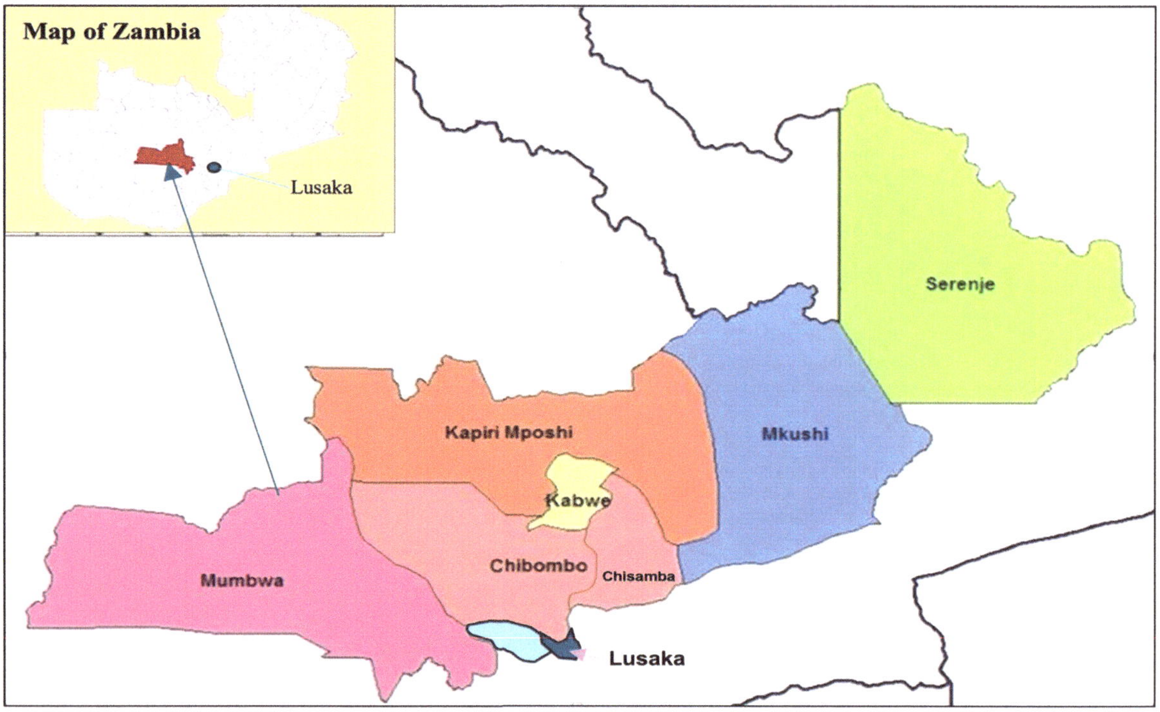

Fig. 3 Location of Mumbwa town. *Source* Google

3.2 The Study Area

The study area, Mumbwa Town Council was selected purposively as it is one of the local authorities that have mainstreamed NBS for climate change mitigation and adaptation actions in its District Development Plan (DDP). Mumbwa Town Council is situated in Mumbwa District, Central Province of Zambia, and approximately about one hundred and fifty kilometers (150 km) from Lusaka, the capital city of Zambia as shown in Fig. 3.

3.3 Data Sources

Both primary and secondary data were used in this study. Interview guides with semi-structured interview questions were used to collect primary data from the key informants. A total of 9 interviews were conducted with the key informants. Due to the coronavirus pandemic, a research assistant was employed to assist with conducting physical interviews with some key informants as the researcher was not able to travel to the study area (Zambia). With the consent of the interviewees, all interviews were recorded. A saturation point in data collection was reached when all the questions were answered based on the research operationalization. The case study strategy was complemented by the desktop research design in which local policies were reviewed to identify the NBS strategies that were mainstreamed (Government of The Republic of Zambia 2016; Government of The Republic of Zambia 2017; Mumbwa Town Council 2017; Mumbwa Town Council 2017).

3.4 Data Analysis

The primary aim of this study was to evaluate the extent to which Nature-Based Solutions (NBS) have been integrated into Local Development Plans within Mumbwa. To achieve this objective, it was first necessary to gather comprehensive information regarding the specific NBS initiatives that have been mainstreamed and implemented, as well as the institutional factors potentially influencing these processes. By reviewing pertinent local policy documents and conducting in-depth interviews with key informants involved in the mainstreaming and implementation of NBS, this study obtained valuable insights into the current state of NBS mainstreaming in Mumbwa.

The collected data were transcribed verbatim (word-for-word) into interview scripts, which were subsequently coded and analyzed qualitatively using Atlas.ti software. A co-occurrence analysis was employed to explore the relationships between various institutional factors and the processes of NBS mainstreaming and implementation. This analysis enabled the identification of key patterns and correlations, thereby informing the conclusions regarding the extent to which institutional factors have influenced the mainstreaming and implementation of NBS in the study area.

4 Results and Discussion

This chapter presents the results and discusses the mainstreaming and implementation of Nature-Based Solutions (NBS) in Mumbwa. It highlights the successful mainstreaming of NBS into local development plans, focusing on sustainable forest management, water resource restoration, wildlife ecosystem restoration, urban greening, and sustainable agriculture. The chapter emphasizes the multi-level governance approach, combining top-down and bottom-up strategies to ensure the sustainability and effectiveness of NBS initiatives. Key projects such as the REDD+ Project, SCRIKA Project, and AWARE Project are discussed, showcasing their contributions to climate mitigation, adaptation, and resilience. The chapter also explores the anticipated environmental, economic, and social-cultural co-benefits of these projects, including carbon sequestration, improved air quality, enhanced agricultural productivity, and social cohesion. Additionally, it examines the influence of institutional factors on NBS mainstreaming and implementation, identifying both facilitating and hindering factors. The proposed conceptual framework illustrates the relationship between mainstreaming and implementation, highlighting that the full benefits of NBS are realized through effective implementation. The findings underscore the importance of integrating NBS into local and national policies to address climate change and support sustainable development.

4.1 NBS Mainstreamed in the Local Development Plans in Mumbwa

The findings from Mumbwa show that the Mumbwa Town Council has successfully integrated Nature-Based Solutions (NBS) for climate mitigation, adaptation, and resilience into its 2017–2021 District Development Plan (DDP) and 2018–2022 Strategic Plan (SP), adapted from Zambia's national policies such as the Zambia National Policy on climate change and the Seventh National Development Plan (Government of The Republic of Zambia 2016; Government of The Republic of Zambia 2017).

The main initiatives mainstreamed are sustainable forest management, water resource restoration, wildlife ecosystem restoration, urban greening, and sustainable agriculture. Most NBS initiatives mainstreamed in Mumbwa fall under Type 2: Ecosystem Management and Restoration, as per Eggermont's typology (Eggermont et al. 2015). This type of NBS focuses on actively managing and restoring ecosystems to enhance their functionality, a crucial strategy in areas affected by climate-induced degradation, such as Mumbwa. Sustainable forest management, water resource conservation, and conservation agriculture all fall predominantly under Type 2, reflecting a focus on restoring degraded ecosystems while maintaining their resilience against future environmental stressors.

The mainstreaming of NBS into strategic documents highlights the increasing recognition of the importance of ecological approaches to address environmental and socio-economic challenges. This is also in line with the literature, which indicates that NBS strategies emphasizing ecosystem restoration are particularly valuable in rural and peri-urban settings, where environmental degradation directly affects local livelihoods (Seddon et al. 2020). Type 1: Ecosystem Protection, Maintenance, and Improvement also play a significant role in Mumbwa, particularly concerning wildlife ecosystem restoration and water resource conservation initiatives mainstreamed. These initiatives focus on protecting existing ecosystems that are critical for biodiversity and ecosystem services, such as carbon sequestration, which has both climate mitigation and adaptation benefits. This dual role of ecosystems, especially forests, in mitigating climate change and supporting local communities aligns with Seddon et al. (2020), who emphasize the multifunctionality of ecosystems in providing both direct (carbon storage) and indirect (livelihood support) benefits to human populations. Urban greening, an example of Type 3: Ecosystem Modification/New Ecosystem Creation represents a forward-looking strategy aimed at modifying urban environments to make them more sustainable. By incorporating the "garden city" concept, Mumbwa follows the global trend of rethinking urban spaces through green infrastructure, aligning with arguments by Kabisch et al. (2016) that urban greening is essential for enhancing ecosystem services in densely populated areas. Under this concept, the council leaves green open spaces in the newly planned areas to be developed as recreational parks. Open spaces are also left along the sidewalks of the road for tree planting purposes as shown in Fig. 4. This strategy not only reduces heat island effects and improves air quality but also provides recreational spaces, which can improve human well-being which is an important co-benefit of NBS (Raymond et al. 2017).

4.1.1 Multi-level Mainstreaming and Governance Approaches

The findings also highlight how the mainstreaming of NBS in Mumbwa follows a multi-level governance approach, cutting across all six strategic levels identified by Wamsler et al.

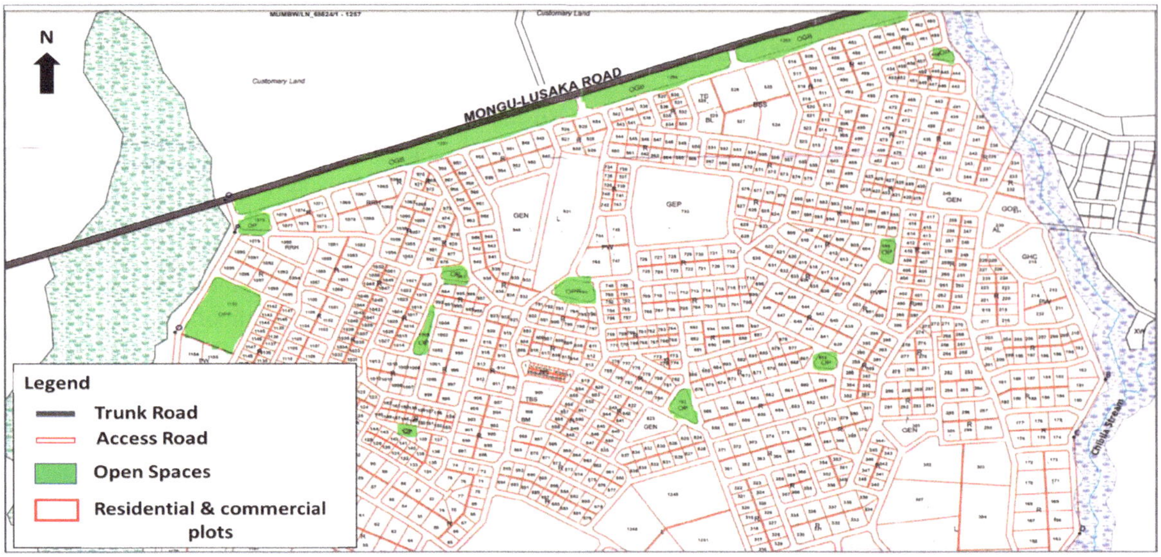

Fig. 4 Extract of the Nambala hill extension land-use plan highlighting green open spaces reserved for recreational parks and green road sideways. *Source* Mumbwa Town Council (2017)

(2017) which are; Strategies I and II (household and community level), Strategy III (institutional), Strategy IV (interdepartmental cooperation), Strategy V (internal council policies), and Strategy VI (interinstitutional collaboration with national and international stakeholders). This multi-level governance approach underscores the importance of integrating NBS across various governance scales, from grassroots to national levels, to ensure their long-term sustainability and effectiveness.

Correspondingly, Mumbwa has a combination of both vertical mainstreaming which entails a top-down mainstreaming involving of adaption of strategies from the interinstitutional level to the local level, and horizontal mainstreaming which entails a bottom-up approach in which strategies are mainstreamed from the local level to interinstitutional levels. The vertical mainstreaming evident in Mumbwa, where national policies are adapted at the local level, presents the top-down approach necessary to align local actions with broader national climate strategies. This is particularly important in the case of sustainable forest management for instance, where local actions such as tree planting and forest conservation are part of Zambia's Nationally Determined Contributions (NDCs). By aligning local and national policies, Mumbwa can leverage national resources and policy frameworks to support its local initiatives. This top-down approach has been emphasized by scholars like Nesshöver et al. (2017), who argue that effective NBS implementation requires coherence between local and national policy frameworks.

Conversely, horizontal mainstreaming, which involves the incorporation of community-driven initiatives into broader institutional frameworks, has also played a significant role in Mumbwa. The community-driven identification of key NBS initiatives, such as conservation agriculture and water resource restoration, demonstrates how bottom-up approaches can enhance the relevance and acceptability of NBS at the local level. This bottom-up process not only ensures that the solutions are tailored to local conditions but also fosters community ownership, which is critical for the long-term success of NBS projects. Wamsler et al. (2017) support this approach, emphasizing that community participation is vital for the adaptive governance of NBS, as it ensures that local knowledge and needs are integrated into broader policy frameworks.

4.1.2 Contributions of NBS to Climate Mitigation, Adaptation, and Resilience

NBS initiatives in Mumbwa contribute to both climate mitigation and adaptation. Sustainable forest management, for instance, serves as a carbon sink, reducing greenhouse gas emissions and contributing to global climate mitigation efforts. At the same time, forests provide important resources for local communities, such as wood, food, and materials, which support adaptation by enhancing the community's resilience to climate variability. This multifunctionality of NBS is well-documented in the literature, with scholars such as Mosca et al. (2023) emphasizing how NBS can simultaneously address environmental challenges and support sustainable development goals (SDGs). Water resource conservation in Mumbwa is another key initiative that addresses both mitigation and adaptation. By restoring local rivers and water catchment areas, the district is improving its capacity to manage water resources during periods of drought and flooding—two critical climate risks in the region. The construction of dams as part of blue-gray infrastructure projects demonstrates how NBS can be integrated with engineered solutions to enhance their effectiveness, a point highlighted by Eggermont et al. (2015) who argue that hybrid solutions combining natural and engineered elements can provide the most robust responses to climate change. Urban greening contributes primarily to climate adaptation by reducing the heat island effect, improving air quality, and providing green spaces that enhance the quality of life in urban areas. According to Kabisch et al. (2016), urban greening is essential in building urban resilience, particularly in rapidly urbanizing areas where natural spaces are limited. By integrating green spaces into urban planning, Mumbwa's approach reflects a shift toward more sustainable land-use planning, as advocated by Mendes et al. (2020), which highlights the importance of incorporating NBS into urban development to meet sustainability and climate goals.

4.1.3 Synergistic and Multi-scalar Approaches to NBS in Mumbwa

The synergistic mainstreaming of NBS across multiple governance levels—from local communities to national policies—enhances the capacity of Mumbwa to adapt to and mitigate climate risks. The combination of both vertical and horizontal mainstreaming in Mumbwa demonstrates a flexible, adaptive governance model that integrates top-down strategies with community-driven initiatives. This model is essential for building long-term climate resilience, as it ensures that local actions are supported by national frameworks, while national policies are informed by on-the-ground realities. Such multi-scalar approaches to NBS implementation are critical in addressing the complex and interconnected challenges posed by climate change (2020).

In conclusion, the successful mainstreaming of NBS in Mumbwa's local development plans reflects a well-structured, multi-level governance approach that integrates local initiatives with national and international frameworks. The emphasis on ecosystem restoration and management, coupled with urban greening and water conservation, highlights the district's commitment to leveraging nature-based approaches to enhance climate resilience. By incorporating both vertical and horizontal mainstreaming, Mumbwa sets a precedent for how adaptive governance models can support the widespread adoption of NBS for climate mitigation, adaptation, and resilience building, echoing the findings of Wamsler et al. (2017) and Seddon et al. (Seddon et al. 2020).

4.2 NBS Projects Implemented

In terms of NBS project implementation, the findings reveal that the projects carried out in Mumbwa were closely aligned with the initiatives mainstreamed into Local Development Plans, focusing on key areas such as forest management, wildlife resource conservation, water resource management, sustainable agriculture, and urban greening. Key projects include the reduce emissions from Deforestation and Forest Degradation (REDD+) Project, funded by the UN, aimed at addressing forest degradation by promoting afforestation, reforestation, and forest carbon stock enhancement, directly contributing to Mumbwa's sustainable forest management efforts. Similarly, the Strengthening Climate Resilience in the Kafue Sub-Basin (SCRIKA) Project (2018–2020), which was collaboratively implemented by the Mumbwa Town Council and local communities and aimed to enhance climate adaptation and resilience through forest tree planting and other adaptive agricultural practices. Funded by the African Development Bank, this project involved peri-urban and rural communities in activities such as gardening, fruit tree planting, and reforestation. Figure 5 shows the nursery trees for the SCRiKA project under the Muleke community.

Additionally, the Accelerating Water and Agriculture Resource Efficiency (AWARE) Project, co-financed by the European Union and GIZ, aimed to increase agricultural resource efficiency through sustainable water management practices, addressing climate change hazards like droughts that are experienced in some parts of Mumbwa and have affected water availability thereby affecting agricultural production as well. This project included activities such as water catchment protection, reforestation, and soil conservation measures. The project also involves setting up demonstration sites for sustainable irrigation and rainwater harvesting through trenching as shown in Fig. 6. The Water Resource Management Project, funded by the World-Wide Fund for Nature (WWF), promoted the natural regeneration of ecosystems in the Kafue River catchment areas, focusing on assisted natural regeneration and river health protection. Last but not the least is the urban greening initiative, which includes the Keep Zambia Clean and Green and Plant a Million Trees Campaigns and aims to develop green spaces and recreational parks in Mumbwa as resources become available. These sister projects are being implemented as national strategies adapted at the local level under sustainable forest management and urban greening initiatives, integrated into the local development plans.

The research findings on the implementation of Nature-Based Solutions (NBS) projects in Mumbwa show a significant alignment with the existing literature on NBS, particularly in terms of ecological restoration, climate adaptation, and urban greening. These initiatives, as demonstrated by projects such as the REDD+ project, the Strengthening Climate Resilience in the Kafue Sub-Basin (SCRiKA) Project, and the Accelerating Water and Agriculture Resource Efficiency (AWARE) Project, reflect the broader theoretical framework discussed by key scholars such as Cohen-Shacham et al. (2016), Eggermont et al. (2015) and Seddon et al. (2020).

Fig. 5 Muleke Community Tree planting project under the SCRiKA Project, Mumbwa. *Source* Field data, 2021

Fig. 6 Trenching demonstration site under the AWARE project, Mumbwa. *Source* Field Data, 2021. *Source* Field Data, 2021

4.2.1 Ecological Restoration and Sustainable Forest Management

The implementation of the REDD+ Project in Mumbwa, which focuses on afforestation, reforestation, and forest carbon stock enhancement, directly supports the principles of forest landscape restoration discussed by Cohen-Shacham et al. (2016). By targeting forest degradation and promoting sustainable forest management, this project aligns with Type 1 NBS in Eggermont et al.'s (2015) typology, which emphasizes the conservation and protection of existing ecosystems. The goal of enhancing carbon sequestration through these efforts directly addresses climate mitigation, a key aspect of NBS highlighted in Seddon et al. (2020). This project also contributes to biodiversity preservation and the recovery of ecosystem services, demonstrating the multifunctional potential of NBS to meet both ecological and societal needs.

In *addition, the involvement* of local communities in these projects, particularly through afforestation and sustainable land management efforts, reflects the participatory and adaptive management principles that Cohen-Shacham et al. (2016) advocate. This community involvement also addresses the need for inclusive governance in NBS implementation, as discussed by Raymond et al. (2017), reinforcing the importance of bottom-up approaches to ensure the long-term success of NBS projects.

4.2.2 Climate Resilience and Water Management

The SCRiKA Project in Mumbwa, which promotes forest tree planting and adaptive agricultural practices, aligns closely with the concept of ecosystem-based adaptation discussed in the literature (Cohen-Shacham et al. 2016). By involving local communities in activities such as reforestation and sustainable agriculture, the project reflects Type 2 NBS, as described by Eggermont et al. (2015), which focuses on the restoration and management of ecosystems to enhance their resilience and functionality. The project's focus on enhancing climate adaptation and resilience through such measures directly ties into Seddon et al. (2020) argument that NBS can build the capacity of ecosystems to buffer climate change impacts while supporting the livelihoods of local communities.

Similarly, the AWARE Project demonstrates the application of ecosystem-based water management approaches, a central theme in Cohen-Shacham et al.'s (2016) categorization of NBS. By focusing on water catchment protection, reforestation, and soil conservation, this project addresses the critical need for sustainable water management in the face of climate-induced droughts. This aligns with Seddon et al. (2020) framework on climate change adaptation, where NBS are positioned to reduce socio-ecological vulnerability by enhancing the adaptive capacity of both ecosystems and human communities. The establishment of demonstration sites for sustainable irrigation and rainwater harvesting under the AWARE project further highlights the practical application of green infrastructure in agricultural systems, which can mitigate water scarcity while improving agricultural productivity.

Overall, the findings echo Raymond et al. (2017), who emphasize that NBS provide co-benefits beyond their primary objectives. In the case of Mumbwa, the projects not only address water scarcity and forest degradation but also contribute to ecosystem resilience, biodiversity conservation, and improved agricultural practices, offering socio-economic benefits to the local communities.

4.2.3 Urban Greening and Sustainable Development

The urban greening initiatives in Mumbwa, particularly the Keep Zambia Clean and Green and Plant a Million Trees Campaigns, align with Type 3 NBS as categorized by Eggermont et al. (2015). These initiatives aim to create new ecosystems through urban greening efforts, which aligns with the green infrastructure approach described by Cohen-Shacham et al. (2016). By developing green spaces and recreational parks, these projects contribute to urban resilience by enhancing biodiversity, improving air quality, and mitigating the urban heat island effect, as noted by Bartlett and Mistry (2022). The multifunctionality of these initiatives underscores the potential of NBS to address a range of urban challenges, from climate adaptation to public health improvements. Urban greening in Mumbwa also reflects the argument by Kabisch et al. (2016) that NBS can provide significant co-benefits for human well-being. These initiatives, originally aimed at enhancing urban resilience and climate adaptation, also foster social cohesion by creating recreational spaces that improve the quality of life for residents. This corresponds to Raymond et al.'s (2017) framework, which highlights the multiple societal areas of concern that NBS can address, including public health, biodiversity conservation, and economic opportunities.

4.3 Co-benefits of NBS Projects in Mumbwa

The findings from the ongoing NBS projects in Mumbwa indicate a wide range of anticipated environmental, economic, and social-cultural co-benefits that resonate with the broader literature on NBS, although no formal evaluation has been conducted yet. While these co-benefits remain potential outcomes at this stage, they provide promising insights into how NBS can address both ecological and socio-economic challenges simultaneously.

4.3.1 Environmental Co-benefits

Carbon sequestration is a key environmental benefit across all the projects, particularly in reforestation efforts under the Integrated Forest Management projects like SCRiKA, REDD+, and the AWARE project. The capacity of forests to act as carbon sinks has been widely acknowledged in climate mitigation literature (Seddon et al. 2020). By enhancing tree cover, these projects contribute significantly to carbon sequestration, which aligns with global climate goals under the Paris Agreement to reduce greenhouse gas emissions. This reinforces the role of forests in contributing to long-term climate stability and ecosystem health, as underscored by Cohen-Shacham et al. (2016).

In addition to carbon sequestration, projects such as urban greening initiatives contribute to urban regeneration and air purification. These activities not only improve air quality, a benefit well-documented in studies of urban NBS (Kabisch et al. 2016) but also help reduce the urban heat island effect, promoting urban sustainability and resilience to extreme temperatures. The Keep Zambia Green and Clean Campaign, for instance, aligns with literature emphasizing the benefits of green infrastructure in improving urban climate adaptation and air purification (Raymond et al. 2017).

The focus on stormwater management, flood control, and groundwater retention in both the urban greening and water resource management projects reflects their multifunctionality in improving water regulation. These benefits are critical, especially in the face of increasing rainfall variability and flooding risks due to climate change, as highlighted by Eggermont et al. (2015). By enhancing water infiltration and reducing surface runoff, these NBS projects contribute to flood risk reduction, an essential adaptation measure, particularly in vulnerable regions like Mumbwa.

Moreover, the prevention of soil erosion is a vital co-benefit in both urban and rural areas, particularly through reforestation efforts. The role of NBS in land stabilization and biodiversity conservation is well-documented in ecosystem-based adaptation literature (Seddon et al. 2020), underscoring how these projects can contribute to both climate adaptation and habitat restoration.

4.3.2 Economic Co-benefits

NBS projects in Mumbwa are anticipated to generate significant economic co-benefits, particularly through their positive impact on agriculture and tourism. The integration of agroforestry and sustainable land management in projects like SCRiKA and AWARE is expected to increase agricultural productivity, aligning with findings by Seddon et al. (2020) that NBS can enhance food security by improving ecosystem services that support agriculture, such as soil fertility and water availability.

Increased agricultural productivity directly links to livelihood improvement and increased household income, as these projects offer farmers improved access to resources such as water and healthier soil conditions, which in turn boost crop yields. This aligns with the broader literature that emphasizes the role of NBS in supporting sustainable agriculture and poverty alleviation, particularly in rural communities (Raymond et al. 2017).

Tourism and game ranching are additional economic benefits associated with these NBS projects, particularly in terms of forest conservation and the promotion of wildlife. The increase in wildlife through habitat restoration can potentially boost ecotourism and foreign direct investment, reflecting the findings of Cohen-Shacham et al. (2016), who note that NBS can generate revenue through nature-based tourism and recreation. This also contributes to national economic development, as NBS offer opportunities for eco-friendly industries, thereby diversifying the local economy.

4.3.3 Social-Cultural Co-benefits

The social-cultural co-benefits of NBS projects are equally noteworthy, as they contribute to social cohesion, recreation, and human health. The creation of green open spaces in urban areas, as part of the Keep Zambia Green and Clean Campaign, provides recreational areas for residents, which promote mental and physical well-being. This aligns with the findings of Kabisch et al. (2016), who highlight the mental health benefits of exposure to green spaces, particularly in urban settings. These spaces also foster social cohesion by providing common areas for community interactions, as noted by Bartlett and Mistry (2022).

Projects that contribute to the expansion of wildlife and national parks also support tourism, which in turn contributes to national food security and infrastructure development. This reflects broader research on NBS projects that demonstrate how ecosystem restoration and biodiversity conservation can foster cultural identity and social well-being by connecting people to nature (Raymond et al. 2017).

The integration of human health benefits through projects that improve air quality, regulate urban heat, and provide access to clean water further strengthens the argument for NBS as a multifunctional solution to societal challenges. These benefits are consistent with the conclusions of Raymond et al. (2017), who emphasize the role of NBS in improving public health outcomes, particularly through the reduction of air pollution and the promotion of active lifestyles in green spaces.

Overall, the co-benefits of NBS projects in Mumbwa align closely with the literature, illustrating how these projects simultaneously address multiple sustainability challenges. The multifunctionality of NBS, ranging from climate mitigation and adaptation to economic development and social well-being, demonstrates their holistic potential, as noted by Seddon et al. (2020) and Eggermont et al. (2015). However, it is important to note that the full realization of these co-benefits depends on the successful and sustained implementation of these projects. While the anticipated benefits are promising, an evaluation of the actual impacts will be necessary to fully understand the effectiveness and scalability of these initiatives.

4.4 Influence of Institutional Factors on the Mainstreaming and Implementation of NBS in Mumbwa

Institutional factors highlighted in the literature review were examined to establish which ones among them have facilitated the mainstreaming and implementation of NBS and which ones are hindering factors. Tables 3 and 4 provide summaries of the facilitating factors and hindering factors, respectively. The findings from Mumbwa Town Council highlight the significant role that institutional factors have played in the mainstreaming and implementation of Nature-Based Solutions (NBS). These factors have either facilitated or hindered the mainstreaming of NBS into local policies and the successful implementation of related projects. This section discusses the extent to which institutional factors have influenced NBS mainstreaming and implementation, aligning these findings with the existing literature to provide a more comprehensive analysis (Tables 3 and 4).

Table 3 Facilitating factors for NBS mainstreaming

Institutions	Institutionalization
	Institutional arrangements
Planning	Holistic, Flexible, and inclusive planning system
	City's focus on both economic and sustainability goals
	Availability of specific NBS planning and design standards
Policy and legal framework	Supportive local policies
	Consistency of local policies with National Policy/legal framework
Actors	Actors' professional knowledge and skills on NBS and their co-benefits
	Lack of Silos mentality among actors (Teamwork and good communication and information sharing)
Governance	Good governance system
	Political will and NBS prioritization
	Stakeholder engagement and collaboration (Co-implementation)
Financial capacity	International support/funding
	Co-financing and risk-sharing

Source Developed by Author, 2021

Table 4 Hindering factors for NBS mainstreaming

Financial capacity	Lack of local financial resources
	Inadequate government incentives
	Lack of participation in the financing of NBS by the private sector and the local citizens
Planning	Lack of available land to accommodate new NBS
	Conflicting land uses
	Land ownership Complexes
Governance	Stakeholders (Community) perception of NBS
Environmental factors	Climate variability-Droughts/flooding

Source Developed by Author, 2021

4.4.1 Facilitating Factors for NBS Mainstreaming and Implementation

Institutionalization and Institutional Arrangements

Institutionalization has emerged as a key factor in the successful mainstreaming of NBS in Mumbwa. The findings indicate that Mumbwa has institutionalized NBS practices, such as tree planting and sustainable forest management, which have become a normative response to climate change mitigation and resilience building. This is in contrast to Nesshöver et al.'s (2017) argument that NBS are often not institutionalized at the local level, creating barriers to mainstreaming. In this case, the strong institutional arrangements in Mumbwa, coupled with consistent support from both local and national policies, have facilitated the mainstreaming and implementation of NBS, underscoring the importance of embedding these solutions in formal structures for long-term success (Maes and Jacobs 2015; Mendes et al. 2020).

Planning Systems

A holistic, flexible, and inclusive planning system has been identified as another critical facilitating factor. Mumbwa's focus on balancing economic growth with sustainability goals aligns with Mendes et al. (2020), who emphasize the need for flexible urban planning frameworks that can accommodate NBS initiatives. The availability of specific NBS planning and design standards in Mumbwa has provided a clear framework for integrating these solutions into local development plans, reflecting the argument made by Frantzeskaki

(2019), who states that well-defined standards are crucial for operationalizing NBS in urban contexts.

Supportive Policy and Legal Framework

The consistency between local policies in Mumbwa and the national legal framework, particularly in areas like sustainable forest management, has been instrumental in mainstreaming NBS. This is in contrast to other findings in the literature, which highlight the absence of supportive policy frameworks as a barrier to NBS implementation (Sarabi et al. 2020). In Mumbwa, the alignment between local and national goals such as Zambia's 2020 Nationally Determined Contributions (NDCs) provides a conducive policy environment for NBS, supporting arguments by Mendes et al. (2020) that coherent policy frameworks are key enablers of NBS mainstreaming.

Actors' Professional Knowledge and Skills on NBS and Their Co-Benefits

In Mumbwa, the successful mainstreaming and implementation of NBS was greatly facilitated by the professional knowledge and skills of local actors. The devolution of forestry, agriculture, and wildlife management departments to the local authority was particularly beneficial, as it ensured the availability of knowledgeable experts. These experts collaborated effectively with the town council, leading to the successful implementation of several NBS projects. This collaboration highlights the critical role that well-informed local actors play in the mainstreaming and success of NBS initiatives.

Governance and Stakeholder Engagement

Good governance and effective stakeholder engagement have also played significant roles in Mumbwa's NBS implementation. The decentralized governance system, in which the Mumbwa Town Council collaborates closely with communities and other stakeholders, reflects the literature's emphasis on co-creation, co-design, and multi-stakeholder collaboration as essential governance models for NBS success (Wamsler et al. 2017; Mahmoud et al. 2021). The study's findings confirm that an inclusive governance system, which promotes collaboration and participation, can significantly enhance the effectiveness of NBS implementation (Kabisch et al. 2016).

International Financial Support and Co-financing Mechanism

International financial support and co-financing mechanisms have been critical in supporting NBS projects in Mumbwa as except for urban greening which is in its initial stage, all the NBS implemented in Mumbwa were financed through international financial support. The international financial support and co-financing mechanisms are vital for NBS projects in regions like Mumbwa. They provide essential resources, technical expertise, and capacity building, ensuring the projects' sustainability and success. These funds help mitigate risks, attract additional investments, and integrate NBS into national policies, promoting global collaboration to address environmental challenges. Without such financial support, most projects in Mumbwa could have not been implemented (Sarabi et al. 2020).

4.4.2 Hindering Factors for NBS Mainstreaming and Implementation

Financial Capacity

Despite the positive impact of international funding, the lack of local financial resources significantly hampers NBS implementation in Mumbwa. Limited government incentives and minimal private sector and local citizen involvement exacerbate this issue. Most projects depend on international funding as local councils and communities lack the financial capacity. The Town Council allocates some budget to NBS, but these funds are often unavailable for project execution. The national government provides general grants but none specifically for NBS. Furthermore, poverty and inadequate income options lead locals to exploit ecosystems for income, such as charcoal production, resulting in deforestation. This overexploitation, coupled with insufficient local funding and low public awareness of NBS benefits, presents significant barriers to effective implementation. The reliance on international funding aligns with the broader literature, which points to limited local financial resources as a major challenge for NBS implementation (Seddon et al. 2020). As Acclimatise et al. (2020) and Cooper (2020) argue, municipalities often face financial constraints, with most of their budgets allocated to social services and infrastructure development. This limits the resources available for NBS projects, which require long-term investments. The study's findings support this, showing that the limited local financial capacity and inadequate government incentives have slowed down NBS implementation in Mumbwa such as the urban greening projects.

Land Availability and Conflicting Land Uses

Another major barrier identified in Mumbwa is the lack of available land for NBS projects, exacerbated by conflicting land uses and complex ownership structures. Population growth and rural–urban migration have led to illegal settlements in areas initially reserved for NBS, such as local stream and river catchment areas. This issue is further complicated by Zambia's dual land tenure system, where most land is under traditional authorities' customary tenure, while a smaller portion is managed by local councils under state or leasehold tenure. As a result, the Town Council must always negotiate with traditional authorities to open new areas for development including NBS projects. While traditional authorities are sometimes cooperative, they can also be resistant. In some instances, they allocate land reserved for NBS to local residents or businesses for residential or commercial use without consulting the local council. This often leads to conflicts, particularly in urban fringes. These challenges align with the literature, which highlights land-related issues as key obstacles to NBS implementation. In urban contexts, where land is contested, the lack of space for NBS initiatives can significantly hinder their adoption, especially when municipal regulations are inconsistent or land ownership issues are unresolved. Frequent monitoring inspections by local authorities aim to maintain these areas, but political circumstances add complexity to the situation.

Governance Challenges and Community Perception

While governance has been identified as a facilitating factor, certain challenges persist. The study found that community perceptions of NBS, particularly the view that these projects are the sole responsibility of local authorities, act as a hindrance to community involvement. This is consistent with the findings of Sarabi et al. (2020), who note that a lack of understanding of NBS benefits among local communities can limit their participation and support. Moreover, Seddon et al. (2020) observe that governance systems at the local level are often not fully established to promote NBS, which further complicates implementation efforts.

Environmental Factors

Environmental factors, including geographical location, ecological rainfall zones, and climate variability, were also identified as significant hindrances to the implementation of NBS. Mumbwa has experienced extreme climate variability, with alternating floods and droughts from year to year. These inconsistent climate patterns have hindered the effective implementation of NBS projects, such as afforestation.

4.4.3 The Extent of Influence of the Institutional Factors on NBS Mainstreaming and Implementation

The co-occurrence analyses performed revealed that, institutional factors have had a generally positive influence on the mainstreaming and implementation of NBS in Mumbwa, with more facilitating factors than hindering ones. The institutionalization of NBS, combined with supportive policies, inclusive planning, and effective governance, has enabled the successful mainstreaming of NBS into local development strategies. This supports the argument made by Nesshöver et al. (2017) and Mendes et al. (2020), who emphasize that strong institutional frameworks are key to the success of NBS projects. Overall, institutional factors have positively influenced NBS mainstreaming and implementation in Mumbwa to a higher extent. However, the study also highlights several challenges that need to be addressed, particularly the local financial constraints and land-related issues. The reliance on international funding underscores the need for greater local financial capacity and government incentives, as argued by Acclimatise et al. (2020). Additionally, resolving land ownership complexities and addressing conflicting land uses will be critical for scaling up NBS in Mumbwa and similar urban contexts (Kalantari et al. 2018).

4.5 Final Consideration

With the lessons learned from the NBS mainstreaming and implementation in Zambia, this study has proposed a conceptual framework to highlight specific indicators of institutional factors facilitating and hindering NBS mainstreaming and implementation in Mumbwa as shown in Fig. 7. In this framework, NBS mainstreaming, and implementation are placed in a staked Venn diagram to indicate an overlapping relationship between the two. With mainstreaming placed in the outer circle and implementation in the inner circle, it indicates that initiatives are effectively implemented when they are part of the mainstream. However, there is also room for implementing projects outside what is mainstreamed in the plans and policies in the case of ad hoc projects. This is why the implementation circle is not a complete subset of the mainstreaming circle. This Conceptual framework also places NBS benefits and co-benefits as a subset of implementation to indicate that, the benefits and co-benefits of NBS are not realized merely by mainstreaming them in the policies. It is only when the initiatives are effectively implemented that their full benefits and co-benefits can be realized. Also, the study discovered that facilitating and hindering factors go beyond institutional factors but include environmental and socio-economic factors as shown in the conceptual framework.

5 Conclusions and Recommendations

This concluding chapter highlights the substantial progress made by the Mumbwa Town Council in Zambia in integrating Nature-Based Solutions (NBS) into local development frameworks to address climate change. Key initiatives such as sustainable forest management, water ecosystem conservation, urban greening, and specific projects like SCRiKA, REDD+, and AWARE demonstrate the Council's commitment to resilience and sustainability. The role of institutional factors, including supportive policies and inclusive planning, has been instrumental in the successful implementation of these solutions. Despite notable challenges such as limited financial capacity and environmental vulnerabilities, the study offers recommendations for enhancing NBS effec-

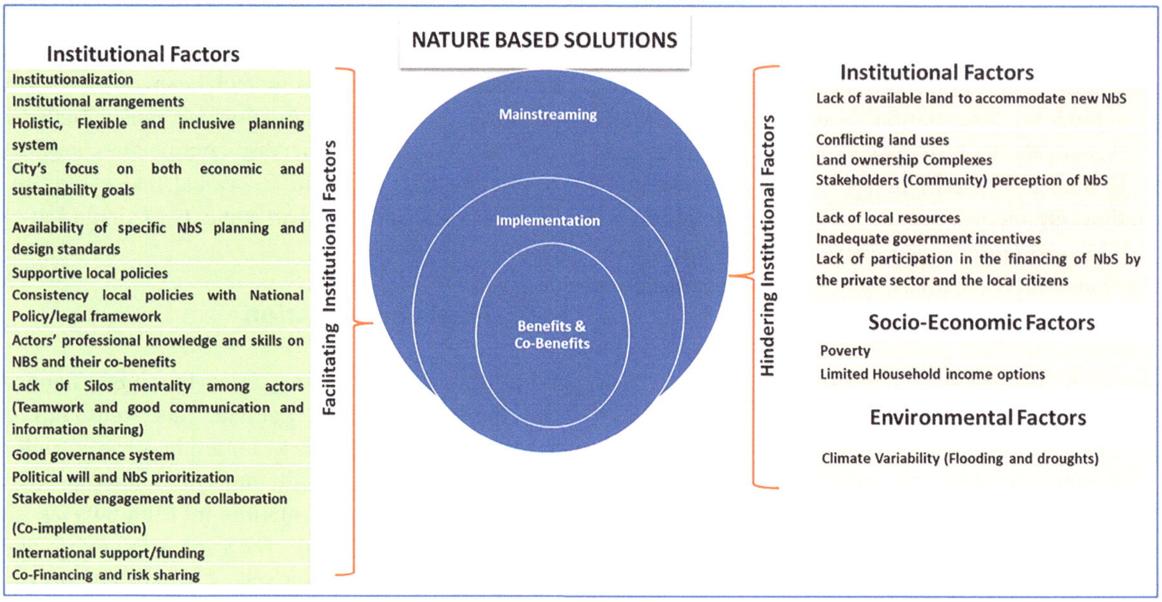

Fig. 7 Conceptual framework of the institutional factors that have influenced the mainstreaming and implementation of NBS in Mumbwa. *Source* Developed by the author, 2021

tiveness through financial capacity building, strengthened government support, international collaboration, community empowerment, and diversification of NBS types. This integrated approach aims to promote sustainable development, environmental conservation, and improved livelihoods in Mumbwa.

5.1 Recommendations

To overcome the NBS mainstreaming and implementation barriers revealed by the findings, and further enhance the effectiveness of NBS in Mumbwa, the following recommendations are made:

5.1.1 Financial Capacity Building

One of the key challenges identified is the limited financial capacity of local authorities. To address this, there is a pressing need for financial capacity building at the local level. The Mumbwa Town Council should actively seek diversified funding sources, including local revenue streams, public–private partnerships, and external grants from international organizations. Additionally, building the capacity of local staff to effectively access and manage climate finance will enhance their ability to secure resources for ongoing and future NBS projects.

5.1.2 Strengthened Government Support

The continued support of the national government and policymakers is critical to the sustained success of NBS initiatives. The Council should advocate for greater national financial and technical assistance, ensuring that NBS remains a national development priority. Enhancing partnerships with international funding agencies, such as the Green Climate Fund and Global Environment Facility, will bolster efforts to secure the necessary financial backing. Government policies should also foster local ownership and accountability in NBS projects, ensuring that local communities can effectively engage in and benefit from these initiatives.

5.1.3 International Collaboration and Networks

To further support the scaling of NBS, the Council should actively participate in international networks and platforms that focus on climate resilience and sustainable development, such as ICLEI-Local Governments for Sustainability and the Global Covenant of Mayors for Climate and Energy. These platforms provide valuable opportunities for knowledge exchange, technical expertise, and funding opportunities, allowing Mumbwa to leverage global best practices and innovations in NBS implementation.

5.1.4 Community Empowerment and Climate Adaptation

With rural communities highly dependent on activities such as charcoal production for livelihoods, it is essential to enhance climate adaptation efforts. Targeted programs, such as SCRiKA, should be expanded to build the adaptive capacity of vulnerable populations. By integrating climate-smart agriculture, alternative livelihood options, and sustainable energy solutions, these initiatives can reduce the dependence on environmentally harmful practices while simultaneously improving community resilience to climate impacts. Empowering local communities through capacity-building programs will ensure they are active participants and beneficiaries of NBS projects.

5.1.5 Exploration of Diverse NBS Types

Current NBS initiatives in Mumbwa primarily focus on Type 1 (natural ecosystems) and Type 2 (managed or restored ecosystems) solutions. To broaden the scope of climate mitigation and adaptation strategies, the Council should invest in research and development to explore Type 3 NBS, which focuses on technological integration with natural systems. This might include exploring innovations such as green roofs, constructed wetlands, and urban farming systems. By adopting a more comprehensive and diversified approach, Mumbwa can address both rural and urban climate challenges more effectively.

5.1.6 Carbon Credits and Rural Communities

A significant opportunity exists in developing community-based carbon projects that allow rural populations to directly benefit from carbon credit revenues. Establishing community carbon sequestration projects can empower rural communities while contributing to global carbon markets. Such projects not only incentivize sustainable land-use practices but also create economic opportunities for local populations, transforming carbon sequestration into a livelihood diversification tool. These initiatives can contribute to Zambia's national climate goals while promoting sustainable development at the grassroots level.

5.2 Conclusion

The study highlights that Mumbwa Town Council in Zambia has made substantial progress in mainstreaming and implementation of NBS, particularly concerning climate change mitigation and adaptation. Initiatives such as sustainable forest management, water ecosystem conservation, urban greening, and projects like SCRiKA, REDD+, and AWARE exemplify the Council's proactive efforts to address the multifaceted challenges posed by climate change. The NBS projects currently underway not only enhance the region's climate resilience but also open up new economic opportunities for rural communities, particularly through the generation of carbon credits. Besides, these NBS interventions are well-aligned with the Council's broader developmental strategies, signaling a strong commitment to fostering resilience and sustainability at the local level. The role of institutional factors in enabling the mainstreaming and implementation of NBS is also notable. Facilitating elements such as the institutionalization of NBS, supportive policies, and inclusive planning frameworks have contributed to the successful deployment of these solutions. Good governance, stakeholder engagement, and cross-sectoral collaboration also emerged as crucial enablers. However, the study also identifies significant challenges, including limited financial capacity, environmental vulnerabilities, and socio-economic constraints, all of which hinder the full-scale implementation of NBS initiatives. However, by fostering strong local governance, enhancing financial capacity, and building international collaborations, the Mumbwa Town Council can overcome these challenges and solidify its position as a leader in NBS implementation in Zambia, ensuring both environmental conservation and improved livelihoods for its residents. Overall, the steps taken by Mumbwa Town Council to mainstream and implement NBS have proven to be a pivotal strategy for addressing climate change impacts while promoting economic development and environmental sustainability.

Acknowledgements I wish to thank the Almighty God who made all things possible. Secondly, I would like to thank the institutions and individuals who gave me support in the course of this study. My gratitude goes to the Orange Knowledge Program (OKP) Scholarship Board for funding my study. My debt of gratitude is also extended to my employer the Local Government Service Commission for granting me the study leave and to Chinsali Municipal Council, in particular Mr. Joseph Zulu, Town Clerk for the recommendations and for facilitating my study leave. My sincere appreciation also goes to my duty station Mumbwa Town Council, for releasing me to further my studies, especially at a time when my presence was greatly needed. To the Institute of Housing and Urban Studies, Erasmus University Rotterdam, thank you for offering me an opportunity to study at the faculty. I am also very grateful to the lecturers and tutors at IHS for introducing me to various concepts of Urban Management and Development and climate change and sustainability.

Permissions Not applicable since this work is not published anywhere other than the University repository, on which outstanding theses are published and where the main thesis is also stored.

Competing Interests The author declares that there are no conflicts of interest regarding the publication of this paper. This research was conducted with the support of the Orange Knowledge Program Scholarship (OKP), which provided funding for overall educational expenses in terms of tuition fees, accommodation, book allowance, meal allowance, and air tickets to and from my home country (Zambia) to study in the Netherlands. The scholarship however did not fund research expenses. Thus, no financial or personal relationships have influenced the outcomes and interpretations of this research.

Ethics Approval Informed consent was obtained from the key informants who participated in the interviews and their identity was kept confidential.

References

Allen, M., Antwi-Agyei, P., Aragon-Durand, F., Babiker, M., Bertoldi, P., Bind, M., et al.: Technical summary: global warming of 1.5°C. An IPCC special report on the impacts of global warming (2019)

Acclimatise: Nature-Based Solutions: Finding Solutions in Nature for Climate Change (Internet). Building Climate Resilience, Acclimatise (2020) (cited 2024 Oct 4). Available from: http://www.acclimatise.uk.com/tag/nature-based-solutions/

Bartlett, R., Mistry, J.: The contributions of NBS to Urban resilience. In: Nature-Based Solutions for More Sustainable Cities—A Framework Approach for Planning and Evaluation. Emerald Publishing Limited, p. 11–20 (2022)

Cohen-Shacham, E., Andrade, A., Dalton, J., Dudley, N., Jones, M., Kumar, C., et al.: Core principles for successfully implementing and upscaling nature-based solutions. Environ Sci Policy **98**, 20–29 (2019)

Cohen-Shacham, E., Walters, G., Janzen, C., Maginnis, S.: Nature-Based Solutions to Address Global Societal Challenges. IUCN, Gland, Switzerland (2016)

Cooper, R.: Nature-Based Solutions and Water Security: Best Practice Examples of Nature-Based Solutions for Water Security. K4D Knowledge, Evidence and Learning for Development Help Desk, UK (2020)

Dunlop, T., Khojasteh, D., Cohen-Shacham, E., et al.: The evolution and future of research on nature-based Solutions to address societal challenges. Commun. Earth Environ. **5**, 132 (2024)

Eggermont, H., Balian, E., Azevedo, J.M., Beumer, V., Brodin, T., Claudet, J., et al.: Nature-based solutions: new influence for environmental management and research in Europe. Gaia **24**(4), 243–248 (2015). https://doi.org/10.14512/gaia.24.4.9

Field, C.B.: Impacts, adaptation, and vulnerability: Part A: global and sectoral aspects. Working Group II contribution to the fifth assessment report of the intergovernmental panel on climate change. Cambridge University Press, New York (2015)

Frantzeskaki, N.: Seven lessons for planning nature-based solutions in cities. Environ Sci Policy **93**, 101–111 (2019)

Government of The Republic of Zambia. Zambia National Policy on Climate Change. Lusaka: PMRC Zambia; (2016)

Government of The Republic of Zambia: The Seventh National Development Plan 2017–2021. The Republic of Zambia, Lusaka (2017)

International Union for Conservation of Nature (IUCN). IUCN global standard for nature-based solutions: a user-friendly framework for the verification, design, and scaling up of NBS. IUCN, Gland, Switzerland (2020)

Kabisch, N., Frantzeskaki, N., Pauleit, S., Naumann, S.: Nature-based solutions to climate change mitigation and adaptation in urban areas: perspectives on indicators, knowledge gaps, barriers, and opportunities for action. Ecol. Soc. **21**(2), 39 (2016)

Kalantari, Z., Ferreira, C.S., Keesstra, S., Destouni, G.: Nature-based solutions for flood-drought risk mitigation in vulnerable urbanizing parts of East Africa. Environ Sci Health. **5**, 73–78 (2018)

Mahmoud, I., Morello, E.: Co-creation pathway for urban nature-based solutions: testing a shared governance approach in three cities and nine action labs (2021)

Mumbwa Town Council. Mumbwa District Development Plan (DDP) 2017–2021. Ministry of Local Government (2017)

Magni, F., Musco, F., Litt, G., Carraretto, G.: The mainstreaming of NBS in the SECAP of San Donà di Piave: the LIFE master adapt methodology. Sustainability **12**(23), 10080 (2020)

Mendes, R., Fidélis, T., Roebeling, P., Teles, F.: The institutionalization of nature-based solutions—a discourse analysis of emergent literature. Resources **9**(1), 6 (2020)

Maes, J., Jacobs, S.: Nature-based solutions for Europe's sustainable development. Conserv. Lett. 1–4 (2015)

Mosca, F., Canepa, M., Perini, K.: Strategies for adaptation to and mitigation of climate change: key performance indicators to assess nature-based solutions performances. Urban Climate. **48**, 101506 (2023)

Mumbwa Town Council: District Development Plan. Mumbwa, Republic of Zambia, Government Printers (2017)

Nelson, D.R., Bledsoe, B.P., Ferreira, S., Nibbelink, N.P.: Challenges to realizing the potential of nature-based solutions. Curr. Opin. Environ. Sustain. **45**, 49–55 (2020)

La Notte, A., Zulian, G.: An ecosystem services-based approach to frame NBS in Urban context. In: La Notte, A., Zulian, G. (eds.) Nature-Based Solutions for More Sustainable Cities—A Framework Approach for Planning and Evaluation, pp. 47–65. Emerald Publishing Limited (2022)

Nesshover, C., Assmuth, T., Irvine, K.N., Rusch, G.M., Waylen, K.A., Delbaere, B., et al.: The science, policy, and practice of nature-based solutions: An interdisciplinary perspective. Sci. Total. Environ. **1**(579), 1215–1227 (2017)

Ochieng, S.B., Charles, O.A., Ang'awa, O.F.: Effect of flooding on the socio-economic livelihood of the farmers in Lower Kano Plain, Kisumu County, Kenya. Int. J. Nov. Res. Interdiscip. Stud. **4**(4), 8–16 (2017)

Raymond, C.M., Frantzeskaki, N., Kabisch, N., Berry, P., Breil, M., Nita, M.R., et al.: A framework for assessing and implementing the co-benefits of nature-based solutions in urban areas. Environ Sci Policy **77**, 15–24 (2017)

Sarabi, S., Han, Q., Romme, A.G.L., de Vries, B., Valkenburg, R., den Ouden, E.: Uptake and implementation of nature-based solutions: An analysis of barriers using interpretive structural modeling. J Environ Manag. **270**, 110749 (2020)

Seddon, N., Chausson, A., Berry, P., Girardin, C.A., Smith, A., Turner, B.: Understanding the value and limits of nature-based solutions to climate change and other global challenges. Philos. Trans. R. Soc. B **375**(1794), 20190120 (2020)

United Nations: Climate change: exacerbating poverty and inequality. In: World Social Report 2020. United Nations Publications, New York, pp. 82–106 (2020)

United Nations Environment Programme (UNEP). Zambia turns to nature to tackle climate change (Internet). Global Adaptation Network; 2021 (cited 2024 Oct 4). Available from: https://www.unep.org/gan/news/press-release/zambia-turns-nature-tackle-climate-change

University of Oxford. Global NDC-Your Nation: Zambia (Internet). Nature-based Solutions Policy Platform; 2021 (cited 2024 Oct 4). Available from: https://www.NBSpolicyplatform.org/your-nation/?country=262

Wamsler, C., Pauleit, S., Zölch, T., Schetke, S., Mascarenhas, A.: Mainstreaming nature-based solutions for climate change adaptation in urban governance and planning. In: Nature-Based Solutions to Climate Change Adaptation in Urban Areas (pp. 257–273). Springer, Cham (2017)

Wang, Z., Huang, L., Xu, M., Wang, S.: Bridging the science-practice gaps in nature-based solutions: a riverfront planning in China. Ambio, pp. 1–19 (2021)

Winsemius, H.C., Jongman, B., Veldkamp, T.I., Bangalore, M., Ward, P.J.. Disaster risk, climate change, and poverty: assessing the global exposure of poor people to floods and droughts. Environ. Dev. Econ. 1355–770X (2018)

Zhao, Y., Weng, Z., Chen, H., Yang, J.: Analysis of the evolution of drought, flood and drought-flood abrupt alternation events under climate change using the daily SWAP index. Water **1**, 23 (2020)

Zingraff-Hamed, A., Hüesker, F., Albert, C., Brillinger, M., Huang, J., Lupp, G., et al.: Governance models for nature-based solutions: seventeen cases from Germany. Ambio, 1–18 (2020)

Sustainable Growth and Development of Towns in a Dual Land Tenure System in Zambia: A Case of Lundazi Town

Kalapula Marrien, Nchito Wilma, and Kajoba Gear

Abstract

Zambia has experienced rapid urban growth, primarily characterized by informal development. This study identifies the dual land tenure system, comprising customary and leasehold tenure, as the principal factor hindering the sustainable growth and development of Lundazi town. The research focuses on establishing how this dual land tenure system impacts Lundazi and other regional capitals facing similar challenges. Qualitative methods were employed for data collection and analysis, including interviews with developers in both state and customary areas and structured interviews with key informants. Qualitative data were analyzed using content analysis, while the geographical information system (GIS) was used to produce maps illustrating the findings. Findings reveal that the dual land tenure system has both negative and positive consequences on Lundazi's growth and development, with the negatives outweighing the positives. The fact that the town is surrounded by customary land impedes smooth expansion, limiting the capacity of spatial planning authorities and resulting in haphazard development and proliferation of informal settlements. Addressing these challenges requires integrated development planning (IDP), enabling collective planning encompassing both leasehold and customary tenure. Additionally, a comprehensive land administration framework is recommended to harmonize the dual land tenure system and integrate them into spatial planning, aligning with the objectives of the Sustainable Development Goal (SDG) 11.

Keywords

Urban growth and development · SDGs · Dual land tenure system · Integrated development planning · Rural–urban fringe · Spatial planning

List of Abbreviations

CSO	Central statistics office
DC	District commissioner
DPPH	Department of physical planning and housing
EPPA	Eastern province planning authority
GIS	Geographical information system
GLTN	Global land tool network
IDP	Integrated development planning
LDC	Lundazi district council
LGSC	Local government service commission
MLGH	Ministry of local government and housing
MLNREP	Ministry of lands, natural resources and environmental protection
SDG	Sustainable development goals
UN	United nations
URP Act	Urban and regional planning act
UNICEF	United nations international children's emergency fund
UNZA	University of Zambia
USAID	United States agency for international development
ZANIS	Zambia news and information services

1 Introduction

Urban growth, which refers to an increase in the proportion of a population living in urban areas and the physical expansion of existing urban centers (Cain 2023), is a global concern. This

K. Marrien (✉) · N. Wilma · K. Gear
Department of Geography and Environmental Studies, School of Natural Sciences, University of Zambia, Lusaka, Zambia
e-mail: kmarrien@yahoo.com

P. Mohammad et al. (eds.), *Geographic Approaches to Climate Change and Mitigation: Urban and Rural Perspectives (Volume 2)*, Advances in Science, Technology & Innovation, https://doi.org/10.1007/978-3-031-92115-5_10

trend of urbanization has steadily increased over the years, with the milestone of more people living in urban areas than rural areas being reached around 2007 (Ritchie et al. 2024; Bloch 2012). In 2010, urban areas housed 3.5 billion people, representing 50.5% of the world's population (Bloch 2012). As of 2024, this number has risen to over 4 billion, which accounts for more than half of the global population (Ritchie et al. 2024). While urbanization rates in Europe and North America have plateaued, rapid urban population growth is projected in Africa and Asia. A 2018 United Nations report predicts that by 2050, these regions will collectively host 5 billion urban dwellers, with 3.5 billion in Asia and 1.5 billion in Africa (United Nations 2019). This rapid urbanization in Africa, including Zambia, has resulted in significant urban spatial expansion as urban centers grow outward to accommodate population increases and infrastructure needs (Yussef et al. 2023; Payne and Paskowitz 2014; Nchito 2013). However, such expansion into peri-urban areas raises critical challenges related to land tenure systems (Taylor and Thole 2015; Lombard 2012; Bloch 2012).

In countries with efficient land tenure and planning systems, such as the United Kingdom, the United States, or Australia, urban planners proactively account for urban growth by facilitating boundary expansions and ensuring adequate service infrastructure (United Nations 2023a, b, c; M'Membe 2013; Bloch 2012). By contrast, in Zambia and other Sub-Saharan African countries, the ambiguous and often conflicting dual land tenure systems comprising leasehold and customary tenure pose significant challenges to effective urban growth and spatial planning (Yussif, Dompreh, and Gasparatos 2023; Chitonge 2018; Weldegebriel et al. 2021). Leasehold tenure governs state land, typically where towns are established, while customary tenure governs peri-urban areas, the natural direction for urban expansion. Local authorities often lack the mandate to manage land under customary tenure, hindering urban planning and development. This dual system has forced many Zambian municipalities to address urban growth pressures by revising municipal boundaries to incorporate peri-urban and rural settlements into urban limits (Bloch 2012; Yussif et al. 2023). However, expanding urban boundaries into customary land introduces complexities that can exacerbate informal urban growth and complicate service delivery, as observed in towns like Lundazi.

In Zambia, informal urban growth is reflected in the proliferation of informal settlements, which house approximately 70% of the urban population and lack adequate housing, water, and sanitation services (UN-Habitat 2023a, b; USAID 2010). Studies in other Sub-Saharan African countries, such as Kampala in Uganda, highlight the critical role of land tenure systems in shaping urban growth patterns (Muinde 2013). Ineffective land tenure systems hinder sustainable urban development by creating barriers to land access and regulatory compliance. Despite these findings, there has been limited empirical research in Zambia specifically linking dual land tenure systems to urban development challenges. Most studies in the country attribute informal urban growth to inefficient land-use planning legislation and practices while overlooking the role of land tenure. This research addresses this gap by examining how the dual land tenure system affects the growth and development of Lundazi town. The study seeks to demonstrate that if the issue of dual land tenure is not addressed, local authorities will lose critical opportunities to manage urban growth effectively. This would result in unplanned urban development characterized by challenges such as slum proliferation, environmental degradation, transportation problems, and poor public health outcomes.

The study aims to establish the effects of the dual land tenure system on the growth and development of towns in Zambia, using Lundazi town as a case study. Specifically, it explores the characteristics of the dual land tenure system in Lundazi, the influence of this system on the town's morphology, and the urban land management challenges faced by local authorities. To address these issues, the study poses the following research questions: What are the characteristics of the dual land tenure system in Lundazi? How does this system affect the morphology of the town? What urban land management challenges do local authorities face in such a system?

By highlighting these challenges, this study contributes to the understanding of how dual land tenure systems hinder sustainable urban development. The findings can inform policymakers in devising strategies to harmonize land tenure systems with urban planning frameworks, thereby promoting inclusive, safe, resilient, and sustainable urban growth in alignment with SDG 11 (United Nations 2023a, b, c). Additionally, the study sheds light on the challenges faced by urban land management authorities, providing insights that could help address the broader urban development issues in Zambia and similar contexts.

2 Literature Review

This chapter explores the historical and contemporary dynamics of Zambia's dual land tenure system and its impact on urban development. Land tenure plays a critical role in shaping social, economic, and political structures, particularly in the context of rapid urbanization. The review examines the evolution of land tenure in Zambia, the challenges posed by the coexistence of customary and statutory systems, and their implications for urban growth and planning. By identifying key gaps in the administration and management of land, this chapter provides the foundation for understanding the unique challenges faced by Lundazi town, addressing the study's research objectives.

2.1 History of Land Tenure System in Zambia

Zambia operates a dual land tenure system consisting of customary tenure and leasehold tenure systems (UN-Habitat 2023a, b; USAID 2018; Sichone 2010; USAID 2010). Before colonial rule, all land in Zambia was managed under the customary land tenure by the chiefs and village headmen on behalf of the community members (Mudenda 2007). With British colonization, the British South Africa Company (BSA Co.) introduced English land tenure principles, creating Crown Lands for settlers and Native Reserves for indigenous people (Mudenda 2007; Sichone 2010). This division limited indigenous land access, segregating Crown Lands for settlers while assigning less fertile Native Reserves to the local population (UN-Habitat 2023a, b; USAID 2018; Sichone 2010; Mudenda 2007). Upon independence, Zambia retained this dual system, converting freehold titles to leasehold and designating Crown Land as state land through the Conversion of Titles Act of 1975. The act also consolidated Native Reserves and Native Trust Lands as customary lands, establishing the 94% customary, 6% leasehold division (UN-Habitat 2023a, b; USAID 2018; Tagliarino 2014; Sichone 2010). As of today, this distribution remains largely unchanged, with approximately 94% of land still under customary tenure and 6% under leasehold tenure despite some customary lands being reclassified to leasehold for urban expansion (UN-Habitat 2023a, b; USAID 2018).

2.2 Current Dual Land Tenure System in Zambia

The current land tenure system in Zambia is still comprised of both leasehold and customary tenure (UN-Habitat 2023a, b; USAID 2018). Leasehold tenure allows for formal registration and enhanced security, facilitating access to loans (UN-Habitat 2023a, b; USAID 2018; Chinene et al. 2010; Adams 2003). Leaseholders are also required to develop the land within a set time and pay annual ground rent, yet all transactions require Presidential consent, restricting tenants' freedoms (Chinene et al. 2010; Mudenda 2007). Customary tenure remains community-based, managed by chiefs and headmen, with rights granted through traditional practices and lacking statutory protection (UN-Habitat 2023a, b; USAID 2018; Tagliarino 2014; Van Loenen 1999; Adams 2003; Sichone 2010).

2.3 Institutional and Legal Framework

The administration of both leasehold and customary tenure is managed by the Ministry of Lands, Natural Resources and Environmental Protection, local authorities, and traditional authorities, with legislation including the Land Policy, Lands Act, the Lands and Deeds Registry Act, and the Urban and Regional Planning Act (URP Act) guiding land management (Umar and Nyanga 2023; USAID 2020; Antonio et al. 2023; M'Membe 2013; Sichone 2010).

2.4 Urban Growth and Development Challenges

Urban areas in Zambia, like many in Africa, experience rapid expansion driven by rural–urban migration and population growth, resulting in urban sprawl and informal settlements (Ritchie et al. 2024; Yussif et al. 2023; Payne and Piaskowy 2014; Tembo 2014; Nchito 2013; Nsiah-Gyabaah 2003). Zambia's town planning systems struggle to keep pace with this growth, leading to unplanned urbanization, especially in peri-urban areas where customary land is converted for state use (UN-Habitat 2015; UNICEF 2012; Taylor and Thole 2015).

2.5 Urban Growth and the SDG 11

The United Nations SDG 11 aims to "Make cities inclusive, safe, resilient and sustainable" (UN-Habitat 2023a, b; Connor 2015). This goal underscores the importance of effective urban planning, land administration, and infrastructure development to manage urban growth responsibly, ensuring cities evolve in a way that supports all residents. In the Zambian context, where rapid urbanization is met with a dual land tenure system, achieving SDG 11 requires addressing challenges associated with both customary and leasehold land tenures and their implications for inclusive and sustainable urban growth. SDG 11 targets the creation of urban areas that support inclusivity, safety, resilience, and sustainability. Key indicators for this goal include the accessibility of safe housing, improved urban planning practices, and sustainable land use that mitigates the risks of informal settlements and urban sprawl (United Nations 2015). Zambia's urban growth, however, often extends into peri-urban and rural areas where land is primarily held under customary tenure, complicating municipal jurisdiction and planning efforts (Tembo 2014; UN-Habitat 2015).

For instance, urban sprawl in Zambia is frequently characterized by unplanned settlements, highlighting the need for urban planning reforms that integrate traditional landholding systems with statutory land administration (Taylor and Thole 2015). The dual land tenure system poses unique challenges for SDG 11's objectives, particularly in creating equitable access to land for housing and infrastructure development. The process of obtaining land for urban expansion often requires negotiations with traditional authorities, which can delay or prevent the transfer of customary land to state-held leasehold tenure, essential for municipal planning. This situation results in the proliferation of informal settlements, limited infrastructure, and spatial inequality in many urban areas (ZANIS 2012; Times of Zambia 2014a, b). In the absence of integrated land-use planning across tenure types, cities struggle to achieve resilient growth patterns that can withstand economic, social, or environmental shocks (Tembo 2014).

Furthermore, SDG 11's emphasis on resilience highlights the role of secure land tenure in mitigating urban vulnerability to risks such as displacement and land insecurity. Customary tenure, which often lacks formal registration, provides limited legal protections, leaving communities vulnerable to eviction or expropriation without compensation (UN-Habitat 2023a, b; United Nations 2015; Adams 2003; Tagliarino 2014). In contrast, leasehold tenure provides formalized rights that enhance security for investment and development (Chinene et al. 2010). However, leasehold tenure's requirement for presidential consent for land dealings can limit urban residents' flexibility in land transactions, reducing the efficiency of land markets and impacting resilience (UN-Habitat 2023a, b; USAID 2018; Chinene et al. 2010; Mudenda 2007). In working toward SDG 11, Zambia could benefit from land administration reforms that harmonize statutory and customary systems, allowing for more consistent urban planning and the provision of secure tenure across both tenure types. Enhanced collaboration between municipal and traditional authorities, as well as policy frameworks supporting land registration and land-use planning in customary areas, could facilitate more inclusive and sustainable urban growth (UN-Habitat 2023a, b; USAID 2018; Bloch 2012). Integrated land management that bridges customary and leasehold systems is essential to ensure that urban expansion supports equitable access to resources and resilient infrastructure, advancing Zambia toward the SDG 11 objectives of sustainable and inclusive cities.

2.6 The Link Between Urban Growth and Land Tenure

Land tenure is integral to urban growth, especially as municipal boundaries expand into customary lands, requiring negotiation between urban authorities and traditional custodians (Deininger and Goyal 2024; Chigbu 2021; Muinde 2013; Lombard 2012; Bloch 2012). These tenure rules shape land access, development regulations, and the security of land rights, all influencing investment and land use in urban fringe areas (Fekade 2000; ZANIS 2012; Times of Zambia 2014a, b). For instance, challenges arise when councils must negotiate with chiefs to acquire customary land for urban projects, as seen in the cases of Kapiri Mposhi and Mkushi (Times of Zambia 2014a, b; ZANIS 2012).

3 Material Methods

This chapter outlines the research methods and materials employed to investigate the effects of a dual land tenure system on the growth and development of towns. A qualitative research approach was adopted, and a case study design was chosen to provide a detailed understanding of the unique challenges and opportunities presented by the dual land tenure system in Zambia, enabling the collection of context-specific data to address the research questions. This approach facilitated an in-depth exploration and analysis of land tenure dynamics and their implications for urban development and Lundazi District, located in Eastern Province, Zambia, served as the study area for this research as shown in Fig. 10.1.

As shown in the figure, Lundazi district borders Malawi to the east and shares boundaries with Chasefu and Lumezi districts, which were part of Lundazi until 2018 when they achieved district status. Lundazi was selected as the focus of this study due to its exemplification of the challenges associated with urban informal growth driven by the complexities of a dual land tenure system. These characteristics make it an ideal context for examining the implications of such systems on urban development in Zambia.

The study targeted key informants from institutions dealing with land administration in Lundazi, as well as developers operating in both state land and customary land. A total of 57 respondents participated in the study, including seven key informants, 10 village headmen from Mphamba Chiefdom, 20 developers in state land, and 20 developers in the rural–urban fringe. Purposive sampling was used to

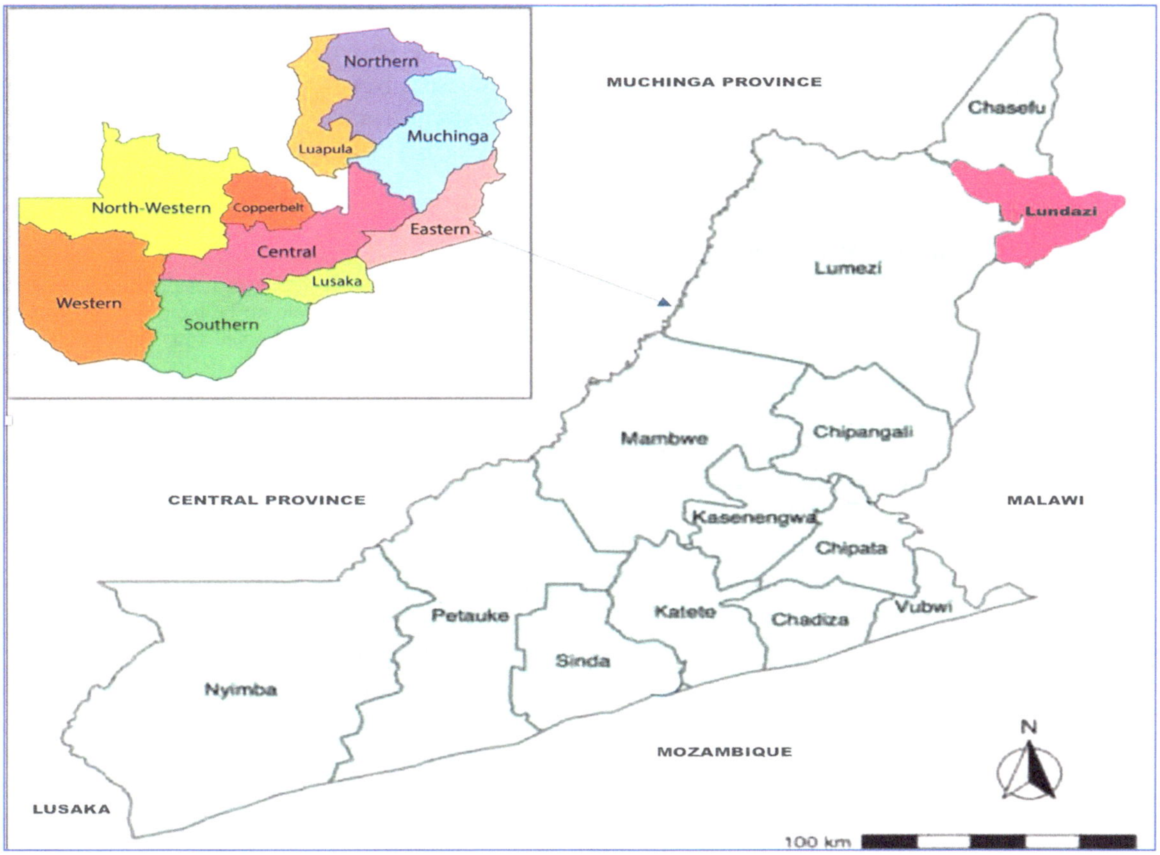

Fig. 10.1 Location map of Lundazi. *Source* Google Image

select key informants and village headmen, while snowball sampling was employed to identify developers.

Data collection methods included semi-structured interviews with experts and officials from various institutions, such as the Lundazi District Council (LDC), Eastern Province Planning Authority (EPPA), the Ministry of Lands, Natural Resources and Environmental Protection (MLNREP), the District Commissioner's Office (DC), the District Department of Agriculture, the Ministry of Chiefs and Traditional Affairs, and the Royal Highness of Mphamba Chiefdom. Interviews were also conducted with developers working in both customary land at the urban fringe and state land within the urban boundary. Semi-structured interview guides with open-ended questions provided a framework for these sessions while allowing respondents to express their perspectives freely.

Additionally, a focus group discussion was held with 10 village headmen from Mphamba Chiefdom, whose villages were affected by the extension of Lundazi's urban boundary. A focus group discussion guide facilitated the session, ensuring a structured but open dialogue. The study also included the review of council records, aerial imagery, and field observations to gather supplementary data on land-use patterns and urban development.

Data analysis was conducted using multiple methods. Content analysis was applied to data from interviews, focus group discussions, and document reviews, allowing for the systematic examination of themes and patterns. Spatial data, derived from interpreting Lundazi Google Earth imagery, topographic sheets, and town layout plans, were analyzed using the GIS software. This combination of qualitative and spatial analysis provided a comprehensive understanding of the effects of the dual land tenure system on urban growth and development in Lundazi.

4 Results

This chapter presents the findings of the study, highlighting the effects of the dual land tenure system on the growth and development of Lundazi town. The research revealed both positive and negative impacts, with the latter being more pronounced. While the dual system has contributed to some modernization in the town, the lack of integrated development planning has emerged as a critical factor driving urban informality and uncoordinated development. The findings delve into key areas such as land rezoning, urban fringe transformation, rural gentrification, informal settlements, and the challenges posed by the absence of a cohesive land administration framework. Details of the findings are as follows.

4.1 Rezoning of Open Spaces into Commercial and Residential Developments

The dual land tenure system in Lundazi comprising of leasehold tenure and customary tenure is predominantly characterized by customary land encircling the town. This leads to a scarcity of urban land. The growing need for residential and commercial buildings has prompted the LDC to convert some open spaces, originally designated for social activities and environmental conservation, such as parks and playgrounds, into areas for commercial and residential development.

Figure 10.2 depicts several open spaces that have been converted due to the lack of urban land. For instance, Vigodo grounds, marked as open space number 1, were traditionally used for district events like political gatherings and various national and international celebrations, including Independence Day and International Women's Day. In 2009, this area was reclassified and divided into commercial and special-use plots.

Open spaces 2 and 3, previously inner-city play areas, were converted into commercial plots for the bank and service stations, respectively. Open space 4, once the district airstrip, was transformed into a mixed-use area for housing, business, and special uses. The Daily Mail Newspaper in 2014 reported:

> Lundazi District is witnessing swift residential development on its undulating hills, a sign of a thriving economy. Council secretary Boyd Kaoma shared plans for Lundazi's future with the author, including a new town center with a mall, bank, and upscale residential area named Kanele, to be built on the current airstrip site. The airstrip will be relocated away from residential zones (Zambia Daily Mail 2014).

Furthermore, open space 5, which served as a buffer for the airstrip and was known as "Prisons' Farm" due to its temporary use by the Lundazi Prisons Service for gardening, was rezoned into 96 upscale residential plots and a special plot for a secondary school and renamed it as "Sunnydale Garden," and open space 6, a local playground in the Dunda informal area, was also rezoned and divided into residential plots, now known as "Dunda Planned Settlement."

Analysis of the findings shows that rezoning of open spaces into residential and commercial uses has affected the liveability of Lundazi town. As indicated by Balogh and Takacs (2011), open spaces provide environmental, economic, and social benefits critical to sustaining the health of the town. Thus, sustainable towns and cities are comprised of more than just buildings and people as they encompass open spaces

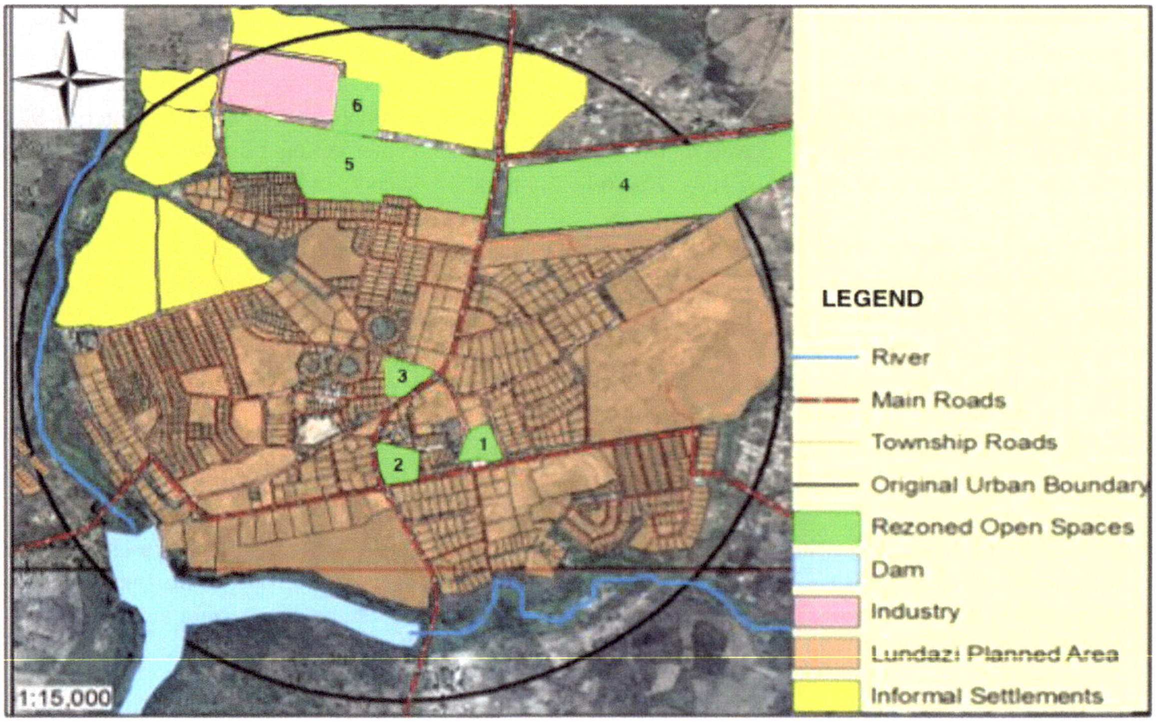

Fig. 10.2 Converted open spaces in Lundazi. *Source* Author 2016

such as playgrounds, play parks, and others. In fact, the most liveable cities and some of the world's most famous cities like Amsterdam, Berlin, Stockholm, Helsinki, New York, and others are known for their open spaces as they provide attraction for both their local and international tourists. The benefits of open spaces in urban areas among others include the provision of space for recreation, sport and cultural facilities, and the preservation of natural environments which are also associated with health benefits (Balogh and Takacs 2011). Therefore, the preservation of open spaces must be a key consideration in urban planning if sustainable growth of Lundazi town is to be achieved.

4.2 Transformation of the Urban Fringe and Uncoordinated Developments

Shortage of urban land has also caused individual developers mainly Lundazi urban residents to resort to acquiring customary land in the rural–urban fringe. This has led to the expansion of Lundazi urban into the surrounding customary lands resulting in the urbanization of the urban fringe areas.

Positively, the expansion of the Lundazi urban and consequently the urbanization of the fringe areas has led to spatial improvement of the rural–urban fringe as modern infrastructure such as housing, lodges, commercial infrastructure, and light industries has developed in these areas as shown in Fig. 10.3 which is a huge transformation from the traditional village structures (Fig. 10.4) and a rural lifestyle that once dominated these areas.

Consequently, this development has led to increased employment opportunities for the residents of the rural–urban fringe areas as their economy diversifies from depending entirely on agricultural activities to various non-farm activities like work at the Lundazi Ginnery located in the fringe area. This has resulted in higher income levels for most residents of the rural–urban fringe areas.

Negatively, the expansion of the Lundazi urban, however, has been largely informal as new developments in the rural–urban fringe have been unregulated resulting in the intermixture of commercial, residential, and agricultural land uses in an uncoordinated and conflicting manner. The study further found that there are no municipal services such as land zoning and layout mapping, roads, and drainages provided in these areas. This is because there has been a lack of integrated development planning exacerbated by the lack of harmony in the administration of the dual land tenure system. While development within the Lundazi urban boundary administered under leasehold tenure has been subjected to urban land-use planning policies and regulations, developments in its surrounding customary land have never been regulated as the customary tenure system that governs its administration has not been in support of planning regulations and controls.

Both the LDC and EPPA officials indicated the lack of a National Land Policy framework as a contributing factor

Fig. 10.3 Modern housing infrastructure in the Lundazi urban fringe areas. *Source* Field Data 2016

Fig. 10.4 Traditional village structure in the Lundazi rural–urban fringe. *Source* Field Data 2016

toward the lack of coordination in the land administration system which has also affected spatial planning in Lundazi. They revealed that, since independence, there has never been a National Land Policy document to guide effective land management and coordinate the land tenure systems. Although numerous drafts and policy pronouncements have been made over the years, the land policy document had been on draft level since time memorial.

4.3 Rural Gentrification and Takeover of Formerly Habited Rural Settlement/ Village

The study found that, as urban developments cross over the official urban boundary into the rural–urban fringe areas of Lundazi, displacement of the typical rural households by the urban elites mainly the civil servants and the business community occurs. This has also come along with the replacement of traditional village structures (huts) and rural lifestyles that once dominated these areas with modern urban housing infrastructure and urban lifestyle. This spatial form can be referred to as rural gentrification characterized by the influx of affluent middle- and high-class populations into the lower-income neighborhoods causing displacement of the lower-income population that had initially occupied the area (Shaw 2008). The term gentrification is often interpreted as a largely urban phenomenon, with urban gentrification being a widely acknowledged research subject. Gentrification however does not only occur in the urban context but in the rural setting as well like in the case of Lundazi. In Lundazi, this kind of development is more evident in the Chimuthyulu settlement due to its proximity to the town which led to its reclassification from a rural settlement into an urban settlement in 2009. Interviews with both the officials from LDC and the chief revealed that Chimuthyulu was initially a village located within the urban fringe area of Lundazi with traditional village structures like any other rural area but was later surrendered to the LDC by the traditional authority following the transformation of the village into a suburban area. Shortage of urban land exerted pressure on Chimuthyulu village for land for development due to its close proximity to the town as some urban elites mainly civil servants and local businessmen started buying some vacant plots from Chimuthyulu and constructed modern housing infrastructure whereas others bought the poor village residents off their land and they too constructed modern houses. As established by the study, these urban elites started converting the land tenure from customary to leasehold in their quest for security of tenure. They also started demanding services such as access roads, piped water, electricity, and a modern market that the village headman and his traditional village council could not provide. The increasing demand for services led the Chimuthyulu village headman and his village council in consultation with the chief to resort to surrendering the entire village to the LDC. The

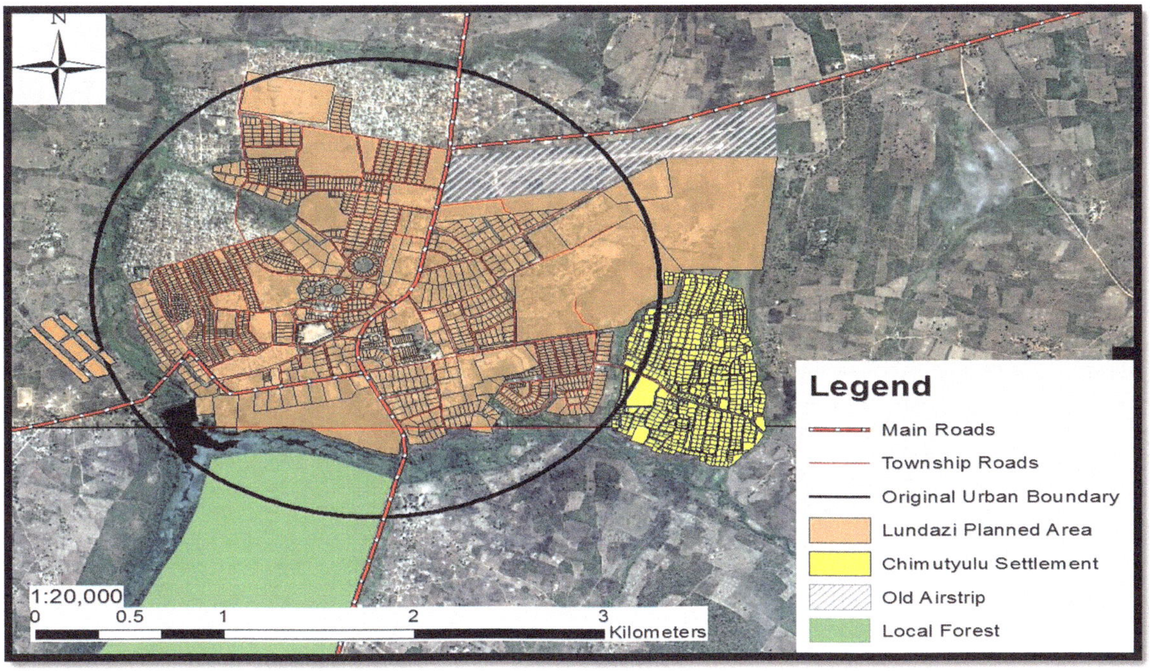

Fig. 10.5 Location of chimuthyulu settlement *Source* Author 2016

local authority later planned the Chimuthyulu settlement and subjected it to urban service provisions such as piped water, electricity, roads, and drainages to mention but only a few. Figure 10.5 below shows the location of the Chimuthyulu settlement.

4.4 Failed Planning Regulation Actions

Further, the findings revealed some failed planning actions that were implemented and were expected to make a difference in the history of the growth and development of Lundazi town. The first action was in August 2015 when the Government enacted the URP Act of 2015 which extended planning controls into customary areas under planning agreements. However, despite the URP Act having provided for planning and control of development in customary land, at the time of data collection, this provision had not yet been implemented in Lundazi as the planning regulations to guide the LDC on how to go about planning agreements were not yet in place.

The second major action was in December 2015 when the Lundazi urban boundary was extended, and the LDC gained authority over rural–urban fringe areas that were initially administered under the customary tenure system. This too could not help improve regulation of development in the rural–urban fringe areas as the traditional authority had continued administering land falling within the new urban boundary based on customary laws. Thus, development in these areas has continued to take place in an unregulated fashion.

4.5 Emergence of Informal Settlements at the Interface of Customary and Leasehold Tenure Systems

The other spatial form resulting from the existence of the dual land tenure system in Lundazi format is the emergence of informal settlements at the interface of customary and leasehold tenure systems. These mostly come about due to the uncertainty and ambiguous nature of existing legal boundaries. Informal settlements such as Gabon, Williams, Riverside, Kalopa, Dunda, and Chinyumba as shown in Fig. 10.6 emerged in state land that fell within the first urban boundary drawn in 1972 but was eventually perceived as customary land when the LDC lost track of the original urban boundary. Because these areas were perceived as customary land, the LDC did not recognize them as an integral part of the town. Thus, development control could not take place in these areas. Neither were they provided with any municipal services such as access roads and drainages, water and sanitation, and waste management. In terms of morphology, informal settlements in Lundazi have affected the orderly development of Lundazi town, therefore making the town look haphazard and unplanned. In addition, most of these informal settlements are located on the banks of the Lundazi River and its tributaries which sites are at high risk of flooding streams. This does not only endangers the lives of the residents of these informal settlements but also causes natural resource degradation as well as deterioration of living conditions within the

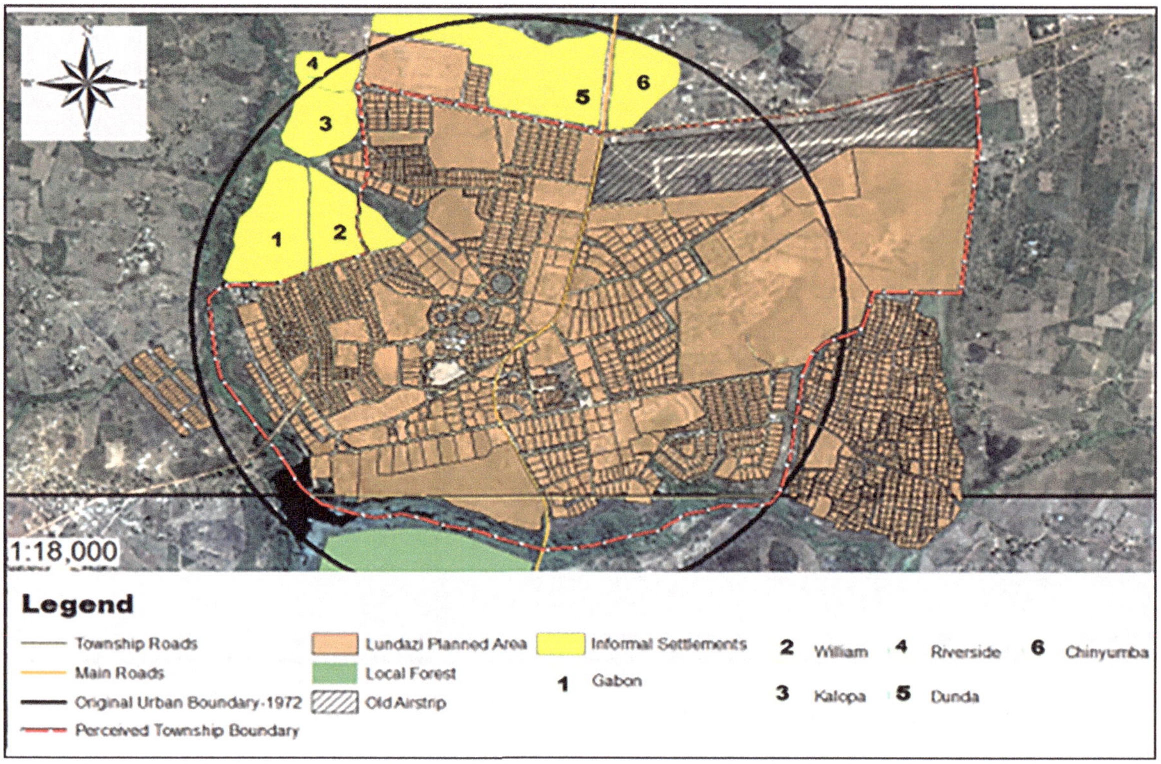

Fig. 10.6 Location of informal settlements in Lundazi town. *Source* Author 2016

urban boundary, problems that threaten the urban sustainability. Encapsulated within the informal settlements is the upsurge of urban poverty levels and social ills such as unemployment, crime, lack of or poor service delivery, and health hazards such that residents are susceptible to poor conditions of living. This is the case with the Lundazi informal settlements.

4.6 Direct Acquisition of Customary Land by Local Authority upon Agreement with Customary Leadership

The study established that the formidable challenges of shortage of urban land led the LDC and the EPPA to resort to engaging the traditional leaders for urban boundary extension. However, urban boundary extension had been challenging as there had been resistance from the traditional authority to surrender part of the land under their custody. It took two years for the traditional leadership of Mphamba Chiefdom to finally consent to the extension of the urban boundary. Even though all land in Zambia is vested in the Republican President who holds it in perpetuity on behalf of the people of Zambia as provided in the Lands Act, traditional authorities in Lundazi have opted to hold on to customary land to the extent of denying the state land for public purposes. Figure 10.7 shows the extended urban boundary. Findings further reveal that the delay in finalizing the urban boundary extension process gave way to illegal selling of land in the rural–urban fringe areas by the traditional leaders and their subjects to the extent that, by the time the process of boundary extension was finalized, there was little land available for greenfield planning. The implication of this situation on the Lundazi urban growth and development is that it has also contributed to the informal growth of Lundazi as no proper green field planning is being done in these areas but rather regularization of the existing unplanned developments. This also has come with its own challenges in that planning standards are compromised to accommodate the majority and avoid demolitions.

5 Analysis and Discussion

This chapter provides an analysis of the study's findings, linking them to the research objectives and existing literature. The discussion critically examines the effects of the dual land tenure system on Lundazi town's growth and development, with a focus on urban morphology, land management challenges, and sustainable planning. By situating the findings within the broader context of Zambia's land administration dynamics and SDG 11, this chapter explores the implications of the dual tenure system on urban

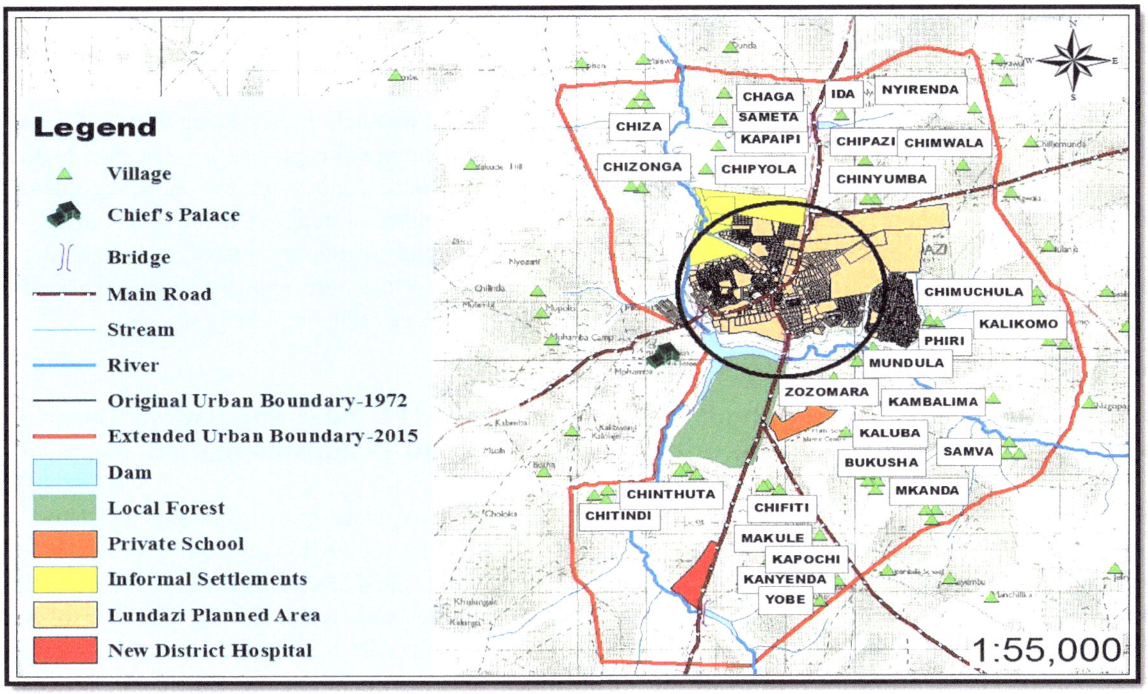

Fig. 10.7 Extension of the Lundazi urban boundary *Source* Author 2016 N.B. Topo sheet NO. 1233-A3 was adapted from Google Earth?

inclusivity, resilience, and sustainability. The analysis highlights the need for harmonized policies and integrated planning to address the challenges identified.

5.1 Lack of a National Land Policy Framework for the Administration of a Dual Land Tenure System

According to the findings, Zambia has historically lacked an approved National Land Policy framework to effectively guide the administration of its dual land tenure system. Although numerous drafts and policy pronouncements have been made over the years, at the time of data collection, the land policy was at draft level. It was not until 2021 that a final, officially approved land policy was published (Ministry of Lands and Natural Resources 2021). This longstanding absence of a formal policy has significantly hindered coordinated land management and development efforts. The lack of policy has contributed to inefficiencies in integrating customary and statutory land administration, leading to challenges such as unregulated developments and conflicts over land use. A comprehensive National Land Policy could address these gaps by defining the roles of customary and statutory systems in land management, as proposed by Tagliarino (2014) and UN-Habitat (2023a, b). The failure to establish this framework has perpetuated informal settlements and unregulated urban sprawl, negatively affecting urban growth and development in Lundazi, as indicated by both the study findings and earlier studies (Taylor and Thole 2015). Sichone (2010) indicates that a defined land policy guidance is a necessary requirement for an effective land management system as it identifies the goals of land and settlement development and identifies the strategies for achieving these goals. It also guides land management institutions in the performance of their duties and provides a means for evaluating their performance.

5.2 Rezoning of Open Spaces: Implications on Access to Public Spaces

The study highlighted the persistent issue of rezoning open spaces in Lundazi for commercial and residential use, significantly reducing access to safe and inclusive public spaces. Examples include the transformation of Vigodo grounds, once a social and cultural hub, into commercial plots, and local playgrounds in the Dunda informal area, converted into residential zones. These changes undermine the town's livability and compromise the benefits associated with open spaces, such as recreation, cultural activities, and environmental preservation.

Public spaces are crucial for fostering community cohesion, supporting recreation, and providing health benefits, as emphasized by Balogh and Takacs (2011). Their loss erodes urban resilience, making it challenging to achieve sustainable urban growth. Zambia's urban planning challenges, driven by weak institutional frameworks and inadequate enforcement, mirror the situation in other rapidly urbanizing regions of sub-Saharan Africa, where the dual land tenure system complicates land allocation for public purposes. The limited statutory regulation in customary land areas often prioritizes short-term economic gains over sustainable urban planning, as evidenced by the conversion of spaces intended for communal use into private developments.

5.3 Transformation of the Urban Fringe: Impact on Housing and Tenure Security

The transformation of the urban fringe in Lundazi has had both positive and negative implications. On the positive side, the development of infrastructure such as housing, electricity, and piped water has modernized the rural–urban fringe, enhancing the quality of life for residents. This aligns with SDG Indicator 11.1, which focuses on ensuring adequate housing, and Indicator 1.4.1, which measures access to basic services (United Nations 2015). For example, areas like Chimuthyulu have seen substantial improvements, transitioning from traditional village structures to modern housing settlements.

However, this expansion has been largely informal, with unregulated housing developments and inadequate municipal services like drainage and zoning. Customary landowners often sell plots without converting them to legally recognized leasehold tenure, resulting in tenure insecurity and uncoordinated urban growth. This duality reflects systemic challenges in Zambia's land administration, where customary practices coexist with statutory regulations. Unregulated fringe development mirrors broader trends across Zambia's peri-urban areas, where urban sprawl is driven by rapid population growth and rural–urban migration. The dual tenure system exacerbates these issues by creating parallel land governance structures, complicating urban planning efforts.

5.4 Rural Gentrification: Erosion of Inclusive Urbanization

Urban expansion into areas like Chimuthyulu illustrates the process of rural gentrification, where traditional rural households are displaced by wealthier urban elites. This shift has modernized housing and service provision, improving outcomes for some residents and aligning with SDG Indicator 11.1. (United Nations 2015). However, it has also exacerbated social inequities, as lower-income residents lose access to their ancestral lands without secure tenure rights, undermining Indicator 1.4.2. The dynamics of rural gentrification in Lundazi are consistent with findings by Shaw (2008), who notes that gentrification often displaces marginalized populations, replacing them with more affluent groups. In Zambia, this process is intensified by the absence of a harmonized framework for managing customary and statutory tenure systems. Chiefs and village headmen, lacking formal regulatory oversight, often allocate land based on informal agreements, leaving displaced residents vulnerable to eviction and tenure insecurity.

5.5 Failed Planning Regulations: Barriers to SDG 11 Implementation

Attempts to regulate growth through legislative measures like the URP Act (2015) and urban boundary extensions have largely failed. Resistance from traditional authorities to relinquish land, coupled with the lack of implementation frameworks, has delayed control over rural–urban fringe developments. These challenges perpetuate informal settlements, where basic services such as water, sanitation, and waste management remain inadequate, falling short of SDG Indicator 1.4.1. (United Nations 2015). Zambia's weak planning systems, as noted by Tembo (2014) and UN-Habitat (2015), are emblematic of broader regional trends, where rapid urbanization outpaces institutional capacity. Customary tenure systems, while deeply rooted in traditional governance structures, lack statutory alignment, preventing effective urban management. This situation reinforces social inequities and limits opportunities for sustainable growth.

5.6 Informal Settlements: Challenges in Slum Upgrading and Service Provision

The emergence of informal settlements in Lundazi, such as Gabon and Riverside, underscores the deficiencies in urban planning and land administration. These settlements, characterized by inadequate housing and infrastructure, fail to meet the benchmarks of SDG Indicator 11.1. Residents often lack access to clean water, sanitation, and waste management, further falling short of Indicator 1.4.1. (United Nations 2015).

This trend reflects broader challenges in Zambia's urban areas, where informal settlements are proliferating due to unclear boundaries between customary and leasehold tenure systems. Similar patterns have been observed across sub-Saharan Africa, where weak planning enforcement and insecure tenure exacerbate the growth of informal settlements. Addressing these issues requires integrating informal settlements into formal planning processes and providing secure land tenure to residents, as emphasized by Mwamba and Peng (2020), Mukumba (2019).

5.7 Delayed Boundary Extensions: Impacts on Public and Private Land Use

The prolonged negotiation process for boundary extensions with Mphamba Chiefdom reveals systemic inefficiencies in Zambia's dual tenure system. Delays have enabled illegal land sales by customary authorities, increasing informal developments and limiting opportunities for greenfield planning. These challenges threaten tenure security, undermine SDG Indicator 1.4.2, and compromise urban planning standards (United Nations 2015).

The resistance of traditional authorities to relinquish land for urban purposes is consistent with findings by Sichone (2010), who highlights the entrenched role of customary leaders in land administration. Without a clear policy framework to guide such negotiations, urban authorities face significant barriers in acquiring land for sustainable development. This dynamic underscores the need for reforms to streamline land acquisition processes and integrate customary land governance into statutory systems.

5.8 Lack of a National Land Policy Framework: Structural Barriers to Progress

The absence of a National Land Policy exacerbates the inefficiencies of Zambia's dual land tenure system, creating uncertainty around tenure rights and hindering sustainable urban planning. Without policy guidance, land administration remains fragmented, perpetuating informal and unsustainable growth patterns. This limitation directly impacts Indicator 1.4.2, as many residents in customary areas lack formal recognition of their land rights.

Sichone (2010) emphasizes that a defined land policy is essential for effective land management, providing a framework for harmonizing customary and statutory systems. Zambia's failure to implement such a policy has stalled efforts to achieve equitable urban growth. Developing a comprehensive National Land Policy would enable coordinated land administration and support the integration of customary tenure into broader urban planning frameworks.

6 Conclusion

This study has revealed that the dual land tenure system in Lundazi town significantly influences its growth and development. The duality, comprising leasehold and customary tenure systems, has both shaped the urban morphology and presented considerable challenges for land management. While leasehold tenure supports spatial planning and development control, customary tenure operates independently, often resisting integration into statutory frameworks. This disjointed administration has resulted in the informal and uncoordinated growth of Lundazi, particularly in its rural–urban fringe areas.

6.1 Achievements and Insights

The research aimed to examine the effects of a dual land tenure system on urban growth, using Lundazi town as a case study. Guided by the objectives to analyze the characteristics of the tenure systems, their influence on urban morphology, and challenges faced by the LDC, the study provided key insights.

Characteristics of the Dual Land Tenure System.

The dual tenure system encompasses a statutory leasehold system that emphasizes formal registration and spatial planning and a customary tenure system managed by traditional authorities. Customary tenure governs most peri-urban and rural land, prioritizing communal practices over statutory requirements. The absence of overlap between the two systems has created governance gaps and hindered sustainable urban growth.

6.2 Influence on Urban Morphology

Lundazi town's morphology reflects the fragmented administration of its land tenure systems. Unplanned developments dominate the rural–urban fringe, where customary land is being sold without statutory regulation. Informal settlements, mixed land uses, and insufficient service provision characterize these areas, resulting in spatial and social inequities. The expansion of the town into customary lands has led to haphazard growth patterns and rural gentrification, displacing traditional households.

6.3 Urban Land Management Challenges

The lack of coordination between leasehold and customary tenure systems has created significant challenges for the LDC. Resistance from traditional authorities to relinquish land for urban planning, delays in implementing urban boundary extensions, and the absence of a National Land Policy have exacerbated informal growth. Municipal services remain inaccessible in many fringe areas, and informal settlements continue to proliferate, undermining urban sustainability.

7 Recommendations

In light of the study findings and the challenges posed by the dual land tenure system, the following recommendations are proposed to address the barriers to sustainable urban development in Lundazi town. These measures aim to harmonize land administration, improve urban planning, and align urban growth with SDG 11 objectives.

Development of a Comprehensive Land Administration Framework A National Land Policy should be developed to harmonize the dual land tenure system, integrating both customary and leasehold tenures into spatial planning. This framework should outline clear guidelines for land allocation, development control, and dispute resolution, ensuring coordinated urban growth and tenure security for all residents.

7.1 Integrated Development Planning

Urban development in Lundazi requires a holistic approach through integrated development planning. Municipal and traditional authorities should collaborate to collectively plan for areas held under both leasehold and customary tenure, ensuring consistency in zoning, infrastructure development, and service provision. This will facilitate smoother transitions of customary land into statutory systems and address uncoordinated urban fringe growth.

7.2 Decentralization of Land Registration

The current centralized system of land registration delays access to land and hampers timely development processes. Decentralizing land administration to local levels would streamline registration, enhance efficiency, and make land management processes more accessible to communities in Lundazi and similar towns.

7.3 Formalization of Informal Settlements

Integrating informal settlements into the formal urban planning system is critical. This includes providing essential services such as water, sanitation, roads, and electricity while ensuring secure tenure for residents. Formalization efforts should align with planning standards to support equitable and sustainable urban growth.

7.4 Preservation and Revitalization of Public Spaces

The protection and revitalization of public spaces must be prioritized to enhance urban livability and resilience. Open spaces like parks and playgrounds provide critical social, economic, and environmental benefits. Strategies should include zoning regulations to safeguard existing spaces and create new ones to meet the needs of a growing population.

7.5 Strengthened Collaboration Between Authorities

Enhanced coordination between municipal authorities and traditional leaders is essential to bridge the governance gap in land administration. Formal agreements and joint committees could facilitate effective negotiations for land transfers, planning approvals, and service delivery in areas under customary tenure.

7.6 Monitoring and Future Studies

Conducting subsequent studies to monitor Lundazi's expansion and evaluate the impact of integrated development planning is crucial. These assessments would provide insights into the effectiveness of implemented policies and guide further refinements to support sustainable growth.

Acknowledgements Firstly, I wish to thank the Almighty God who made all things possible. Secondly, I would like to thank the institutions and individuals who gave me support in the course of this study. To the University of Zambia (UNZA), I wish to extend my debt of gratitude for offering me an opportunity to study with the faculty. I am also very grateful to the lecturers in the Department of Geography and Environmental Studies for introducing me to various aspects of Spatial Planning and in particular Dr. Wilma Nchito and Mr Gear Mumena Kajoba, under whose supervision this study was undertaken, for their guidance, support, and insightful comments that helped in shaping this study. My thanks are also extended to my classmates; the spatial planning class of 2014/2016 with whom I found support, encouragement, and laughter amidst challenges.

My gratitude goes to the Ministry of Local Government and Housing (MLGH), the Department of Physical Planning and Housing (DPPH) for funding my study. My sincere thanks are also extended to my employer the Local Government Service Commission for granting me study leave. My appreciation also goes to my workstation the Lundazi District Council (LDC), for releasing me to further my studies, especially at a time when my presence was greatly needed.

I send special thanks to members of my family, in particular my mother Rabbeca Chanda Kalapula, My father Dennyson Mwansa Kalapula, my two elder sisters Hope Kalapula and Brendah Kalapula, and my one and only brother Dennyson Kalapula Jr for your unconditional support, encouragement, prayers, and sharing in my journey. May God continue blessing you all.

Many thanks also go to my special friends Naomi Katongo Chikwanda, Paul Magiya Manda, and Felix Mulilo for your constant encouragement and support. Special thanks are also extended to my colleagues Mwanta Matthews and Makalani Mukuni for the encouragement you gave me when I almost gave up processing my study leave when things were really hard for me. A friend in need is a friend indeed. I am very grateful.

My sincere gratitude also goes to all the research participants. I am grateful for your contribution.

Dedication To Mum and Dad, for your unconditional love and for being great parents to me, love you always.

Competing Interests The authors declare that they have no conflicts of interest relevant to the content of this article.

References

Adams, M.: Land Tenure Policy and Practice in Zambia: Issues Relating to the Development of the Agricultural Sector. Mokoro Ltd., Oxford (2003)

Aka, E.O.: Gentrification and socioeconomic impacts of neighborhood integration and diversification in Atlanta, Georgia. Natl. Soc. Sci. J. **35**(1), 1–13 (2010)

Antonio, D., Gitau, J., Sylla, O.: Innovative customary land governance in Zambia: experiences, lessons learned and emerging Impacts. UN-HABITAT GLTN (2023). Retrieved from Academia.edu.

Van Asperen, P.: Evaluation of Pro-poor Land Administration from an End-user Perspective: A Case Study from Peri-urban Lusaka (Zambia)'. Morocco: Marrakech (2011)

Bloch, R.: The future of water in African cities: why waste water? Integrating urban planning and water management in Sub-Saharan Africa: background report. The World Bank Group, New York (2012)

Balogh, P. I., Takács, D.: The significance of urban open spaces and green areas in urban property developments. In: First International Conference "Horticulture and Landscape Architecture in Transylvania" Agriculture and Environment Supplement, vol. 110, p. 121. Sapientia Hungarian University of Transylvania, Cluj-Napoca, Romania (2011)

Cain, L.P.: Urban growth. In: Diebolt, C., Haupert, M. (eds) Handbook of Cliometrics. Springer, Berlin, Heidelberg (2023). Available at: https://doi.org/10.1007/978-3-642-40458-0_92-1

Central Statistics Office (CSO): Zambia National Population and Housing Census. Government Printers Lusaka (2010)

Chigbu, U.E.: Urban-Rural Land Linkages: A Concept and Framework for Action. United Nations Human Settlements Programme (UN-Habitat). HS/003/21E. Nairobi, Kenya. (2021) Available at: https://www.unhabitat.org (Accessed: 21 November 2024).

Chinene, V.R., Maimbo, F., Banda, D.J.: A Comparison of Customary and Leasehold Tenure: Agriculture and Development in Zambia. The University of Zambia, Lusaka (2010)

Chitonge, H.: Customary land in Zambia: the new scramble and the evolving socio-political relations. In: Reclaiming Africa: Advances in African Economic, Social, and Political Development (2018)

Connor, R.: The United Nations world water development report 2015: water for a sustainable world, vol. 1. UNESCO publishing (2015)

Deininger, K., Goyal, A.: Land policies for resilient and equitable growth in Africa. International Bank for Reconstruction and Development/The World Bank. Washington, DC (2024). Available at: https://www.worldbank.org

Fekade, W.: Deficits of formal urban land management and informal responses under rapid urban growth: an international perspective. Habitat Int. **24**, 127–150 (2000)

Kajoba, G.M.: Land Use and Land Tenure in Africa: Towards an Evolutionary Conceptual Framework. The University of Zambia, Lusaka (2002)

Lombard, M.: Land Tenure and Urban Conflict: A Review of the Literature. The University of Manchester Press, Manchester (2012)

Ministry of Lands and Natural Resources: National Lands Policy. Lusaka: Ministry of Lands and Natural Resources (2021)

M'Membe, F.: The Zambian Land Tenure System and the Process of Land Alienation: The Need for Land Reform in Zambia. University of Zambia, Retrieved from UNZA DSpace (2013)

Mudenda, F.S.: Land Law in Zambia: Cases and Materials. The University of Zambia Press, Lusaka (2007)

Muinde, D.K.: Assessing the Effects of Land Tenure on Urban Development in Kampala. University of Twente, Enschede (2013)

Mukumba, C.P.: Enablement approaches to the upgrading of informal settlements: a case study of Misisi Compound in Lusaka, Zambia. (2019) Retrieved from Academia.edu.

Mwamba, J.S., Peng, Z.W.: Analysis of Informal Urban Settlement Upgrading: The Case of Ng'ombe Slum Upgrading in Zambia. Curr. Urban Stud. **8**, 509–532 (2020)

Nchito, W.: The Growth and Functions of Small Urban Centres in Zambia: A Case of Mazabuka and Kalomo. Unpublished Thesis. Lusaka: The University of Zambia (2013)

Nsiah-Gyabaah, K.: Urbanization, Environmental Degradation and Food Security in Africa. Prepared for Poster Presentation, Sunyani, Ghana: Sunyani Polytechnic (2003)

Payne, G., Piaskowy, T.: Land Tenure in Urban Environments. LandLinks (2014)

Post: Shortage of land is inhibiting government plans to expand Choma, The Post Newspaper, 3 March (2014)

Ritchie, H., Samborska, V., Roser, M.: Urbanization. Our world in data (2024). Available at: https://ourworldindata.org/urbanization (Accessed: 2 November 2024)

Shaw, K.: Gentrification: What it is, why it is, and what can be done about it. Geogr. Comp. **2**(5), 1697–1728 (2008)

Sichone, F.: The System of Land Alienation in Zambia: A Critical Analysis of the Legal and Institutional Framework. The University of Zambia, Lusaka (2010)

Tagliarino, N.K.: Zambia: Conversions of Customary Land to Leasehold Title. Focus on Land in Africa (2014)

Taylor, T.K., Thole, C.B.: Rethinking Town and Country Planning Practice in Zambia. Copperbelt University, Kitwe (2015)

Tembo, Y.: Rapid Urbanization in Zambia: The Challenges Facing Our Cities and Towns. The University of Zambia, Lusaka (2014)

Times of Zambia: Zambia: Luano on Economic Growth Path. Times of Zambia [online]. (2014) Available at: https://www.timesofzambia.com (Accessed: 11 February 2015).

Times of Zambia: The shortage of land is inhibiting government's Plans to Expand Choma'. [Online] (2014)

Umar, B.B., Nyanga, P.H.: Customary land certification, governance, and natural resource use in Zambia: a social learning approach. Eur J Dev Res **35**, 997–1027 (2023). Available at: https://doi.org/10.1057/s41287-022-00555-9

UN-Habitat: The state of planning in Africa: an overview. African Planning Association Nairobi (2015)

UNHabitat: Global Land Tool Network. (n.d.) Zambia Country Context (2023). Retrieved from GLTN , https://gltn.net/country-work/zambia/.

UN-Habitat: Urbanization in Zambia: Building inclusive and sustainable cities (2023). Retrieved from UN-Habitat.

UN-Habitat: Rescuing SDG 11 for a Resilient Urban Planet. United Nations Human Settlements Programme (UN-Habitat). Nairobi (2023)

UNICEF: The State of World's Children. UNICEF, New York (2012)

United Nations Department of Economic and Social Affairs. Population Division (2019) World Urbanization Prospects, The 2018 Revision. New York: United Nations. ST/ESA/SER.A/420.

United Nations: Land Governance and Local Tenure Systems: The Role of Government Policy (2023)

United Nations: Sustainable Cities and Communities. Retrieved from UN Zambia (2023).

United Nations: Transforming Our World: The 2030 Agenda for Sustainable Development. United Nations New York (2023). Available at: https://sdgs.un.org/goals/goal11.

USAID: USAID Country Profile: Property Rights and Resource Governance, Zambia. USAID Lusaka (2010)

USAID: Zambia land and resource tenure sector overview. Retrieved from LandLinks (2018), https://www.land-links.org/wp-content/uploads/2019/05/Zambia-Land-and-Resource-Tenure-Sector-Assessment.pdf.

USAID: Zambia Resource Tenure and Governance Review (2020). Available at: https://www.land-links.org/wp-content/uploads/2023/08/Zambia-2020-Resource-Tenure-and-Governance-Review.pdf.

Weldegebriel, A., Assefa, E., Janusz, K., Tekalign, M. and Van Rompaey, A.: Spatial analysis of intra-urban land use dynamics in Sub-Saharan Africa: the case of Addis Ababa (Ethiopia). Urban Science, **5**(3), 57. Retrieved from MDPI (2021)

Yussif, K., Dompreh, E.B., Gasparatos, A.: Sustainability of urban expansion in Africa: a systematic literature review using the Drivers–Pressures–State–Impact–Responses (DPSIR) framework. Sustain. Sci. **18**, 1459–1479 (2023)

Zambia Daily Mail: Lundazi: Small town with big dreams. Zambia Daily Mail [online] (2015). Available at: https://www.dailymail.co.zm (Accessed: 18 November 2015).

ZANIS: Chief Nkole of the Swaka speaking people says the claim for part of the land of his chiefdom by the council is illegal. ZANIS News, 18 April (2012)

Design of a Land Use–Land Cover Index Using Fuzzy Logic for Territorial Planning in the Andean Basins of Colombia

David Alejandro Munõz, Yesid Carvajal Escobar, and Diego F. Tarapuez Bolanõs

Abstract

Territorial planning emerges as an essential tool to address environmental degradation and to guarantee the comprehensive sustainability of ecosystems, considering the preservation of natural reserves as a fundamental approach for the conservation of water resources in Andean basins and the promotion of sustainable practices in Colombia for climate risk. This study was carried out in the municipality of Santander de Quilichao, department of Cauca, by identifying the land covers using the CORINE (Coordination of information on the Environmental) Land Cover (CLC) methodology and the MOLUSCE (Modules for Land Use Change Evaluation) to estimate the potential for change in land use through specialized program for GIS, QGIS. Using these techniques, maps of change in coverage for the period from 2018 to 2021 were generated using a territorial transition matrix and a prediction model of a possible future scenario for the year 2030. The results obtained were used as input for the design of a land use coverage index (LUCI) using fuzzy logic. 60 coverage points of the Andean Quilichao river basin were classified into three categories: high, medium, and low which are the results of LUCI. From these results, it is possible to see that the most significant changes in coverage are: another transitional crop: medium corresponding to 35% and mosaic of pastures with natural spaces: low with 12%. This research is pioneering in the department of Cauca and is expected to be replicated in Colombia, since these types of methodologies, through a land use coverage index (LUCI), allowed for the identification of territorial conflicts in areas of environmental importance. Therefore, it can provide good references for the improvement of the ecological environment in future which is necessary for a territorial ecological planning framework. Finally, the LUCI values are similar to the coverage prediction for the year 2030.

Keywords

Climate risk · Water resources · GIS · Fuzzy logic · Watershed · Land use

Y. C. Escobar and D. F. Tarapuez Bolanõs: These authors contributed equally to this work

D. A. Munõz (✉)
Faculty of Natural and Exact Sciences—PhD Student in Environmental Sciences, Universidad del Valle, Cali, Valle del Cauca, Colombia
e-mail: david.a.munoz@correounivalle.edu.co

Y. C. Escobar
School of Natural Resources and Environmental Engineering, Universidad del Valle, Cali, Valle del Cauca, Colombia
e-mail: yesid.carvajal@correounivalle.edu.co

D. F. T. Bolanõs
Faculty of Natural and Exact Sciences, Universidad del Valle, Cali, Valle del Cauca, Colombia
e-mail: diego.tarapuez@correounivalle.edu.co

1 Introduction

Climate change has been generating growing global concern in society, due to the impacts it is causing on ecological, social, and territorial dynamics. This is why societies frequently face extreme events associated with climate variability (CV) and climate change (CC), such as storms, flash floods, droughts (Birkmann et al. 2009), cyclones, and hurricanes, among others (Webster et al. 2005) affecting coastal areas (Wehner et al. 2018; Mafi-Gholami et al. 2020), cities (Reckien 2016; Nasiri and Khosravani 2019; Santos et al. 2019), biodiversity (Souza et al. 2010; Daly 2014; Xu et al. 2020), and hydrographic basins (Pabon-Caicedo et al. 2018; Rosero Mier 2031; Giri et al. 2020).

P. Mohammad et al. (eds.), *Geographic Approaches to Climate Change and Mitigation: Urban and Rural Perspectives (Volume 2)*, Advances in Science, Technology & Innovation, https://doi.org/10.1007/978-3-031-92115-5_11

Therefore, efficient territory management is essential to guarantee sustainable development and the preservation of biodiversity in various regions of the world. In Colombia, the territorial planning of basins is regulated by various regulations that seek to promote sustainable development, the conservation of natural resources, and adequate water management, under a legal regulatory protocol called Hydrographic Basin Planning and Management Plans—POMCA. (DNP 2013). Therefore, conservation and territorial planning processes are fundamental for the capacity to respond to socio-environmental and climatic stressors (Tagliari et al. 2023). Given this reality, it is important to consider that the impact of land cover and land use change represents a crucial indicator that reflects the effect of human actions on natural ecosystems (Liu et al. 2021).

In addition, a greater frequency of extremely low flows has been recorded in recent decades, a phenomenon that does not coincide with times of drought. These findings suggest that the variability in trends may be related to changes induced by human activities within the watersheds (Bedoya and Ramırez, B.H. 2023). Therefore, it is necessary to make watershed ecosystems in the Andean region more resilient to global climate change, since an increase of 3 °C is expected by 2070, the disappearance of intertropical glaciers due to temperature increase. These unique and diverse high mountain ecosystems play a fundamental role in water regulation, biodiversity conservation, and the livelihoods of local communities, especially in the Andes of South America (IUCN 2020).

The objective of this research was to design a land use and land cover index (LUCI) using fuzzy logic in the Andean Quilichao river basin of Colombia and to study how this index impacts on the processes of improving territorial planning in these areas. This basin is important because of the presence of the last relict of forest in the north of the department of Cauca, according to Holdridge, Premontane Humid Forest in Climate change adaptation plan (2019); the LUCI is compared with the European CORINE Land Cover classification in the time period (2019–2021) in which the MOLUSCE (Modules for Land Use Change Evaluation) complement is used to evaluate changes in land use dynamics. Finally, it is expected that the use of the LUCI will facilitate the implementation of conservation strategies, sustainable management of resources, and mitigation of environmental impacts, thus contributing to the adequate and sustainable management of the territory in the Andean basins of Colombia.

1.1 Statement of the Problem and Objective of the Study

Currently, various studies indicate that the intensity and change of land use impact water patterns and flows (Castellanos-Tapia et al. 2022). This modification, combined with climate change, could affect the availability of water resources in the Andean basins of Colombia (Romero Hernandez 2023). The country has adopted European standards since 2005 to understand the dynamics of land cover, contributing to territorial planning (IDEAM 2007). Local communities express concerns about agricultural sustainability, water accessibility, and resource deterioration (Departmental plan for climate change 2016). This research was carried out in the municipality of Santander de Quilichao, department of Cauca, a region that faces significant challenges, including pressure for water demand and vulnerability due to shortages in more than 40% of the Cauca basins, pollution, and deforestation (Departmental plan for climate change 2016). Watersheds are considered essential for human subsistence, providing environmental services such as the supply of drinking water (Vargas-Franco and Restrepo-Tarquino 2018). In the face of unprecedented climate risks, natural reserve zones and protected areas are presented as essential components to confront climate change. These areas not only reduce greenhouse gas emissions, but also conserve ecologically important ecosystems (Figueroa et al. 2011). Given their relevance, it is crucial to have integrated information on these areas to make strategic decisions in territorial planning and strengthen the response capacity of natural resources to climate change. In this context, the design of a land use index using fuzzy logic is presented as an innovative tool. This index will allow more precise evaluation and understanding of changes in land use, especially in critical areas such as the Andean basins of Colombia. Fuzzy logic provides flexibility and precision in modeling complex phenomena, such as land use changes, providing a valuable tool for territorial planning and water resources management (Wang et al. 2020). The use of GIS and artificial intelligence techniques such as fuzzy logic is still a novel and unexplored field in Colombia. It is therefore expected that this research will become a baseline for decision-making at the basin scale and resource management from the basis of science, technology, and innovation CTeI. In the world, some applications have become information to provide good references for the improvement of the ecological environment in future to prevent the degradation of environmental systems (Hussain et al. 2022) and delimit potential areas of water bodies. In this way, the proposed research seeks to comprehensively address the challenges associated with climate change, watershed management, and the conservation of key natural areas in Colombia. Finally, the current situation of the municipality is derived from the torrential avenue that occurred last March 2023, and the alarms were set off to pay maximum attention to these risk-generated events and that have put municipal government entities on alert to the creation of information, through reports and technical visits to the most affected communities 47 families, mostly in the rural area of the municipality, upper part of the Munchique-Los Tigres Natural Reserve, where the Quilichao river basin is located (Munoz and Baro Saurez 2023). It is necessary to generate knowledge and information whose results become a baseline for decision-making, providing a technical and scientific approach and input for the response capacity of rural and urban populations within the framework of land use, allowing a more accurate and adaptable representation of reality in various contexts and contributing to effectiveness in resource management.

1.2 Literature Review

The planet Earth has presented various changes over the years, due to the intervention of civilizations which have modified the different environments, such as the earth's surface, which has been directly affected in order to satisfy human needs reflected in the considerable change in land cover globally (Cuevas Caamanõ 1986–2020). Loss has been suffered mainly in forests, savannahs, and other ecosystems, transforming them into productive systems for agricultural use. This significant change has led to human activity as the main responsible for the alterations in the flows and states of the biosphere (Vergara et al. 2021). Understanding land cover and use is crucial for territorial planning, as it influences soil formation and the provision of ecosystem services (Flores and Olmos-Martınez 2020), in addition, understanding land use and occupancy changes is a crucial tool for natural resource management. The processes of agricultural expansion and decline have generated an emerging global debate due to their socio-environmental consequences (Vliet et al. 2017). These alterations become an important driving force for changes in ecosystem services, as they affect key ecological processes such as the water cycle, having a significant impact on water availability. At the regional level, alterations in land cover and use influence the water cycles of the basins, temperature patterns, and biodiversity, increasing the vulnerability of human communities. At the local level, these changes cause soil degradation and adjustments in ecological niches, conditioning the biological composition of the soil, and affecting its biological quality (Saavedra Briones and Sepulveda-Varas 2016). Given the importance of the hydrological cycle in the functioning of ecosystems, hydrographic basins are presented as an appropriate spatial framework to study the effects of change in land cover and use. The characterization of this change is essential for the integrated management of ecosystem goods and services (Munevar Quintero and Ramırez Rıos 2021). Within climate risks, protected areas and natural reserves are presented as crucial elements to address climate change, reducing greenhouse gas emissions and preserving strategic areas in ecosystem terms. In particular, the Cerro Munchique-Los Tigres Natural Reserve in the Municipality of Santander de Quilichao, Colombia, stands out for its environmental importance and water potential, since three rivers originate here, which are essential for the ecosystems and water supply of this area in the north of the department of Cauca (Climate change adaptation plan 2019). In 1965, Lotfi Zadeth proposed the theory of fuzzy logic, also known as fuzzy logic. However, its roots go back approximately 2500 years, when Greek philosophers explored the existence of varying degrees of truth and falsehood. These thinkers proposed a perspective that departed from the rigidity of traditional logic, recognizing the diffuse and nuanced nature of many concepts and propositions in reality. This historical context sheds light on the subsequent development of fuzzy logic as a conceptual framework that more faithfully reflects the complexity inherent in many systems and phenomena (Baez and Duran). Currently, fuzzy logic is applied in different fields of science and is considered an important tool within artificial intelligence (AI). Finally, in this study, the researchers present fuzzy logic (FL) as a rule-based model that represents various types of knowledge about the research problem and models a compendium of structures in a specific situation. Furthermore, they highlight that one of its main advantages consists of its tolerance to different types of uncertainty, as well as its ability to process numerical and linguistic information, simultaneously; therefore, FL can be used for various practical applications in terms of modeling and control of complex systems (Vadiati et al. 2019). For this reason, the FL is proposed as the instrument to generate the land use coverage index (LUCI), as a management tool for territorial planning in Andean basins in Colombia.

2 Methods

2.1 Study Area

The Quilichao river basin is located north of the department of Cauca, between elevations 960 and 2600 m above sea level and the coordinates 3° 03' and 3° 02' north latitude and 76° 15' and 76° 18' west longitude. It is located southeast of the municipality of Santander de Quilichao, with

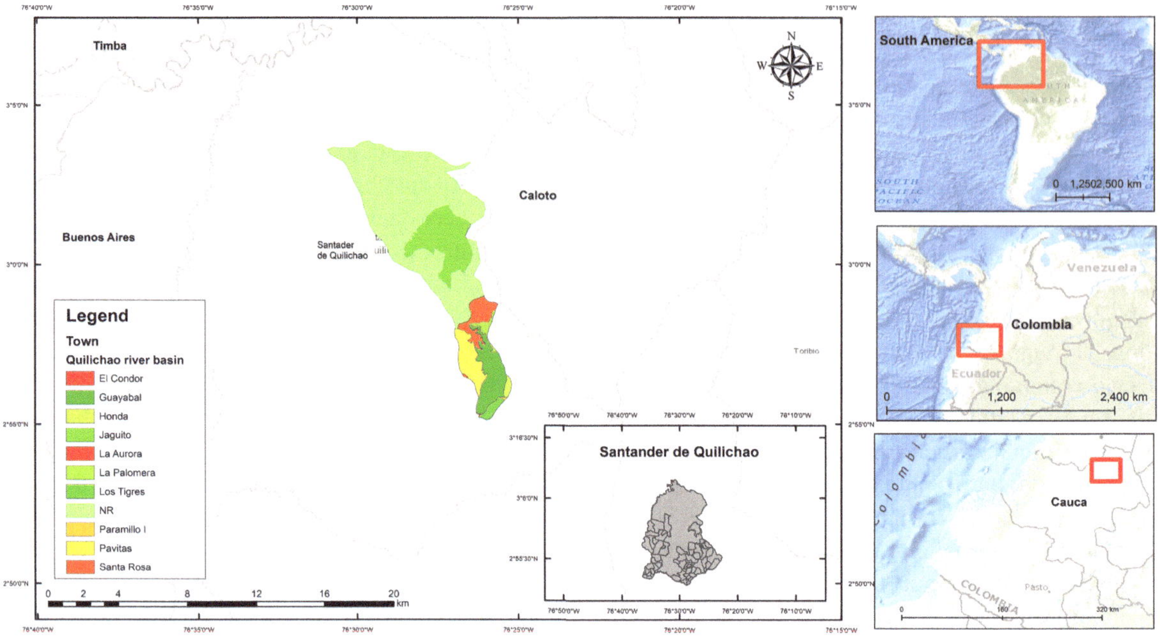

Fig. 1 Location of the study area, Quilichao river basin, Cauca, Colombia. *Source* Self made

limits between the municipalities of Caloto, Buenos Aires, and Timba. It is classified as a premontane humid forest according to Holdridge; the basin is impacted by various environmental stressors because it is the main source of water supply for human consumption of the population that is settled in the urban area, more than 70,000 inhabitants. (Climate change adaptation plan 2019; EMQUILICHAO 2016). Figure 1 shows the map of the Quilichao river basin in the department of Cauca, Colombia. The figure shows administrative divisions, main localities, and sub-basins, as well as the geographical location of the basin at the level of South America, Colombia, and Cauca.

2.2 Supervised Classification, Cartography, and Field Work—CORIN Land Cover (CLC) Methodology

To meet the objective of this study, mapping of the vegetation cover in the Quilichao river basin was carried out using satellite images. This process was carried out using the supervised classification method, in order to create a digitized cartographic database that could distinguish different types of vegetation cover, providing detailed quantitative information. The main purpose of this mapping was to obtain a clear view of the current state of vegetation cover, which is essential for making informed decisions. This information would allow us to analyze environmental impacts in the basin, evaluate anthropogenic activities, implement forest monitoring and restoration programs, and address issues such as deforestation and territorial planning. This is due to the close relationship that exists between land use, the different vegetation covers, and the quality and quantity of water in the region.

For this purpose, a satellite image from 2019 from the Sentinel 2 platform was selected. This image has 13 bands and a spatial resolution of 10, 20, and 20 m, respectively. In addition, it has a cloud coverage percentage of 20%. This image was processed at a 1C processing level, which includes radiometric and geometric corrections, as well as orthorectification and spatial registration.

Regarding the classification of land covers and uses in the Quilichao river basin, the CORINE Land Cover (CLC) methodology was used. This classification was carried out at a scale of 1:25.000, in the MAGNA-SIRGAS Colombia West, Quilichao coordinate system, covering an area of approximately five thousand hectares (5.000 ha). This process was carried out through digital processing of remote sensors, that is, satellite images, combined with the collection of field information using GPS.

To adapt the legend of the national CORINE Land Cover (IDEAM 2010) to the scale of 1:25.000, a legend was proposed that adjusted the available information and eight land use classes were defined. These classes include: body of water, dense vegetation, natural grasslands, flooded agriculture, crops, shrubs, urban areas, and bare soil. This detailed and precise classification provides a comprehensive view of the different land coverage and uses in the Quilichao river basin, thus facilitating territorial analysis and planning based on solid and quantifiable data. This classification can be defined as a process in which pixels of known identity, located within the training areas, are used to classify pixels of unknown identity (http://www.teledet.com.uy/tutorial-imagenessatelitales/classification supervised.html). Supervised classification in ArcGIS allows the raster to be classified by means of each pixel of a cell; it is necessary to have prior knowledge of the classes of the study area (current land use). You must mark points and assign a numerical value for each class that you want to separate. The signatures are created, and then, the probability classification is performed, and with this, a raster file containing the classification performed is finally generated.

2.2.1 CORINE Classes Use

- Photointerpretation on Sentinel 2 satellite images of the current vegetation cover. For mapping purposes, the CORINE Land Cover classification was used, adapted to a scale of 1:25.000.
- Based on the CORINE Land Cover legend (IDEAM 2010) on land cover that was adopted interinstitutionally by Colombia at a scale of 1:100.000, a proposal was made for a legend at a scale of 1:25.000, where the national CORINE Land Cover legend of IDEAM (2010) was adjusted to the available information.
- Taking into account the information that could be obtained from the images, the interpretation of the CORINE classes found and shown in the coverage changes was considered.

Finally, it was necessary that materials and equipment to carry out the verification were programmed, including visits that were made to the area using GPS for coverage points. Therefore, it was considered necessary to map the coverage in the Quilichao river basin, using satellite images, through the supervised classification method, to build a digitalized cartographic database that distinguishes vegetation coverage, with its corresponding quantitative information, to know the current state of the vegetation cover that constitutes basic information for decision-making, which allows analyzing the environmental impacts on the basin of anthropogenic activities, implementing monitoring programs, forest restoration, deforestation, and for planning and territorial planning, given the close relationship between use, land cover, and quality and quantity of water.

2.3 Fuzzy Logic

Fuzzy logic has concepts such as fuzzy sets, membership functions, degrees of membership, or degrees of veracity; in this way, it is recognized that reality is not linear and complex dynamic systems are incorporated, allowing it to be evident in turn that there will always be a degree of uncertainty regarding the analyzed phenomenon. Although fuzzy logic is a field of research of great importance today, both for its mathematical or theoretical implications and for its practical applications, there are three reasons to use this tool which are: In complex processes if there is no model simple solution, in nonlinear processes, and in case it is necessary to include the experience of an expert operator based on misconceptions acquired from his experience (Fleites Ramos et al. 2021). In this research, the three reasons mentioned above were considered sufficient to implement the fuzzy logic methodology. This is because a complex model that adequately integrates land cover and use to identify vulnerable areas has not yet been developed. On the other hand, the adoption of a linear model in this research would imply assuming a directly proportional relationship between the variables, which may be too simplistic and not take into account the complexity and nonlinearity of the processes that influence the coverage and the land use. Therefore, we seek to implement a nonlinear model that can more accurately capture these complex relationships. Furthermore, to enrich this study, it is necessary to have the experience and knowledge of experts in the field, who can provide valuable perspectives to design a model that more closely approximates reality. According to Zadeth, a system based on fuzzy logic is structured in three main phases: fuzzifier (fuzzification), inference mechanism (rules) (fuzzy rule inference), and the defuzzifier (defuzzfication) as shown in the following Fig. 2. On the other hand, Fig. 2. is the diagram of the fuzzy modeling process applied in the Cerro Munchique-Los Tigres Natural Reserve. The process includes the creation of a database, fuzzification of land cover and land use data, inference based on fuzzy rules, and finally de-fuzzification to obtain a quantitative land use indicator.

- Fuzzifier: At this stage, membership functions are established in line with the analysis of each variable, such as land cover and land use. The inputs to this phase are specific values of the variables under consideration, and the outputs are the degrees of membership. Each input value is transformed into a number within the range from zero to one, using a predefined membership function, which is configured in advance for this purpose.
- Inference mechanisms (rules): It relates the input and output fuzzy sets through predetermined rules for the system, which can be formulated by experts in the respective field or generated by computational algorithms. These rules are essential for determining how fuzzy input variables are translated into fuzzy output variables, and their formulation can derive from both specialized human knowledge and algorithmic approaches.
- Defuzzifier: In this phase, the fuzzified values, which are in the range from zero to one, are transformed to the scale used by the study variables. In this procedure, mathematical methods are applied, such as the maximum method

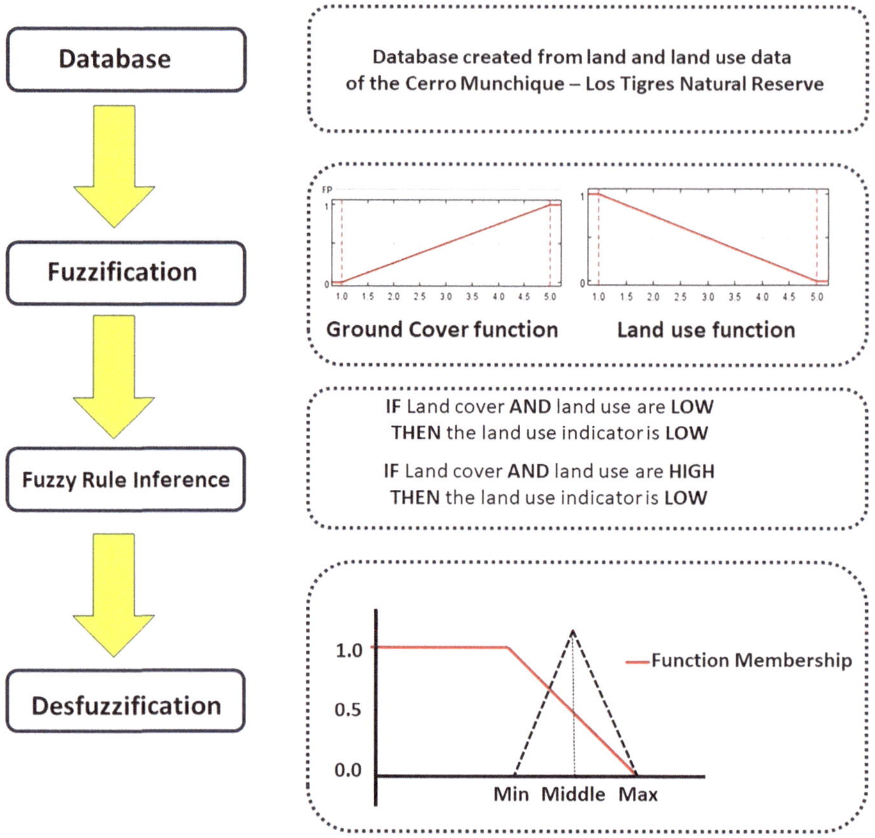

Fig. 2 Diagram of a system with fuzzy logic. *Source* Self made

(using the maximum value), the centroid method (based on the center of gravity of the function), and the height method (using a weighted average). These methods allow adequate rescaling of the fuzzy values to adapt them to the specific magnitude and context of the analysis variables.

On the other hand, the degrees of membership are calculated for any $x \in [0, +\infty)$ according to Eq. 1:

$$\deg(\mathrm{x}) = \begin{cases} \dfrac{j-x}{j-i} & \text{if } x \le i \text{ if } i \le x \le j \text{ if } \mathrm{x} \le j \end{cases} \quad (1)$$

The most commonly used characteristic functions due to their mathematical simplicity and manageability are: triangular, trapezoidal, and GAMMA. Conceptually, there are two approaches to determine the characteristic function associated with a set: The first is based on the knowledge of experts, and the second is through the application of direct methods that use the collection of data to design the function (Baez and Duran).

2.3.1 Advantages and Applications

- Flexibility: Fuzzy logic allows us to handle uncertainty and imprecision in data.
- Intuitive Interpretation: Fuzzy categories can be easily related to real-world concepts.
- Complex Modeling: It can capture nonlinear relationships and consider multiple variables simultaneously.

This index can be used to monitor changes in land cover and use in the Quilichao river basin over time and to identify areas of deforestation, urban expansion, changes in agricultural patterns, among others. Furthermore, it can provide valuable information for land use planning, biodiversity conservation, and sustainable management of natural resources in this specific region.

Figure 3 is a workflow for land use and land cover analysis using GeoFIS and ArcGIS. The figure shows the steps from the creation of the database with longitude, latitude, land use and land cover data, to the loading in GeoFIS to define functions and aggregation rules, and finally the visualization of results in ArcGIS.

In Sect. 1, the database is made up of geospatial information including longitude, latitude, land cover and land use. It contains a total of 60 points, which were collected during field work. During this process, coverage verifications were carried out in rural areas, considering the current land use, with the purpose of assigning them in the fuzzy model.

Fig. 3 Methodological protocol for the fuzzy model. *Source* Self made

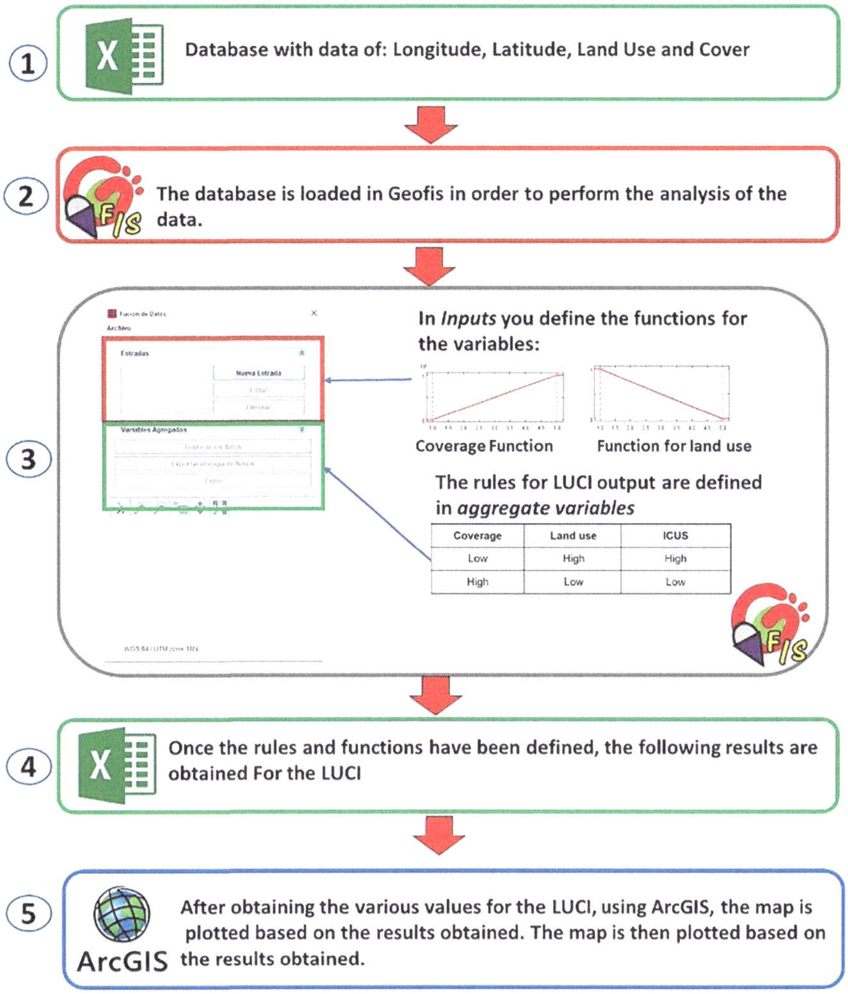

In Sect. 2, as a fundamental instrument in data analysis, it was decided to use the open-source software GeoFIS. This program, known for its accessibility, constitutes a friendly tool for developing decision-making support systems. Unlike platforms like R or Python, which require programming skills, GeoFIS stands out for its simplicity, and a distinctive advantage is its focus on incorporating expert knowledge in data analysis, a feature that is not present in other processing tools space. This aspect is particularly relevant, since it facilitates a more contextualized and specific analysis. (Leroux et al. 2018).

In Sect. 3, the membership functions for the land cover and land use variables are defined. These functions are used to model and manage uncertainty and imprecision in the data for each of our study variables. There are various membership functions that allow us to model the different variables used in a study, such as triangular, trapezoidal, Gaussian, upper semi-trapezoidal, lower semi-trapezoidal functions, among others. In GeoFIS, some pre-established functions have been defined (Guillaume et al. 2020). This shows that the membership function (land use–land cover) assigns a membership value to each point in the range of the input variable, allowing the uncertainty and imprecision inherent in the data to be modeled of land use. Basically, it describes how each input value belongs to a fuzzy set that represents a specific land use or land cover category. In this study, for the coverage variables, a higher semi-trapezoidal function was assigned because the higher the value of the degree of membership of the function, the better the coverage and for the land use variable. There is a lower semi-trapezoidal function since the higher the value of the degree of membership of the function, the better the land use.

The zoning and assignment of values for the variables that make up the LUCI were based on specific criteria related to the dynamics of land use in the basin. Factors such as the intensity of agricultural activity, urban expansion, deforestation, preservation of natural areas, among others, were taken into account. These criteria allowed us to establish a scale of values for the LUCI that reflected the magnitude and direction of changes in land use in each of the districts.

In summary, the LUCI of the Quilichao River was carried out through a process that integrated the georeferencing and zoning of the paths present in the ecosystem. This methodology allowed us to obtain an accurate and detailed evaluation of changes in land use in the basin, providing valuable information for the management and conservation of this important water and environmental resource. For the coverage variables, a higher semi-trapezoidal function was assigned because the higher the value of the degree of membership of the function, the better the coverage, and for the land use variable, there is a lower semi-trapezoidal function, since between the higher the value of the degree of membership of the function, the better the land use, as shown.

On the other hand, once the functions are defined, the next step is to define the fuzzy rules which are the set of IF–THEN propositions that model the problem to be solved. The behavior of the entire system is determined by the union or sum of all of them and is expressed with statements of the type:

IF input conditions THEN action to execute

Rules establish connections between fuzzy sets and values, providing the ability to express in linguistic language various problem considerations that would otherwise be difficult to model or require considerable time and effort to formulate. Once the functions are defined, the rules are made for each variable, having three levels: low, medium, and high. Possible combinations between coverage levels and percentage of land use are generated. It is important to keep in mind that for the research, the coverage variable has a greater degree of importance with respect to land use, and this is because soil coverage in the Quilichao river basin is of utmost importance, due to its direct influence on the health and functioning of the entire ecosystem. Some key points of its importance are: conservation of biodiversity, regulation of the hydrological cycle, protection of water quality, mitigation of climate change, sustainability of natural resources; it is essential for ecological balance, and its conservation and proper management are essential to ensure the health and resilience of the entire ecosystem of the premontane humid forest. These are some of the rules that were defined, and LUCI index in Table 1 corresponds to "Rules for the Land Use Index," that is, the relationship between coverage and land use so that the fuzzy model defines the range of the index (Table 1).

In Sect. 4, once we have obtained the final results, the GeoFIS program gives us the ability to export these results in format shape and *csv. To analyze the resulting values for each of the 60 points that we have collected, we chose to export the data in *csv format to facilitate analysis and interpretation of the results. Finally, in Sect. 5, once the results have been obtained, we proceed to visualize them in ArcGIS, in order to create the map of our study area from the values obtained in fuzzy logic. To obtain the ranges of: high, medium, and low, we used a natural classification (Jenks), which is a data classification method that seeks to divide the data into classes according to natural groups present in the distribution of the data. Looking at the histogram of the data, you look for natural breaks at the low points of the valleys. The assignment of the order of the slices is done based on the size of the valleys, so that the first natural slice is assigned to the largest valley identified in the histogram. This approach allows an intuitive classification based on the intrinsic structure of the data, which facilitates its interpretation and analysis. (https://pro.arcgis.com/). The land use coverage index (LUCI) map is presented in Fig. 4, showing the various coverage labels in the basin with distinctive colors. In this scheme, green represents the low category, yellow the medium category, and red the high category.

Table 1 Rules for the land use index

Coverage	Land use	Land use index
Low	High	High
High	Low	Low

3 Results

Below are the most abundant covers in the basin, along with their respective label based on the results generated through fuzzy logic from the LUCI (land use coverage index):

Table 2 presents the main land use coverages identified in the Quilichao river basin, along with their respective uses and classification labels. The agricultural zone includes sugar cane crops and is classified as low density. Permanent herbaceous crops include other transitional crops and are classified as medium density. Finally, the artificialized areas represent the discontinuous urban fabric, classified as high density.

The land use coverage index (LUCI) map is presented in Fig. 4, showing the various coverage labels in the basin with distinctive colors. In this scheme, green represents the low category, yellow the medium category, and red the high category. Figure 4 is the map resulting from the classification of land use in the Quilichao river basin, Cauca, Colombia. The figure shows different land use categories with a color scale, indicating critical areas, agricultural zones, and vegetation areas. It also includes the geographical location of the watershed in South America, Colombia, and the department of Cauca.

The output of the land use coverage index is fuzzy (Fig. 4, which is characterized by its ability to manage uncertainty by assigning degrees of membership to different fuzzy sets). When coverage is low and land use is high, a high value is obtained for the LUCI, and this is because there is a

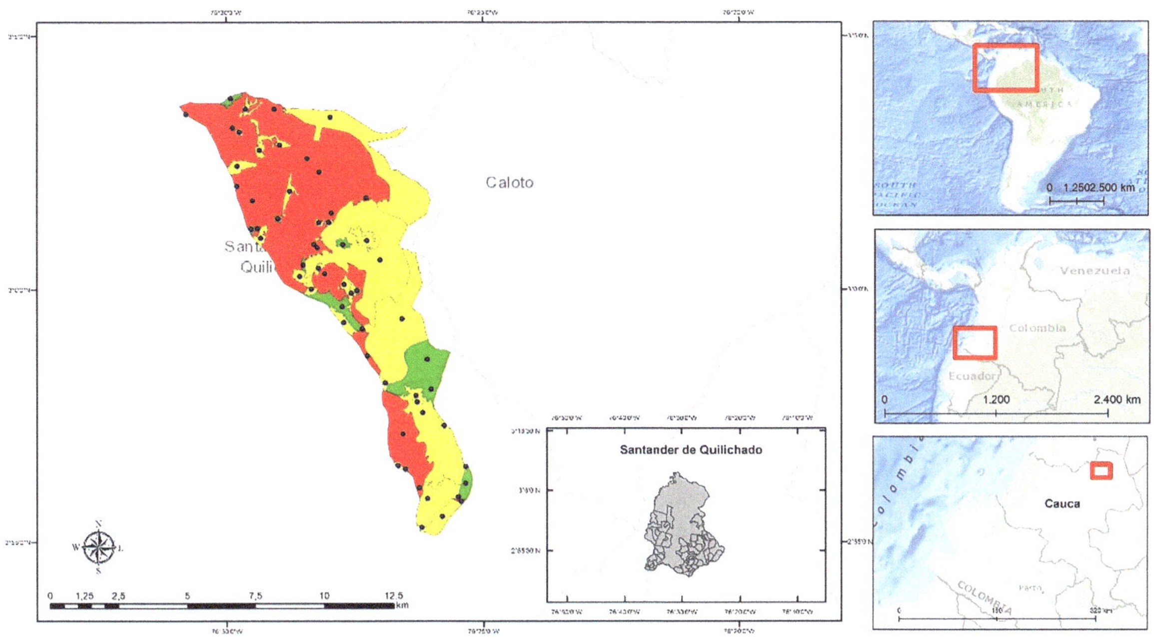

Fig. 4 Map of the land use coverage index (LUCI). *Source* Self made

Table 2 More abundant coverage in the basin

Coverage	Land use	Labels
Agricultural zone	Sugar cane	Low
Herbaceous permanent crops	Other transitional crops	Average
Artificialized	Discontinuous urban fabric	High

greater impact, that is, a conflict arises over land use when the ideal conditions of the ecosystem change. In the opposite case, having a low LUCI classification in the Quilichao river basin can have several significant implications for the coverage and health of the river, as well as for the ecosystem in general. Here are some key implications: loss of natural plant cover, increased vulnerability to erosion, risks of floods and droughts, decreased water quality, reduced wildlife habitats, less resilience to climate change.

3.1 LUCI and CLC Classification 2018–2021

The use of the land cover and land use index through fuzzy logic in the Quilichao river basin proposes a more detailed and specific classification of land use classes in the Quilichao river basin. This is due to its ability to capture subtle nuances and variations in land use patterns, especially in transitional or mixed areas. Therefore, for the purpose of this research, the CORINE Land Cover classification, although it may be more easily interpretable and comparative at an international level due to its standardization and familiarity at the European level, may lose details and subtleties in the representation of land cover in specific areas of the Quilichao river basin. With the classification, the area of each of the years of study was established, and in this way, a comparison was made of the results obtained from the LUCI and the changes in coverage in these time periods, where obtained the following: the transition matrix for the years 2019–2020 (Table 3). Dense vegetation with −119 ha and shrubs with −95.91, respectively, represent those areas that changed with respect to their initial coverage. Table 3 shows the changes in the different land use classes in the Quilichao river basin between the years 2019 and 2020. The areas in hectares for each class in both years are included, as well as the absolute (Δ) and percentage variation. Categories include water bodies, dense vegetation, sparse grasslands, crops, shrublands, buildings, and clouds. Negative values indicate a decrease in area, with a notable reduction in dense vegetation and buildings, while water bodies and crops show an increase.

The relationship between LUCI and CLC classification is that LUCI can be used as a measure to validate and improve the accuracy of land cover classification carried out using CLC. By comparing the results of both methods, it is possible to identify in the period from 2019 to 2020 that the impacts on dense vegetation and shrub coverage are related to conflicts over land use from the upper part of the basin. Therefore, urban expansion affects ecosystem services due to the absence of frameworks that guarantee territorial planning. If there is no adequate planning toward sustainability, it would have a negative impact on the livelihoods of the communities that inhabit the territory in rural

Table 3 Variation analysis of soil changes between the years 2019 and 2020

Classes	Use class	2019 (ha)	2020 (ha)	Δ	2019%	2020%	Δ (ha)
1	Water bodies	22.28	26.63	4.35	0.368	0.440422	0.0719
2	Dense vegetation	2467.85	2348.27	−119.58	40.815	38.837	−1.9776
3	Scarce pastures	574.45	607.72	33.27	9.501	10.05	0.5502
4	Crops	1804.38	2097.93	239.55	29.842	34.6968	4.85
5	Shrubbery	436.11	340.20	−95.91	7.213	5.6264	−1.5862
6	Buildings	740.84	624.84	−116.00	12.252	10.3339	−1.9184
7	Clouds	0.55	0.87	0.32	0.009	0.01438	0.005292

Note Δ represents the change between years. Negative values indicate decrease in area

and urban areas. Given the conditions in Colombia, where the implications of climate change are due to the geographical, ecological, and socioeconomic particularities of the territory, it requires the implementation of adaptive measures in response, as established in the National Plan for Adaptation to Climate Change (PNACC) of the National Planning Department (DNP 2013).

The variation analysis of soil changes between the years 2020 and 2021 (Table 4), showing a decrease in its order of −1.95 ha for bodies of water. In this context, the change in the coverage of the dynamics of the Quilichao river basin can have serious consequences for biodiversity, climate, local communities, and ecosystem services, which highlights the critical importance of conserving and managing sustainable these valuable natural resources. It is necessary in this context that initiatives be aimed at reducing the vulnerability of natural and human systems against the real or anticipated effects of climate change, as indicated by the Ministry of Environment and Sustainable Development (MADS 2017). According to the results of the LUCI, the high classification for the discontinuous urban fabric shows changes from the upper zone of the basin; therefore, tools are required to organize the territory given the population growth that is extending to rural areas and that would generate conflicts in the Quilichao river ecosystem.

The description of Fig. 5 is made in the methodology to establish the importance of the information with the classification, the area of each of the years of study was established, and in this way, a comparison of this sequence was made. Transition matrix and variation analysis: This is with the objective of comparing sequential changes in the 2019–2021 time period. The CLC classification is established also considering the year 2018. This is with the objective of validating whether the output of the fuzzy model for the land cover and land use index is comparable with the classes obtained from the CORIN database. Once the coverage has been established, detailed maps have been generated that show the spatial distribution of land cover at different time points, allowing patterns and trends to be identified throughout the period studied. In addition, coverage classification techniques have been applied to categorize areas based on their uses, such as agriculture, forests, urbanized areas, bodies of water, among others (Fig. 5). This is the land use and land cover (LULC) classification base map in the Quilichao river basin for 2018. The figure shows the different cover categories, including water bodies, dense vegetation, cultivated areas, built-up areas, clouds, and scattered grasslands, allowing

Table 4 Variation analysis of soil changes between the years 2020 and 2021

Classes	Use class	2020 (ha)	2021 (ha)	Δ	2020%	2021%	Δ (ha)
1	Water bodies	26.63	25.10	−1.53	0.4404	0.41511	−0.0253
2	Dense vegetation	2348.27	2416.46	68.19	38.83	39.9648	1.1277
3	Scarce pastures	607.72	590.80	−16.92	10.0508	9.771	− 0.2798
4	Crops	2097.93	2007.12	−90.81	34.696	33.194	−1.501
5	Shrubbery	340.20	527.78	187.58	5.626	8.728743	3.1023
6	Buildings	624.84	478.31	−146.53	10.333	7.9105	−2.423
7	Clouds	0.87	0.89	0.02	0.0143	0.0147	0.00033

Note Δ represents the change between years. Negative values indicate decrease in area

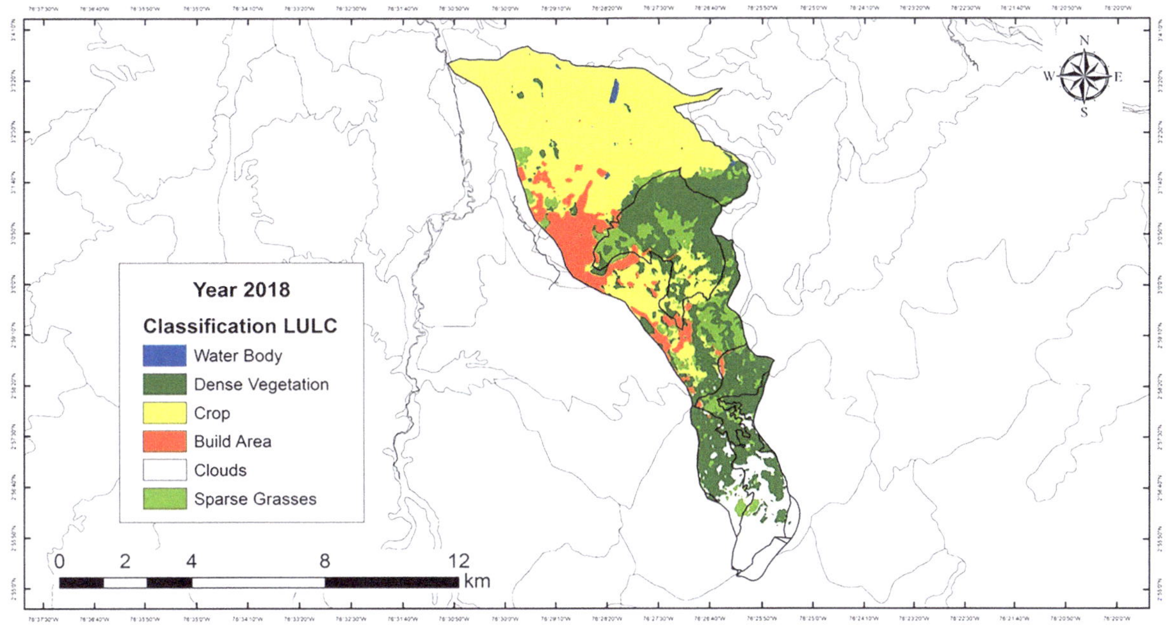

Fig. 5 CLC classification, year 2018. *Source* Self made

to evaluate the spatial distribution of these elements in the region.

The result of the land use and land cover (LULC) classification analysis in the Quilichao river basin for the year 2019, using the MOLUSCE complement, is shown in Fig. 6. The results show the land cover categories, including water bodies, dense vegetation, crops, built-up areas, and scattered pastures. This classification allows identifying the main areas of land use and cover, which is fundamental to understand the territorial dynamics and environmental changes affecting the watershed.

It is important to highlight that the territorial change matrix is represented in Fig. 7, as a result of the land use and land cover (LULC) classification analysis in the Quilichao river basin for the year 2020, obtained through

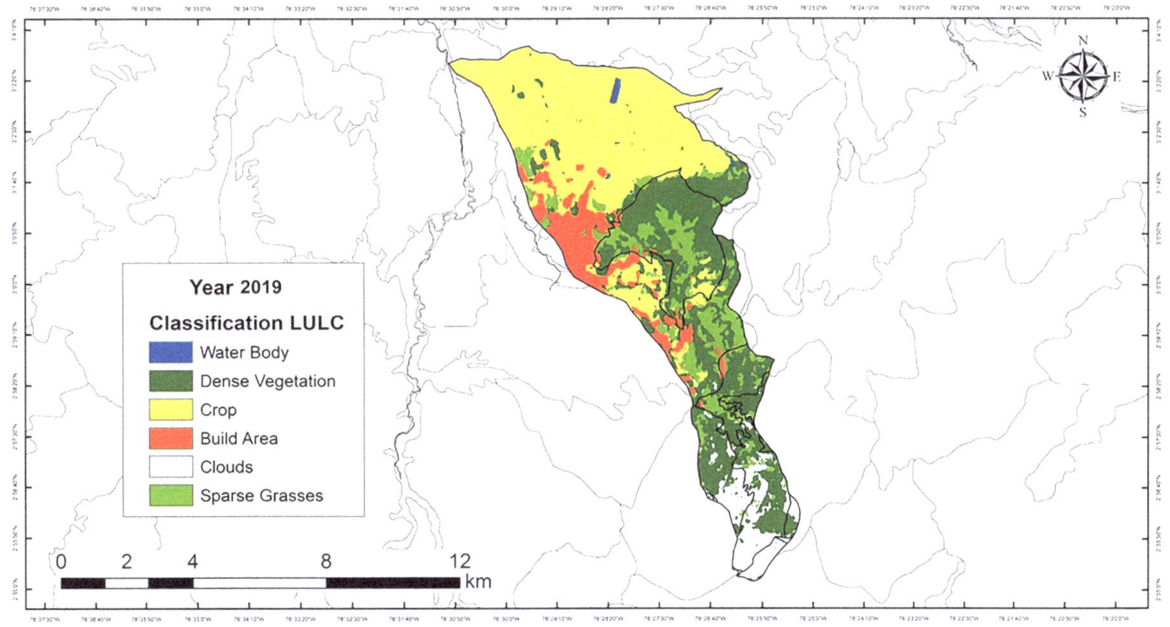

Fig. 6 CLC classification, year 2019. *Source* Self made

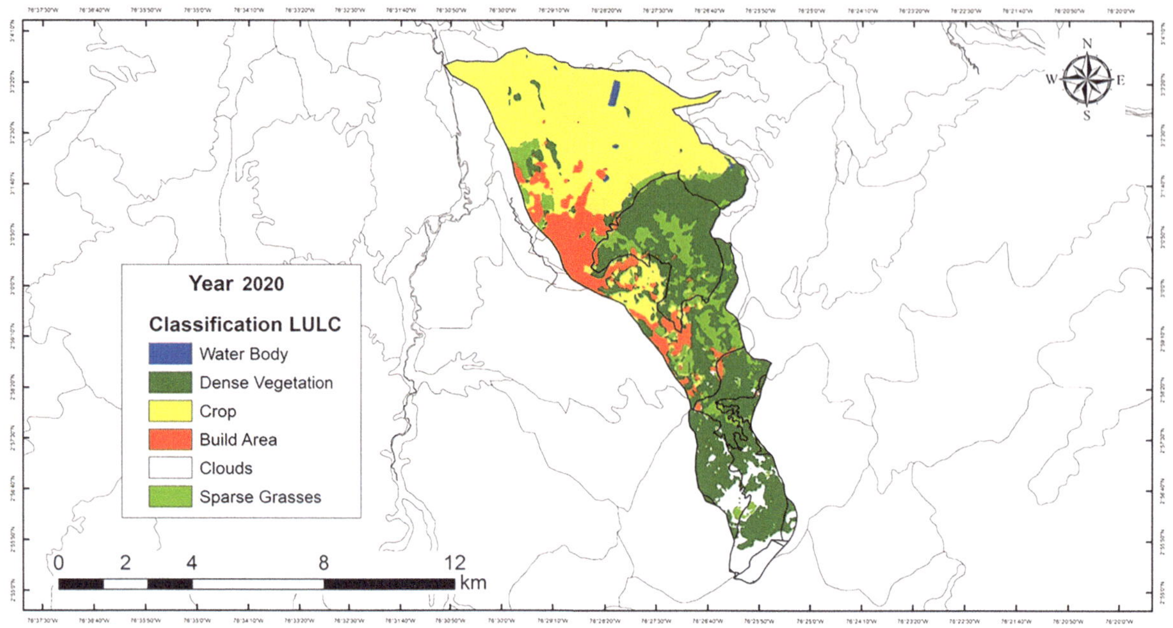

Fig. 7 CLC classification, year 2020. *Source* Self made

the MOLUSCE complement, allowed identifying and quantifying the spatial distribution of the different land cover categories, including water bodies, dense vegetation, crops, built-up areas, and dispersed pastures. The results highlight the dynamics of land use change compared to the previous year (Fig. 6), providing critical information for assessing spatial and temporal trends of land change in the watershed. These changes have important implications for the sustainable management of the territory, as they allow identifying areas of agricultural expansion, decrease in natural vegetation and urban development, which directly affect the resilience of the ecosystem to anthropogenic pressures and climate change.

Finally, the land use and land cover (LULC) classification analysis was done in the Quilichao river basin for the year 2021, using the MOLUSCE add-on. Figure 8 shows the spatial distribution of the different land use and land cover categories, such as water bodies, dense vegetation, cultivated areas, built-up areas, and scattered pastures. This analysis provides a detailed view on the evolution of the landscape in the watershed and allows us to identify key patterns in land use change, highlighting the reduction in natural vegetation cover and the expansion of built-up areas. These results are fundamental for assessing the cumulative impacts of anthropogenic activities and their relationship with ecosystem sustainability, as well as for the design of territorial management strategies that promote resilience to climate change and the conservation of natural resources.

4 Discussion

The use of LUCI logic in the Quilichao river basin proposes a more detailed and specific classification of land use classes in the Quilichao river basin. This is due to its ability to capture subtle nuances and variations in land use patterns, especially in transitional or mixed areas. Therefore, for the purpose of this research, the CORINE Land Cover classification, although it may be more easily interpretable and comparative at an international level due to its standardization and familiarity at the European level, may lose details and subtleties in the representation of land cover. Based on the categories established at the high level, there is coverage of fragmented forest, discontinuous urban fabric, continuous urban fabric, and broadleaf riparian forest. Given this scenario, it is necessary to prioritize the preservation of existing forest areas in the water sources that feed the Quilichao River through their conservation. This measure would contribute significantly to the conservation and increase of biodiversity in both terrestrial and aquatic ecosystems. This has a high degree of environmental importance because according to Drenkhan et al. (2023) the regions of the tropical Andes stand out as priority areas. The interpretation of the results of the land cover and land use index will require a more detailed and specific analysis of the classes obtained. This can provide valuable information on specific land use patterns, such as intensive agricultural areas, degraded forests, or expanding urban areas.

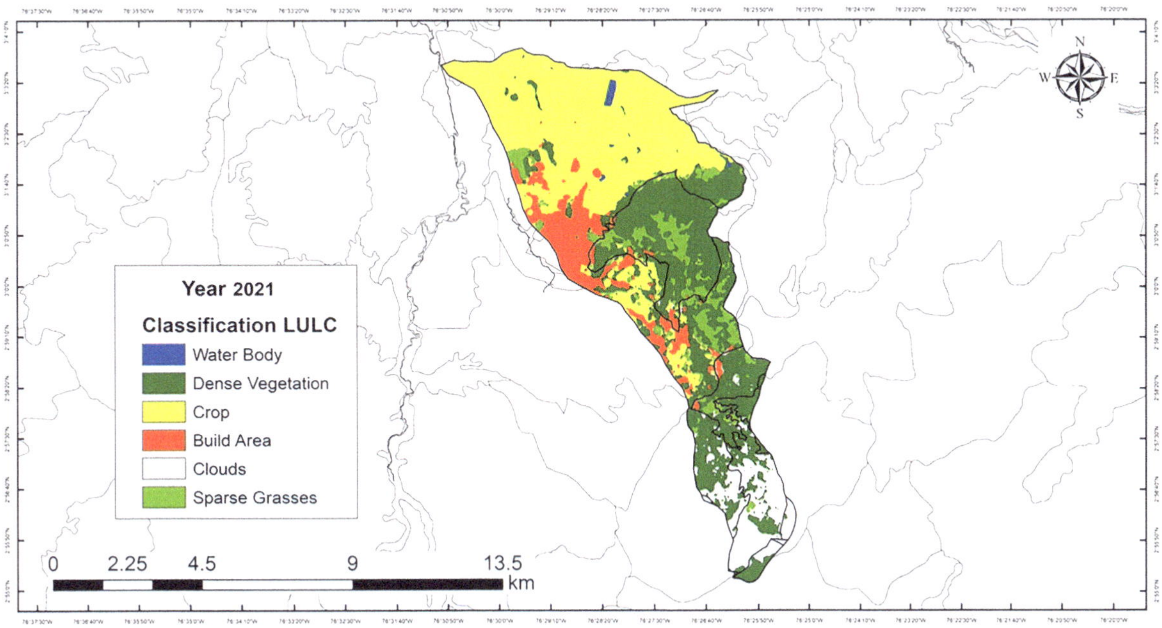

Fig. 8 CLC classification, year 2021. *Source* Self made

At the medium level, it is shown that the coverage of other permanent tree crops, other transitional crops, high secondary vegetation, wooded pastures, low dense terra signature forest, and mixed riparian forest are the most predominant. This reality shows that agricultural areas and permanent productive systems impacted the coverage of the basin. The absence of forest vegetation in the protection areas of the Quilichao River and the use of land in agricultural activities impact the retention of sediments and chemicals derived from agriculture; it is necessary to carry out actions in a collaborative and participatory manner to guarantee a comprehensive and sustainable approach to environmental conservation. (Arias-Arevalo and Pacheco-Valles 2022). At the low level, there are: landfills, mosaic of crops, pastures and natural spaces, sugar cane, coniferous plantation, clean grass, swampy area, weeded pastures, mosaic of pastures with natural spaces. They represent the land uses that still exist preserve most of its coverage, therefore, in the presence of the last relict of premontane humid forest life zone seven according to Holdridge, in the Munchique nature reserve territory of the Quilichao river basin. It is a key piece of conservation to establish the dynamics of the most important water system in northern Cauca.

5 Conclusions

The land cover and land use index using fuzzy logic can be especially useful for decision-making at the local or regional level in the Quilichao river basin. It can provide detailed information for territorial planning, natural resource management, and the identification of critical areas for conservation or restoration.

The CORINE Land Cover classification, for its part, may be more suitable for comparative studies at a national or international level, as well as for monitoring long-term changes in land cover in a given region, since when analyzing these impacts in coverage should be a reason for planning since biological diversity, landscape connectivity, soil stabilization, and ecosystem services must be guaranteed (Rodrıguez et al. 2023; Rincon-Ruiz et al. 2014; Bennett et al. 2023; Schmera et al. 2023).

A significant advantage of the fuzzy logic land use and cover index is its ability to capture uncertainty and imprecision in the data, allowing for more detailed and accurate classification of land use patterns. This is especially useful in areas where there are transition zones or mixing of different types of land use (Zhou et al. 2022; Ji and Wu 2022); fuzzy logic is useful in land use classification due to its ability to model the inherent uncertainty in the data and provide more accurate classification in transition areas.

A disadvantage that could be presented by the land cover and land use index is that it can result in an excessively detailed or complex classification, which makes interpretation and comparison with other standardized classifications difficult (Wu et al. 2020).

6 Declarations

The authors declare that they have no known competing financial interests or personal relationships that could have appeared to influence the work reported in this paper.

Acknowledgements This research is funded within the framework of the internal call to support the training of doctoral students #149 of 2023, Universidad del Valle; the Water and Soil Resource Engineering Research Group - IREHISA, of the Universidad del Valle, Cali, Colombia; the Interinstitutional Doctorate in Environmental Sciences assigned to the Faculty of Natural and Exact Sciences of the Universidad del Valle, Cali, Colombia; the Indigenous Authorities of the Ancestral Territory of Munchique Los Tigreskwekwe nejwesx, Santander de Quilichao, Cauca, Colombia.

References

Arias-Arevalo, P., Pacheco-Valles, N.: Implementation of payments for environmental services in the Cali River Basin, Colombia: a look from socioecological systems. Revista de Economia e Sociologia Rural **61** (2022)

Baez, A., Duran, V.: Classification of the applications of fuzzy logic in accounting sciences

Bedoya, M.A., Ramırez, B.H.: The effects of climate and forest cover variability on the hydrological regulation of an eastern Andean Cusiana river sub-basin. Ecohydrology & Hydrobiology **23**(4), 569–587 (2023)

Bennett, A.C., Rodrigues de Sousa, T., Monteagudo-Mendoza, A., EsquivelMuelbert, A., Morandi, P.S., Coelho de Souza, F., Phillips, O.L., et al.: Sensitivity of South American tropical forests to an extreme climate anomaly. Nat. Climate Change **13**(9), 967–974 (2023)

Birkmann, J., et al.: Addressing the challenge: recommendations and quality criteria for linking disaster risk reduction and adaptation to climate change, Bonn (2009)

Busby, J.W., Cook, K.H., Vizy, E.K., Smith, T.G., Bekalo, M.: Identifying hot spots of security vulnerability associated with climate change in Africa. Clim. Change **124**(4), 717–731 (2014)

Castellanos-Tapia, F., Fernandez, G., Ruiz-Luna, A., Cervantes-Escobar, A., Berlanga-Robles, C.A.: Landscape changes in a critical subtropical coastal wetland in northwestern Mexico: Is shrimp farming a driver of concern? Land (2022)

Chen, H., Xie, W., Chen, X., Han, J., A¨ıt-Ahmed, N., Zhou, Z., Benbouzid, M.: Fractional-order PI control of DFIG-based tidal stream turbine. J. Mar. Sci. Eng. **8**(5), 309 (2020)

Climate change adaptation plan. Decree 23, Municipal Mayor's Office, Santander de Quilichao (2019)

Cuevas Caamanõ, F.R.: Analysis of the effects of land cover change on water erosion processes in the Coyanco estuary basin (1986–2020)

Daly, C.: A framework for assessing the vulnerability of archaeological sites to climate change: theory, development, and application. Conservat. Manage. Archaeol. Sites **16**(3), 268–282 (2014)

Departmental plan for climate change, department of Cauca. Technical report, PDCC CAUCA, Popayan (June 2016)

Departmental Government Responds to the Emergency in Santander de Quilichao in a Timely Manner. https://www.magazincnc.com/gobierno-departamental-atiende-de-manera-oportuna-emergencia-en-santander-de-quilichao/

DNP: National development plan. Working document, National Planning Department, Colombia (2013)

Drenkhan, F., Buytaert, W., Mackay, J.D., Barrand, N.E., Hannah, D.M., Huggel, C.: Looking beyond glaciers to understand mountain water security. Nat. Sustain. **6**(2), 130–138 (2023)

EMQUILICHAO: Environmental management plan, Munchique-Los Tigres Natural Reserve. Environmental management plan, Municipal Company of Quilichao, Santander de Quilichao, Colombia (2016). https://emquilichao.gov.co/forestal/

Figueroa, F., Sanchez-Cordero, V., Illoldi-Rangel, P., Linaje, M.: Evaluation of the effectiveness of protected areas to contain processes of change in land use and vegetation: Is an index enough? Mexican J. Biodivers. **82**(3), 951–963 (2011)

Fleites Ramos, Y., Vizcaıno Escobar, A.E., Crespo Hurtado, E.: Logica difusa: metodologıa cuantitativa util para determinar el consenso entre expertos de ciencias estomatologicas. Medicentro **25**(2), 305–314 (2021)

Flores, A.R., Olmos-Martınez, E.: Ecosystem services in the Coyuca Lagoon, Guerrero, Mexico, a tourism perspective. PASOS J. Tour. Cultural Heritage **18**(5), 779–793 (2020)

Giri, S., Lathrop, R.G., Obropta, C.C.: Climate change vulnerability assessment and adaptation strategies through best management practices. J. Hydrol. **580**, 124311 (2020)

Guillaume, S., Bates, T., Lablee, J.-L., Betts, T., Taylor, J.: Combining spatial data layers using fuzzy inference systems: Application to an agronomic case study. In: Proceedings of the 6th International Conference on Geographical Information Systems Theory, Applications and Management (2020)

Hussain, S., Lu, L., Mubeen, M., Nasim, W., Karuppannan, S., Fahad, S., Aslam, M., et al.: Spatiotemporal variation in land use land cover in the response to local climate change using multispectral remote sensing data. Land **11**(5), 595 (2022)

Hydrographic basin planning and management plans – POMCA. Technical report, National Development Plan (2011)

IDEAM: Second communication from Colombia to the United Nations Framework Convention. Technical report, Institute of Hydrology, Meteorology and Environmental Studies of Colombia (2010). http://www.pnud.org.co//imgupload/36353463616361636163616361636163/2%C2%AAComunicaci%C3%B3nCap5.pdf

IDEAM: Global circulation models for Colombia. Technical report, Institute of Hydrology, Meteorology, and Environmental Studies, Bogota DC (2007)

IDEAM: National water study 2014. Technical report, Institute of Hydrology, Meteorology and Environmental Studies of Colombia (2015)

IUCN: Global standard for nature-based solutions. a user-friendly framework for the verification, design and scaling up of nbs. first edition. Technical report, IUCN, International Union for Conservation of Nature, Gland, Switzerland (2020). https://doi.org/10.2305/IUCN.CH.2020.08.en

Ji, M., Wu, Z.: Automatic detection and severity analysis of grape black measles disease based on deep learning and fuzzy logic. Comput. Electron. Agric. **193**, 106718 (2022)

Leroux, C., Jones, H., Pichon, L., Guillaume, S., Lamour, J., Taylor, J., Naud, O., Crestey, T., Lablee, J., Tisseyre, B.: Geofis: an open source, decision-support tool for precision agriculture data. Agriculture **8**(6) (2018)

Liu, B., Pan, L., Qi, Y., Guan, X., Li, J.: Land use and land cover change in the Yellow River Basin from 1980 to 2015 and its impact on the ecosystem services. Land **10**(10), 1080 (2021)

MADS: National and departmental inventory of greenhouse gases - Colombia. Third national communication on climate change, Ministry of Environment and Sustainable Development, Bogota DC (2017)

Mafi-Gholami, D., Jaafari, A., Zenner, E.K., Kamari, A.N., Bui, D.T.: Vulnerability of coastal communities to climate change: thirty-year trend analysis and prospective prediction for the coastal regions of the Persian Gulf and Gulf of Oman. Sci. Total. Environ. **741**, 140305 (2020)

Magazine CNC: Departmental Government Responds to the Emergency in Santander de Quilichao in a Timely manner. https://www.magazincnc.com/gobierno-departamental-atiende-de-manera-oportuna-emergencia-en-santander-de-quilichao. Accessed 2023

Metodos de Clasificacion de datos. https://pro.arcgis.com/es/pro-app/latest/help/mapping/layer-properties/data-classification-methods.htm. Accessed 13March 2024

Mun˜oz, D.A., Baro Saurez, J.E.: Identificacion de los factores subyacentes de la vulnerabilidad institucional: una reflexion sobre los procesos generadores de riesgoa escala local. In: Geologıa Ambiental Y Recursos Hıdricos **1**, 151–163 (2023)

Munevar Quintero, C., Ramırez Rıos, M.: The National System of Protected Areas in Colombia: Conceptual developments from the socio-legal and environmental doctrine. Magaz. Legal **18**(2) (2021)

Nasiri, S., Khosravani, M.R.: Faults and failures prediction in injection molding process. Int. J. Adv. Manuf. Technol. **103**(5), 2469–2484 (2019)

Pabon-Caicedo, J.D., Pilar Ycaza, R., Friend, F., Espinoza, D., Fenzl, N., Apostolova, M.: Vulnerability of the Amazon basin to extreme hydroclimatic phenomena. Geography Notebooks: Colombian Geography Magaz **27**(1), 27–49 (2018)

Rai, S., Srinivas, R., Magner, J.: Using fuzzy logic-based hybrid modeling to guide riparian best management practices selection in tributaries of the Minnesota River Basin. J. Hydrol. **608**, 127628 (2022)

Reckien, D.: Assessing social vulnerability in New York City: Methods, mapping and implications for adaptation. In: North American Symposium on Climate Change Adaptation: Fostering Resilience and the Regional Capacity to Adapt (2016)

Rincon-Ruiz, A., Echeverry-Duque, M., Pineros, A.M., Tapia, C.H., David, A., Arias-Arevalo, P., Zuluaga, P.A.: Valoracion Integral de la Biodiversidad Y Los Servicios Ecosistemicos: Aspectos Conceptuales Y Metodologicos. Instituto de Investigacion de Recursos Biologicos Alexander von Humboldt, Bogota, Colombia (2014)

Rodrıguez, T., Reu, B., Bolıvar-Santamarıa, S., Cortes-Aguilar, A., Buendıa, C.: A framework for participatory scenario planning to guide transitions towards sustainability in mountain social-ecological systems: a case study from the Colombian Andes. Land Use Policy **132**, 106817 (2023)

Romero Hernandez, C.P.: Analysis of the impact of the growth of megacities on the hydrological cycle under climate change scenarios. Application to the Bogota River Basin (Colombia). PhD thesis, Universitat Polytechnic of Valencia (2023)

Rosero Mier, M.M.: Multi-temporal analysis of land use and vegetation cover of the Tahuando River basin and projection of changes to the year 2031, in the Ibarra canton, province of Imbabura. Master's thesis (2018)

Saavedra Briones, P., Sepuveda Varas, A., Esse Herrera, C.: Analysis of land cover and use change in a Chilean pre-Andean subbasin: Tool for the productive sustainability of a territory (2022)

Saavedra Briones, P., Sepulveda-Varas, A.: Systematic transitions in land use and land cover in a pre-Andean sub-watershed with high human intervention in the Araucania Region, Chile. Ciencia e Investigacion Agraria **43**(3), 396–407 (2016) https://doi.org/10.4067/s0718-16202016000300006

Santos, R.B.D., Menezes, J.A., Confalonieri, U., Madureira, A.P., Duval, I.D.B., Garcia, P.P., Margonari, C.: Construction and application of a human vulnerability index to climate change for the Brazilian context: the experience of the state of Espırito Santo. Health Soc. **28**, 299–321 (2019)

Schmera, D., Boschi, C., Baur, B.: The type of forest edge shapes snail assemblages at forest-pasture transitions. Sci. Rep. **13**(1), 16761 (2023)

Souza, L., Norby, R.J., Weltzin, J.F., Classen, A.T.: Climate change effects on plant biomass alter dominance patterns and community evenness in an experimental old-field ecosystem. Glob. Change Biol. **16**(10), 2676–2687 (2010)

Tagliari, M.M., Bogoni, J.A., Blanco, G.D., Cruz, A.P., Peroni, N.: Disrupting a socio-ecological system: could traditional ecological knowledge be the key to preserving the Araucaria Forest in Brazil under climate change? Clim. Change **176**(2), 2 (2023)

Tutorial Introduccion a la Percepcion remota. http://www.teledet.com.uy/tutorial-imagenes-satelitales/clasificacion-supervisada.htm

Vadiati, M., Nalley, D., Adamowski, J., Nakhaei, M., Asghari-Moghaddam, A.: A comparative study of fuzzy logic-based models for groundwater quality evaluation based on irrigation indices. J Water Land Dev. (2019)

Van Vliet, J., Eitelberg, D.A., Verburg, P.H.: A global analysis of land take in cropland areas and production displacement from urbanization. Glob. Environ. Chang. **43**, 107–115 (2017)

Vargas-Franco, V., Restrepo-Tarquino, I.: Construction of index with artificial intelligence to evaluate vulnerability to climate change in Andean tropical microwatersheds. Study case in Colombia. DYNA **85**(204), 194–203 (2018). https://doi.org/10.15446/dyna.v85n204.67048

Vergara, X., Carmona, A., Nahuelhual, L.: Spatial coupling and decoupling between ecosystem services provisioning and benefiting areas: Implications for marine spatial planning. Ocean Coast. Manag. **203**, 105455 (2021)

Wang, Y., Shen, X., Jiang, M., Lu, X.: Vegetation change and its response to climate change between 2000 and 2016 in marshes of the Songnen Plain Northeast China. Sustainability **12**(9), 3569 (2020)

Webster, P., Holland, J., Curry, G.J., Chang, A.: Changes in Tropical cyclone number, duration, and intensity in a warming environment. Science **309**(5742), 1844–1846 (2005)

Wehner, M., Stone, D., Mitchell, D., Shiogama, H., Fischer, E., Graff, L.S., Krishnan, H., et al.: Changes in extremely hot days under stabilized 1.5 and 2.0 C global warming scenarios as simulated by the HAPPI multi-model ensemble. Earth Syst. Dynam. **9**(1), 299–311 (2018)

Wu, B., Cheng, T., Yip, T.L., Wang, Y.: Fuzzy logic based dynamic decisionmaking system for intelligent navigation strategy within inland traffic separation schemes. Ocean Eng. **197**, 106909 (2020)

Xu, Z., Bambrick, H., Frentiu, F.D., Devine, G., Yakob, L., Williams, G., Hu, W.: Projecting the future of dengue under climate change scenarios: progress, uncertainties and research needs. PLoS Negl. Trop. Dis. **14**(3), 0008118 (2020)

Zhou, J., Liu, L., Leng, Y., Yang, Y., Gao, B., Jiang, Z., Yuan, Q., et al.: Both cross-patient and patient-specific seizure detection based on self-organizing fuzzy logic. Int. J. Neural Syst. **32**(06), 2250017 (2022)

The Remorse of the Earth. A Comparison of Agro-Reforestation Practices on the Ruins of the Salento Olive-Growing Landscape

Gustavo D'Aversa

Abstract

Xylella fastidiosa's rapid olive tree desiccation disease in Salento, along with its landscape repercussions, epitomizes an ecological crisis and a cultural apocalypse, linking landscapes and local communities to climate change. New reforestation and agro-reforestation projects are being built and tested on the rubble of the olive-growing landscape. These will be compared and contrasted. On the one hand, super-intensive exploitation practices, linked to governmental microbiopolitical prescriptions to combat the epidemic, will be confronted. On the other hand, we will compare two coexisting case studies of experimental practices related to postdisaster reforestation/agro-reforestation: the Sylva Foundation and Manu Manu Riforesta. On a local scale, in the present case of the agro-reforestation movement in Salento, this apocalyptic moment becomes a field of experimental practices in which a more than social movement resists the bio-financialization of the ecological transition, rejecting the carbon credit market.

Keywords

Climate change · Xylella fastidiosa · Postdisaster reforestation · Bio-security in mediterranean farming · Ecological regeneration

G. D'Aversa (✉)
Department of Human and Social Sciences, University of Salento, Lecce, Italy
e-mail: gustavo.daversa@unisalento.it

1 Introduction

The contribution intends to analyze the current implications of the *Xylella* phenomenon in the Salento region, which has been hardest hit by the proliferation of the bacterium since 2013. We will analyze its multiple effects in detail from a geographical perspective in two stages: the first stage focuses on the phenomenon in fieri, starting from the history of olive cultivation as a monoculture, the identity and landscape implications (Alliegro 2017), the valences of this capitalist ecological regime (Haraway and Tsing 2019), and the biosecurity fallout (Bandiera 2020; Ciervo 2020, 2023); the second related to the phenomenon *ex post*, in which the rapid olive tree desiccation is configured as an epiphenomenon of the global ecological crisis, in the general framework of climate change.

In particular, we will look at how the break in the Nature–Culture schema is linked to cultural apocalypses (De Martino 1977) and the Anthropocene (Latour 2009, 2014) from the point of view of hybrid geography (Whatmore 2003), which studies the more-than-human geographies of other nonhumans such as bacteria, trees, and plants (Lorimer 2012; Bingham 2006; Clark 2003; Franklin 2008; Griffin 2010; Jones and Cloke 2008).

As a result, Salento has become an experimental field for ecological adaptation techniques. Therefore, we will examine some practices that have been developed on the ruins of Salento olive growing, which, by declaring an ongoing state of emergency, have enabled governmental biological policies to combat the epidemic, including the replanting of resistant varieties. Huge portions of land have been freed up by emergency measures and compulsory eradication, opening it up to new markets and energy-oriented profitability, making Salento attractive to multiple interests (Ciervo 2023). These include the reindustrialization of the countryside through

P. Mohammad et al. (eds.), *Geographic Approaches to Climate Change and Mitigation: Urban and Rural Perspectives (Volume 2)*, Advances in Science, Technology & Innovation, https://doi.org/10.1007/978-3-031-92115-5_12

the installation of solar panels and wind turbines, or reforestation practices, specifically agroforestry, which propose profoundly different models. These models include the biofinancing of the ecological transition through the carbon credit market, or the more-than-social movements (Ghelfi 2023) that aim at renaturalization, inspired by the principle of biodiversity.

In order to offer a cross-sectoral view of what is happening on a local scale while exemplifying active policies and practices at a global level, we will comparatively analyze two case studies in the Salento area, Fondazione Sylva, and Manu Manu Riforesta, as model of these dichotomous reforestation practices The comparative study will focus on objectives, strategies, methodologies and techniques; assessment of environmental impacts; community and stakeholder involvement; monitoring tools and evaluation of long-term effectiveness.

2 Problem Statement and Study Aim

This contribution seeks to reflect on the effects of climate change by examining the intersection between global phenomena and local practices in reorganizing the natural environment, landscape, and reterritorialization following a catastrophic event. The rapid drying of olive trees in the southeastern territory of Apulia, specifically in Salento, due to the *Xylella fastidiosa* bacterium, stands out as a particularly significant case. This phenomenon has been the subject of numerous studies since 2013, when the presence of the bacterium was first identified.

2.1 The Emerging Phenomenon (in Fieri)

Up until 2013, approximately eleven million olive trees dominated the agricultural land in Salento, specifically in the Province of Lecce, accounting for over 55% of the utilized agricultural area (SAU) (Ciervo 2020).

These trees, protected by law, were not uprooted, reflecting their deep connection to the identity of the local community. However, the epidemic caused by *Xylella fastidiosa* since 2013 has become the most severe phytosanitary crisis in Italy in recent years, with devastating economic, landscape, and social consequences (EPPO 2024). Beyond the economic impact, which has had significant repercussions on the olive oil production chain and nursery activities, the olive groves represent a natural heritage closely tied to the identity of the community. Public alarm intensified in 2020, driven by the threat to the monumental olive trees, ancient specimens that were proposed for UNESCO World Heritage status (Martella 2022).

Xylella fastidiosa, identified in Puglia as the *pauca* subspecies in 2014, is the causal agent of the disease known as "olive quick decline syndrome" (CODIRO). The symptoms of CODIRO include leaf scorch, scattered twig, and branch dieback that starts from the top of the tree canopy and spreads downward, stunted growth, desiccation, and ultimately, the death of the plants. The pathogen multiplies within and colonizes the plant's vascular system, inducing physiological alterations in response to the infection. The bacterium is naturally transmitted by vector insects, primarily *Philaenus spumarius*, which feed on the raw sap circulating in host plants.

In 2013, about 8000 hectares were affected, compromising approximately one million olive trees. Twelve months later, the affected area had expanded to 20,000 hectares, and a few years later, the disease was declared endemic and considered no longer eradicable, leading the European Commission to adopt a "containment" strategy. *Xylella* is an environmental catastrophe that has profoundly altered the landscape, with over 750,000 hectares of territory affected, accounting for 40% of the region. Moreover, the so-called buffer zone encompassed the provinces of Bari and Barletta-Andria-Trani (BAT), which, with their approximately 132 thousand hectares, represented 12 percent of Italy's usable olive-growing agricultural land (Camera dei Deputati 2024). The contagion, which started in the province of Lecce, covered much of the province of Brindisi, first spreading to that of Taranto and then, by early 2018, reaching the borders of the province of Bari (Ciervo 2020).

The spread of *Xylella* is closely linked to certain climatic areas, including the Mediterranean basin (Purcell 1997), and has significant implications for capitalist ecological impacts and biosecurity (Bandiera 2020; Ciervo 2023). Recent studies have identified four subspecies of the bacterium: *Fastidiosa, Multiplex, Sandy*, and *Pauca*. The strain found in Italy is believed to have originated from ornamental coffee plants imported from Costa Rica, identified as ST53 (Saponari 2013). A crucial factor in the proliferation of *Xylella* is the practice of monoculture planting. The development of this bacterium is intricately linked to monoculture, as this cultivation method creates ideal conditions for the bacterium's spread.

The history of olive cultivation in Southern Italy, especially in Apulia, reveals a longstanding monocultural tradition, with significant implications for the region's identity and landscape (Alliegro 2017).

Starting in the eighteenth century, various notable figures and enlightened scientists showed interest in olive tree diseases and techniques to increase production, also aiming to improve the poor conditions of laborers. During these same decades, King Charles III of Bourbon, an enlightened ruler, promoted the intensification of olive cultivation in Puglia through tax exemptions and land reforms, aligning this period with the cycle of material and financial capital accumulation associated with the English industrial activity (Arrighi 1994). The demand for oil to lubricate the cotton-spinning machines of the emerging Northern European textile industry incentivized the planting of new olive trees, adopting more advanced cultivation techniques for the time. This period significantly shaped the modern landscape, influencing both the ecological and socio-economic structures in the region (Trizio 2000).

This era, characterized by significant olive production, was marked by a highly polarized social system, relying on extremely low-cost labor from farmhands. However, the fundamental change was not a better redistribution of wealth from olive cultivation but rather the control of these trades and the transfer of the "oil" product to English hands (Spedicato 2018). Scholars like Donna Haraway (2019) have described this period as a plantation system, characterized by the emergence of numerous pathogens and the subsequent efforts to counteract their production threats.

2.2 Postoutbreak Effects (Phenomenon *Ex Post*)

The rapid desiccation of olive trees in Salento must be viewed within the broader context of the global ecological crisis, within the framework of climate change. The spread of *Xylella fastidiosa* serves as further evidence of hybrid geography, which encompasses the geographies of nonhuman entities such as trees, plants, and bacteria (Whatmore 2003; Lorimer 2012; Bingham 2006; Clark 2003; Franklin 2008; Griffin 2010; Jones and Cloke 2008).

Latour (2014) notes that these hybrids are both local and global, similar to how *Xylella* spreads by relying on the global transportation of plants via the nursery market. This brings to mind the concept of the end of Nature (Latour 2009) and the idea of "multinatural," co-evolutionary systems (Lorimer 2012). In the Anthropocene, humanity has more power but less control, in fact as humans and their products move, so do bacteria and other entities, usually unnoticed until they become problematic, like *Xylella fastidiosa.*

Ernesto De Martino's exploration of the rupture of the Nature–Culture schema is deeply embedded in his anthropological and philosophical work, where he examined how cultural practices and beliefs shape human experiences, especially in moments of crisis or "cultural apocalypses." De Martino (1977) focused on the tension between nature and culture as a fundamental aspect of human existence. He argued that culture serves as the means by which humanity confronts and manages the uncertainties of nature. However, this tension is starkly revealed when cultural systems fail to address the challenges posed by nature, leading to a rupture in the established schema. This rupture exposes the fragility of cultural constructs and the underlying chaos that culture seeks to control, resulting in what De Martino called a "cultural apocalypse," where the protective layer of culture is stripped away, leaving humanity vulnerable to the raw forces of nature.

This theme is particularly relevant in the context of modern ecological crises, such as the one unfolding in Salento with the spread of *Xylella fastidiosa.* Here, the breakdown of the Nature–Culture relationship is evident, as the disease ravages olive groves that are not only economically vital but also culturally symbolic of the region. In studying *Xylella fastidiosa*, it becomes clear that the forms of knowledge embedded in agricultural practices, and the shifting perspectives on how to address the crisis, reflect a deeper tension between nature and culture. Agriculture, central to the ecological regime of capitalism, plays a significant role in greenhouse gas emissions and the exploitation of natural resources. Yet, it also offers a crucial perspective for rethinking human interaction with the earth. The process of "cultural domestication" transforms physical space into a personal and usable landscape, but when disease strikes, it represents a crisis in the organization of this landscape, evoking fears of "losing the world" and "being lost in the world."

The ecological apocalypse caused by *Xylella* highlights the intricate link between human and non-human health and the harmful effects of contemporary production realities in Puglia. The impact of plantations and industry is inscribed not only in the visible landscape but also invisibly in the soil, waters, plants, and eventually in humans and animals. A perspective beyond the human is necessary to understand health as a property emerging from the interconnections between humans and nonhumans.

The olive plantations in Puglia, while appearing homogeneous, actually encompass diverse agricultural practices and work organization systems. Applying Anna Tsing's concept of a "patchy" landscape, the situation in Salento illustrates the reactualization of capitalist dynamics and the peasant landscape in response to the disease and death of olive trees.

Salento has thus become a testing ground for ecological adaptation techniques, where the loss of these olive groves has not only transformed the landscape but has also triggered broader discussions about climate change, biosecurity, and the ongoing intersection of nature and culture. De Martino's work provides a lens through which to understand these developments, emphasizing the vulnerability of cultural systems in the face of natural forces and the potential for crisis when these systems fail to mediate the relationship between humanity and nature.

In a remarkably short period, the olive landscape of Salento has dissolved, yet from its ruins, new forms of valorization and historical reintegration have emerged, reshaping the territory in profound ways. The traumatic event of the olive tree desiccation has not only connected Salento to the hyper-object of "climate change" but has also introduced a complex array of responses, varying among different stakeholders.

The regional regulations aimed at mitigating the ecological crisis require the eradication of infected plants. However, farmers often resort to burning to dispose of the carcasses of dead or infected trees, adding another problem to the existing epidemic: that of wildfires.

Additionally, the replanting of resistant varieties, such as Leccino and FS-17 (also known as "Favolosa"), cannot be overlooked. The Region of Puglia has actively promoted replanting, leading to numerous concerns about the project's feasibility. While Leccino is a classic but low-yield variety, the Favolosa (FS-17) is a Frantoio variety combined by Giuseppe Fontanazza, designed to be suitable for medium and high-density planting.

Simultaneously, the desiccation caused by *Xylella fastidiosa* has facilitated the repurposing of abandoned inland areas for renewable energy production, such as wind and solar power, aligning with a broader vision of a "techno-rural" scenario in the era of the green economy (Patel 2012). This transformation echoes the historical monocultural practices that once dominated the region, now recast in the service of the "long green revolution."

This contribution intends to investigate analytically and comparatively the experimental reforestation practices, which are a direct response to a catastrophic phenomenon like the one that occurred in the Salento region. These practices refer to vast and deeply contrasting horizons of meaning, capable of deeply affecting the future of the Salento territory, its landscape, communities, and development model. These implications can have both local and global significance, with reference to current green policies being implemented. Studies on the carbon credit market (West et al. 2023), in fact, highlight numerous critical issues regarding these practices, both in terms of methodologies and effectiveness. Therefore, the author intends to verify whether there are grounds to assert that what is happening on a local scale could be construed as a form of greenwashing. Similarly, the author intends to assess the effectiveness of experimental agroforestry regeneration practices, characterized by a critical evaluation of past agricultural practices and the questioning of the boundaries between nature and culture, which are gaining significant social relevance. These practices are pushing to scrutinize the economic and political agendas of industry associations, the Puglia Region, and speculators (Papadopoulos 2018). The challenge of reterritorialization not only lies in the communities engaged in building a new landscape, but also in the effort to imagine alternative landscapes.

3 Methodology

The aim of this research is to obtain a picture of the present, useful for a process of reterritorialization, by combining recent perspectives of academic relevance with the objective reality of the territory in its complexity.

Numerous institutions and ONG are involved in forestation and agroforestation in Puglia. Although driven by the same purpose, they adopt extremely different practices. The Puglia Region addresses forestry issues (including the Xylella emergency) through the Department of Environment, and A.R.I.F. Puglia, the agency for irrigation and forestry activities. The Alta Murgia National Park and other regional parks, along with the Forestry Corps of the Carabinieri, are also actively involved in this issue.

In 2022, the Puglia Region issued a call for proposals (PSR PUGLIA 2014–2022 Rural Development Program 2014–2022. "CALL 2022—Measure 8—Investments in the development of forest areas and in the improvement of forest profitability—Submeasure 8.1 'Support for forestation and afforestation'") with the aim of increasing the use of agricultural and nonagricultural land, not yet forested, throughout the region with fast-growing species for forestry purposes. Public and private landowners, along with their consortia, are the beneficiaries. For state-owned land, support is only available if the land's management body is either a private entity or a municipality.

In terms of NGOs, Legambiente Puglia and WWF Puglia have dedicated decades of effort, particularly in protected areas, alongside recently established organizations like OlivaMi ETS, which operates through certified carbon credits for the Carborea spin-off. We have also selected Fondazione Sylva and Manu Manu Riforesta as case studies.

The methodology used is that of the comparative analysis of Fondazione Sylva and Manu Manu Riforesta, NGOs animated by the same mission but characterized by considerable differences We conducted numerous interviews with the contact persons identified by the two associations to gather relevant data for each case study. The data provided will be subjected to comparative analysis to identify similarities and differences. After that, the results will be analyzed to find patterns, trends, and the real reasons behind the differences that were seen. This will be done by looking at both real-world data and relevant academic theories, and it will involve comparing and reviewing the results.

Manu Manu Riforesta (MMR) is a nonprofit organization that in 2020 started a pilot agroforestry project, acquiring the land called 'Kurumuni', sprout in griko. The mission is the recovery of the Belvedere Forest, a forest that until the end of the nineteenth century extended for about 7 thousand hectares and 32 km^2, within an area involving sixteen municipalities of the lower Salento region. The NGO relies in the support of a team of experts: biologist, ecologist, and agronomist. The fundamental characteristic of MMR is the time span of its intervention, which is configured as permanent. In fact, the assets are acquired by the association with an act of donation, sealed by private deed, to compensate for the inability of private individuals to manage the agricultural fund. Once the property has been acquired, the organization proceeds to a mapping of the land (identification of the SAU type) and to a design of the reforestation intervention, conducted according to the guidelines of biodiversity protection. The aim of the Manu Manu reforestation intervention is to preserve the identity of the still-fertile farmland, with particular reference to traditional agricultural practices, which involve a mix of areas subject to replanting trees and areas that can be cultivated according to seasonal crops. MMR therefore embraces a practice of agro-reforestation, implemented through native biological material, with a technique that generates new plants directly from the mother plant, located on site. The acquisitions of portions of agricultural land, also distributed in a disjointed manner over the intervention areas, are aimed at building ecological corridors. They connect even distant areas and are fundamental for the biological balance, with the aim of renaturalising disused areas, upgrading rural landscapes, protecting food quality, as well as helping to mitigate the effects of climate change.

The Sylva Foundation also is a nonprofit organization, a foundation born in Lecce in March 2021 with the goal of landscape regeneration through reforestation, now headquartered in Tricase but operating nationally. The organization board is a mix of local and nonlocal personalities, starting by its board. The organization's intervention spans 5 years, focusing primarily on public areas, with very few exceptions in the private sector. It also selects lands that have a public destination or vocation.

Since these are public spaces, Sylva solely oversees the design phase and implements a change of land use to woodland. This is in accordance with Article 3, paragraph 3 of the Consolidated Act on forests and forest supply chains (Legislative Decree April 3, 2018, no. 34 and subsequent amendments—T.U.F.), which defines a forest as areas covered by arboreal forest vegetation, whether associated or not with shrub vegetation, of natural or artificial origin in any stage of development and evolution, with an extension of not less than 2000 m^2, an average width of not less than 20 m, and an arboreal coverage of greater than 20%.

In fact the minimum intervention area for Sylva Foundation is at least 2 hectares. Indeed, the legal definition of a forest captures the inherent spatial and temporal dynamism of the woodland as an ecological system, subjecting the asset to forestry constraints of equal rank with other landscape and environmental constraints of territorial significance, such as landscape, hydrogeological, and protected area constraints, which impose a land use restriction.

Considering the area as a social resource, Sylva remains open to collaborating in parallel with other projects, effectively turning the forest into a capacity-building project. Under the supervision of a technical-scientific committee, the foundation engages external forestry and agronomy technical consultants to map the land and design the reforestation intervention. The aim of the Sylva reforestation intervention is to plant native trees, guided by the specificity of the terrain and surrounding flora, in a mix that involves selecting 80% trees and 20% shrubs from the Mediterranean scrub. The plant supply comes from local nurseries, generating a small local supply chain, considering that the floriculture-nursery sector has suffered heavy losses in the years of Xylella. The reference person for the Fondazione Sylva, states that the organization does not rely on the carbon credits market for its fundraising, but rather on voluntary donations from corporations, which invest in this project for reasons of reputation and social responsibility. Management report of Sylva Foundation can help to understand the funding structure of the foundation. A Restricted Assets Reserve was established in 2021 with the contribution of Treedom s.p.a., an Italian company that manages the online e-commerce platform of the same name that allows companies or individuals to plant trees in various countries around the world (https://fondazionesylva.com/wp-content/uploads/2023/07/RELAZIONE-DI-GESTIONE-2022.pdf). For the year 2022 in fact the management report shows a total income of €465,328.69. The heading "Revenues, income and income from activities in the general interest" highlights the largest liberal donations from corporate donors, such as traditional and digital companies, banking institutions, insurance companies, international law firms. There is no mention of the use of carbon credits or carbon offsets in any documents that can be found on the foundation.

4 Results

The data provided by the two NGOs were subjected to comparative analysis (as shown in Table 1) in order to identify similarities and differences regarding a specific reforestation initiative: the Municipality of Minervino di Lecce for the Sylva Foundation and at the Kurumuni estate (in the Municipality of Ruffano) for MMR. The relevant data required are exemplified in the following table:

Table 1 Comparative analysis of the two case studies

General information	SYLVA foundation	Manu Manu Riforesta (MMR)
Name of the reforestation area	Land located in the agro of Minervino di Lecce, owned by the Province of Lecce	Municipality: Ruffano Land Name: Kurumuni Surface Area: 60 are Coordinates: 39°58′51.83″ N - 18°16′55.02″ E Ownership: Acquired on a loan for use basis granted by Merico Company
Date of reforestation	Planting: January 2022 Maintenance: 2022–2026	First intervention: 02/02/2020–18/10/2020 Existing olive trees: 72 Planting: 02/02/2020—92 trees planted Fire in summer 2023—12 olive trees survived, and approximately 80% of the trees planted in 2020 Planting: 10/03/2024—15 new trees planted (mastic, holm oak, carob)
Project coordinator	Agronomist Dr. Francesco Marra (Vice President of Sylva Foundation); Agronomist Dr. Salvatore Cavallo (Project Director)	n/a
Type of soil	Calcarenitic clay loam	Clayey with water stagnation during the rainy season
Climatic conditions	Mediterranean	Mediterranean
Reforestation project		Requalification intervention: Creation of a buffer zone with Mediterranean scrub species along the northern side of the field Strengthening of the oak core (Quercus robur s.l.) and mastic (Pistacia lentiscus) in the N-NE quadrant of the field Planting fruit trees in the S-SW quadrant of the field without a predefined planting pattern, considering the exposure and soil nature at each transplant point
Tree species planted	Holm oak—Kermes oak—Carob—Aleppo pine + shrubs	The transplants were carried out along the borders and in the spaces between the 64 existing olive trees. The species transplanted are as follows: 26 Oaks (Q. cfr robur) 19 Mastic (Pistacia lentiscus) 5 Carob (Ceratonia siliqua) 5 Fig (Ficus carica) 2 Quince (Cydonia oblonga) 6 Mulberry (Morus alba and nigra) 2 Myrtle (Myrtus communis) 1 Buckthorn (Rhamnus claternus) 1 Viburnum (Viburnum tinus) 2 Strawberry tree (Arbutus unedo) 1 Azarole (Crataegus azarolus) 3 Pomegranate (Punica granatum) 1 Jujube (Ziziphus jujuba) 4 Walnut (Juglans regia) 1 Field elm (Ulmus minor) 1 Hackberry (Celtis australis) 1 Olive (rooted shoot from the area) (Olea europea) 2 Lippia (Argentine mint Lippia polystachya) 1 Rosemary (Rosmarinus officinalis)

(continued)

Table 1 (continued)

General information	SYLVA foundation	Manu Manu Riforesta (MMR)
Planting details	Soil preparation – Drilling holes with a 40 × 40 cm auger—Water-retentive agent—Support stake	The planting was not carried out according to predefined planting patterns but in small clusters. The tree species are planted about 3 m from the boundary line, and the shrub species are planted 1.5–2 m apart. The buffer zone occupies a final width of 2 m; here, tree species (spaced 5–6 m apart) such as virgiliana (Q. virgiliana), farnia (Q. robur s.l.), and shrub species with lower xeric characteristics: viburnum (Viburnum tinus), mastic (Pistacia lentiscus), common phillyrea (Phillyrea latifolia), privet (Ligustrum vulgare), terebinth (Pistacia terebinthus), and bush honeysuckle (Lonicera implexa) can be planted
Total number of trees planted	9.500	92 specimens belonging to 19 plant species planted on 6,000 m^2
Size of the seedlings	1 year	Age less than 3 years. The intervention was spontaneous, as the enthusiasm to repopulate the destroyed lands led many participants to donate (and plant themselves) any plant they had in their garden/field
Planting method	Manual	The transplants, holes were dug (an auger was used for the tree species), then partially filled with a mix of agricultural soil, chipped branches, and shredded prickly pear cladodes. After placing the plant with its root ball, the holes were filled with fine agricultural soil, compacted around the root, and the surface around the collar was covered with mulch for protection
Maintenance	5 years	No time limits
Date of the last maintenance	14/06/2024	10/03/2024
Maintenance activities performed	Weeding—Mowing—Emergency irrigation—Replacement of failed transplants—Firebreak maintenance	Fire in summer 2023—12 olive trees survived, and approximately 80% of the trees planted in 2020 Planting: 10/03/2024—15 new trees planted (mastic, holm oak, carob)
General condition of the plants	Pines show the best vigor, high mortality in the first year of planting, now all in good vegetative condition	Currently, the project has seen the planting of 92 plants, of which about 80% survived a fire in 2023
Problems or challenges encountered during reforestation	Severe frosts after planting; need for emergency irrigation up to 16 times per season	Previous damage: Fire that destroyed most of the century-old olive trees, with only the trunk and suckers surviving
Ideas for improving future reforestation projects	Useful irrigation system installation, use of mulch	n/a

4.1 Critical Comparative Analysis Results

4.1.1 Time Span and Area of Focus

Manu Manu Riforesta's intervention is designed as a permanent project, with a strong emphasis on creating and maintaining ecological corridors. This long-term approach allows the organization to focus on restoring the landscape over extended periods, which is crucial for achieving deep ecological integration and resilience.

In contrast, Fondazione Sylva operates within a more constrained timeframe of 5 years, focusing on the rapid reforestation of large public areas. The foundation's approach is effective in terms of quickly transforming public lands into forested areas, but the short duration of the intervention may limit the long-term ecological benefits compared to the more gradual and sustained efforts of MMR.

4.1.2 Species and Planting Techniques

Both organizations prioritize the use of native species, but their methods of propagation and planting differ significantly. Manu Manu Riforesta emphasizes traditional techniques, with a strong reliance on propagating plants directly on-site from existing mother plants. This method ensures that the plants are well-adapted to the local environment, promoting greater ecological stability.

Fondazione Sylva, on the other hand, employs external forestry and agronomy consultants to guide their planting efforts. The use of local nurseries for plant supply supports the local economy but may introduce variations in plant adaptation due to differences in nursery conditions versus the actual planting site. The foundation's structured approach ensures consistency and scale but may not offer the same level of ecological integration as the more localized methods used by MMR.

4.1.3 Maintenance and Challenges

Manu Manu Riforesta adopts a long-term maintenance strategy that focuses on minimal human intervention once the ecosystem has been established. The organization's goal is to create self-sustaining environments that require little external input, aligning with their broader mission of ecological balance. (As shown in Fig. 1, the regeneration efforts on the damaged areas. Additionally, Fig. 2 highlights the first stages of planting conducted by Manu Manu Riforesta in February 2020, while Fig. 3 provides insights into the progress made over a year.)

Conversely, Fondazione Sylva implements an intensive maintenance regime for the first 5 years, ensuring that the newly planted forests are well-established. However, this approach requires significant resources and may not be sustainable in the long run without continued intervention. The challenges faced by Fondazione Sylva, particularly severe frosts and high-water requirements, highlight the difficulties of managing large-scale reforestation projects in Mediterranean climates.

4.1.4 Collaboration, Legal Framework, Funding, and Resource Management

Collaboration with local community and others local NGOs is a key strength of Manu Manu Riforesta.

On the other hand, Fondazione Sylva operates under a strict legal framework that guides its reforestation efforts. By adhering to the Italian Consolidated Act on forests and forest supply chains, the foundation ensures that its interventions meet national standards for forest creation. This legal adherence provides a strong foundation for their projects but may also limit flexibility in adapting to local ecological conditions.

Funding is another area where the two organizations diverge. Manu Manu Riforesta depends on donations and academic collaborations, which may constrain the scope

Fig. 1 Kurumuni's front band in 2020 (*Source* MMR Archive)

Fig. 2 Kurumuni's front band first planting in February 2020 (*Source* MMR Archive)

of their operations but guarantee the alignment of their interventions with ecological and community objectives.

In contrast, Fondazione Sylva is heavily funded by corporate donations, which provides significant resources for large-scale projects. However, the reliance on corporate funding may introduce pressures to meet donor expectations, potentially influencing project priorities.

4.2 Discussion

The comparative analysis of Manu Manu Riforesta and Fondazione Sylva reveals two distinct approaches to landscape regeneration and reforestation in the Salento region. MMR's long-term, ecologically integrated approach is well-suited for areas where biodiversity and traditional agricultural practices are priorities, as illustrated in Fig. 4, which shows the Kurumuni project plan. Their focus on agroforestry and ecological corridors ensures that their interventions contribute to the restoration of the region's natural heritage.

On the other hand, Fondazione Sylva's more structured, short-term approach is effective for large-scale reforestation, particularly in public areas where rapid land transformation is needed. However, the challenges associated with their intensive maintenance strategy and the reliance on external consultants suggest that this approach may benefit from further adaptation to local ecological conditions.

By comparing these two case studies, this research highlights the importance of tailoring reforestation strategies to the specific needs of the land and communities involved. Both short-term and long-term interventions have their place in landscape regeneration, and the most effective strategies will likely involve a combination of both, adapted to the unique challenges and opportunities of each site.

Fig. 3 Kurumuni's front band after a year (*Source* MMR Archive)

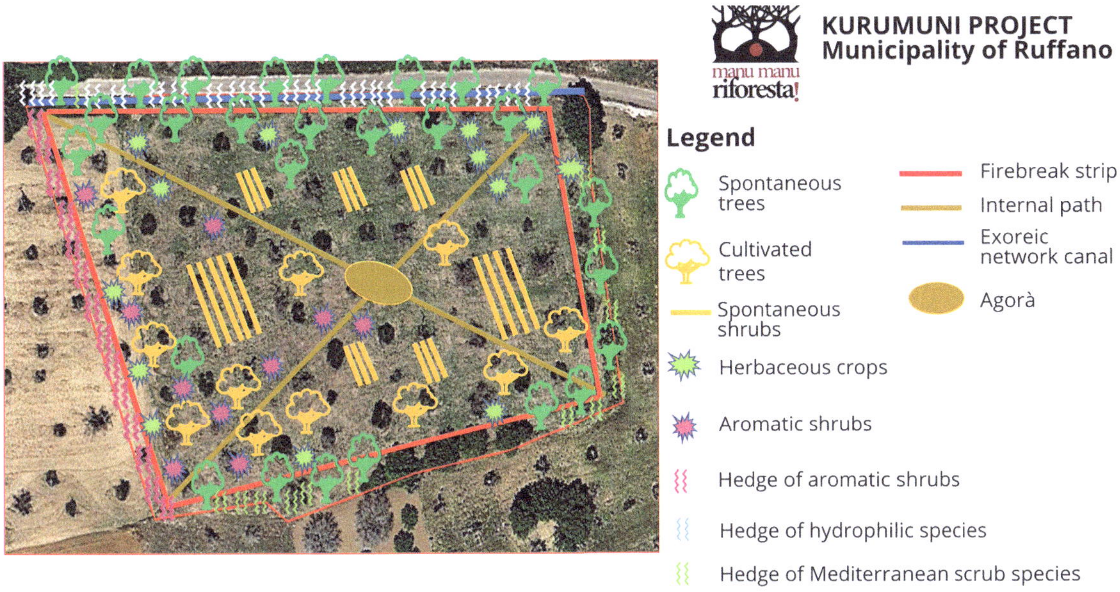

Fig. 4 Kurumuni project—Manu Manu Riforesta (*Source* Adapted from MMR Archive, with modifications by the author)

5 Conclusion

The *Xylella* has led to the implementation of various emergency measures that, by removing olive trees from the land, have transformed Salento into an area of interest for soil profitability. The regional regulations regarding the replanting of olive trees permit the planting of only two varieties of olive trees: one non-native and self-sterile (Leccino), and the other patented (FS-17, also known as "Favolosa"), both suitable for intensive and super-intensive cultivation systems.

The spread of *Xylella fastidiosa* has transformed Salento into a laboratory for experimental reforestation and agroforestation practices. The various strategies adopted, such as those by the Sylva Foundation and MMR, reveal contrasting approaches that reflect different views on land use, resource management, and community involvement. The first employs a forestry approach, while the second has opted for agroforestry. In summary, Sylva has a top-down approach, offering itself as an intermediary capable of providing an efficient response to the demands for sustainability in the corporate world, but also to public entities struggling to manage large portions of demesne territory, representing a significant cost. Manu Manu Riforesta instead embraces a bottom-up approach, in which the qualitative indicator of interventions does not lie in an extensive intervention or in the quantity of trees planted, but in the ecosystemic quality of the interventions, in the constant maintenance, safeguarding the ecosystem and the phenomenon of wildfires, which are the main antagonists of reforestation actions. Due to this profound difference, the scope of their respective interventions is significantly different.

Although the carbon credits market is a legal practice, there are contrasting opinions on the effectiveness and ethics of using carbon credits. Scientific studies have demonstrated the ineffectiveness of some of these practices, which, through ineffective environmental compensations, provide the opportunity to continue polluting instead of reducing direct carbon emissions (West et al. 2023). Others believe that if reforestation projects financed through carbon credits are well-structured and monitored, they can effectively contribute to forest conservation and mitigating climate change. In this context, the effort of the contribution has been to carefully consider the specific context, thoroughly evaluating the pros and cons of each approach before drawing definitive conclusions. The need of a regulation of Carbon Farming is evident, in fact the European Commission introduced, on November 30, 2022, the EU's inaugural shared regulatory framework, which aimed at certifying emissions offsetting through the acquisition of carbon credits. This pivotal regulatory instrument is poised for replication across all sectors aligned with the green economy. While the utilization of carbon credits/offsets might appear to be ruled out, inquiries persist concerning their efficacy, ethical implications, and the engagement of the community.

However, to fully understand the impact of these initiatives, it is crucial to consider the role of public policies and local socio-economic dynamics. These factors influence not only the implementation but also the sustainability and social acceptance of reforestation practices. Furthermore, the long-term assessment of the effectiveness of these interventions must go beyond merely counting the surviving plants, incorporating measures of ecological resilience, community well-being, and contributions to climate change mitigation. Therefore, we should recommend continuous and participatory monitoring of reforestation initiatives, integrate local communities into project management, and use multidimensional indicators that reflect both ecological and socio-economic success.

Only through a synergy of sustainable practices, coherent public policies, and community participation can the lasting and meaningful regeneration of the Salento landscape be ensured.

References

Journal

Alliegro, E.V.: Identità territoriale, crisi ambientale e processi di identizzazione. Alcuni esempi nel Mezzogiorno d'Italia. In: Zola, L. (ed.), Ambientare. Idee, saperi, pratiche, pp. 143–161. Franco Angeli, Milano (2017)

Bandiera, M.: Biosicurezza nella Puglia del disseccamento. *Geotema*, Supplemento, 97–107 (2020)

Bingham, N.: Bees, butterflies, and bacteria: biotechnology and the politics of nonhuman friendship. Environ Plan A **38**(3), 483–498 (2006). https://doi.org/10.1068/a38436

Ciervo, M.: Emergenza Xylella in Salento (Puglia): una questione geopolitica. Quaderni del Semestrale di Studi e Ricerche di Geografia, Speciale 17–37 (2023)

Clark, N.: Feral ecologies: Performing life on the colonial periphery. In: Szerszynski, B., Heim, W., Waterton, C. (eds.) Nature performed: environment, culture and performance, pp. 163–182. Blackwell, Oxford (2003)

Franklin, A.: A choreography of fire: a posthumanist account of Australians and eucalypts. In: Pickering, A., Guzik, K. (eds.) The Mangle in Practice: Science, Society, and Becoming, pp. 17–45. Duke University Press, Durham, NC (2008)

Griffin, C.: More-than-human histories and failure of grand state schemes: Sylviculture in the New Forest, England. Cultural Geograph. **17**(4), 451–472 (2010). https://doi.org/10.1007/978-3-031-04958-3_63

Haraway D., Tsing, A.: Reflection on the plantationocene. a conversation with donna Haraway and Anna Tsing. In: Mitman, G. (ed.) Edge Effects Magazine. University of Wisconsin-Madison, pp. 1–20 (2019). Retrieved August 25, 2024 from https://edgeeffects.net

Latour, B.: Agency at the time of the Antropocene. New Literary History **45**(1), 1–18 (2014)

Lorimer, J.: Multinatural geographies for the Anthropocene. Prog. Hum. Geogr.. Hum. Geogr. **36**(5), 593–612 (2012). https://doi.org/10.1177/0309132511435352

Patel, R.: The long green revolution. J. Peasant Stud. **40**(1), 1–63 (2012). https://doi.org/10.1080/03066150.2012.719224

Purcell, A.H.: Xylella fastidiosa, a regional problem or global threat to agriculture? Comprehensive Rev. Food Sci. Food Safety **41**, 1245–1250 (1997)

Saponari, M., Boscia, D., Altamura, G., D'Attoma, G., Cavalieri, V., Loconsole, G., Zicca, S., Dongiovanni, C., Palmisano, F., Susca, L., Morelli, M., Potere, O., Saponari, A., Fumarola, G., Di Carolo, M., Tavano, D., Savino, V., Martelli, G.P.: Pilot Project on Xylella fastidiosa to Reduce Risk Assessment Uncertainties, in EFSA supporting publication, pp. 1–60 (2013)

West, et al.: Action needed to make carbon offsets from forest conservation work for climate change mitigation. Science **381**(6660), 873–877 (2023). https://doi.org/10.1126/science.ade3535

Book

Arrighi, G.: Il lungo XX secolo: Denaro, potere e le origini del nostro tempo. Il Saggiatore, Milano (1994)

De Martino, E.: La fine del Mondo: Contributo all'analisi delle apocalissi culturali. In: Gallini, C. (ed.) Einaudi, Torino (1977)

Ciervo, M.: Il disseccamento degli ulivi in Puglia Evidenze, contraddizioni, anomalie, scenari. Un punto di vista geografico (2020). Retrieved August 25, 2024 from http://societageografica.net/wp/wp-content/uploads/2020/11/ebookuliviHD.pdf

Ghelfi, A.: New peasantries in Italy: eco-commons, agroecology and food communities. In: Papadopoulos, D., de la Bellacasa, M.P., Tacchetti, M. (eds.) Ecological Reparation: Repair, Remediation and Resurgence in Social and Environmental Conflict, pp. 194–207. Bristol University Press, Bristol (2023)

Jones, O., Cloke, P.: Non-human agencies: trees, relationality, time and place. In: Knappett, C., Malafouris, L. (eds.) Material Agency: Towards a Non-anthropocentric Approach, pp. 79–96. Springer, Guilford (2008)

Latour, B.: Non siamo mai stati moderni. Elèuthera, Milano (2009)

Martella, S.: La morte dei giganti. Il batterio Xylella e la strage degli ulivi millenary. Meltemi, Milano (2022)

Papadopoulos, D.: Experimental Practice: Technoscience, alterontologies, and more-than-social movements. In: Papadopoulos, D., de la Bellacasa, M.P., Tacchetti, M. (eds.) Ecological Reparation: Repair, Remediation and Resurgence in Social and Environmental Conflict, pp. 194–207. Bristol University Press, Bristol (2018)

Spedicato, G.: La parabola dell'olio salentino. Milella, Lecce (2018)

Trizio, B.: Il paesaggio del Salento: Storia e trasformazioni. Congedo Editore, Lecce (2000)

Whatmore, S.: Hybrid Geographies. Natures Cultures Spaces. Routledge, London (2003)

Web Sources

Camera dei Deputati. (2024). Gli interventi per il contrasto al batterio della Xylella fastidiosa. Retrieved August 25, 2024 from https://www.camera.it

European and Mediterranean Plant Protection Organization (EPPO): Distribution of Xylella fastidiosa. Retrieved August 25, 2024 (2024) from https://gd.eppo.int

Urban Territory, Transportation, and Technological Innovations

Preferences Toward Shared Automated Electric Vehicles: Evidence from Mixed Logit and Bayesian Inference Analysis

Pooja Rao, Mohammed Quddus, and Washington Y. Ochieng

Abstract

This study investigates the factors influencing the public's acceptance of shared automated electric vehicles (SAEVs) as a practical alternative to public transit during peak travel times in London, UK. Advanced statistical models, including mixed multinomial logit (MMNL) and Bayesian multinomial logit (BMNL), analyze mode preferences and determinants. The MMNL model underscores the pivotal role of cost, travel attributes and demographic considerations in shaping preferences, emphasizing SAEVs as a sustainable alternative to private vehicles and public transport. Incorporating random coefficients and case-specific variables enhances the model's accuracy, revealing distinct preferences among individuals with driver's licenses and certain gender groups during high-demand periods. The BMNL model explores intricate relationships between gender, driver's licenses, employment variables, and mode choices. Despite limitations of Bayesian uninformative priors, the study highlights its potential for future research and encourages further exploration of random effects in Bayesian modeling using the framework's flexibility by incorporating priors, utilizing new information to update subjective beliefs over time. Research findings predict a substantial 63% likelihood of individuals choosing SAEVs during peak travel, providing valuable guidance for policymakers aiming to reduce private vehicle use. The study emphasizes the importance of addressing public concerns about safety and privacy through targeted educational campaigns, advocating for a holistic approach to promote SAEV adoption. Overall, this research contributes to a comprehensive understanding of the factors influencing mode choices in urban transportation using advanced MMNL and BMNL statistical models, and providing insights for transportation planning interventions when scaling new automated transport technologies.

Keywords

Ridesharing · Shared automated electric vehicles · Bayesian methods · Mixed logit methods · Mode preferences

1 Introduction

The emergence of automated vehicles (AVs) presents a significant opportunity to revolutionize urban transportation, promising enhanced traffic flow, safety, and efficiency. This potential is particularly evident in the concept of shared automated electric vehicles (SAEVs), which offer adaptable on-demand services through mobile applications (Maeng 2022). SAEVs represent a convergence of three transportation revolutions: automation, electrification, and shared mobility, aiming to provide a sustainable and efficient alternative to private vehicle ownership and traditional public transit systems. SAEVs have the potential to increase passenger-miles-traveled (PMT) and reduce vehicle-miles-traveled (VMT), therefore reducing the number of vehicles while benefiting from lower generalized costs associated with private AV ownership. When integrated with enhanced value of travel time savings (VTTS), Shared Automated Electric Vehicles (SAEVs) offer significant advantages in terms

P. Rao (✉)
Imperial College London, London, UK
e-mail: pooja.rao16@imperial.ac.uk

M. Quddus
Intelligent Transport Systems, Imperial College London, London, UK
e-mail: m.quddus@imperial.ac.uk

W. Y. Ochieng
Positioning and Navigation Systems and Civil and Environmental Engineering, Imperial College London, London, UK
e-mail: w.ochieng@imperial.ac.uk

P. Mohammad et al. (eds.), *Geographic Approaches to Climate Change and Mitigation: Urban and Rural Perspectives (Volume 2)*, Advances in Science, Technology & Innovation, https://doi.org/10.1007/978-3-031-92115-5_13

of productivity during in-vehicle travel time, compared to conventional human-driven vehicles. The VTTS is inherently contingent upon an individual's value of travel time (VOTT) and the duration of their trip (Choi 2023), thus influencing the rationale for adopting SAEVs.

The cost-effective on-demand shared mobility service can contribute to a more sustainable transition away from private vehicle use. Certain scholars propose a drastic reduction of up to 90% in greenhouse gas emissions for shared automated vehicles (SAVs) compared to non-automated vehicles (Greenblatt 2015). Although this may seem like an overestimation, renewable energy generation powering SAVs may likely contribute to substantial GHG reductions, and it is projected that SAEVs will become a personal mobility choice for many travelers in London. SAEVs can overcome dynamic ridesharing challenges through optimized vehicle repositioning and rider pickup capabilities, which can reduce passenger wait times and significantly improve accessibility (Faisal 2019). For example, first-mile-last-mile (FMLM) services offer user flexibility in providing multi-modal travel options, allowing users to make relevant mode-choice decisions (Fagnant 2015). Better education campaigns are needed to ensure the successful deployment of SAEVs in the long term, given public concerns around safety, privacy, and data security, and these further vary depending on demographics and respective policy issues (Lee 2022).

The benefits of SAEVs include the potential to transform urban development through anticipated reductions in parking needs within the city center (Stein 2020). This opens up opportunities for the revitalization of existing building stock and encourages the adoption of active mobility options, such as walking and cycling. Lower operating costs of $0.50 per mile reduce congestion and fleet size with a vehicle replacement rate of 4, meaning one SAEV can potentially replace four private vehicles (Liu 2022). Ridesharing on-demand services can be attractive to those with and without mobility access, but limited work exists measuring variation in preferences using multivariate statistical models, specifically the mixed multinomial logit model (MMNL) and Bayesian multinomial logit model (BMNL) that aim to measure the heterogeneity in model parameters. In the context of London, where peak travel times pose significant challenges to existing transportation infrastructure, further research is required to assess the factors influencing decisions about mode choice behavior using advanced quantitative discrete choice methods.

This paper discusses the preference for SAEVs using two advanced multivariate statistical models. Specifically, and firstly, the objective is to show how the mixed multinomial logit model (MMNL) allows individual-level heterogeneity in the model parameters by introducing random effects and individual variation. These random effects capture unobserved variability across variables and individual effects across individuals. This study enhances the multinomial logit model employed in prior research by adopting the mixed multinomial logit (MMNL) model. This adjustment is made to accommodate unobserved preference heterogeneity among individuals corresponding to random coefficients in the presence of alternative invariant (case-specific) variables in mode selection. Secondly, the Bayesian multinomial logit model (BMNL) is a Bayesian approach that is used in this study to estimate the parameters of the multinomial logit model (MNL) by incorporating prior information with experimental results specific to demographic and travel attributes and treating the model parameters as random variables with probability distributions.

Analyzing the influence of individuals' transportation mode preferences utilizing advanced modeling techniques that recognize individual-level heterogeneity in preferences can enhance transportation planning and policies related to shared automated electric vehicles (SAEVs).

The remainder of the paper is organized as follows: Sect. 2 provides a brief literature review on mode choice preferences toward shared automated vehicles. Section 3 provides details on the methods, including survey design, data collection, and model formulation for MMNL and BMNL. Section 4 contains the results of the study, followed by concluding remarks in Sect. 5.

2 Summary of Related Work

The proliferation of shared transportation has the potential to substantially alter the rate of vehicle ownership. According to estimates, each car-sharing vehicle is anticipated to replace approximately 7–10 individual vehicles two years after implementation (Kolleck 2021). This shift is attributed to the fact that a majority of private cars are typically utilized for less than their total available time, meaning that an average vehicle is parked for 95% of the time (Barter 2019). Moreover, shared mobility stands to enhance the overall efficiency of the transport sector by reducing the number of vehicles and journeys made, resulting in lower vehicle-miles-traveled (VMT) (Greenblatt 2015). Additionally, shared transport is envisaged to offer a more cost-effective alternative compared to private transportation (Shaheen 2018). While cost savings, convenience, and other non-monetary factors are significant drivers, the success of shared mobility also depends on factors such as regulatory support, infrastructure development, and public acceptance. As technology and societal norms continue to evolve, the landscape of shared mobility is likely to undergo further transformations.

Automated driving technology is gradually being introduced and is positioned to accelerate the adoption of this emerging transport mode, particularly through its integration with shared mobility. This adaptation aims to enhance convenience and bring about significant changes in lifestyles.

Automated vehicles have the capability to reach riders for pickup and dropoffs with minimal relocation efforts and without driver-related costs (Tscharaktschiew et al. 2022). The success of comparably convenient, albeit relatively costly, driver-enabled services underscores the potential of automated transport technologies. Extensive and ongoing testing is underway, projecting that 25% of the commercial fleet will be operational by the year 2035 (Davidson 2015; Ng 2023). Other gains include fuel efficiency improvements with the potential to significantly reduce greenhouse gas emissions given a 60–80% AV penetration rate (Massar et al. 2021).

Shared automated vehicles (SAVs) represent a nascent concept of mobility and have recently witnessed a surge in academic and commercial interest dedicated to exploring and understanding SAVs. Extensive studies have been undertaken to scrutinize the potential impacts, operational intricacies, and anticipated penetration of SAVs in the market.

Scaling the worldwide deployment and use of electric vehicles (EVs), SAVs naturally play a pivotal role in advancing the electrification of automated vehicles. Commercially operated shared automated electric vehicles (SAEVs) alleviate range anxiety and charging management by dynamically handling range and charging activities in response to real-time trip demand and utilization of location-based charging stations, as demonstrated by an agent-based microsimulation model using a gridded city in Austin, Texas (Chen 2016).

Previous work on SAEVs is limited to MATSim modeling of fleet size, wait time, and vehicle repositioning (Liu et al. 2017; Spieser et al. 2014; Fagnant 2014; Iacobucci 2019). Other studies focus on operational aspects, including algorithmic optimization in solving rebalancing strategies using mixed integer linear programming (MILP) techniques (Kessler 2020). Additionally, there is a concentrated effort to measure public sentiment regarding mode preferences of SAEVs in hypothetical scenarios using stated preference surveys and various methodological methods.

Research indicates that sociodemographic factors, encompassing age, education, and gender, alongside trip-related elements like travel time, travel cost, and waiting time, play pivotal roles in shaping the adoption of SAVs. Additionally, factors such as possessing a driver's license (DL) and owning a private vehicle have been identified as influencers in users' preferences for both AVs and SAVs incorporating either multinomial logit, nested logit, parameter probit, or regression models (Haboucha 2017; Winter 2017; Menon et al. 2019; Cools 2017). Few studies consider advanced statistical methods such as mixed logit models to capture the random effects of unobserved variability across individuals when modeling influencing factors such as travel time, cost, and trip type in the choice of mode alternatives (Yap 2016). The literature on mixed logit models is even sparse for shared automated vehicles, as researchers focus mainly on AVs and specifically on travel time, with limited understanding of cost variation, demographics, and travel attributes on mode preferences, although some work has been done on SAV preference given cost, wait time, and multimodal utilization (Krueger 2016).

In efforts to include other multivariate statistical techniques, studies utilizing Bayesian methods for multinomial logit have varying use cases for shared mobility in densely populated cities. Investigating residents' travel behavior and the adoption of shared mobility using Markov Chain Monte Carlo and Bayes-Mixed Logit algorithms is used to emphasize weather impact, travel time influence on car pooling, and safety concerns of female users (Zhou et al. 2022). Bayesian network structural learning (BNSL) approaches model probabilistic relationships among variables in making predictions of carsharing propensity based on uncertain information (D'Urso et al. 2021). However, there are several disadvantages and challenges to this technique, including computational complexity and assumptions of faithfulness, meaning that the observed statistical dependencies reflect causal relationships. The handling of latent variables and non-linear relationships can appear more difficult to handle, along with limiting causal relationships in which Bayesian networks may not always capture causal relationships accurately, and remains a challenge (Guo 2020; Kitson et al. 2023). Estimation techniques have been proposed to advance Bayesian estimation mixed multinomial logit models (MMNL) in expanding existing Variational Bayes (VB) methods. Despite the operational efficiency and scalability enhancements offered by these techniques, the requirement to fine-tune parameters for convergence, may result in biased estimates and increase the likelihood of model misspecification. This introduces an additional layer of complexity, making it an improbable method of choice. In the effort to contribute to the wider scholarly work, research focusing on Bayesian inference multinomial logit methods for analyzing mode preferences in the context of shared automated electric vehicles (SAEVs) is lacking to the best of the author's knowledge, specifically when datasets containing informative priors from previous studies are unavailable. Given the uniqueness of this study's data, additional research is needed to study the random effects captured by unobserved variability among individuals in a unique choice experiment, given a set of discrete alternatives in the context of mixed multinomial logit (MMNL) and Bayesian multinomial logit (BMNL) models.

3 Methods

The methods section provides a detailed overview of the survey design, data collection process, and the methodologies employed in the analysis, including the mixed multinomial logit (MMNL) and Bayesian multinomial logit (BMNL) model estimation.

The section initiates with an extensive exposition on London's urban dynamics, encompassing its topography, transport network, demographic characteristics, and socio-economic resilience.

3.1 Integrated Analysis of London's Urban Dynamics: From Topography to Socio-economic Resilience

To provide context, London's topography exhibits significant variations across its metropolitan expanse and is up to 30 m above sea level. The Thames River basin dominates the central region, carving through the landscape and influencing land use patterns, in which urban spaces are analyzed through land cover mappings, indicating a diverse distribution.

The transport network is a critical aspect of London's urban fabric. The London Underground, a complex system comprising 11 lines, covers approximately 402 km, facilitating efficient intra-city travel. An analysis of bus routes, which total over 700, identifies key corridors supporting multi-modal road transport. Illustrated in Fig. 1, the transportation demand in London has consistently increased throughout 2022, reaching current levels that have remained stable in 2023. Additionally, an examination of cycling infrastructure highlights the expansion of dedicated lanes, particularly in response to the increased popularity of cycling during the COVID-19 pandemic. London's transport system has left an indelible mark on its mobility patterns, reporting decreased public transport ridership, a surge in active mobility, and changes in travel behavior (Fig. 2). Naturally, as other global cities emerged from the pandemic, London adopted hybrid work models, which influenced commuting patterns, necessitating strategies in urban planning.

In the aftermath of the global pandemic, London's transport dynamics have undergone significant adaptations to ensure public safety. The London Underground, for instance, has implemented enhanced air circulation systems, reducing airborne particle concentrations in stations. Spatial analysis of ridership data indicates shifts in travel patterns, with an increased emphasis on off-peak hours and a surge in cycling commuters, a rise of 40% compared to 2019 (London 2022).

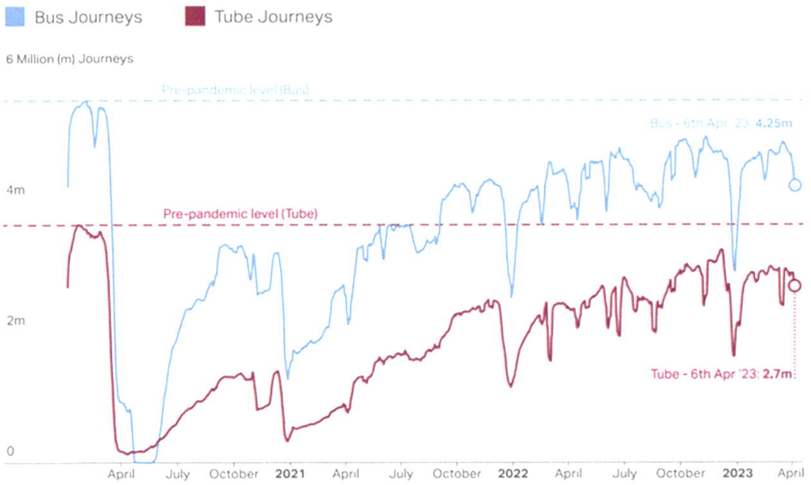

Fig. 1 Number of journeys by bus and tube on the TfL network during 2021–2023. *Source* London Datastore

Fig. 2 Number of daily journeys by bus and tube on the TfL network during 2020–2023. *Source* London Datastore

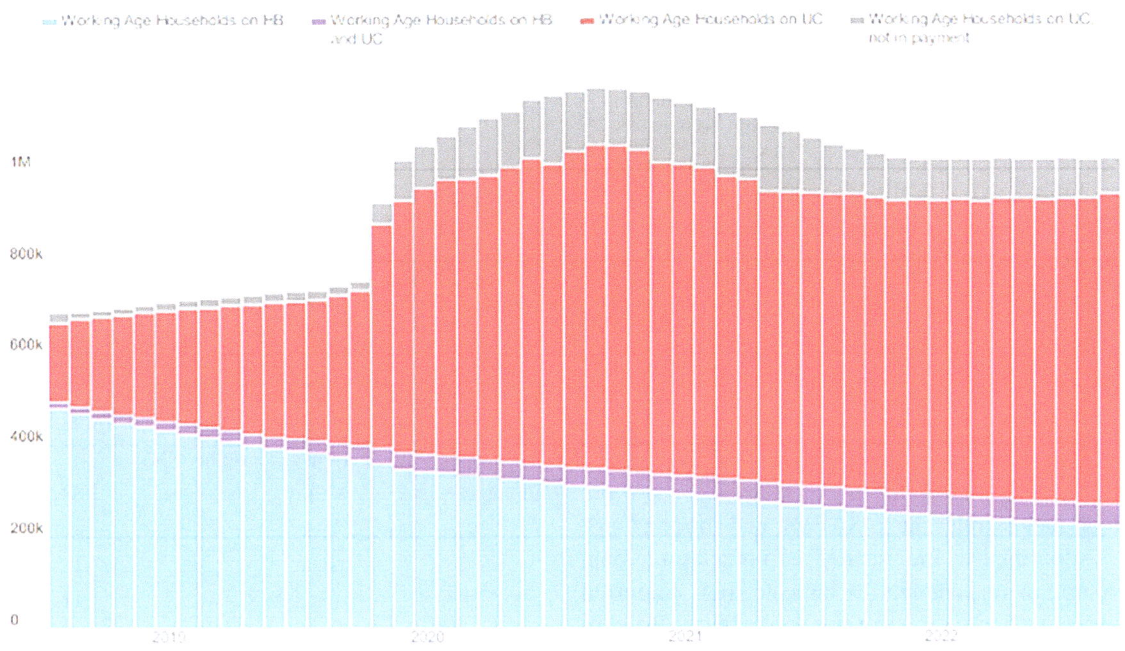

Fig. 3 Working age London households on means tested benefits between 2019 and 2023. *Source* London Datastore

Demographic analysis revealed that London, with a population exceeding 9 million (MacroTrends 2023), continues to experience steady growth. Gender dynamics indicate a balanced distribution, with ongoing efforts to address gender disparities in employment and access to education aimed at promoting inclusivity in various sectors. A continued understanding of the spatial distribution of population growth across London boroughs is needed to allow for targeted urban planning initiatives to accommodate this influx. For instance, an exploration of residential patterns in relation to employment centers provides insights into commuting behavior.

London's GDP growth and income distribution show that economic growth remains robust, although discernible disparities in income remain, with certain boroughs experiencing higher poverty rates (Fig. 3).

London's transport ecosystem is scrutinized for its impact on socio-economic dynamics. Accessibility metrics, considering proximity to transport nodes, showcase the uneven distribution of accessibility across neighborhoods. The road infrastructure analysis examines traffic patterns, congestion hotspots, and the integration of smart technologies to enhance traffic management given London's complex road network (London-City-Hall 2016).

In conclusion, the city's topography and complex transport network are used to model key factors that influence the preference for shared automated mobility during peak travel times in London. Advanced multivariate statistical methods are employed to capture individual choices from a discrete set of alternatives. The evolution of automated vehicles encompasses factors such as policy development, infrastructure adjustments, and public opinion, with the latter being particularly scarce. This research offers valuable insights into public sentiment, shedding light on the city's readiness to embrace automated vehicles as part of its strategy for sustainable urban mobility.

3.2 Survey Design and Data Collection

The Stated Preference Discrete Choice Experiment (SPDCE), a cross-sectional study, employs a non-probabilistic convenience-based sampling technique distributed online. Executed in London, this choice experiment aims to unravel insights into individual travel behavior for transport-related decision-making. Targeting adults across all London boroughs and categorized into four regions, North, South, East, and West London, an IP mechanism ensures participation, accompanied by location requests that are carefully mapped based on responses.

The research questionnaire centers on discerning public attitudes and mode-choice perceptions related to shared automated electric vehicles (SAEVs). Respondents are provided with a comprehensive understanding of SAEVs through both a textual definition and a visual representation of a rideshare pickup. For example, inquiries about travel consider the main reason for traveling to the current activity and their frequency of travel. The comprehensive questionnaire encompasses four sections dedicated to capturing demographic characteristics, travel attributes,

Table 1 Situational factors for each mode choice alternative in SP experiment

	London bus	Private car	Shared automated electric vehicle (SAEV)
Wait-time (min)	15	0.00	5
Travel time (min)	58	20	33
Travel cost (£)	1.65	8.02	4.82
What is your choice?	☐	☐	☐

psychological factors relevant to SAEVs, and mode choice questions. Additionally, the survey explores situational factors, prompting respondents to indicate their preferred mode of transportation based on various attributes. Psychological questions probe technological acceptance of SAEVs and users' likelihood of using them for their main or leisure activity, where responses are meticulously collected using a 5-point Likert scale. Preferences for travel modes, which are the primary focus of this paper and include London buses, private vehicles, and SAEVs during peak travel periods, are provided in Table 1. The questionnaire asks respondents to select their preferred mode of travel when traveling a maximum distance of 12.8 km or 8 miles based on their attributes.

While acknowledging inherent biases in sample selection, the distribution of the online questionnaire follows a non-probabilistic approach, a widely accepted method in the research community. This approach ensures a well-rounded distribution of respondents across London, maximizing data capture with an impressive 45% completion rate, surpassing the average by 5%.

Data cleaning processes involve the identification and removal of incomplete, straight-line, patterned, and Christmas tree responses (incremental increasing or decreasing), as well as unreasonably fast responses (less than 5 min). The survey successfully collected responses from 446 participants, a sample size computed based on Cochran's formula (Adam 2020) maintaining a 95% confidence interval and a 5% margin of error. Notably, the completion rate of 45% attests to the robustness of the dataset, ensuring its suitability for in-depth analysis.

Key demographic and travel characteristics are listed in Table 2.

Table 2 Key demographic and travel attributes of sample base

Variable	Category	Frequency tabulation (%)
Age	18–24	< 0.5
	25–34	1
	35–44	2
	45–54	6
	55–64	28
	65–74	40
	> 75	20
Gender	Male	68
	Female	28
	Transgender/other/PNS	4
Ethnicity	White	88
	Black	< 0.5
	Asian	1
	Mixed/PNS	11
Education	Postgrad	10
	Undergrad	18
	Vocational training	65
	High school degree	28
	None to limited education	9
Employment	Full time/S.E./P.T./contract	39
	Retired	55
	Student	< 0.5
	Unemployed	6
Drivers license	No	5
	Yes	95
Number of vehicles	0	9
	1	53
	2 or more	38
Mode used for current activity	Private car	79
	Taxi/Uber	2
	Public transport	16
	Active mobility	3

3.3 Variables Framework

For conciseness, only the specific variables pertinent to this paper are presented in Table 3, as they constitute the key components employed in the mixed logit and Bayesian analysis models, respectively.

Most cost and choice of mode during peak travel times based on an individual's age and if they hold a driver's

Table 3 Key variables used in MMNL and BMNL analyses

Variable name	Description	Scale/data type	Model used
Mode cost	Cost of alternative	Discrete	MMNL
Mode options peak	Bus/private car/ SAEV	Nominal	MMNL
Driver's license	Yes/no	Dichotomous	MMNL/BMNL
Age	18–24...above 75	Ordinal	MMNL
Gender	Male/female/ other	Nominal	BMNL
Employment	FT/SE/PT/ contract/retired/ student/ unemployed	Nominal	BMNL

license (DL) are variables used in the MMNL model specification. The final mode choice during peak travel times is the response variable based on the three indicator variables. Likewise, driver's license, gender, and employment are indicators used in the BMNL model specification. These key variables are used to model heterogeneity in model parameters in the MMNL estimation and uncertainty of parameters observed from prior distributions using BMNL estimation.

3.4 Mixed Multinomial Logit Model (MMNL)

Discrete choice models, assuming a random utility-maximizing framework, are widely used among researchers for investigating travelers' preferences in the context of mode selection. This framework posits that individuals choose the alternative with the highest utility among a set of alternatives, and various discrete choice models, including the multinomial logit model (MNL), the nested logit model (NL), and the mixed logit model (MMNL) have been applied to analyze preferences for automated vehicles (AVs). In this part of the study, the mixed multinomial logit (MMNL) model is employed.

The MMNL is an extension of the standard multinomial logit model (MNL). It overcomes the limitation of the MNL model by allowing for individual-level heterogeneity in the model parameters by introducing random effects. These random effects capture unobserved variability across individuals within a population. MNL models, on the other hand, assume homogeneity across individuals, treating all individuals as having the same preferences. This can lead to model misspecification and inflexibility in parameterization if there is significant individual-level heterogeneity that is not captured by the fixed coefficients. The MMNL model used in this study is to account for individual-level variation in mode preferences, including random effects from unobserved variables, leading to improved model fit.

3.4.1 Model Specification

In discrete choice analysis, utility, considered a latent variable, represents a measurable indicator of satisfaction, desirability, or preference that an individual associates with various alternatives in a decision-making scenario. This pivotal concept serves as a fundamental tool for modeling and comprehending the decision-making process, illustrating how individuals choose among a set of distinct options. Essentially, utility is a function of observed individual attributes, coefficients, and random components.

In the development of the MMNL model, an individual i receives from alternative j the utility U_{ij} given by,

$$U_{ij} = x_{ij}\beta_i + w_{ij}\alpha + z_i\delta_j + \varepsilon_{ij} \tag{1}$$

where β_i are the stochastic coefficients that vary across individuals across the sample population. The vectors x_{ij} and w_{ij} encompass alternative-specific variables, while δ_j represents fixed coefficients specific to alternative based on case-specific variable z_i. The stochastic component of utility denoted by $\in_{ia}$, follows a Type 1 Extreme Value Gumbel distribution—a commonly employed distribution in discrete choice models examining extreme tail behavior of a dataset. This is to help understand the likelihood of events that fall far outside the typical range of observations. The fixed coefficients α and δ_j are derived from the distribution $\beta_i \sim \mathbb{N}(\mu, \sigma^2)$, where the MMNL method estimates the mean and variance of the model parameter.

By emphasizing the significance of the rank order of utilities for each alternative, the location and scale of utility becomes inconsequential. To address this, normalizing the location by considering differences with respect to the base or reference alternative is crucial. This normalization process assumes that the distribution for the difference in errors follows a logistic distribution, ensuring proper scaling.

The choice probabilities of the MMNL model for an individual i selecting alternative j are determined by the integral of the conditional probability of choosing j, considering the individual-specific parameters and random variation (utility) provided by,

$$\wp_{ij} = \frac{\int e^{U_{ij}}}{\sum_j e^{U_{ji}}} f(\beta_i)\mathrm{d}\beta_i \tag{2}$$

where U_{ij} denotes the utility of individual i for a specific alternative j, and $\sum_j e^{U_{ij}}$ signifies the summation of exponentiated utilities across all available alternatives. $f(\beta_i)$ represents the probability density function (pdf) of individual-specific parameters β_i.

Equation (2) indicates that the integration is conducted across the distribution of individual-specific parameters. These parameters account for the diversity among individuals in their preferences. Each person may possess distinct parameter values that impact their evaluation of different alternatives. As these parameters introduce variability in utility across individuals, they are treated as random during model estimation.

3.5 Bayesian Multinomial Logit Model (BMNL)

Bayesian methods are a vast and complex set of advanced multivariate statistical models used by academic scholars in discrete modeling analysis. For brevity, this paper considers the Bayesian multinomial logit model (BMNL) which is a Bayesian approach to estimating the parameters of a multinomial logit mode (MNL). It allows for the incorporation of prior information and treats model parameters as random variables with probability distributions.

In a Bayesian framework, uncertainty about parameters is expressed through probability distributions, and Bayesian methods update these distributions based on observed data that is useful in parameter estimation.

In terms of the two models used in this study, the mixed multinomial logit (MMNL) model and the Bayesian multinomial logit (BMNL) model are two statistical techniques designed for analyzing discrete choice data. Both models extend the multinomial logit (MNL) framework and incorporate individual-level heterogeneity in preferences, acknowledging variations in individual preferences. However, they diverge in several key ways.

Notably, their dissimilarities arise in terms of estimation methods, parameter distribution, handling of uncertainty, and computational complexity. MMNL primarily employs maximum likelihood estimation (MLE) to estimate model parameters (Molloy 2021), accommodating random coefficients to capture heterogeneity. In contrast, BMNL utilizes Bayesian methods, specifying prior distributions for parameters and updating them with observed data to obtain posterior distributions (van de Schoot 2021). Bayesian methods are advantageous for handling complex models and limited data, yielding entire probability distributions for parameter estimates.

Regarding parameter distribution, MMNL assumes the distribution of individual-specific parameters is known or follows a specific parametric form, often a normal or log-normal distribution for random coefficients. In contrast, BMNL explicitly models the parameter distribution within the Bayesian framework, assigning prior distributions to parameters and deriving posterior distributions through Bayes' theorem.

The treatment of uncertainty further distinguishes the two models. MMNL provides point estimates and standard errors based on the MLE approach, offering a frequentist perspective on uncertainty. Conversely, BMNL delivers a complete distribution of parameter estimates, providing a more comprehensive understanding of uncertainty.

Lastly, computational complexity varies between the two models. MLE in MMNL may involve solving complex optimization problems, especially with numerous alternatives and random coefficients. Bayesian methods in BMNL often employ Markov Chain Monte Carlo (MCMC) techniques, which can also be computationally demanding. Nevertheless, advancements in computational methods have made Bayesian estimation more accessible, and there is rich literature on this.

In summary, while MMNL and BMNL both extend the multinomial logit model and address individual-level heterogeneity, they differ in estimation methods, treatment of parameter distributions, uncertainty handling, and computational complexity. Both models are therefore used in this study based on the available data.

3.5.1 Model Specification

The model specification is an extension of an MNL model specification. To maintain brevity and simplicity, a concise overview of the BMNL model specification is provided. For a more in-depth review of Bayesian methods, interested readers are encouraged to refer to Gelman (1995) and Polymeropoulos (2020). STATA 18 is employed for implementing the complex statistical model using MCMC methods. Additionally, OpenBUGS, an open-source program for Bayesian modeling, is a viable alternative for model implementation. Consider a BMNL model with N observation, K classes, and D input variables. Let x_{ni} denote the value of input variable i for observation n, where $y_n = 1, 2, \dots, K$. The model posits a linear relationship between log odds, expressed as,

$$\log\left(\frac{p_{nk}}{p_{nK}}\right) = \beta_{k0} + \beta_k^T x_n \tag{3}$$

for $k = 1, 2, \dots, K-1$. Here β_{k0} is the intercept for class k, and β_k is the coefficient vector for class k.

The probabilities p_{nk}, representing the probability of observation n belonging to class k, are obtained using the softmax function (or normalized exponential function). This function transforms a vector of numbers into a vector of probabilities,

$$p_{nk} = \frac{e^{\beta_{k0}+\beta_k^T x_n}}{1 + \sum_j^{K-1} e^{\beta_{j0}+\beta_j^T x_n}} \tag{4}$$

The likelihood of the observed data follows a multinomial distribution,

$$P(y_n = k \mid x_n, \beta_{k0}, \beta_k) = p_{nk} \tag{5}$$

The mathematical formulation underscores that, in the Bayesian framework, priors for the parameters β_{k0} and β_k are explicitly defined. Subsequently, posterior inference is conducted to estimate these parameters, integrating observed data with prior beliefs, as previously discussed.

4 Results and Discussion

4.1 Descriptive Analysis

The stated preference experiment gathered 446 observations after cleaning the initial dataset, in which only relevant key demographic and travel statistics used in this paper are recorded in Table 2. To expand further, the sample population consists of individuals with distinct sociodemographic characteristics. The age distribution of the sample shows that the majority of participants are aged between 65 and 74, comprising 40% of the total, with a marked representation of individuals aged 55 and older. Males represent approximately 68% of the sample, while females constitute around 28%. The ethnic composition is predominantly White Caucasian (88%), with other ethnic groups accounting for a minimal 2% of the population. Notably, 81% of the respondents reported having disabilities, a proportion that corresponds with the older age profile of the participants. Geographically, the sample is moderately distributed, with the largest segments located in East London (30%) and South London (34%) (Rao 2023).

Regarding socioeconomic factors, the sample predominantly consists of individuals with limited educational attainment (35%), with a smaller, but similar proportion possessing postgraduate qualifications (10%) or minimal to no education (9%). In terms of employment status, the majority are retired (55%), while 33% are either in full-time employment or self-employed. Essential workers make up 19% of the sample, leaving 81% categorized as non-essential workers. Income levels are relatively balanced across groups, with 35% earning between £10 K and £24.9 K annually, reflecting a higher prevalence of retirees and those in moderate-skill jobs or contract positions (Rao 2023).

An analysis of travel behaviors within the sample reveals that leisure (39%) and work-related travel (33%) are the main reasons for transportation. Private vehicles are the dominant mode of travel (79%), consistent with patterns observed before the COVID-19 pandemic, indicating a return to pre-pandemic levels of car usage. This preference for private vehicles is further supported by the fact that 77% of vehicle owners do not pay for parking. Addressing parking policies is essential to curb the trend of owning multiple vehicles. Older drivers with higher vehicle densities show an elevated incidence of severe car accidents, as highlighted by a high crash rate (49%). Moreover, participation in dynamic ridesharing is low, involving less than 10% of the sample, primarily younger individuals, indicating the need to increase adoption among older adults who are at higher risk of accidents. Promoting ridesharing could enhance social inclusion, given that 91% of respondents do not currently use these services. This low adoption rate can be attributed to pre-existing safety concerns and heightened apprehensions due to the pandemic, which have discouraged ridesharing with unfamiliar individuals (Rao 2023). It is therefore particularly useful to determine the heterogeneity in preferences using MMNL and BMNL models.

Mode preferences from Fig. 4 suggest that SAEVs are a likely choice of mode during peak travel periods for more than 60% of respondents in the sample.

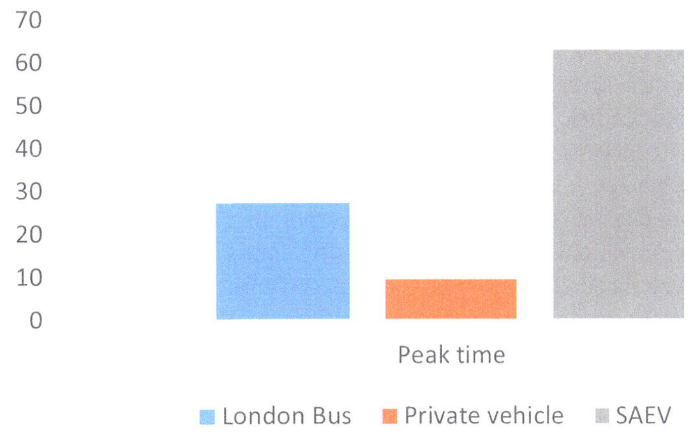

Fig. 4 Preferences for mode choice by respondent sample during peak travel in London. *Source* Author

4.2 MMNL Model Estimates

As outlined in Sect. 3.5.1, the mixed multinomial logit (MMNL) model specification is used for estimating the model parameters and analysis, including calibration for predicted probabilities. Given the absence of a closed-form solution for the integral in Eq. (2), analytical methods are employed, specifically simulation-based techniques to approximate the equation. Subsequently, the model parameters are approximated through the process of maximum simulated likelihood (MSL). The MSL estimator necessitates a substantial number of random draws within the simulation procedure to ensure consistency. While a greater number of draws enhances estimation accuracy by reducing approximation discrepancies, this improvement comes at the cost of increased computational time.

This section focuses on the estimation of the mixed multinomial logit (MMNL) model, which incorporates random coefficients and two constant case-specific (alternative invariant) variables.

4.2.1 Random Intercept Mixed Logit Model Incorporating Constant Case-Specific Variables and a Random Coefficient

Model fitting involves relaxing the independence from irrelevant (IIA) assumption inherent in the multinomial logit (MNL) model. The rationale behind this is the belief that alternative mode choices may exhibit correlations, constituting the primary advantage of the mixed logit model over its MNL counterparts due to its high flexibility.

Mode cost serves as an alternative-specific variable that can vary across alternatives and individuals, specified as a random coefficient. The driver's license and gender variables vary among individuals and are thus case-specific variables in the MMNL model specification. In the model specification, the likelihood is computed using Monte Carlo integration, with default integration points set for numerical robustness.

The model assesses the impact of bus, private car, and shared automated electric vehicle (SAEV) costs and the effect of gender and owning a driver's license on mode choice alternatives. Alternative-specific costs are presumed to be random across individuals and the sample population, and the results are presented in Table 4.

The Random Intercept Mixed Logit Model, incorporating constant case-specific variables and a random coefficient, is employed to analyze mode choices among 446 individuals. The Wald chi-squared test, used to assess the significance of coefficients, revealed the model's overall significance.

The model's significance, based on 446 cases representing individual mode choices, is established through the Wald chi-squared test. For key variables—*mode cost*, *drivers license*, and *gender*, the *p*-values are highly significant within the 90% confidence interval, indicating their statistical significance as predictors of mode choice. The small standard errors associated with each coefficient enhance the reliability of uncertainty estimates.

Table 4 MMNL regression specifying mode cost as a random coefficient and gender as case-specific

Final mode choice	Coefficient	Std. err.	$P > \|z\|$
Mode choice peak			
Mode cost	− 0.3883215	0.1220114	0.001
sd (mode cost)	4.83e−06	0.3222291	
London bus			
Drivers license	− 3.504839	0.7762647	0.000
Gender	0.1550966	0.2391875	0.517
SAEV (base alternative)			
Private car			
Drivers license	− 3.037244	0.4388134	0.000
Gender	− 0.745358	0.4099532	0.069

The estimated coefficient for the case-specific variable *gender* was found to be significant at the 90% confidence interval. This suggests that, relative to males and with other variables held constant, females and other gender categories exhibit a decreased likelihood of choosing private vehicles, indicating a preference for shared automated electric vehicles (SAEVs).

Possession of a driver's license emerges as a distinctive and significant factor with considerable importance. Individuals with a driver's license demonstrate a reduced inclination toward public transport (such as buses) and private vehicles, indicating a higher preference for shared automated electric vehicles (SAEVs). This finding is promising for the acceptance and adoption of shared automated transportation technologies.

The estimated mean coefficient for the normally distributed mode cost is − 0.39, with a small standard deviation of 0.000872. Despite the small standard error, mode cost remains highly significant. This implies a negative relationship, indicating that as the cost of a mode increases, the log odds of choosing that mode decrease. The results suggest substantial heterogeneity in individual preferences for the cost variable, emphasizing its significance over travel time and efficiency, particularly during peak travel periods. This finding aligns with the existing literature indicating that choice-based preferences and predictions using mixed logit models, incorporating individual heterogeneity, lead to improved prediction accuracy (Krueger 2016, 2021; Bai 2017; Paulssen 2014).

The overall model fit was assessed using the likelihood ratio (LR) test, yielding a chi-square test statistic of 187.76 with 5 degrees of freedom. The very small *p*-value ($p \leq 0.1$)

even at the 95% confidence interval suggests that the model significantly improves upon a model with fixed parameters.

4.2.2 Post-estimation Analysis

Post-estimation in the context of MMNL refers to the analysis conducted after estimating the model parameters, which includes computing the marginal effects. The marginal effects provide insights into how changes in the independent variables impact the probability of choosing a particular alternative. It provides insights into individual heterogeneity, the impact of covariates, and model validation, thereby enhancing the usefulness and reliability of the estimated model.

In this study, employing the recently calibrated model and assuming the existence of a randomly selected sample of individuals, marginal analysis is employed as a post-estimation procedure to estimate the proportion of individuals opting for each mode of transport.

The outcomes presented in Table 5 are highly significant, indicating that 63% of individuals are anticipated to choose to ride in SAEVs during peak travel. This optimistic prediction provides valuable insights for planners, enabling them to formulate recommendations and policies aimed at reducing the use of private vehicles during peak travel periods.

In review, this section presents and discusses the results of the MMNL model estimates, which involve the estimation of model parameters and subsequent analysis, including calibration for predicted probabilities.

In terms of including random coefficients and constant case-specific variables, the MMNL model application reveals significant insights into the factors influencing mode choices during peak travel times. By introducing random coefficients and considering individual-specific variables, this paper departs from the limitations of the multinomial logic (MNL) model, acknowledging that alternative mode choices may exhibit correlations.

The model outcomes underscore the pivotal role of cost considerations in individuals' mode preferences. The estimated mean coefficient on mode cost, along with its standard deviation, points to substantial heterogeneity in how individuals in the sample population respond to cost variations. An increase in costs is shown to reduce the likelihood of selecting a mode, underscoring users' prioritization of cost considerations over travel time and efficiency during peak travel. This aligns with existing literature, emphasizing the necessity of accounting for individual heterogeneity for more accurate predictions. The findings specifically highlight the importance of addressing cost-related concerns to enhance the future adoption of shared automated electric vehicles (SAEVs) during busy travel periods.

Moreover, the inclusion of the driver's license and gender coefficient as case-specific variables adds depth to the existing literature. Individuals with driver's licenses exhibit a decreased likelihood of choosing buses over SAEVs during high-demand periods. Similarly, private vehicles become less likely to be chosen as the number of individuals with driver's licenses increases, as do those by women and other gender categories. These findings present promising implications for the gradual introduction and scaling of SAEVs in London, particularly in the context of addressing peak travel demand and possible congestion. However, it addresses concerns about the scope of public transport use, particularly buses, but this is unsurprising given the general decline in ridership due to significant cuts in service (Transport Investment Limited 2018).

The model's post-estimation analysis delves into the marginal effects, providing crucial insights into how changes in the independent variable impact the probability of choosing a specific mode. The anticipated outcome, signifying that 63% of individuals are likely to choose SAEVs during peak travel, is highly significant. This optimistic prediction serves as a valuable guide for planners, offering actionable recommendations and policies to reduce private vehicle use during high-demand travel periods.

In summary, the MMNL model results underscore the significant influence of cost and gender, particularly among women among the surveyed population in London who favor ride-sharing in SAEVs (Shared automated electric vehicles) during peak travel times. Despite the fact that fewer than 10% currently engage in ridesharing with strangers, the overall hesitancy toward SAEVs in the future emphasizes the imperative to enhance accessibility and trust, particularly among older individuals and across genders. The findings suggest that addressing cost-related concerns has the potential to drive increased adoption of SAEVs, and the incorporation of individual-specific variables such as driver's license and key demographic attributes adds nuance to the study's understanding, paving the way for more informed transportation planning strategies in London. The mixed logit model better reflects real world behavior and is suitable for situations where individuals exhibit varying degrees of sensitivity to attributes influencing their choices.

Table 5 Predictive margins; proportional of people in the sample that will select each mode

	Delta-method		$P > \|z\|$
	Margin	Std. err.	
Outcome			
London bus	0.273801	0.012988	0.000
SAEV	0.630301	0.014269	0.000
Your own private car	0.095899	0.009536	0.000

4.3 BMNL Model Estimates—A Two-Step Approach

In the exploration of the mode choice dataset consisting of 446 observations, this study employs a BMNL model with Bayesian multinomial logit (BMNL) model with uninformative priors to unravel the intricate relationship between the response variable (e.g., SAEV, bus, private vehicle) and covariates (employment, gender, driver's license). This involves generating samples from the posterior distribution and merging prior information about parameters (e.g., intercept, employment, gender, and driver's license as coefficients) with the likelihood from the fitted model.

Step 1: The initial step involves obtaining MLEs by fitting a multinomial logistic (ML) regression model. The results (Table 6) underscore the widespread use of MNL models in transportation research due to their utility in decision-making based on random utility theory. This is out of the scope of this paper, but readers interested in transport economics can consult (Cascetta 2001) to further their understanding. However, these models, grounded in robust economic theory, exhibit impressive empirical performance.

Step 2: The subsequent step involves estimating the Bayesian MNL model using Markov Chain Monte Carlo (MCMC) methods (Table 7). The Bayesian approach, treating data and parameters as random variables, offers flexibility in model development and specification. It updates subjective beliefs about parameters based on observed data using Bayes theorem.

STATA 18 utilizes default priors chosen to be fairly uninformative. Regression coefficients are assigned an independent normal with a zero mean and a variance of 10,000. The MCMC engine is employed with a sample size of 50,000, a burn-in period of 8000, and a thinning of 3. Informative priors have a greater impact on the posterior distribution or given sample size. However, precise estimates can be obtained with smaller sample sizes when using more informative priors. In this study, due to the lack of relevant previous datasets, inference analysis relies on uninformative priors.

To further aid the interpretation of the results, Table 7 demonstrates the close convergence of posterior mean parameter estimates to MLEs, confirming the reliability of the BMNL model. Diagnostic plots, especially for London buses and SAEVs, indicate low autocorrelation and acceptable levels of convergence. Posterior summaries are presented as relative risk ratios (RRRs) instead of regression coefficients. This transformation simplifies comprehension by applying the exponential function to simulated values. RRRs indicate how the probability of an outcome belonging to a comparison group changes in response to changes in the variable.

Table 6 MNL reference model specification

Mode choice peak	Coefficient	Std. err.	$P > \|z\|$
London bus			
Gender	– 0.1208969	0.1409173	0.391
Employment	0.1688013	0.0824804	0.041
DL			
Yes	– 3.788356	1.037641	0.000
Cons	– 3.788356	1.083894	0.024
Private car (base outcome)			
SAEV			
Gender	– 0.7747434	0.3622895	0.032
Employment	0.3679137	0.1274786	0.004
DL			
Yes	– 3.400051	1.145076	0.003
Cons	1.206889	1.320815	0.361

4.3.1 Public Transport (Bus) Relative to Private Vehicles

The relative risk ratios (RRRs) obtained through the BMNL regression outlined in Table 8 provide insights into the relationship dynamics between the utilization of London buses and private vehicles. This analysis specifically examines variables such as gender, driver's licenses, and employment.

In terms of gender influences, an increase in gender by one unit corresponds to a decrease in the relative risk of opting for buses over private vehicles by a factor of 0.881, assuming all other variables remain constant. In essence, a rise in the gender category indicates a higher likelihood of preferring private vehicles. Gender influences on travel behavior are aligned with (Si 2023; Ng 2018).

The impact of driver's licenses reveals that ownership of a driver's license is associated with a decrease in the relative risk of choosing buses over private vehicles by a factor of 0.0246 for each unit increase. This aligns with expectations and mirrors maximum likelihood estimates (MLEs).

For employment propensity, the relative risk ratio for employment type signifies a propensity to favor bus transport over private vehicles while controlling for all of the variables in the model. Specifically, the relative risk increases by a factor of 1.190, indicating a preference for buses over private vehicle ownership within this employment category.

Upon further scrutiny, Table 9 dissects the impact of different unemployment categories, revealing significant variations. Retirees exhibit a substantial relative risk of 2.335 for preferring buses, while students demonstrate an even higher relative risk of 6.5. This aligns with expectations, considering demographic composition and uninformative priors.

Retirees and students, likely characterized by lower income, show preferences for public transportation over private vehicles, consistent with the literature (Ravensbergen et al. 2022), possibly due to cost savings and environmental

Table 7 Bayesian MNL (BMNL) parameter estimate

	Mean	Std. dev.	MCSE	Median
London_bus				
Driver's license				
Yes	− 4.29296	1.264996	0.038779	− 4.113657
Gender	− 0.1365533	0.1442595	0.001667	− 0.1310692
Employment	0.1708014	0.082706	0.00102	0.1706739
Cons	2.954632	1.3058	0.038469	2.786495
SAEV				
Driver's license				
Yes	− 3.821668	1.371382	0.041134	− 3.682176
Gender	− 0.8620291	0.3711361	0.004634	− 0.8361726
Employment	0.378278	0.1293404	0.001563	0.3750099
Cons	1.663909	1.527459	0.042868	1.564353

Table 8 BMNL estimated relative risk ratios

	RRR	Std. dev.	MCSE	Median
London_bus				
Driver's license				
Yes	0.0246415	0.0263316	0.000543	0.0163479
Gender	0.8813922	0.1259022	0.001442	0.8771571
Employment	1.190324	0.0987912	0.001214	1.186104
Cons	95.2895	1363.973	26.3271	16.22406
SAEV				
Driver's license				
Yes	0.0465688	0.0723893	0.001293	0.0251682
Gender	0.4507844	0.1593398	0.001905	0.433366
Employment	1.472104	0.19296	0.002341	1.455006
Cons	31.44246	385.1732	7.57297	4.779584

Table 9 BMNL estimated relatives risk ratios for various employment categories

		RRR	Std. dev.	MCSE	Median
London_bus					
	Employment				
	Self-employed	1.010628	0.4615606	0.009962	0.9240442
Part time I contract I	Temporary	1.322585	0.7477338	0.006879	1.160644
	Retired	2.180615	0.6590654	0.011105	2.08326
	Student	2.1E+171			3.33E+34
	Not working	1.068513	0.7115303	0.009006	0.8944169
	Cons	0.2871963	0.0735237	0.001205	0.279477
SAEV					
	Employment				
	Self-employed	3.84349	3.451289	0.05686	2.890197
Part time I contract I	Temporary	2.810914	3.283858	0.055295	1.859983
	Retired	4.30651	3.232034	0.05245	3.427623
	Student	3.30E+101	7.30E+103	3.00E+101	1.69E−14
	Not working	10.2375	9.52719	0.129974	7.584982
	Cons	0.0568558	0.0292973	0.000483	0.0519049

considerations. In contrast to full-time employment, unemployment leads to a relative risk increase of 1.069, suggesting a preference for public transport, which aligns with logical reasoning.

In essence, the BMNL regression provides nuanced insights into mode choice dynamics, revealing how various factors influence the preference for buses relative to private vehicles. These findings contribute to a comprehensive understanding of the intricate relationships within the surveyed population.

4.3.2 SAEVs Relative to Private Vehicles

Table 8 also provides insights into the comparison between shared automated electric vehicles (SAEVs) and private vehicles, demonstrating consistent magnitude across covariates, akin to the bus versus private vehicle analysis earlier. Notably, individuals with a driver's license (DL) exhibit a trend where a one-unit increase in DL ownership results in a decrease in the relative risk of preferring SAEV ridesharing over driving a private vehicle by a factor of 0.0466. This aligns with expectations and echoes the convergence seen in maximum likelihood estimates (MLEs) utilizing uninformative priors. However, the results contrast those from the Random Intercept MMNL model earlier, where observed factors reveal variation in preferences from individuals holding driver's licenses preferring ridesharing in SAEVs relative to driving private vehicles. This indicates a possible future acceptance of the nascent technology.

Table 9 extends the investigation, unveiling a consistently higher RRR in opting for SAEVs over private vehicles across various employment categories, surpassing preferences for both buses and private vehicles mentioned earlier. Notably, unemployed individuals emerge with the highest RRR within the SAEV category, signifying a substantial increase in the relative risk of choosing SAEV ridesharing over private vehicle use by a factor of 10.238 for a one-unit increase in the unemployment category while keeping other variables constant. This implies that unemployed individuals are significantly more inclined toward ridesharing in SAEVs compared to their counterparts in full-time employment.

Similarly, the analysis reveals that individuals with DLs generally prefer their own vehicles over ridesharing, consistent with the widely accepted notion that those without driving capabilities are more likely to opt for SAEVs. This can also be due to social reasons evident from psychobehavioral ridesharing preferences, consistent with (Carol 2019; Si 2023). It is important to highlight that the study cannot derive informative priors from independent datasets. The justification for constructing informative priors to improve estimation accuracy exists only when available datasets align with the current data of the study. These findings from the Bayesian estimation of the multinomial logit model (BMNL) using uninformative priors contribute valuable insights into the nuanced dynamics influencing the preference for SAEVs over private vehicles, particularly concerning DL ownership and employment status.

5 Conclusions

In summary, this study explores the transformative impact of shared automated electric vehicles (SAEVs) on urban mobility. SAEVs, exemplifying the convergence of automation, electrification, and shared mobility, have the potential to reduce vehicle-miles-traveled, lower greenhouse gas emissions, and reshape urban development.

The environmental benefits and cost-effective on-demand services of SAEVs position them as a sustainable alternative to private vehicle use for many travelers in London. Despite their potential benefits, challenges such as public concerns about safety and privacy need to be addressed through targeted educational campaigns.

The adoption and scaling of SAEVs can only be possible by capturing and modeling individual-level heterogeneity in preferences and uncertainty. The study employs advanced statistical models, namely the mixed multinomial logit (MMNL) and Bayesian multinomial logit (BMNL) to analyze complex individual decision-making and diverse individual preferences.

The Random Intercept Mixed Logit Model incorporating a random coefficient along with two case-specific variables, offers valuable insights into factors like mode cost, gender, and possessing a driver's license that impact mode choices during peak travel times in London. The incorporation of random coefficients and constant case-specific variables in the MMNL model allows for a departure from the limitations of the multinomial logit model, acknowledging correlations in alternative mode choices.

The study consistently emphasizes the pivotal role of cost considerations in shaping individuals' mode preferences, highlighting the potential of SAEVs as a preferred alternative during peak travel times. The findings underscore the importance of addressing cost-related concerns to enhance the adoption of SAEVs.

Moreover, the inclusion of case-specific variables, such as driver's licenses, adds depth to the understanding of mode choices. Individuals with driver's licenses exhibit a decreased likelihood of choosing buses over SAEVs during high-demand periods, and private vehicles become less likely to be chosen as the number of individuals with driver's licenses increases. These insights present promising implications for the gradual introduction and scaling of SAEVs in London.

The post-estimation analysis, focusing on marginal effects, further reinforces the significance of cost considerations and predicts that 63% of individuals are likely to choose SAEVs during peak travel. This optimistic prediction provides valuable guidance for planners to formulate recommendations and policies aimed at reducing private vehicle use during high-demand travel periods.

Moving on to the BMNL model estimates, the two-step approach involving maximum likelihood estimates (MLEs) and the Bayesian multinomial logit (BMNL) model adds depth to the understanding of mode choice dynamics. The BMNL model sheds light on the intricate relationships between gender, driver's licenses, employment variables, and mode choices, providing nuanced insights into the surveyed population's preferences for buses and SAEVs relative to private vehicles.

The utilization of uninformative priors by the model, considering the study's originality and the absence of comparable datasets, enables the data to dominate the inference, particularly with regard to unique items. In this instance, the question to be asked: is the BMNL model using uninformative priors worth the effort in Bayesian modeling?

If only uninformative priors are available and no specific prior information can be incorporated, estimating a Bayesian multinomial logit model may not provide substantial advantages over a standard multinomial logit model. In such cases, both approaches may yield similar results as revealed in this study, when determining the relationship between the response variable and covariates. The resulting posterior distribution is closely similar to the likelihood function of the MNL reference model specification, and the Bayesian model might not offer additional insights.

The decision to use a Bayesian approach often depends on the context and availability of prior information. If informative priors are lacking, the standard MNL model could be simpler and computationally more efficient. However, this study serves as a precursor to utilizing Bayesian inferences without informative priors, given the future prospect of incorporating priors to update subjective beliefs from knowledge observed in new datasets in future work. This will provide the opportunity to update beliefs over time, and a Bayesian framework remains flexible for incorporating new information as it becomes available.

While recognizing limitations, this research contributes valuable insights, emphasizing the need for further exploration of random effects in the context of BMNL models. However, both the MMNL and BMNL model estimates contribute to a comprehensive understanding of the factors influencing mode choices in the context of urban transportation and new technologies. The findings have implications for transportation planning, emphasizing the need to consider individual and unobserved variability that affects decision-making using a mixed logit model, and by providing a more comprehensive understanding of uncertainty using Bayesian methods with the future scope of updating beliefs from this study's observed data. This study relies on key predictors, including cost considerations and the distinct influence of factors like driver's licenses, gender, and employment status, in shaping mode preferences during peak travel times in London. Future research could involve incorporating random effects at different levels of analysis, such as individual, or household levels, to better understand how contextual factors influence mode choice preferences. Longitudinal studies tracking changes in mode choice behavior over time can provide valuable insights into the dynamics of adoption and the effectiveness of transportation policies and interventions. By combining sophisticated Bayesian modeling techniques with longitudinal data collection, future research can yield deeper insights into mode choice behavior and inform more effective transportation planning and policy decisions.

Overall, the study makes significant contributions to the existing literature by employing advanced and robust modeling techniques to analyze mode choice preferences and the adoption of shared automated electric vehicles (SAEVs) in urban settings, particularly London. It offers insights into factors influencing mode choices, such as cost considerations, gender, driver's licenses, and employment status, while addressing challenges like safety and privacy concerns associated with SAEVs. Moreover, the study identifies avenues for future research, including further exploration of random effects in BMNL models and the incorporation of new data. Overall, it provides valuable guidance for researchers, policymakers, and practitioners in urban transportation planning and policy.

Acknowledgements This research was funded by the Imperial College London scholarship.

Declaration This research serves as an extension to a previous paper titled "Exploring SAEVs in the UK: A Behavioral Attitudes and Ridesharing Analysis for Sustainable Urban Mobility," which was presented at FSC 2023. The research builds upon the survey datasets utilized in that initial study.

References

Adam, A.M.: Sample size determination in survey research. J. Sci. Res. Rep. 90–97 (2020)

Bai, T.L.: Effects of cost adjustment on travel mode choice: analysis and comparison of different logit models. Transp. Res. Procedia 2649–2659 (2017)

Barter, P.: Cars are parked 95% of the time. Let's check!, May 2019. Retrieved from Reinventing Parking: https://www.reinventingparking.org/2013/02/cars-are-parked-95-of-time-lets-check.html

Carol, S.E.: Who can ride along? Discrimination in a German carpooling market. Popul. Space Place (2019)

Cascetta, E.: Random utility theory. Transp. Syst. Eng. Theory Methods 95–173 (2001)
Chen, T.D.: Operations of a shared, autonomous, electric vehicle fleet: implications of vehicle & charging infrastructure decisions. Transp. Res. Part A Policy Pract. 243–254 (2016)
Choi, S.K.: Value of travel time savings of autonomous vehicle commuters: a segmented valuation for local and inter-city travel. Int. J. Urban Sci. 1–25 (2023)
Cools, M.R.: Can autonomous vehicles reduce car mobility? Evidence from a stated adaptation experiment in Belgium. In: Transportation Research Board 96th Annual Meeting (2017)
D'Urso, P.G.: To share or not to share? Determinants of sharing mobility in Italy. Soc. Indic. Res. 647–692 (2021)
Davidson, P.: Autonomous Vehicles: What Could This Mean for the Future of Transport. Australian Institute of Traffic Planning and Management (AITPM), Brisbane, Queensland (2015)
Fagnant, D.J.: The travel and environmental implications of shared autonomous vehicles, using agent-based model scenarios. Transp. Res. Part C Emerg. Technol. 1–3 (2014)
Fagnant, D.J.: Preparing a nation for autonomous vehicles: opportunities, barriers and policy recommendations. Transp. Res. Part A Policy Pract. 167–181 (2015)
Faisal, A.K.: Understanding autonomous vehicles. J. Transp. Land Use 45–72 (2019)
Gelman, A.C.: Bayesian Data Analysis. Chapman and Hall/CRC (1995)
Greenblatt, J.B.: Automated vehicles, on-demand mobility, and environmental impacts. Curr. Sustain./Renew. Energy Rep. 74–81 (2015)
Guo, R.C.: A survey of learning causality with data: problems and methods. ACM Comput. Surv. (CSUR) 1–37 (2020)
Haboucha, C.J.: User preferences regarding autonomous vehicles. Transp. Res. Part C Emerg. Technol. 37–49 (2017)
Iacobucci, R.M.: Optimization of shared autonomous electric vehicles operations with charge scheduling and vehicle-to-grid. Transp. Res. Part C Emerg. Technol. 34–52 (2019)
Kessler, T.E.: Linear differential games for cooperative behavior planning of autonomous vehicles using mixed-integer programming. In: 59th IEEE Conference on Decision and Control (CDC), pp. 4060–4066. IEEE (2020)
Kitson, N.K.: A survey of Bayesian network structure learning. Artif. Intell. Rev. 1–94 (2023)
Kolleck, A.: Does car-sharing reduce car ownership? Empirical evidence from Germany. Sustainability (2021)
Krueger, R.R.: Preferences for shared autonomous vehicles. Transp. Res. Part C Emerg. Technol. 343–355 (2016)
Krueger, R.B.: Evaluating the predictive abilities of mixed logit models with unobserved inter- and intra-individual heterogeneity. J. Choice Model. (2021)
Lee, D.: Public concerns and connected and automated vehicles: safety, privacy, and data security. Humanit. Soc. Sci. Commun. 1–13 (2022)
Liu, J.K.: Tracking a system of shared autonomous vehicles across the Austin, Texas network using agent-based simulation. Transportation 1261–1278 (2017)
Liu, Z.L.: Effects and feasibility of shared mobility with shared autonomous vehicles: an investigation based on data-driven modeling approach. Transp. Res. Part A Policy Pract. 206–226 (2022)
London, T.f.: New TfL data shows continued boom in walking and cycling, with almost twice as many now living near a high-quality cycle route. Retrieved from Transport for London | Every Journey Matters, 30 Nov 2022
London-City-Hall: Investing in an accessible transport network. Retrieved from London City Hall: https://www.london.gov.uk/programmes-strategies/transport/improving-londons-roads/investing-accessible-transport-network, 7 Mar 2016
MacroTrends: London, UK Metro Area Population 1950–2023. Retrieved from MacroTrends: https://www.macrotrends.net/cities/22860/london/population (2023)
Maeng, K.: Who will want to use shared autonomous vehicle service and how much? A consumer experiment in South Korea. Travel Behav. Soc. 9–17 (2022)
Massar, M.R.-I.: Impacts of autonomous vehicles on greenhouse gas emissions—positive or negative? Int. J. Environ. Res. Public Health (2021)
Menon, N.B.: Shared autonomous vehicles and their potential impacts on household vehicle ownership: an exploratory empirical assessment. Int. J. Sustain. Transp. 111–122 (2019)
Molloy, J.B.: mixl: an open-source R package for estimating complex choice models on large datasets. J. Choice Model. (2021)
Ng, W.S.: Understanding urban travel behaviour by gender for efficient and equitable transport policies. International Transport Forum Discussion Paper (2018)
Ng, A.: When will self-driving cars be available to consumers. Retrieved from Imagination: https://www.imaginationtech.com/future-of-automotive/when-will-autonomous-cars-be-available/, 4 May 2023
Paulssen, M.T.: Values, attitudes and travel behavior: a hierarchical latent variable mixed logit model of travel mode choice. Transportation 873–888 (2014)
Polymeropoulos, A.: Objective variable selection in multinomial logistic regression: a conditional latent approach. Ph.D. thesis. University of Milano-Bicocca, Milan, 2020
Rao, P., Quddus, M., Ochieng, W.Y.: Exploring SAEVs in the UK: a behavioral attitudes and ridesharing analysis for sustainable urban mobility. In: International Conference on Future Smart Cities, pp. 123–140. Springer Nature Switzerland (2023)
Ravensbergen, L., Van Liefferinge, M., Isabella, J., Merrina, Z., El-Geneidy, A.: Accessibility by public transport for older adults: a systematic review (2022)
Shaheen, S.: Impacts of shared mobility. ITS Berkeley Policy Briefs (2018)
Si, H.S.: What influences people to choose ridesharing? An overview of the literature. Transp. Rev. 1–26 (2023)
Spieser, K.T.: Toward a systematic approach to the design and evaluation of automated mobility-on-demand systems: a case study in Singapore. In: Road Vehicle Automation, pp. 229–245 (2014)
Stein, G.M.: The impact of autonomous vehicles on urban land use patterns. Fla. St. U. L. Rev. (2020)
Transport Investment Limited: TfL is planning significant cuts in bus services and now acknowledges cycling as a cause of London ridership decline. Retrieved from Transport Times: https://www.transporttimes.co.uk/news.php/TfL-is-planning-significant-cuts-in-bus-services-and-now-acknowledges-cycling-as-a-cause-of-London-ridership-decline-333/, 26 Oct 2018
Tscharaktschiew, S.R.: Repositioning of driverless cars: is return to home rather than downtown parking economically viable? Transp. Res. Interdiscip. Perspect. (2022)
van de Schoot, R.D.: Bayesian statistics and modelling. Nat. Rev. Methods Primers (2021)
Winter, K.C.: A stated-choice experiment on mode choice in an era of free-floating carsharing and shared autonomous vehicles. Transp. Res. Board (2017)
Yap, M.D.: Preferences of travellers for using automated vehicles as last mile public transport of multimodal train trips. Transp. Res. Part A Policy Pract. 1–16 (2016)
Zhou, T.Z.: Exploring the determinants of public transport usage and shared mobilities: a case study from Nanchang, China. Sustain. Cities Soc. (2022)

Overview of Transport Policies and Climate Change Mitigation: Toward an Integrated Approach in the Mediterranean Region

Elona Karafili, Sadmira Malaj, and Flora Krasniqi

Abstract

Transporting plays a crucial role in global efforts to climate change mitigation climate, as it is a significant contributor to greenhouse gas emissions. As climate change increasingly impacts the Mediterranean as well, the region's approach to transport policies becomes a crucial instrument in achieving sustainability and resilience. In order to connect the two main such as transport climate change, various policy frameworks have been developed. However, so far, no consistent framework addresses these two areas in an integrated approach. Therefore, this study explores the intersection of transport policies, urban development, and climate changer mitigations, through a methodology of bibliographic analyses. The study highlights the unimportance of understanding the Mediterranean's complex transportation dynamics. Moreover, it will review the specific drivers and barriers influencing the formulation and execution of integrated transport policies in the Mediterranean. These factors, starting from economic considerations and technological advancements to the diverse political, social, and environments lie through the region. In conclusion, this review highlights the relevance of the importance of adopting an integrated approach to transport policies within the Mediterranean context, through a systematic literature review of articles.

Keywords

Sustainable transport · Climate change mitigation · Sustainable strategies · Bibliometric analysis · Integrated policies

1 Introduction

This study highlights the lack of a consistent integrated framework that effectively connects transport with climate change mitigation in the Mediterranean region. Through a bibliographic analysis, the study aims to examine the intersection of transport policies, urban development, and climate change mitigation, with a specific focus on the Mediterranean region.

As climate change increasingly impacts the Mediterranean as well, in this impact, transport plays a crucial role in global efforts to mitigate climate change, as it is a significant contributor to greenhouse gas emissions. Therefore, the region's approach to transportation policy becomes a crucial instrumental in achieving sustainable and resilient development.

In order to connect the two main pillars such as transport with climate change, various policy frameworks have been developed. However, so far, no framework addresses these two areas in an integrated approach. Therefore, the study aims to provide a conceptual framework, highlighting the importance of understanding the Mediterranean's complex transportation dynamics and delving into the specific drivers and barriers influencing the formulation and execution of integrated transport policies in the Mediterranean.

E. Karafili · S. Malaj (✉) · F. Krasniqi
Polis University, Tirana, Albania
e-mail: sadmira_malaj@universitetipolis.edu.al

E. Karafili
e-mail: elona_karafili@universitetipolis.edu.al

F. Krasniqi
e-mail: flora_krasniqi@universitetipolis.edu.al

P. Mohammad et al. (eds.), *Geographic Approaches to Climate Change and Mitigation: Urban and Rural Perspectives (Volume 2)*,
Advances in Science, Technology & Innovation, https://doi.org/10.1007/978-3-031-92115-5_14

2 Literature Review

Importance of Addressing Climate Change in the Context of Transport Policies

Many authors, discuss the significant impact of the transport sector on global gas emissions and emphasize the need to align transport policies with climate change objectives to mitigate its effects (Bakker et al. 2014; Bakker and Huizenga 2010; Black Carbon Mitigation and the Role of the Global Environment Facility: A STAP Advisory Document, n.d.; Creutzig 2016; De Abreu et al. 2022; Gota et al. 2019; Kadir 2022; Marí-Dell'Olmo et al. 2022; Stone et al. 2018). The importance of integrating climate change principles with Sustainable Development Goals (SDG) is highlighted to address the current climate change impact. Additionally, scholars highlight the importance of considering the scholars considering the environment's capacity for regeneration in resource extraction and emission processes to seek a more resilient approach. In this aspect, the role of government is defined as a crucial aspect in addressing climate change challenges, with a focus on whether policy processes or stakeholder involvement are necessary (De Abreu et al. 2022). Bakker et al. (2014) also discusses the challenges posed by different countries' varying contributions, capacities, and challenges in addressing climate change, leading to a complex matrix for integrating climate change aspects into transport sector policies (Bakker et al. 2014). Some authors, do align with sustainable and resilient concepts, while others express skepticism regarding the challenges posed by different countries' capacities and contributions to addressing climate change (Bakker et al. 2014; Black Carbon Mitigation and the Role of the Global Environment Facility: A STAP Advisory Document, n.d.; De Abreu et al. 2022; Häll et al., n.d.; Marí-Dell'Olmo et al. 2022). De Abreu et al. (2022) defines the need for a coherent overview of integrating climate change aspects into transport sector policies, recognizing the varying perspectives and challenges in this area (De Abreu et al. 2022). Additionally, the importance of aligning transport policies with climate change objectives to mitigate its impact brings together all scholars acknowledging both supportive and skeptical perspectives. It also highlights the challenges posed by differing countries regarding this matter, addressing climate change, and stressing the need for a comprehensive approach to integrating climate change aspects into transport sector policies.

This discussion emphasizes the need for further research on the implementation and best practices of low-energy mobility innovations is a complex and precarious issue (Jaeger et al. 2015). As mentioned above, geographical differences, stakeholder involvement, and political complexities play a significant role in shaping the nature and scale of these innovations (Schwanen 2015). Despite the challenges, the Paris Agreement has provided a significant impelling force for developing instruments aimed at mitigating the impact of climate change and enhancing sustainability and resilience (Schwanen 2015). However, some scholars criticize the limited impact of current international instruments on emissions reductions in the transport sector (Bakker and Huizenga 2010).

The analysis conducted by Bakker and Huizenga (2010), reveals that existing climate instruments are not effectively integrated into the framework of the transport sector on a global, regional, or local scale (Bakker and Huizenga 2010). These instruments have limited impact assessment on emissions reductions specifically in transport sector (Bakker et al. 2014; Bakker and Huizenga 2010; Challa 2017). Additionally, limited funding for climatic finance tools in transportation sectors worsens the situation! It is pointed out that specifying analytic framework and normalized presumptions is crucial for effective communications and understanding of these challenges. Moreover, on a global scale, these challenges might inaccurately grasp the implications of localized resolution tactics. This underlines the necessity to consider the localized context in formulating and executing low-power mobility options. Overall, the complexity of the problem and the need for more research and developing prime methods to tackle challenges connected with low-power mobility novelties (Creutzig 2016; Jaeger et al. 2015).

Some authors argue that most policies implemented are slow rather than radical, making it difficult to achieve significant changes in the transport sector ((Bakker et al. 2014; Bakker and Huizenga 2010; Gota et al. 2019). Understanding the radical systemic changes to transform transport systems, recognizing that such changes are infrequent, intricate, and encountering significant resistance due to extensive pay-back times and limited tangible economic advantages. The paper further highlights that investments in extensive, systemic, and communal transport systems frequently confront obstacles from the rapid evolution of conventional individual car-dependent transport systems within identical geographical regions (Gota et al. 2019). Additionally, the lack of success of existing policy instruments is attributed to various factors such as fast urbanization and motorization, inaccessibility of finance, costs of alternative fuels, technological challenges, technological lock-in, sunk costs, and inertia of transport systems, vested interests, low-price elasticity of mobility, diversity of stakeholders, and behavioral barriers (Bakker and Huizenga 2010). Overall, the need for simplified methodological requirements and increased funding for the transport sector to facilitate the transition to low-carbon sustainable transport (Bakker and Huizenga 2010). It cautions against the limitations and barriers associated with incremental policies and highlights the importance of addressing complex social, economic, and environmental considerations in efforts to decarbonize the transport sector and mitigate climate change

(Bakker, n.d.-a; Bakker et al. 2014; Marí-Dell'Olmo et al. 2022).

Environmental impacts and greenhouse gas emissions attributed to the transport sector, emphasizing the need to transition toward sustainable transport to mitigate climate change and lessen resource depletion (Shastry and Pai, n.d.). In the context of sustainable transportation, the paper discusses various determinants of climate change, its environmental and social consequences, and the direct and indirect implications on health. Continuing, the importance of mitigation and adaptive policies in addressing social and health inequalities, and acknowledging the role of individual and collective vulnerability in forming the effects of climate change on health (Challa 2017; Häll et al., n.d.).

The concept of transportation in this context, is quite complicated, including emission and waste reduction, inclusive accessibility, promoting health and boosting the economy. The diverse definitions of sustainability in transportation, emphasizing differences in terms of stringency, inclusion of intergenerational equity, specific targets, and focus on accessibility or mobility (Bakker et al. 2014). Some definitions explicitly encompass objectives related to safety, affordability of transport, and its contribution to development and economic growth, whereas others maintain a more general perspective. Additionally, in some authors' definitions is emphasized the quality of life, while others prioritize economic and social aspects (De Abreu et al. 2022; Angelidou et al. 2022; Bakker et al. 2014; Bakker and Huizenga 2010; Häll et al., n.d.; Marí-Dell'Olmo et al. 2022).

Furthermore, it is necessary to address social and health inequalities in the context of sustainable transportation, advocating for policies that consider the diverse impacts of climate change on various communities. According to Bakker et al. (2014), importance is all about capturing the essence of understanding the individual and combined vulnerability in influencing the impacts of weather changes on well-being, thus emphasizing the necessity for specific actions to minimize these consequences. In line to this context Bakker and Huizenga (2010), accentuates the complex interplay between sustainable transportation and health, underscoring the necessity of integrated approaches that acknowledge the diverse socioeconomic and environmental factors at play.

Emerging Technologies in Transportation and Their Potential Impact on Climate Change Mitigation

According to Gota et al. (2019), it is crucial the establishment of a specific target within the transport sector to achieve significant reductions in emissions and contribute to the 1.5 °C target (Gota et al. 2019). This approach does highlight the importance of adopting low-carbon transport trajectories and implementing mitigation measures to reduce greenhouse gas emissions in different income countries (Stone et al. 2018). Moreover, the authors address the significance of addressing short-lived climate pollutants (SLCPs) and unlocking their mitigation potential through policy action (Bakker et al. 2017; Gota et al. 2019; Nugroho and Zusman 2018).

To have a successful assessment indicator of a model for vehicle population and fuel demand, indicating a steady dieselization trend in the light vehicle fleet (Stone et al. 2018). Additionally, transport sector models are emphasized for their focus on efficiency measures (Creutzig 2016). These finding underline potential for policy intervening a technological advancement within transport sector driving emission reductions. Besides, author finding define significance limiting gas emitting, present opportunity bridging emissions gap. They suggestion that early reductions in gas emitting can have immediate benefit by improve health, mitigate climate changing in short term, and contribute achieving long-term temperature goal outlined in Paris Agreement (Bakker and Huizenga 2010; Creutzig 2016; Stone et al. 2018).

Therefore, they define the role of transport sector, very critical, in achieving deep emission reduction pathways. By implementing low-carbon transport trajectories and mitigation measures, countries can substantially contribute to reducing greenhouse gas emissions and working toward the global temperature goals outlined in the Paris Agreement.

Innovations in urban mobility in these cities are mostly small improvements rather than big changes, and the role of small cities in driving major transformations should not be overestimated. The innovations in urban mobility in various cities emphasize that these innovations are often small improvements rather than major changes (Schwanen 2015). He cautions against overestimating the role of small cities in driving transformative changes. The potential for reusing electric vehicle (EV) batteries in residential areas and integrating renewable energy generation technologies, alongside addressing traffic congestion, economic loss, and pollution in local level, is also highlighted (De Abreu et al. 2022). The introduction of intelligent transportation systems is noted as a way to enhance efficiency, reduce operating costs, and mitigate the negative impact of traffic congestion on economic productivity and the environment (Sajovic and Boh Podgornik 2022). The negative effects of traffic congestion, such as economic losses and fuel waste, are presented as key issues that need to be addressed (Schwanen 2015). Therefore, its crucial to emphasizes the need for measures to reduce dependency on fossil fuels and control CO_2 emissions in order to combat pollution and its associated consequences (Challa 2017). Moreover, the potential of intelligent transportation systems to improve traffic flow efficiency, particularly in the face of rising levels of pollutants and road accidents that disrupt traffic (Challa 2017). Additionally, the importance of introducing the technology as the main tool for merging the gap of transport sector and climate change aspect is crucial. Overall, the authors' findings from an in-depth analysis of the current state of urban mobility, they conclude the challenges posed by traffic congestion and pollution, and the potential for innovative solutions to address these issues.

In contrary from the above paragraph, other authors, mention that the low-energy mobility innovations and their impact on transportation can be noticed more in small- and medium-sized cities but has limitations on measuring that in a bigger scale (Gota et al. 2019; Häll et al., n.d.; Jaeger et al. 2015; Kadir 2022). Most of low-energy mobility innovations are incremental rather than radical, suggesting that major transformative changes are limited in some territorial scale. The authors, highlight in their research, the fact that small and medium-sized cities have limited autonomy in bringing about transformational change, which shapes the nature of mobility innovations in a bigger scale (Gota et al. 2019; Häll et al., n.d.; Jaeger et al. 2015; Kadir 2022). Furthermore, the dominance of private car use as the most important form of transport in these cities (Schwanen 2015). Therefore, the efforts to promote low-energy mobility solutions may face significant challenges in shifting away from car-dependent transportation systems. Meeting this situation with a solution approach, some authors, bring into discussions the potential for a horizontally structured assessment design to improve modeling efforts related to mobility innovations (Angelidou et al. 2022). This approach could offer a more comprehensive and effective way to evaluate the impact of different transportation strategies and technologies on urban mobility (Häll et al., n.d.).

Another key aspect comes in discussions for the consideration of smart cities, transport, and energy as interconnected worlds. This perspective highlights the need for holistic approaches to urban development and planning transportation, emphasizing importance integrating energy-efficiency and smart mobility solutions into broader framework of initiative for smart cities (Angelidou et al. 2022; Creutzig 2016). Moreover, the discussion leads to the role of the importance of the simulation process in introducing integrated transport services and predicting the attractiveness of new mobility solutions (Angelidou et al. 2022). This analytical tool can help operators and planners assess the potential impact of different transport services and technologies, enabling them to make informed decisions about the implementation of low-energy mobility innovations (Häll et al., n.d.). These here findings carry some important implication for them policymakers, city planners, and transportation peeps looking to improve sustainable and efficient moving around in them small urban spots! According to Creutzig (2016), local approaches could mitigate urban transport emissions by 20–50% (Creutzig 2016).

Current State of Transport Policies in the Mediterranean, Analysis of Their Effectiveness in Addressing Climate Change

The Mediterranean countries climate, including the cities in the Mediterranean Basin, are particularly affected by climate change, which can exacerbate health inequalities within these cities (Marí-Dell'Olmo et al. 2022). According to Kadir (2022) in the Mediterranean region, the global warming has raised by 20% already (Kadir 2022). This emerges the importance of considering the specific regional and demographic impacts of climate change when developing policies and initiatives related to sustainable transport. The Mediterranean region, stands as a very complex context when it comes to transport policies. This is caused as the in this territory, the development of the transport is quite disbalanced, cause of many political issues mostly (Jaeger et al. 2015). Many authors, brought in attention discussions, challenges and opportunities this region does face, regarding promoting sustainable paratransit in the context of local and national and regional policies. In different countries, are different concept and approaches regarding transport sector as well as climate change impacts (Bakker, n.d.-b). For example, local governments often prefer less costly and ambitious actions, which can hinder the implementation of sustainable transport initiatives. Therefore, is need the role of policymakers is crucial to get involved to the process of enabling policies that can facilitate the promotion of sustainable transport policies (Nugroho and Zusman 2018). The implementation of such policies and measures should be discussed and coordinated at different levels of government, including local, regional, and national levels (Shastry and Pai, n.d.). Another interesting aspect pointed out is that national transportation policies have traditionally focused on large cities, which can be problematic for smaller cities and towns that also need support in promoting sustainable transport. Therefore, there should be a shift into more inclusive and diverse national transport policies, in order to include all diverse needs of cities and towns of varying sizes.

Identification of Challenges Hindering Effective Climate Change Mitigation

The main aim of this research is to explore the relation between sustainable transport visions and climate change mitigation. As noticed from many authors, that visions on sustainable transport often include environmental considerations, climate change mitigation is not explicitly mentioned (Creutzig 2016). Therefore, through the methodology of the bibliographic analysis, we aim to analyze the gap between global and local effects of transport sector within the framework of climate change mitigation.

Therefore, the research gap identified in the literature, shown in Fig. 1, reveals the lack of a comprehensive and integrated framework connecting transport policies and climate change mitigation, particularly in the Mediterranean region. While many studies emphasize the need to align transport policies with climate change goals, current efforts remain fragmented and often limited in scope. Despite the recognition of transportation's significant contribution to

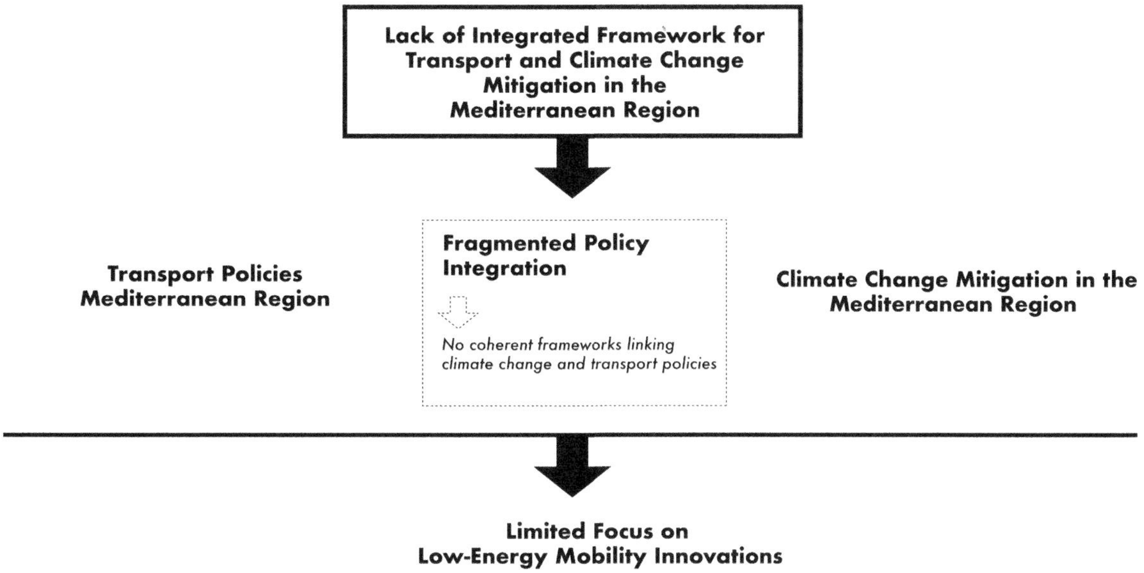

Fig. 1 Conceptual diagram of the research gap. *Source* Author

greenhouse gas emissions, existing policy instruments are either incremental or ineffective at achieving transformative changes. Furthermore, there is limited exploration of the specific challenges and drivers affecting the formulation and implementation of integrated transport and climate policies in the Mediterranean, including political, economic, and geographical differences. This gap highlights the need for more robust frameworks that account for regional complexities and encourage radical systemic changes, particularly in light of the growing urgency to meet global climate goals, such as those outlined in the Paris Agreement. Consequently, further research is required to develop, implement, and assess policy innovations that promote low-energy mobility, improve policy coherence, and address the unique environmental and socio-political challenges of the Mediterranean region.

3 Methodology

This section describes the procedure used to conduct the literature review, focusing on relevant and current studies on the subject, as well as highlighting the main knowledge gaps that the study intends to address. To explore a more in-depth analysis, of the relationship between sustainable transport visions/concepts and climate change mitigation, we did our research by selecting the documents on the Semantic Scholar bibliometric by using three keywords per different territorial level. The aim of this methodology was more to understand, how interlinked are these documents according to the main keyword search by using two platforms, Paper connectivity/ Research Rabbit (as an extension of Semantic Scholar), and VOS viewer software to conduct and analyze networks of different items such as documents; authors; sources; references (Sajovic and Boh Podgornik 2022). This methodology does provide us with bibliographic maps, where shows the linkage among the articles/document (Sajovic and Boh Podgornik 2022).

The literature search was performed in Semantic scholar/ Research Rabbit web databases by using a search guided by two topics "*transport/policies/sustainability*" in European level, and "*transport policies/climate change/Mediterranean region*", shown in Fig. 2. The collected data, was processed on VOSviewer, based on title and abstracts of the relevant titles selected aiming to show the connectivity between keywords used mostly on the articles. The application of VOSviewer facilitated a comprehensive analysis of the keyword link strengths, ultimately contributing to a deeper understanding of the central concepts and relationships within the literature.

4 Results and Discussions

4.1 Transport/Policies/Sustainability

4.1.1 Phase 1—Paper Connection

The analysis of the identified papers revealed several key findings. Firstly, it was observed that there is an increasing scholarly interest in the intersection of transport policies and climate change in the Mediterranean region, as evidenced by the significant number of papers retrieved. Furthermore, the study found that there is a diverse range of perspectives and approaches adopted in the literature concerning the incorporation of climate change considerations into transport policies. Some papers may focus on the direct impact

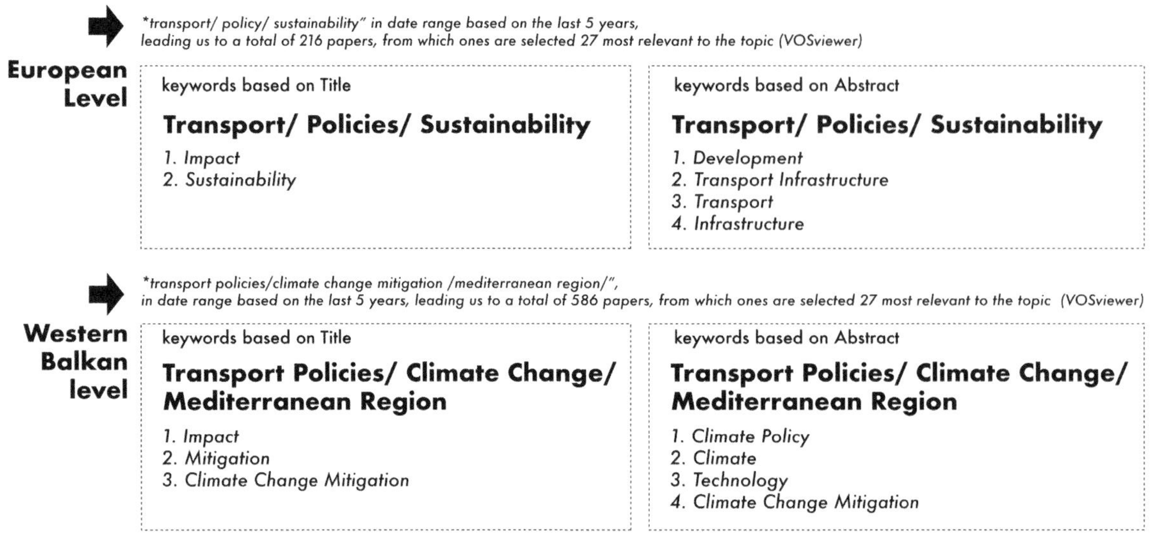

Fig. 2 Conceptual diagram, explaining the implemented methodology. *Source* Author

of climate change on transportation infrastructure and operations, while others may emphasize the role of transportation policies in contributing to or mitigating climate change effects. This diversity reflects the complexity of the issue and indicates the multifaceted nature of addressing climate change within the transportation sector. Therefore, in this phase, we focused more on finding the search topic closer to transport policies, climate change mitigation, and the Mediterranean region. This level aims to understand more how much is considered climate change in transport policies without having the keyword search "Climate change mitigation." By searching in the Paper Connectivity/Research Rabbit platform, for papers that are relevant to the main preselected keywords, we concluded 260 totals of papers, where most relevant were 22. A contribution to this enormous number of papers was the general keywords.

As shown in Fig. 3, the papers were published between 2013 and 2022. On the connection map, we do not see any network among the papers, although we selected the most relevant ones to the keywords.

On the connected papers analysis shown on Fig. 4, we notice that the selected papers do connect to other relevant papers, except other papers connected to each other. However, based on this collection, we continued with the bibliographic analysis after extracting the data in this version.

4.1.2 Phase 2a—Bibliographic Map/Titles Data

This research focused on analyzing the link strength of the main keywords used in the titles of documents related to a specific topic. The authors collected data from a platform and inputted it into VOS viewer, a software tool, to understand the relationships between the keywords and other properties. Table 1 is identified a total of three clusters containing 28 keywords, which exhibited a 65% link strength.

Through these analyses, we found that certain keywords had a higher correlation compared to others. The keywords with the highest link strength were identified as "sustainability," "transport infrastructure," "impact," and "international trade." These keywords were shown to be particularly significant in the context of the topic under investigation, based on the strength of their associations with other keywords. Additionally, we used VOSviewer to examine the main properties of the collected data. We utilized the software to detect and visualize the links between keywords, providing a comprehensive overview of the relationships and patterns within the dataset. The keyword strength link table helps to understand the interconnectedness and the relevance of the keywords, offering insights into the key themes and concepts within the documents.

The overview of the keyword's strength link Table 2, provides valuable insights into the central themes and focal points within the literature, understanding the key areas of focus and relevance in the field. By using VOSviewer, we were able to effectively extract and visualize the essential connections and patterns within the dataset, enhancing their understanding of the underlying associations between the keywords and the overall topic. In Figs. 5 and 6, we can notice the main two keywords such as "impact" and "Sustainability" stand well correlated to the "transport infrastructure."

The bibliographic map interprets each core of the cluster, to which other keywords are connected. As our main aim in this study is to explore the relationship between climate change and the transport sector, we notice that in this analysis, there is a mediated connection between them. The strength link is more between Sustainability—Impact,

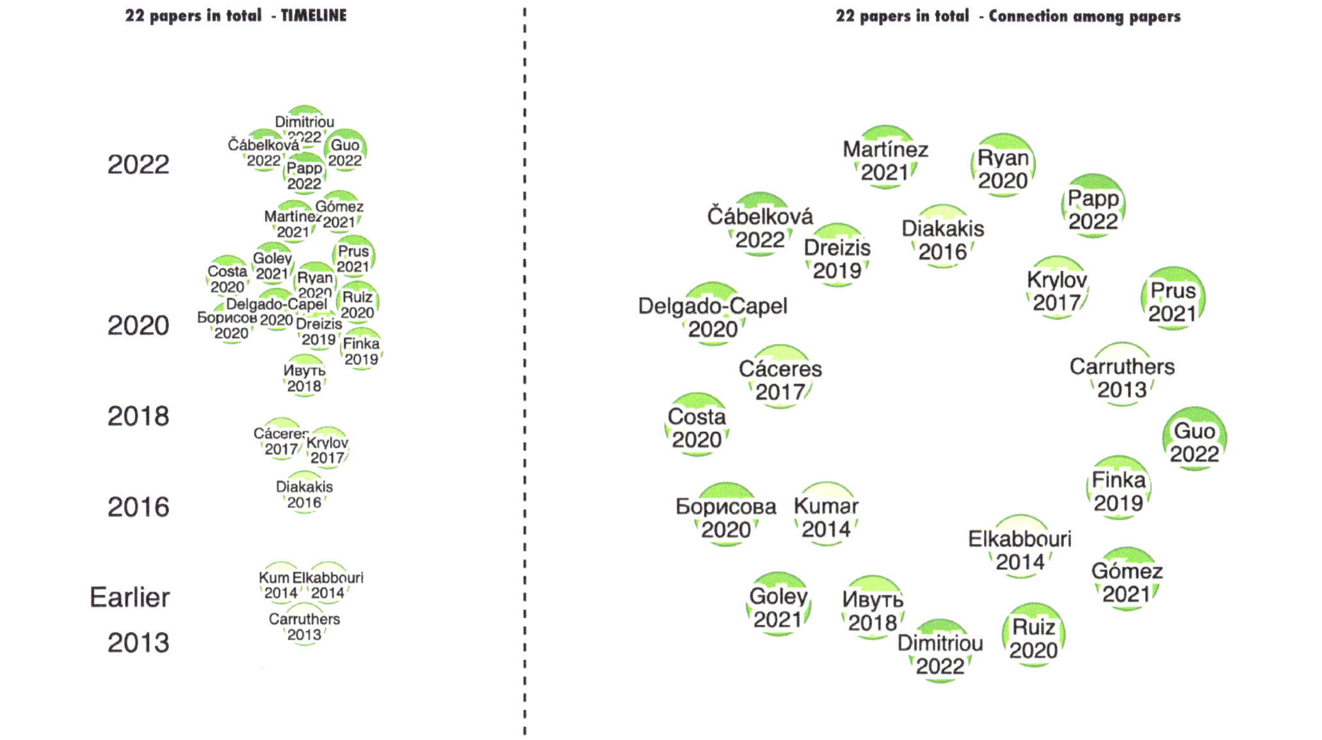

Fig. 3 Map of connection and timeline. *Source* Author

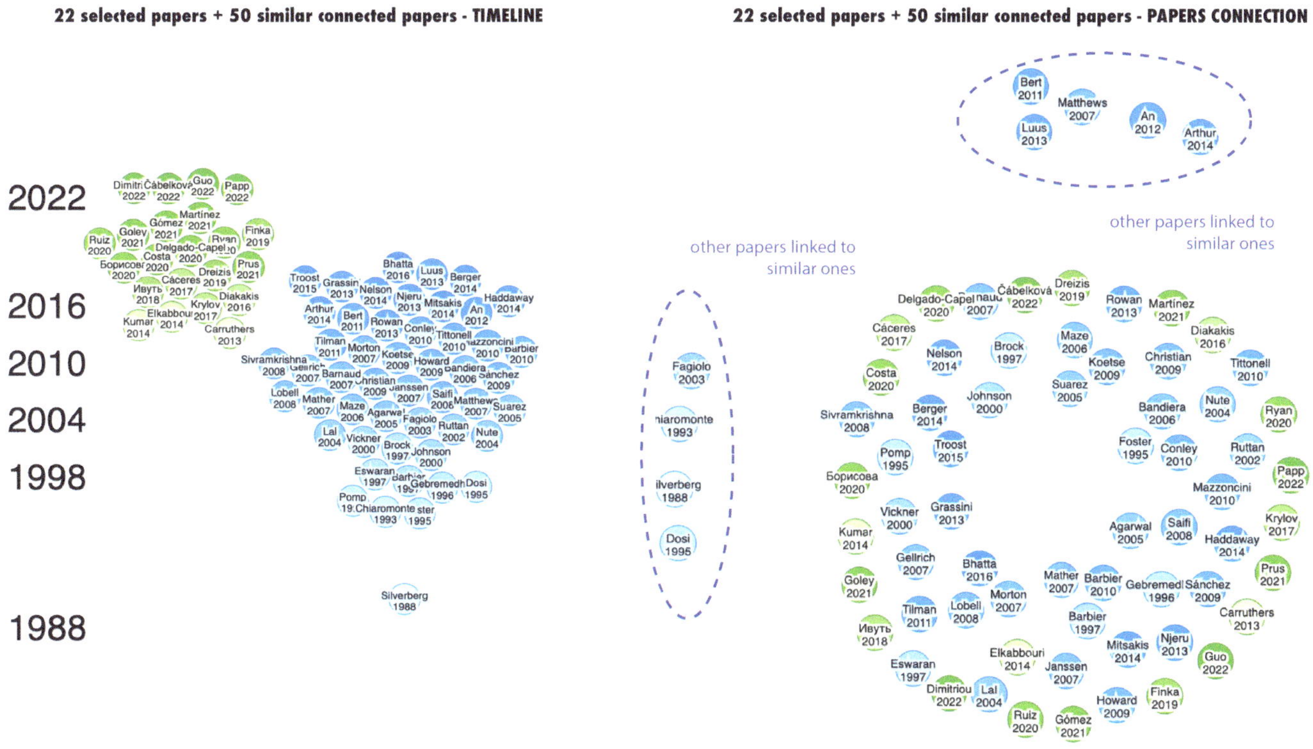

Fig. 4 Map of selected papers connected to other papers. *Source* Author

Table 1 Data input in VOSviewer

Source	Properties
Selection field	Title
Counting method	Full counting
Minimum number of occurrences of a term	1
From 104 terms in total, are selected 64 which do meet the threshold For each of the 64 terms, a relevance score will be calculated Based on this score, the most relevant terms will be selected. The default choice is to select the 60% most relevant terms	

Clusters	Items	Links	Total link strength
3	28	63	65

Source Author

Table 2 Data processing/ clusters core keyword

Cluster	Keyword	Links	Occurrence	Total strength link
2	Impact	12	3	13
3	Sustainability	13	4	13
1	Transport infrastructure	11	5	12
2	International trade	9	2	10

Source Author

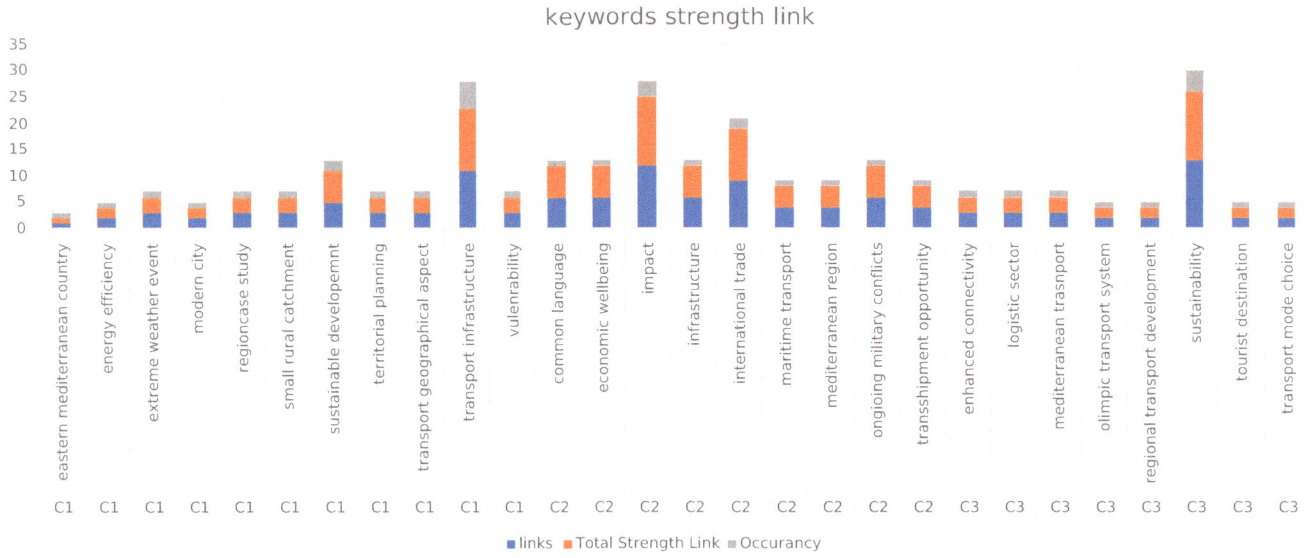

Fig. 5 Overview of the keyword's strength link. *Source* Author

Sustainability International Trade, and Transport Infrastructure—International Trade. The overlay visualization utilizes keywords from article titles to map the connections between clusters and identify predominant keywords used over a specific period. The research finds that the keywords linking clusters in network visualization are moderately connected to the main association between transport and climate change. Furthermore, in overlay visualization, sustainability is identified as a predominant keyword from 2020 to 2022, particularly in relation to "transport," "infrastructure," and "regional connectivity."

The overlay visualization utilizes keywords from article titles to map the connections between clusters and identify predominant keywords used over a specific period. The research finds that the keywords linking clusters in network visualization are moderately connected to the main association between transport and climate change. Furthermore, in overlay visualization, sustainability is identified as a predominant keyword from 2020 to 2022, particularly in relation to "transport," "infrastructure," and "regional connectivity." Despite the keyword collection being derived from article titles, the study demonstrates a high relevance to the

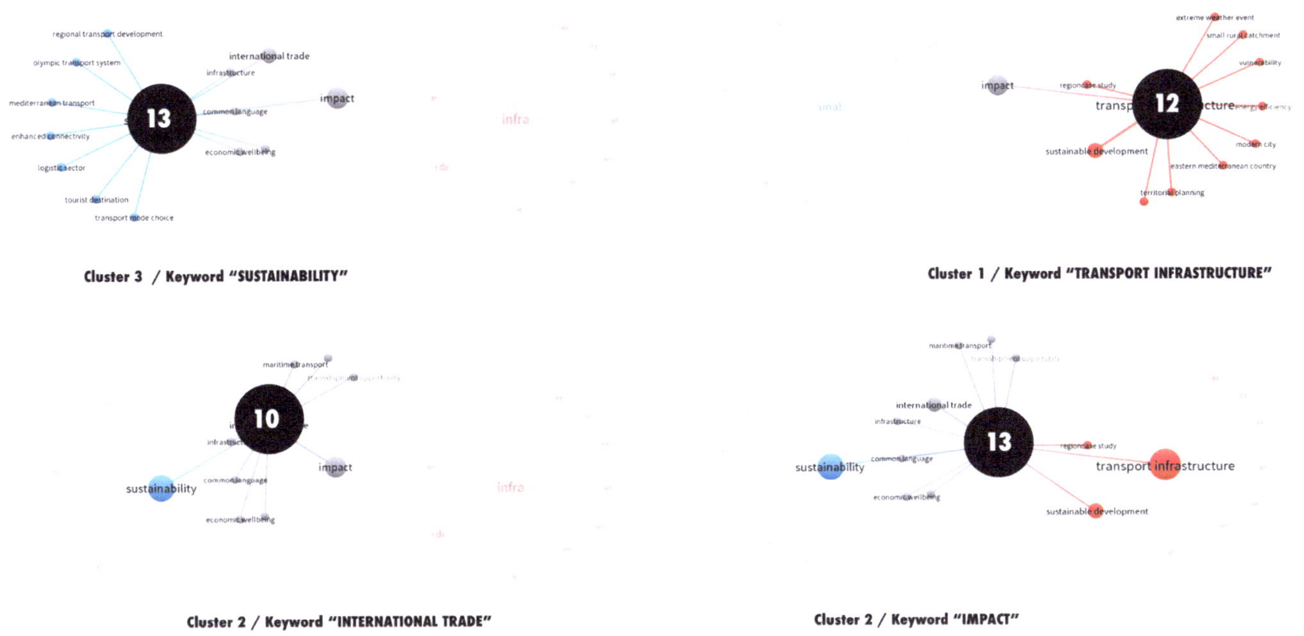

Fig. 6 Bibliographic map of each cluster, within the linkage to the other keywords. *Source* Author

main topic search and strong linkages among the identified keywords. The findings suggest that there is a noticeable link between transport and climate change in network visualization, shown on Fig. 7. This connection is crucial as it highlights the intersection between these two crucial areas. Moreover, the prominent use of sustainability as a keyword in the context of transport, infrastructure, and regional connectivity emphasizes the growing significance of sustainable practices within these domains, particularly in recent years. The findings regarding the significance of these interlinkages, demonstrate the relevance of identified keywords.

4.1.3 Phase 2b—Bibliographic Analysis/ Abstracts Data

After collecting the data from the platform, we did enter it into the VOSviewer, to understand the link strength of the main keywords used on the titles of the documents, and other properties such as shown below on Table 3.

After proceeding with setting the parameters for analyzing the data takes from the abstracts of the articles, we concluded in a total of 5 Clusters, containing 43 keywords which had a 962 total link strength. On Table 4, we can easily identify the core keywords, which has the highest rate of connection to other keywords. The main for keywords are:

- Development
- Transport infrastructure
- Transport
- Infrastructure.

While focusing on the abstracts of articles related to transport infrastructure, development, sustainability, and climate change, the findings shown on Figs. 8 and 9 only four main keywords, which are only transport-related. Despite sustainability and climate change being the main keywords in the search, the analysis revealed that the abstracts mainly centered on transport infrastructure and development, with a lower correlation to sustainability and climate change.

Meanwhile, on the map of network and overlay visualization represented on Fig. 10, we can notice the strength connection among main core cluster keywords, while sustainability, has less than half of the strength connection. The main clusters are from articles which they are published from 2018 to 2020, while the most recent keywords used more, are "Mediterranean," "sustainable transport," and "ecological connectivity." This represents, that most of the articles which are focused on transport sector, do not have a strong correlation to climate change aspects and sustainability, although the tittle it does show as.

The network mapping and overlay visualization can examine the strength of connections among the core cluster keywords. Additionally, the analysis showed that the main clusters were from articles published between 2018 and

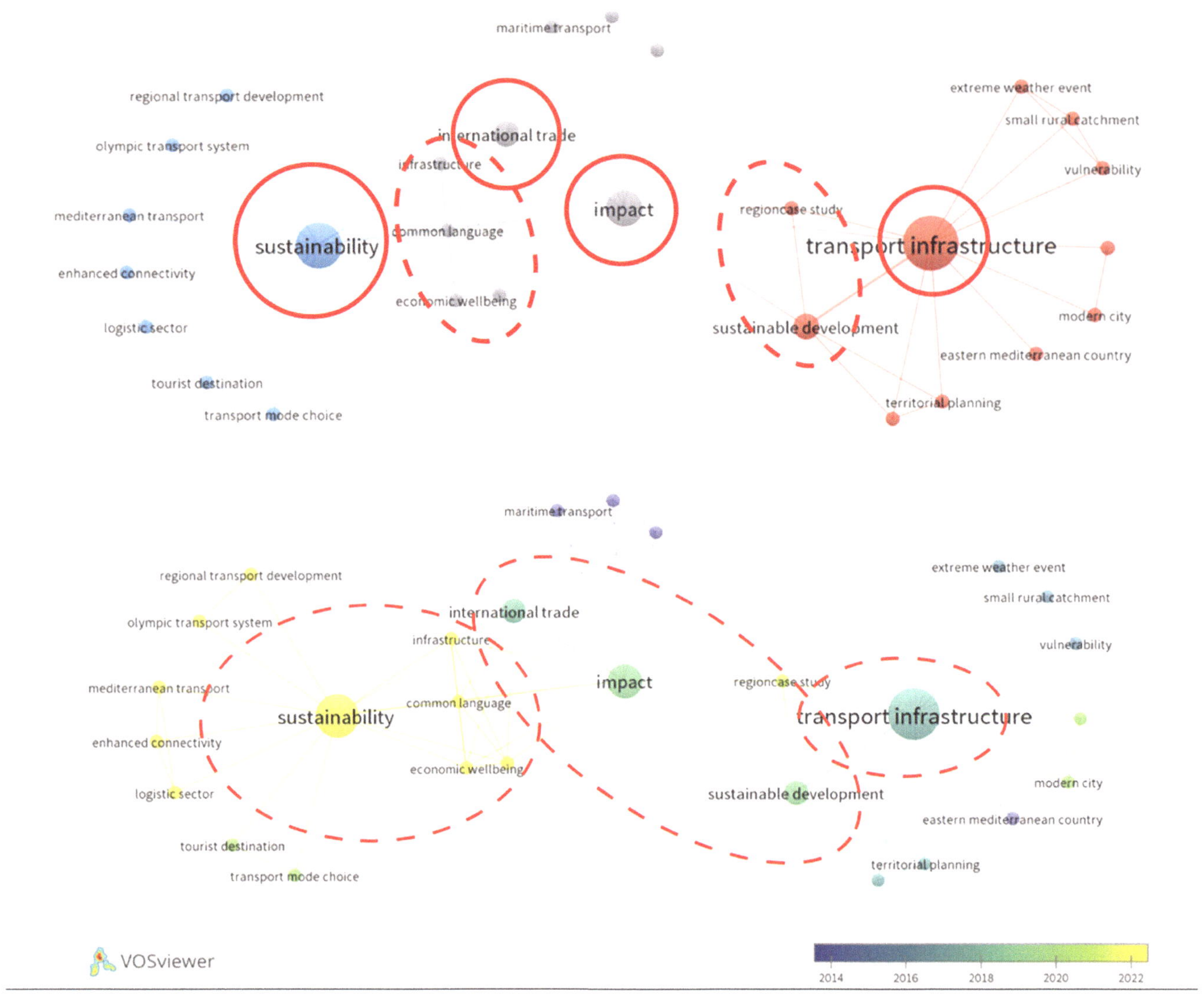

Fig. 7 Network visualization and overlay visualization. *Source* Author

Table 3 Data input in VOSviewer

Source	Properties		
Selection field	Abstract		
Counting method	Full counting		
Minimum number of occurrences of a term	2		
From 1067 terms in total, are selected 204 which do meet the threshold For each of the 204 terms, a relevance score will be calculated. Based on this score, the most relevant terms will be selected. The default choice is to select the 60% most relevant terms			
Clusters	Items	Links	Total link strength
5	43	208	962

Source Author

2020, and the most recently used keywords included "Mediterranean," "sustainable transport," and "ecological connectivity." The findings suggest that the field of transport infrastructure and development outweighs the emphasis on sustainability and climate change in the abstracts analyzed. This indicates a potential *gap in the literature* where the prioritization of transportation and development is not interconnected to sustainability and climate change in the discourse. The network mapping and visualization offered insights into the interconnectedness of core cluster

Table 4 Data processing/clusters core keyword

Cluster	Items	Keyword	Links	Occurrence	Total strength link
1		Development	30	28	251
5		Transport infrastructure	29	24	230
3		Transport	26	18	152
2		Infrastructure	27	18	148
5		Sustainable development	17	9	96
4		Sustainable	19	11	75

Source Author

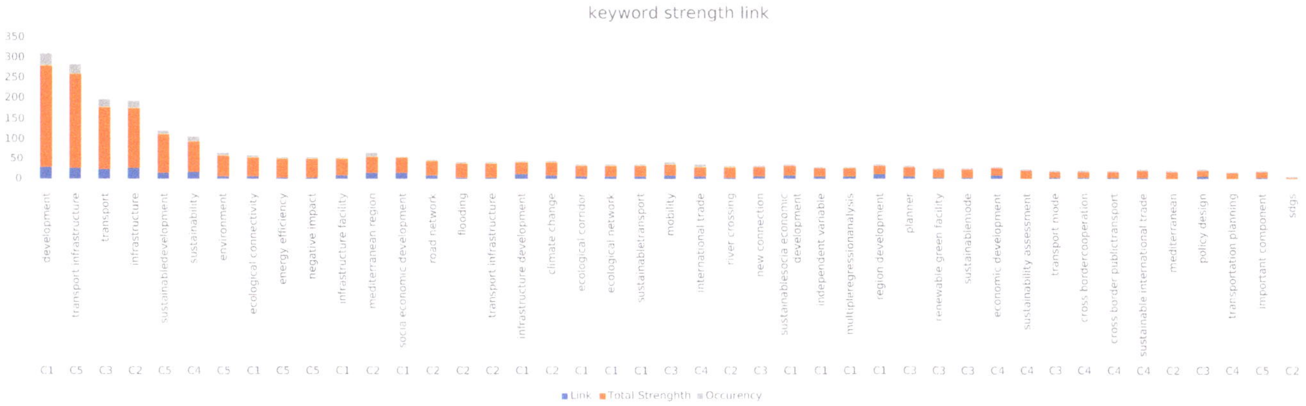

Fig. 8 Overview of the keyword's strength link. *Source* Author

keywords, highlighting the dominance of certain themes over others. Furthermore, the identification of the most recent keywords used, highlights another emerging topic in the field, such as "Mediterranean," "sustainable transport," and "ecological connectivity," indicating potential shifts in research focus and interests.

However, in this section, we can notice the irrelevance of the keyword network in two levels.

The bibliographic map regarding the data collected from the titles of the articles does have a higher correlation a total link strength according to the main topic, while in the case of the bibliographic map of the abstracts, we found an irrelevance to the main topic.

4.2 Transport Policies/Climate Change Mitigation/Mediterranean Region

4.2.1 Phase 1—Paper Connection

The analysis of the identified papers revealed several key findings. Firstly, it was observed that there is an increasing scholarly interest in the intersection of transport policies and climate change in the Mediterranean region, as evidenced by the significant number of papers retrieved. Furthermore, the study found that there is a diverse range of perspectives and approaches adopted in the literature concerning the incorporation of climate change considerations into transport policies. Some papers may focus on the direct impact of climate change on transportation infrastructure and operations, while others may emphasize the role of transportation policies in contributing to or mitigating climate change effects. This diversity reflects the complexity of the issue and indicates the multifaceted nature of addressing climate change within the transportation sector. Therefore, in this phase, we focused more on finding the search topic closer to transport policies, climate change mitigation, and the Mediterranean region. Different from the previous section, the aim of this level, is to understand more how much is considered climate change in the transport policies in the Mediterranean region. By searching in the Paper Connectivity/Research Rabbit platform, for papers that are relevant to the main preselected keywords, we concluded 232 totals of papers, where most relevant were 37, shown on Fig. 11.

As shown on the map of connection (Fig. 11), the 37 papers, published between 2013 and 2022, highlight only two connected papers. Additionally, the table provides

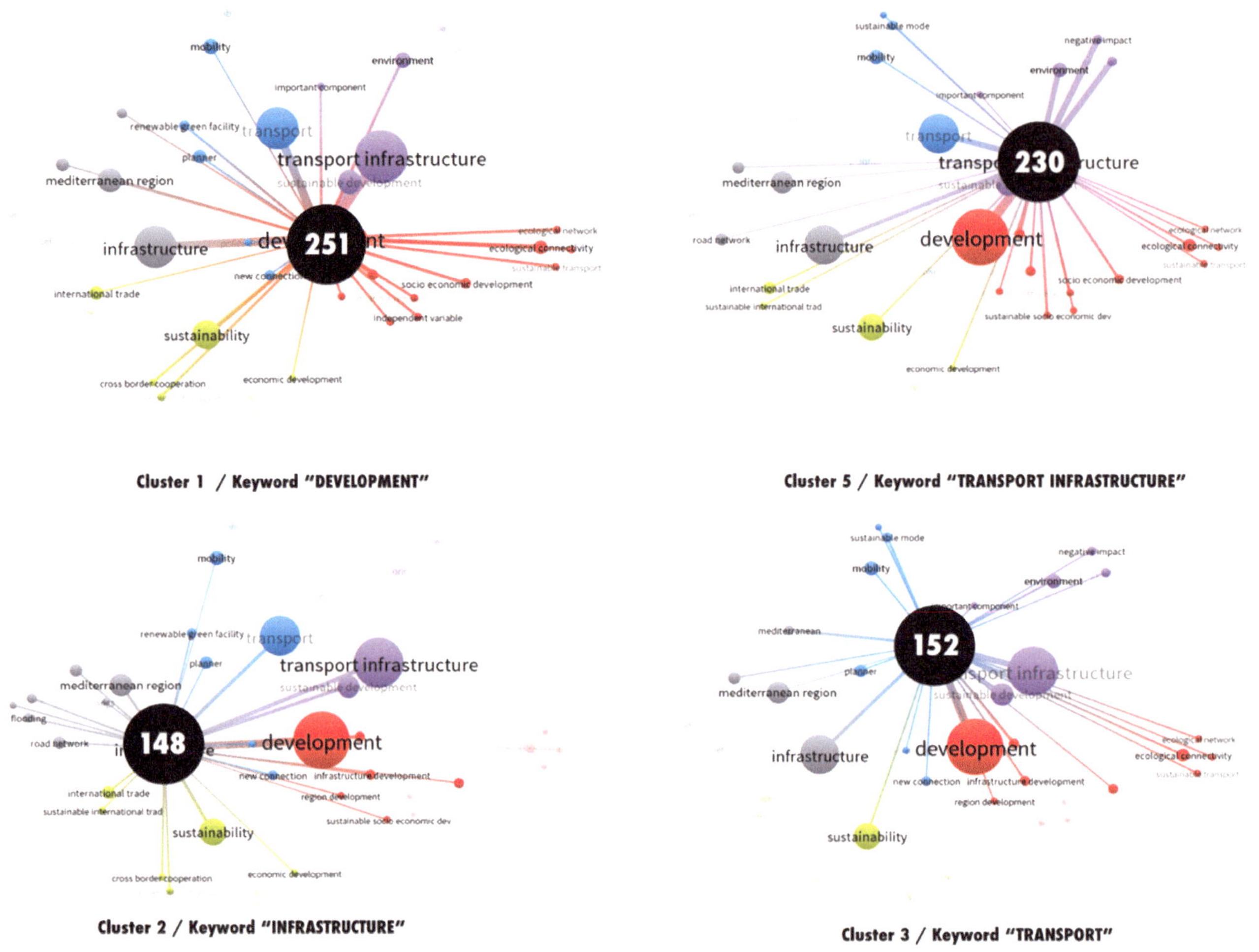

Fig. 9 Bibliographic map of each cluster, within the linkage to the other keywords. *Source* Author

detailed information about the connected articles, specifically focusing on climate change mitigation keywords during the same year. The analysis of these connected papers reveals that they also connect to relevant papers, particularly from 1956 to 2017. Two distinct clusters emerge from the analysis. The first cluster comprises papers whose titles are highly relevant to the main keywords of interest, i.e., climate change mitigation (Table 5). On the other hand, the second cluster primarily delves into the impact of transportation on climate change.

On the connected papers analysis (Fig. 12), we notice that the selected papers do connect to other relevant papers, mostly in the past years from 1956 to 2017.

Furthermore, the first cluster delves into a deeper discussion around the central theme of climate change mitigation (Table 6). This suggests that within the scope of the selected papers, there is a significant focus on addressing the challenges posed by climate change through various mitigation strategies. Conversely, the second cluster is more relevant to the critical role of transportation in contributing to climate change. This illustrates a growing emphasis on understanding and mitigating the environmental impact of transportation systems, positioning it as a pivotal area of research and intervention.

4.2.2 Phase 2a—Bibliographic Map/Titles Data

Based on the data shown on the first phase, where the articles have high relevance, this phase of the analyze is to understand how relevant does stand the title comparison to the abstract of the articles. Table 7 represents the attributes and the results after processing the data.

After proceeding with setting the parameters for analyzing the data takes from the abstracts of the articles, we concluded in a total of 3 Clusters, containing 23 keywords which had a 80 total link strength. On Table 8, we can easily identify the core keywords, which has the highest rate of connection to other keywords. The main for keywords are:

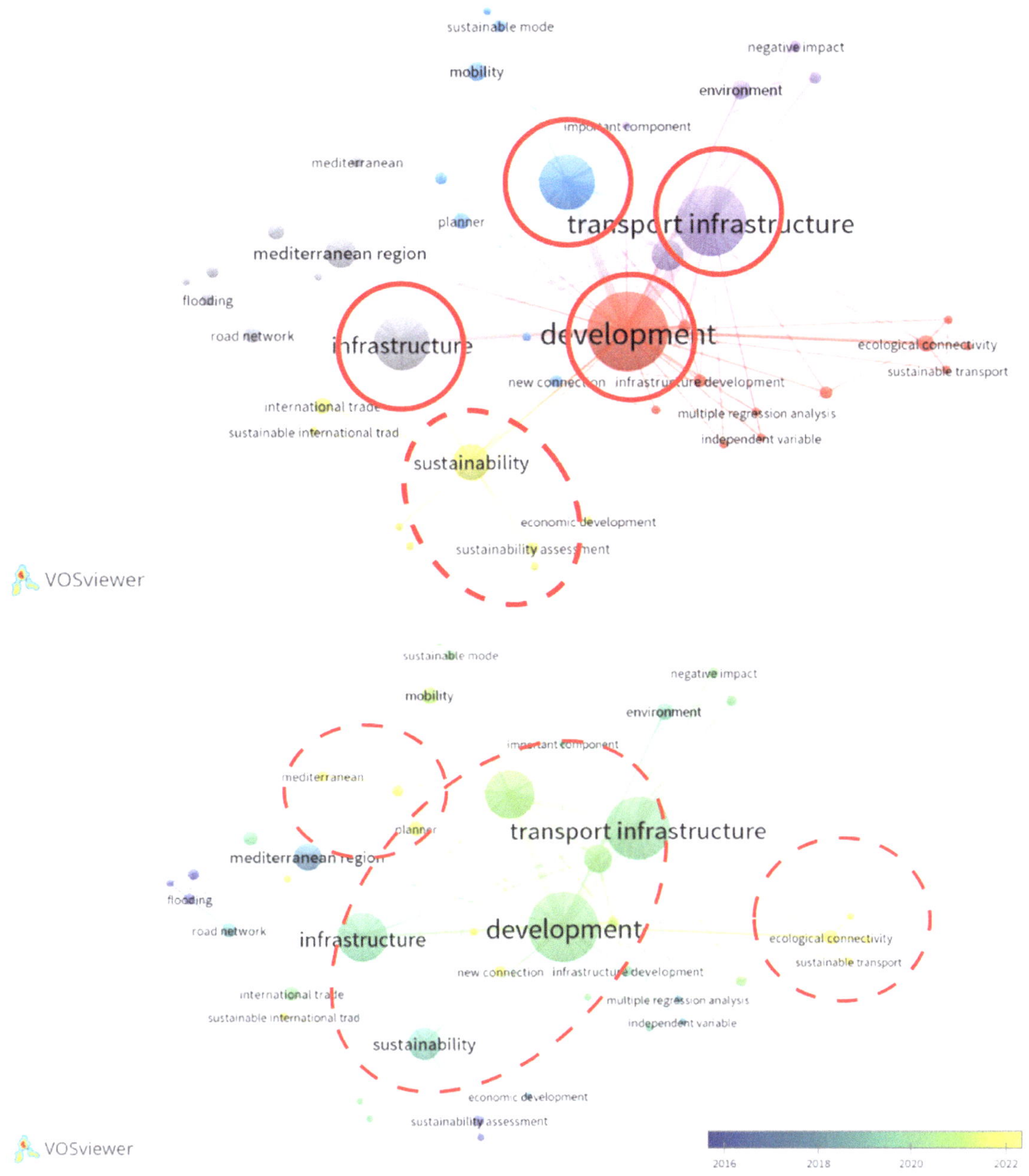

Fig. 10 Network visualization and overlay visualization. *Source* Author

- Climate change
- Mitigation
- Impact
- Climate change mitigation.

As shown in Fig. 13, there is a prevalence of certain keywords in the titles of articles related to climate change and climate change mitigation. The study reveals that the most commonly used keywords in these titles are "Climate change" and "Climate Change Mitigation." However, it is noted that the keyword "Climate Change Mitigation" has a relatively weaker association compared to other main keywords. The authors found that when "climate change mitigation" is added as a keyword in the search, the selected article titles predominantly focus on climate issues, with less emphasis on the transport sector (Table 8).

This finding indicates that "climate change" and "climate change mitigation" are highly prominent topics in the literature, but there is a noticeable disparity in the strength of association with the keyword "Climate Change Mitigation." This observation suggests that the mitigation aspect

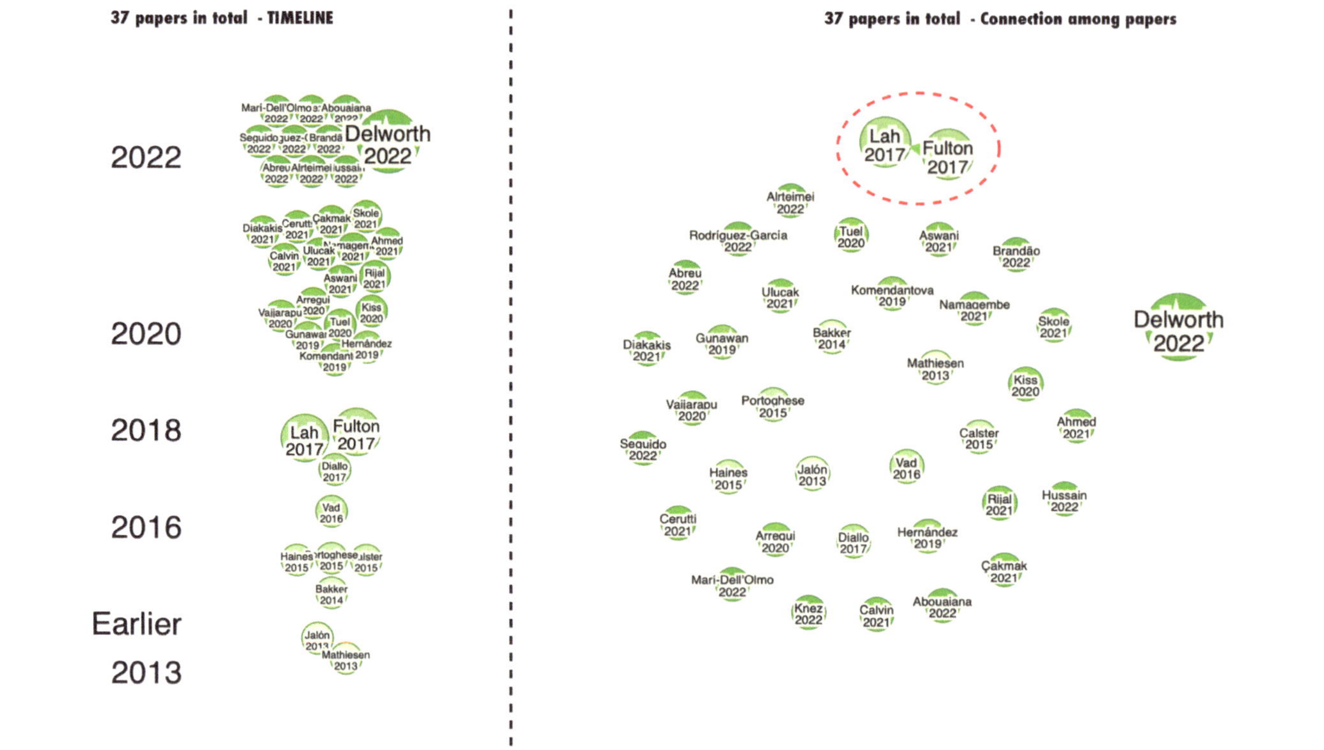

Fig. 11 Map of connection and timeline. *Source* Author

Table 5 Overview of connected papers

Clus.	Author	Cit.	Title
1	Lah (2017)	12	Continuity and change: dealing with political volatility to advance climate change mitigation strategies—examples from the transport sector (Lah 2017)
	Fulton et al. (2017)	47	Climate change mitigation pathways for Southeast Asia: CO_2 emissions reduction policies for the energy and transport sectors (Fulton et al. 2017)

Source Author

may not be as extensively covered or emphasized in the titles of articles compared to the broader topic of climate change. The disparity in keyword strength and the shift in focus when specific keywords are included in the search prompt considerations about the representation and visibility of various sub-topics within the broader field of climate change research (Figs. 14 and 15). This underscores the importance of critically evaluating keyword selection and its potential influence on the prominence and representation of specific facets of climate change and mitigation efforts in academic literature.

4.2.3 Phase 2b—Bibliographic Analysis/ Abstracts Data

The analysis below, aims to show how relevant do the keywords extracted from the abstracts, are to the main topic of search. The tool of VOS viewer has collected 2984 terms in total, from which ones are selected 758 most relevant, shown on Table 9. The processing of this data, derived to 4 clusters, containing 39 keywords, with a strength link of 1320, based on a 60% of most relevant selection.

The results of this step, concludes with 4 most relevant keywords, the same are the core keyword per cluster. As we can see in Fig. 16, the most predominant keywords with a total strength are, "Climate policy," "Climate," "Technology," and "Climate Change Mitigation."

The equivalent keywords to the climate change topic, are the ones, which to show a strong network among each other, while the keywords related to transport and Mediterranean region, has a low interlink and occurrence.

On the bibliographic map and on Fig. 17, we can read the properties of each keyword as a core of the cluster. The cluster number 2, on Table 10, has two main common keywords, which ones predominant on the total strength interlink.

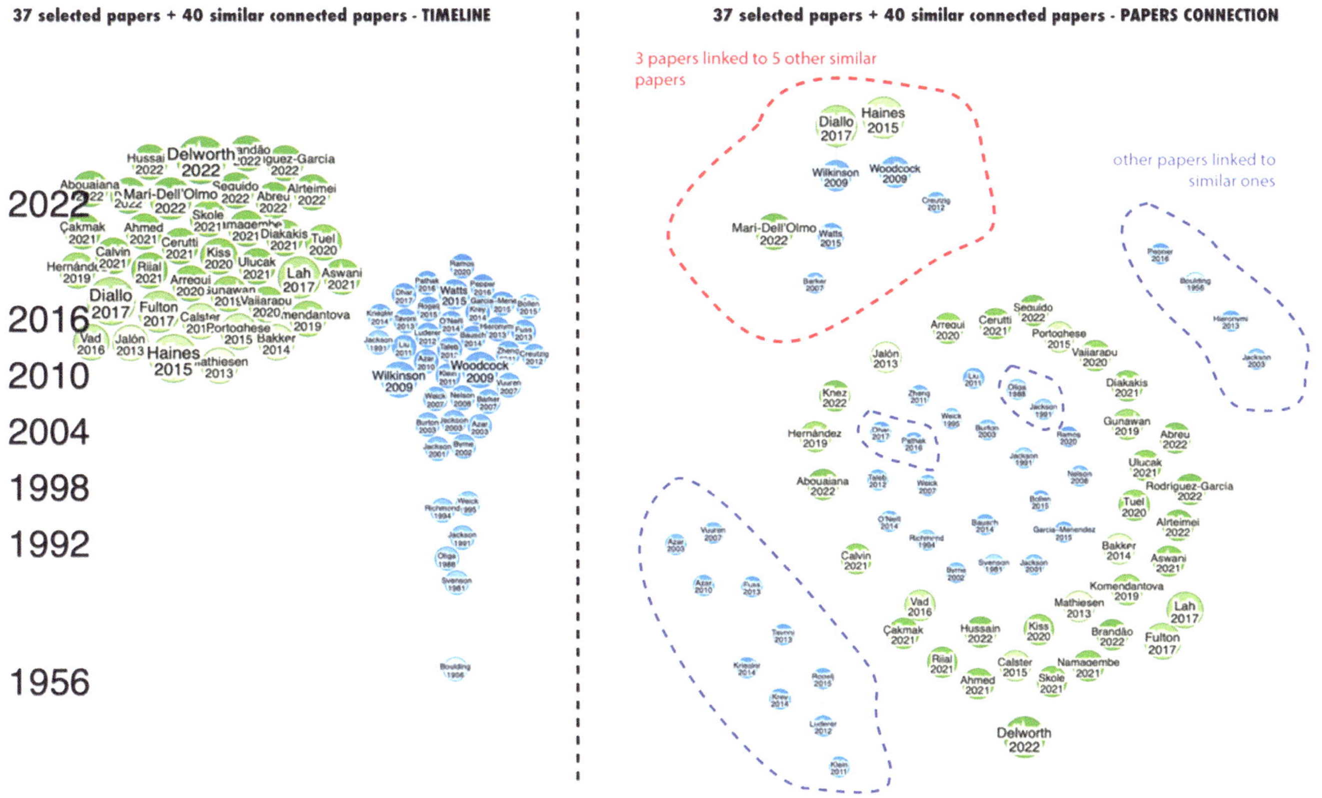

Fig. 12 Map of selected papers connected to other papers. *Source* Author

Table 6 Overview of connected papers

Clus.	Author	Title	Cit.	Author	Title	Cit.
1	Marí-Dell'Olmo et al. (2022)	Climate change and health in urban areas with a Mediterranean climate: a conceptual framework with a social and climate justice approach (Marí-Dell'Olmo et al. 2022)	5	Watts et al. (2015)	Health and climate change: policy responses to protect public health (Watts et al. 2015)	1316
				Barker et al. (2007)	Summary for Policymakers (Barker et al. 2007)	9563
	Diallo et al. (2017)	Is HIA the most effective tool to assess the impact on health of climate change mitigation policies at the local level? A case study in Geneva, Switzerland (Diallo et al. 2017)	8	Wilkinson et al. (2009)	Public health benefits of strategies to reduce greenhouse gas emissions: household energy (Wilkinson et al. 2009)	337
				Woodcock et al. (2009)	Public health benefits of strategies to reduce greenhouse gas emissions: urban land transport (Woodcock et al. 2009)	918
	Haines (2015)	Health co-benefits of climate change mitigation policies (Haines 2015)	0	Creutzig et al. (2012)	Decarbonizing urban transport in European cities: four cases show possibly high co-benefits (Creutzig et al. 2012)	113
2	Dhar et al. (2017)	Electric vehicles and India's low-carbon passenger transport: a long-term co-benefits assessment (Dhar et al. 2017)	127	Pathak and Shukla (2016)	Co-benefits of low-carbon passenger transport actions in Indian cities: case study of Ahmedabad (Pathak and Shukla 2016)	50

Source Author

Table 7 Data input in VOSviewer

Source	Properties
Selection field	Title
Counting method	Full counting
Minimum number of occurrences of a term	2

From 292 terms in total, are selected 46 which do meet the threshold
For each of the 46 terms, a relevance score will be calculated. Based on this score, the most relevant terms will be selected. The default choice is to select the 60% most relevant terms

Clusters	Items	Links	Total link strength
3	23	55	80

Source Author

Table 8 Clusters strength link

Cluster	Keyword	Links	Occurrence	Total strength link
2	Climate change	12	24	29
2	Impact	10	11	18
1	Mitigation	9	8	14
3	Climate change mitigation	9	22	14
1	Policy	9	11	11
1	Transport	7	6	7

Source Author

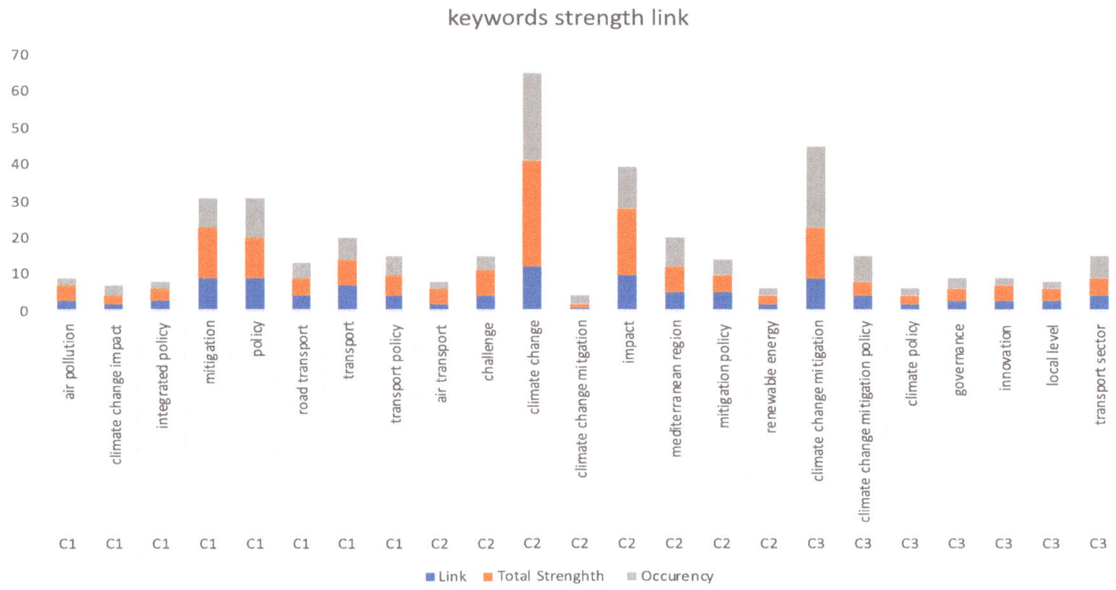

Fig. 13 Overview of the keyword's strength link. *Source* Author

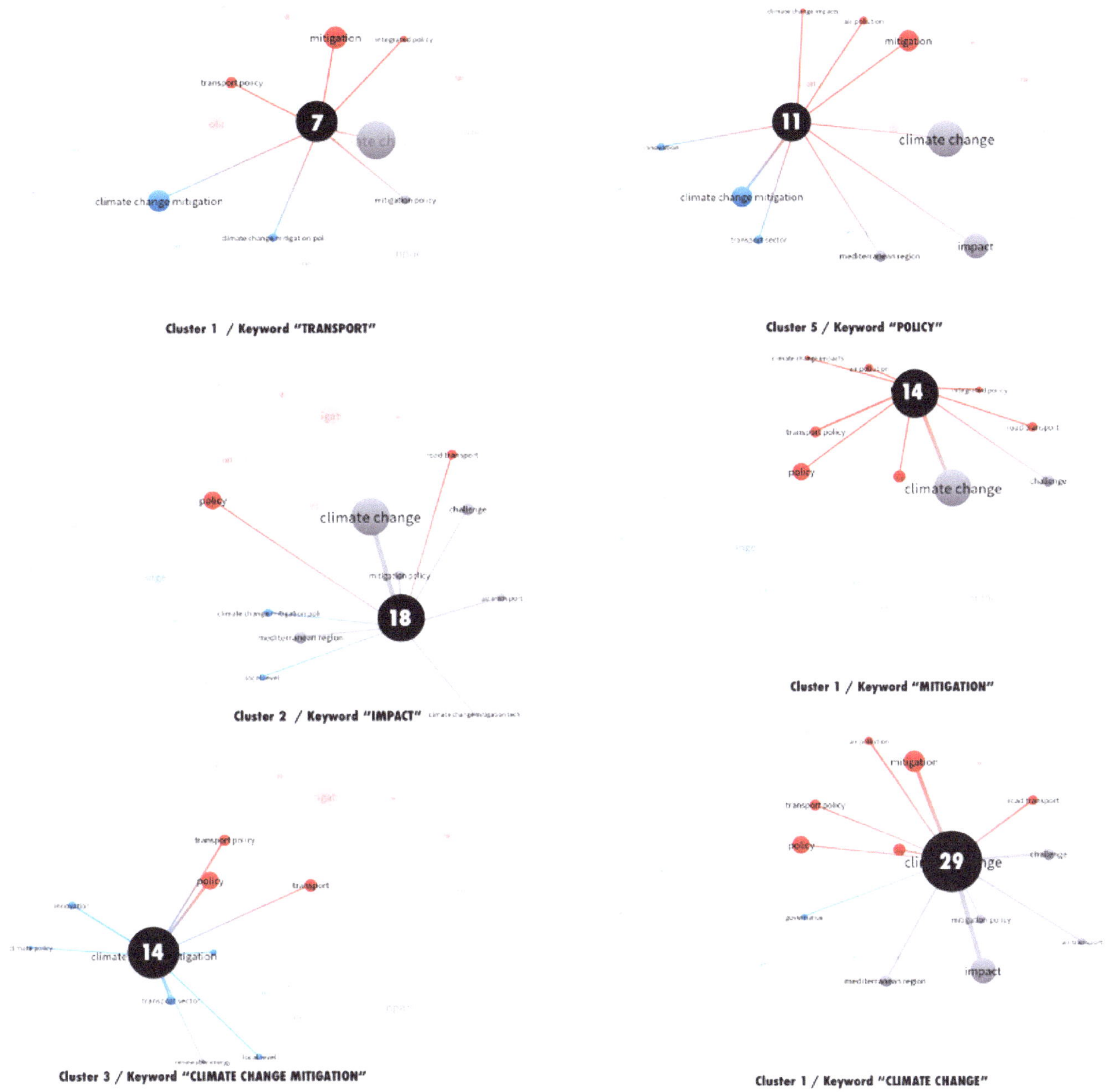

Fig. 14 Bibliographic map of each cluster, within the linkage to the other keywords. *Source* Author

Referring Table 10, the findings show an interesting point, where cluster number 1, with the core keyword of "Technology," has a connection with the "Mediterranean region" and "transport sector."

This finding indicates that "climate change" and "climate change mitigation" are highly prominent topics in the literature, but there is a noticeable disparity in the strength of association with the keyword "Transport Policy and Mediterranean region." This observation, suggests that the keywords. "Climate," "Climate Policy," and "Climate change mitigation" have a strong connection with the keyword of technology. Meanwhile, with the keyword of transport sector, stands a gap in between, in the context of the abstracts. Meaning, although on the title articles, have a strong connection to the main topic of this research, while the content does not reflect the same strength of rele-

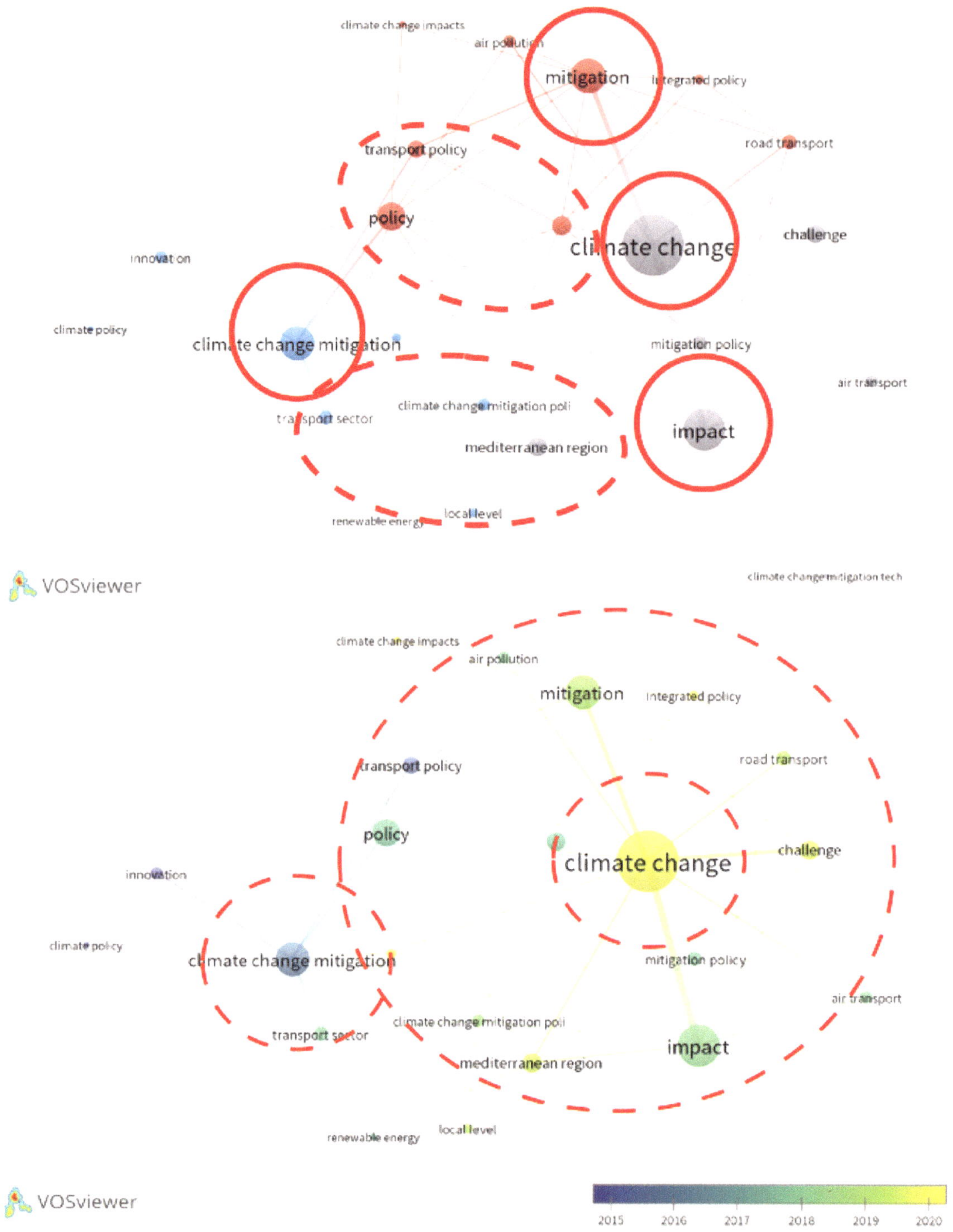

Fig. 15 Network visualization and overlay visualization. *Source* Author

Table 9 Data input in VOSviewer

Source	Properties
Selection field	Abstract
Counting method	Full counting
Minimum number of occurrences of a term	2

From 2984 terms in total, are selected 758 which do meet the threshold
For each of the 758 terms, a relevance score will be calculated. Based on this score, the most relevant terms will be selected. The default choice is to select the 60% most relevant terms

Clusters	Items	Links	Total link strength
4	39	211	1320

Source Author

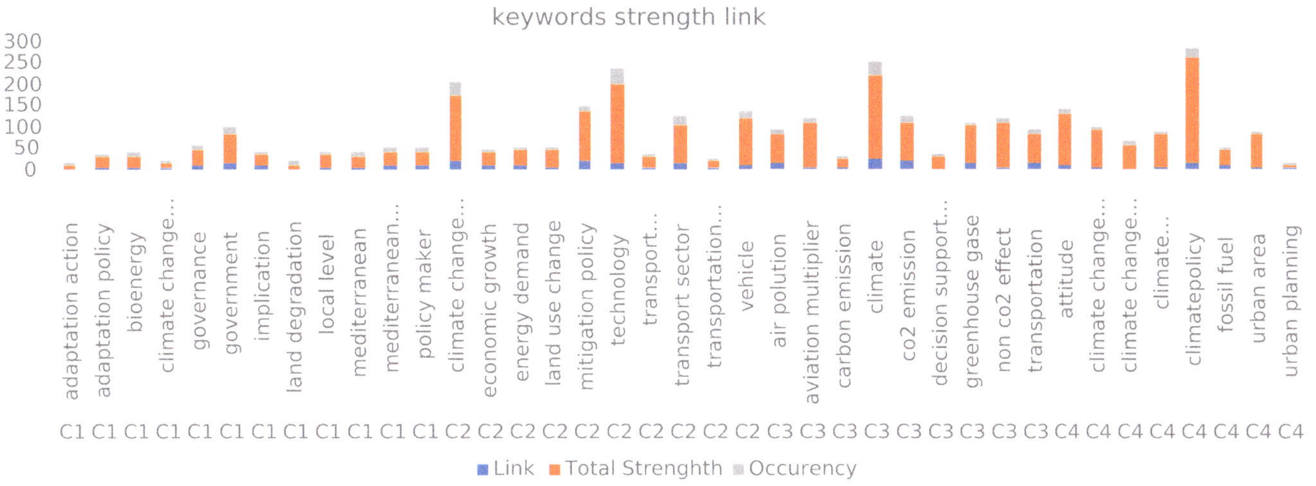

Fig. 16 Overview of the keyword's strength link. *Source* Author

vance, leading to a not very clear and explicit integration between transport policies, to the climate change policies, specifically in the context of Mediterranean basin (Fig. 18).

In this frame, the aim of intersection of transportation and climate change mitigation, in the Mediterranean region, reveals significant insights regarding the challenges and opportunities presented by current policies and frameworks. Aparicio (2017), critically examines the existing institutional framework within Europe, identifying the barrier to developing an integrated approach that reconciles adaptation and mitigation strategies in transport. This path-dependency approach is underlined as crucial by the author, given that short-term actions might either facilitate or further hinder the transition to low-carbon mobility (Aparicio 2017).

Building on this foundation, Moretti and Loprencipe (2018) further emphasizes the lack of an integrated framework that effectively links transport with climate change mitigation in the Mediterranean context. The overview of the bibliographic analysis highlights the critical role transportation plays in climate change, particularly its role on contributing significantly to greenhouse gas emissions. The authors argue that the Mediterranean's challenges necessitate a comprehensive approach to transportation policy, which is essential for achieving sustainable and resilient development in order to face the climate change impacts. This study resonates with the findings of Aparicio (2017) reinforcing the need for cohesive policies aiming to bridge the gap between transport and climate change mitigation. Following more to this approach, Cramer et al. (n.d.) expand upon these themes by exploring the interconnected risks that climate change poses to sustainable development in the Mediterranean. The authors advocate for the development of various policy frameworks that not only addressing the immediate challenges posed by climate change but also integrate long-term sustainability goals within the transport sector.

All these articles selected to this study, collectively underscore the critical need for an integrated framework that connects transport policies with climate change mitigation efforts in the Mediterranean region.

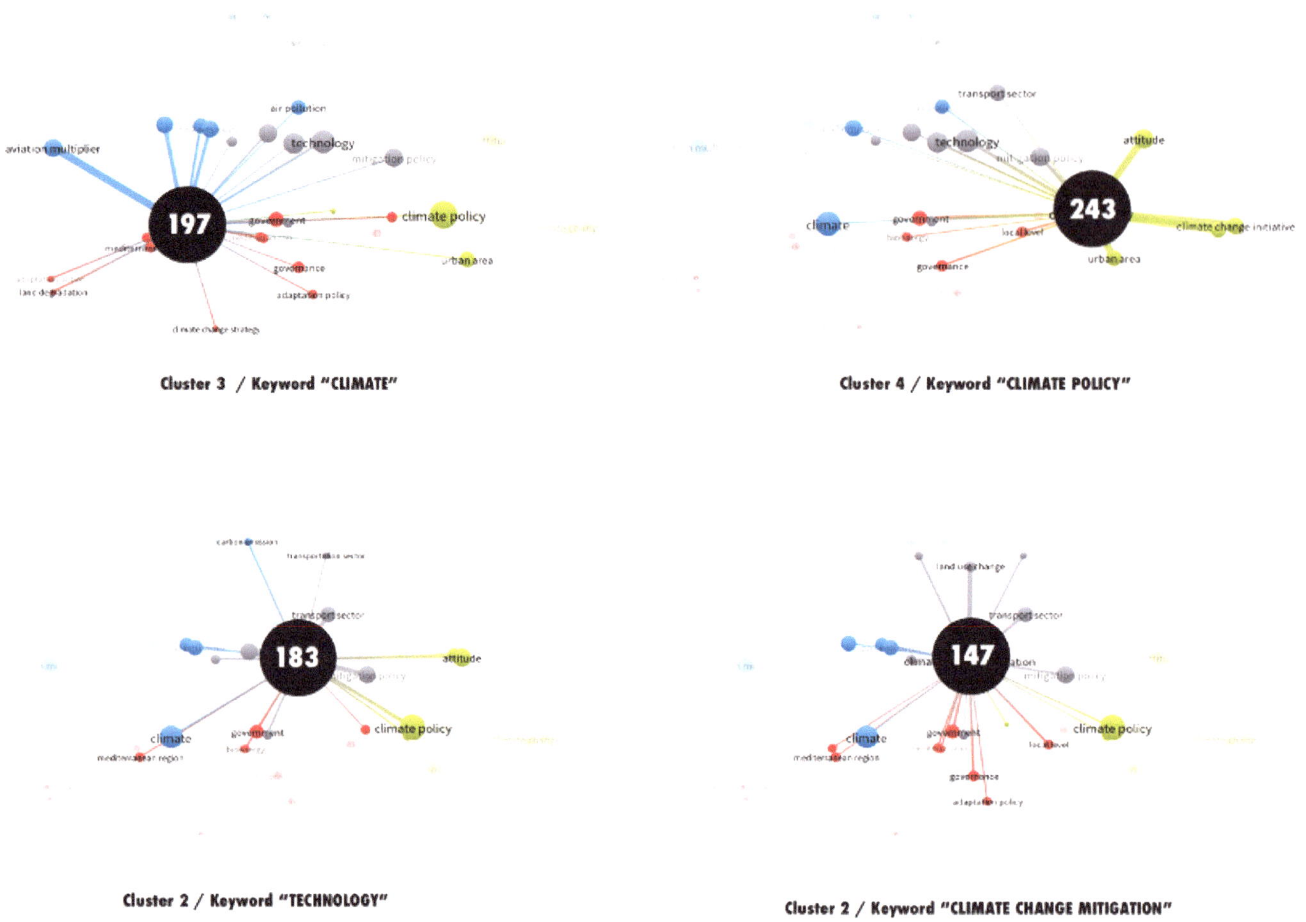

Fig. 17 Bibliographic map of each cluster, within the linkage to the other keywords

Table 10 Data processing/clusters core keyword

Cluster	Items	Keyword	Links	Occurrence	Total strength link
2		Climate policy	19	21	243
2		Climate	26	32	197
1		Technology	19	35	183
3		Climate change mitigation	25	33	147

Source Author

5 Conclusions

The main aim of this research is to explore the relationship between sustainable transport visions and climate change mitigation. The literature review emphasizes that visions of sustainable transport often include environmental considerations, but climate change mitigation is not explicitly mentioned. Although many authors emphasize that addressing this gap is crucial for effective climate change mitigation within the transport sector. They still face barriers to the importance of considering both global and local impacts when formulating sustainable transport policies and strategies.

The results of this study show that climate change mitigation should be explicitly integrated into the transport policy for the visions of sustainable transport to take effect. The findings from the bibliographic analysis show the gap between many articles; according to the title, there is a very strong relation between transport policies and climate policies, while the content, does not reflect explicit integration between these policies.

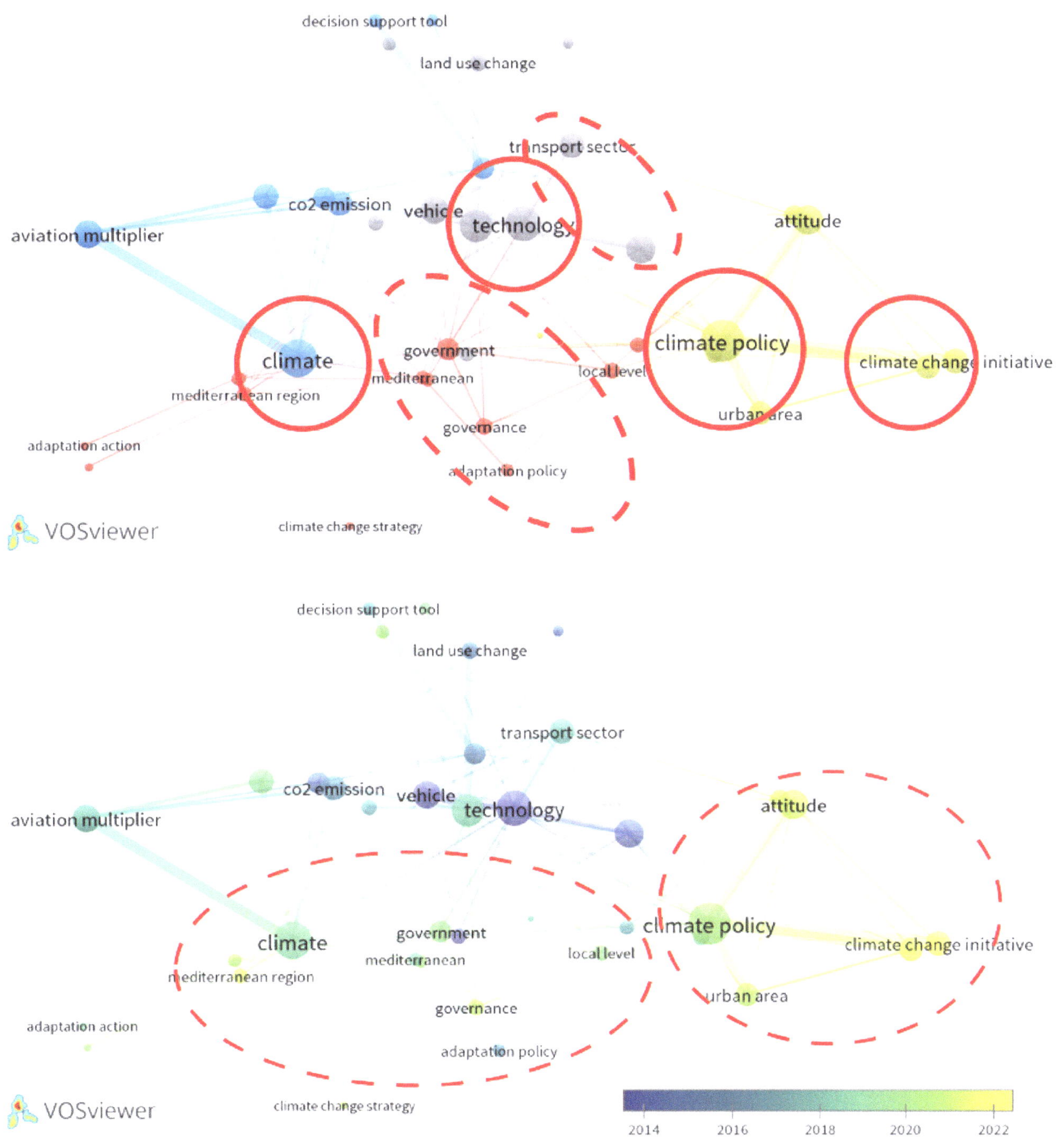

Fig. 18 Network visualization and overlay visualization. *Source* Author

As mentioned in the previous section, while understanding the linkage of the strong connection between climate change mitigation with technology and technology with transport, we do keep this keyword as the main node in developing an integrated framework of transport policies and climate change in the Mediterranean region. Therefore, this study emphasizes the importance of environmental considerations in transport, which might be directed more toward technology/innovation applied to the transport infrastructure. As many authors emerge the necessity for an approach that combines transformative change with incremental progress toward low-carbon transport to

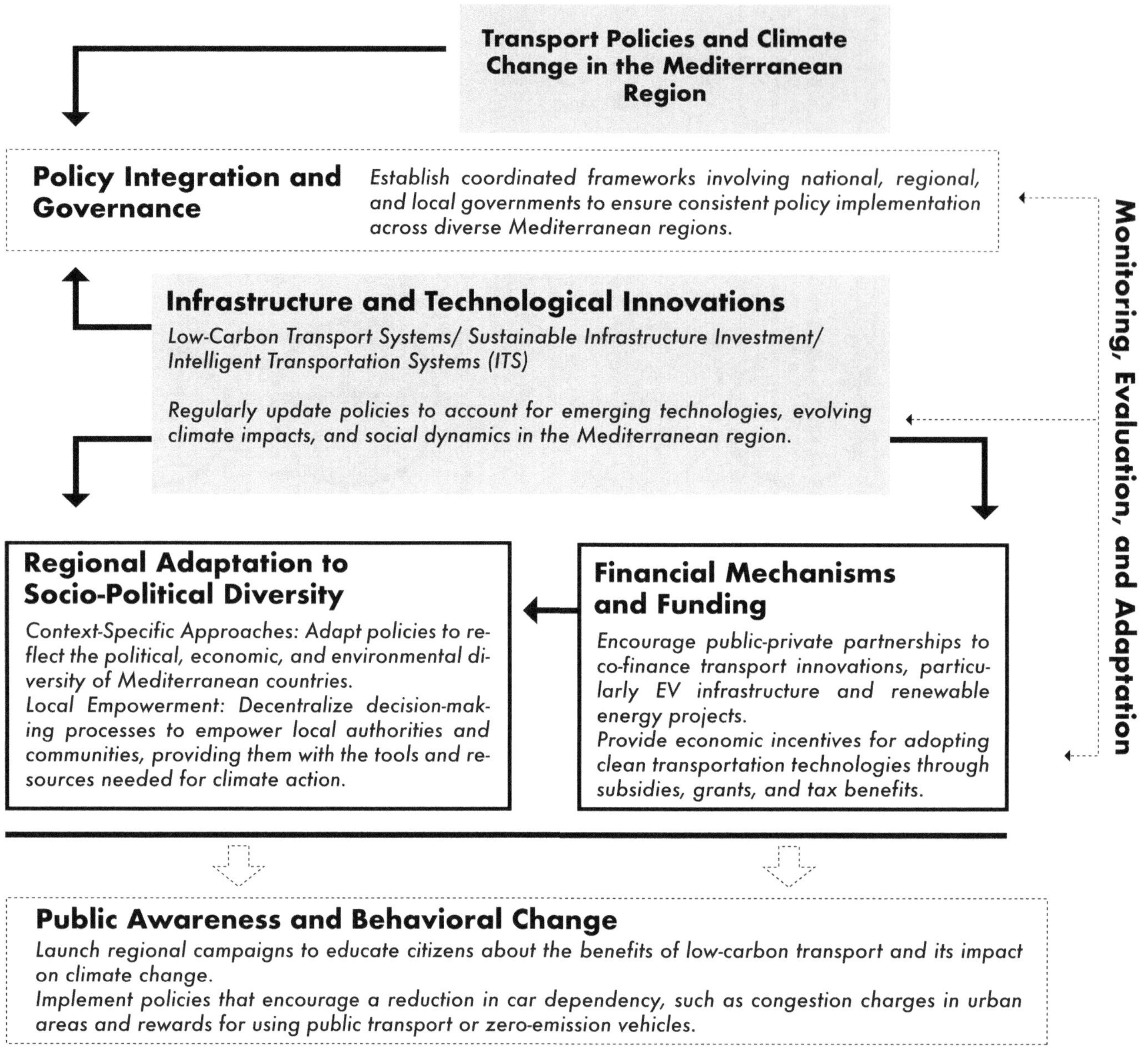

Fig. 19 Conceptual framework on integrating transport and climate change policies. *Source* Author

meet sustainability objectives. Figure 19 emphasizes the need for dynamic and adaptive strategies that would be able to respond effectively to evolving challenges linked to climate change within the transport sector, specifically within the Mediterranean context, by following a more integrated approach in governance and policymakers from different stakeholders. It is crucial to provide a regularly updated overview of policies so that emerging technologies may raise awareness of climate impacts in the region through financial mechanisms and funding.

In conclusion, the study highlights the importance on the link between sustainable transport visions and climate change mitigation. The need for sustainable transport strategies to explicitly incorporate climate change mitigation as a fundamental element it is emergent (Bakker et al. 2014). The paper also emphasizes the importance of addressing the global and local effects of sustainable transport and advocates for a balanced approach that combines transformative change with incremental progress. Additionally, the findings showed the gap between global and local effects of sustainable transport, and how important is to bridge this gap, with adaptive strategies and visions.

Author Contributions E.K.—data analysis, concept, writing draft, and final manuscript; S.M.—data collection, concept, writing draft, and final manuscript; F.K.—concept, writing draft, and final manuscript. All authors have read and agreed to the published version of the manuscript.

Conflicts of Interest The authors declare no conflict of interest.

References

Angelidou, M., Politis, C., Panori, A., Barkratsas, T., Fellnhofer, K.: Emerging smart city, transport and energy trends in urban settings: results of a pan-European foresight exercise with 120 experts. Technol. Forecast. Soc. Change **183** (2022). https://doi.org/10.1016/j.techfore.2022.121915

Aparicio, Á.: Transport adaptation policies in Europe: from incremental actions to long-term visions. Transp. Res. Procedia **25**, 3529–3537 (2017). https://doi.org/10.1016/j.trpro.2017.05.277

Bakker, S.: Shifting to low-carbon transport in ASEAN: policy development in a rapidly motorising region (n.d.-a)

Bakker, S.: Transport NAMA toolbox. www.transport-namas.org. Tool: World of Sustainable Transport Mitigation actions (Version 2, November 2016). www.transport-namas.org/resources/toolbox (n.d.-b)

Bakker, S., Huizenga, C.: Making climate instruments work for sustainable transport in developing countries (2010)

Bakker, S., Zuidgeest, M., de Coninck, H., Huizenga, C.: Transport, development and climate change mitigation: towards an integrated approach. Transp. Rev. **34**(3), 335–355 (2014). https://doi.org/10.1080/01441647.2014.903531

Bakker, S., Contreras, K.D., Kappiantari, M., Tuan, N.A., Guillen, M.D., Gunthawong, G., Zuidgeest, M., Liefferink, D., van Maarseveen, M.: Low-carbon transport policy in four ASEAN countries: developments in Indonesia, the Philippines, Thailand and Vietnam. Sustainability (Switzerland) **9**(7) (2017). https://doi.org/10.3390/su9071217

Barker, T., et al.: Summary for Policymakers. Aberdeen Centre for Environmental Sustainability (2007)

Black Carbon Mitigation and the Role of the Global Environment Facility: A STAP Advisory Document. http://www.thegef.org/documents/gef-6-programming-directions (n.d.)

Challa, P.: A review on integrated transportation system for Bangalore city. Int. J. Res. Appl. Sci. Eng. Technol. **V**(X), 1260–1268 (2017). https://doi.org/10.22214/ijraset.2017.10183

Cramer, W., Guiot, J., Fader, M., Garrabou, J., Iglesias, A., Lange, M.A., Lionello, P., Carmen Llasat, M., Paz, S., Peñuelas, J., Snoussi, M., Toreti, A., Tsimplis, M., Xoplaki, E.: Environmental change represents multiple risks for 1 sustainable development in the Mediterranean Basin 2 (n.d.)

Creutzig, F.: Evolving narratives of low-carbon futures in transportation. Transp. Rev. **36**(3), 341–360 (2016). https://doi.org/10.1080/01441647.2015.1079277

Creutzig, F., et al.: Decarbonizing urban transport in European cities: four cases show possibly high co-benefits. Environ. Res. Lett. (2012)

De Abreu, V.H.S., Da Costa, M.G., Da Costa, V.X., De Assis, T.F., Santos, A.S., de Almeida D'Agosto, M.: The role of the circular economy in road transport to mitigate climate change and reduce resource depletion. Sustainability (Switzerland) **14**(14). MDPI (2022). https://doi.org/10.3390/su14148951

Dhar, S., Pathak, M., Shukla, P.R.: Electric vehicles and India's low carbon passenger transport: a long-term co-benefits assessment. J. Clean. Prod. **146** (2017)

Diallo, T., et al.: Is HIA the most effective tool to assess the impact on health of climate change mitigation policies at the local level? A case study in Geneva, Switzerland. Sage J. (2017)

Fulton, L., Mejia, A., Arioli, M., Dematera, K., Lah, O.: Climate change mitigation pathways for Southeast Asia: CO_2 emissions reduction policies for the energy and transport sectors. Sustainability (2017)

Gota, S., Huizenga, C., Peet, K., Medimorec, N., Bakker, S.: Decarbonising transport to achieve Paris Agreement targets. Energy Effic. **12**(2), 363–386 (2019). https://doi.org/10.1007/s12053-018-9671-3

Haines, A.: Health Co-Benefits of Climate Change Mitigation Policies. ANU Press (2015)

Häll, C.H., Lundgren, J.T., Värbrand, P., Häll, C.H.: Evaluation of an integrated public transport system: a simulation approach. http://www.degruyter.com/view/j/aotrhttp://urn.kb.se/resolve?urn=urn:nbn:se:liu:diva-51813 (n.d.)

Jaeger, A., Nugroho, S.B., Zusman, E., Nakano, R., Daggy, R.: Governing sustainable low-carbon transport in Indonesia: an assessment of provincial transport plans. Nat. Res. Forum **39**(1), 27–40 (2015). https://doi.org/10.1111/1477-8947.12066

Kadir, N.A.: Promoting utility cycling: prospects and challenges from Penang Island, Malaysia. Pertanika J. Soc. Sci. Humanit. **30**(3), 1017–1035 (2022). https://doi.org/10.47836/pjssh.30.3.05

Lah, O.: Continuity and change: dealing with political volatility to advance climate change mitigation strategies—examples from the transport sector. Sustainability (2017)

Marí-Dell'Olmo, M., Oliveras, L., Barón-Miras, L.E., Borrell, C., Montalvo, T., Ariza, C., Ventayol, I., Mercuriali, L., Sheehan, M., Gómez-Gutiérrez, A., Villalbí, J.R.: Climate change and health in urban areas with a Mediterranean climate: a conceptual framework with a social and climate justice approach. Int. J. Environ. Res. Public Health **19**(19) (2022). https://doi.org/10.3390/ijerph191912764

Moretti, L., Loprencipe, G.: Climate change and transport infrastructures: state of the art. Sustainability (Switzerland) **10**(11) (2018). https://doi.org/10.3390/su10114098. MDPI

Nugroho, S.B., Zusman, E.: Low carbon paratransit in Jakarta, Indonesia: using econometric models to improve the enabling environment. Case Stud. Transp. Policy **6**(3), 342–347 (2018). https://doi.org/10.1016/j.cstp.2018.03.006

Pathak, M., Shukla, P.R.: Co-benefits of low carbon passenger transport actions in Indian cities: case study of Ahmedabad. Transp. Res. Part D Transp. Environ. (2016)

Sajovic, I., Boh Podgornik, B.: Bibliometric analysis of visualizations in computer graphics: a study. SAGE Open **12**(1) (2022). https://doi.org/10.1177/21582440211071105

Schwanen, T.: The bumpy road toward low-energy urban mobility: case studies from two UK cities. Sustainability (Switzerland) **7**(6), 7086–7111 (2015). https://doi.org/10.3390/su7067086

Shastry, S., Pai, M.: The Role of Transportation in the Future of Urban Developing Asia: A Case Study of India (n.d.)

Stone, A., Merven, B., Maseela, T., Moonsamy, R.: Providing a foundation for road transport energy demand analysis: a vehicle parc model for South Africa. J. Energy South. Afr. **29**(2), 29–42 (2018). https://doi.org/10.17159/2413-3051/2018/v29i2a2774

Watts, N., et al.: Health and climate change: policy responses to protect public health. Lancet (2015)

Wilkinson, P., et al.: Public health benefits of strategies to reduce greenhouse-gas emissions: household energy. Lancet (2009)

Woodcock, J., et al.: Public health benefits of strategies to reduce greenhouse-gas emissions: urban land transport. Lancet (2009)

Leveraging Urban Air Mobility (UAM) for Sustainable Urban Development: A Comprehensive Analysis of Impact and Congestion Mitigation

Yi Chen and Zhihao Li

Abstract

Since the 1920s, in response to the challenge of mediating between excessive cars and inadequate infrastructure, urban spaces have been transformed to accommodate transportation needs while also providing areas for residents' activities. However, this has led to congestion issues that need to be addressed, along with associated problems such as air pollution, high noise levels, and increasing greenhouse gas emissions that negatively impact quality of life. Simply expanding and densifying urban networks is not a long-term solution; alternative transportation tools must be explored. As experiments have shown that relying solely on one transportation plan is unfeasible, exploring transportation options across different layers presents an alternative solution—urban air mobility (UAM). Urban air mobility offers a promising approach by utilizing small and highly automated aircraft at lower altitudes in urban and suburban areas for efficient and environmentally friendly passenger and cargo transport. This article aims to introduce the strategy of developing UAM in cities not only to alleviate congestion but also to mitigate its side effects. The focus of this paper is on analyzing the benefits brought about by integrating UAM into existing transit systems through data analysis.

Keywords

Alternative transportation tools · Urban air mobility · Efficient and environmentally friendly · Alleviate congestion · Environmental benefits

Y. Chen (✉) · Z. Li
Politecnico di Milano, Milano, Italy
e-mail: yi3.chen@mail.polimi.it

Z. Li
e-mail: zhihao1.li@mail.polimi.it

1 Background

Since the invention of cars, it has significantly facilitated human mobility. It has enhanced travel efficiency by allowing individuals to reach their destinations directly, without the need for transfers or waiting times. Furthermore, it has increased the freedom of mobility, enabling people to travel at their convenience without constraints. They have expanded the scope of human activities by facilitating travel to more distant locations. Consequently, urban development has transitioned from being primarily pedestrian-oriented to car-centric transportation.

As a result, many large cities have embarked on the construction of urban expressways, connecting city centers with suburbs, as well as various commercial and residential areas, thereby reshaping the geographical layout of cities (Joshi et al. 2005). These roads are like vessels, enhancing regional accessibility, promoting the integration of urban spaces, and optimizing land utilization. Planners believe this transport network could be able to efficiently manage the flow of urban populations.

However, due to the continuous growth in population and the increasing ownership and utilization of private vehicles in large cities, the demand for urban transportation continues to escalate. As a result, traffic congestion has transitioned from being limited to peak hours to becoming a persistent issue throughout the day. The expansion of infrastructure fails to keep pace with the rapid growth in vehicular traffic demands (Ng et al. 2010).

The pressures confronting urban areas persist, encompassing issues such as sustainable development, air quality degradation, noise pollution, etc.

For example, by the end of 2023, the total number of automobiles in Shanghai reached 5.479 million, reflecting a yearly increase of 6.1%. Among these, the number of passenger cars stood at 5.04 million, representing an annual growth rate of

P. Mohammad et al. (eds.), *Geographic Approaches to Climate Change and Mitigation: Urban and Rural Perspectives (Volume 2)*,
Advances in Science, Technology & Innovation, https://doi.org/10.1007/978-3-031-92115-5_15

7.2% (Shanghai Municipal Transportation Commission). The traffic jam index (the ratio of actual travel time to free-flow travel time from 8:00 to 20:00) increased by 8.16%, which illustrates that people spend more time commuting (Baidu Map).

2 Challenges

2.1 Unmet Road Needs

Many large cities commonly opt for widening existing roads or constructing new elevated highways to alleviate urban congestion (Figs. 1 and 2). While these measures enhance traffic volume, they are not pedestrian-friendly. Broad and swift roads may impart a sense of difficulty in crossing for pedestrians. The imposing structure of elevated highways obstructs sights, while the shadows cast below create a subconscious sensation of boundary (Carlà-Uhink 2022). Consequently, roads visually and psychologically divide the areas on either side.

Connecting various urban spaces, enhancing regional accessibility, and promoting the integration of urban spaces have led to the constant widening of urban roads or the construction of new elevated highways or underground (Samaniego and Moses 2008). These initiatives aim to link city centers with suburbs and different functional zones. However, such planning primarily prioritizes cars, resulting in the continual expansion of street dimensions, diverging from pedestrian scale. Consequently, these streets generate spaces that are unsuitable for pedestrian movement.

Fig. 1 Construction of Shanghai Inner Ring Elevated Road by Lu (1991)

Fig. 2 View under Shanghai Inner Ring Elevated Road by Author (2020)

As a result, urban roads are under significant pressure, with current capacities failing to accommodate peak-hour traffic volumes, exacerbating congestion. Simply increasing road construction (either through building new roads or widening existing ones) cannot meet the ever-growing demand for vehicular travel. Moreover, expanding road infrastructure often prompts a higher reliance on private automobiles for transportation, subsequently resulting in renewed congestion. This situation perpetuates the cycle where roads are widened to accommodate more vehicles, yet congestion persists, unable to effectively address the root cause of traffic congestion.

Large cities development process, such as Shanghai, Suzhou, and Hangzhou, have built wider roads, elevated highways, and tunnels to serve as major urban thoroughfares (Wei 2015). While these initiatives have improved traffic flow, they have also diminished green spaces and encroached upon pedestrian zones, altering the urban landscape.

2.2 Environmental Pollution

The energy consumption and environmental pollution caused by vehicles present a key challenge. Conventional vehicles rely on engines, burning gasoline to generate power. This process produces numerous harmful gases and a substantial amount of carbon dioxide. These pollutants are major contributors to urban air pollution. Additionally, the significant emissions of carbon dioxide contribute to global climate change as a greenhouse gas (Fitzpatrick 2014). Furthermore, congested traffic exacerbates this phenomenon. Through an analysis of Beijing's air quality from 2002 to 2012, utilizing the air pollution index as a metric, comprehensive quantitative and qualitative assessments were conducted. The study revealed a significant correlation between Beijing's air quality and the number of automobiles in circulation (Xu 2014).

2.3 Parking Challenge

Parking challenges are an inescapable concern for several reasons. Firstly, the persistent increase in the number of motor vehicles outpaces the city's ability to accommodate this change effectively. Secondly, urban areas face constraints in land resources, which impede the provision

Fig. 3 Parking issue [online image] (2018)

of sufficient parking facilities. Thirdly, many neighborhoods were inadequately planned and constructed without due consideration for future parking requirements, resulting in existing facilities falling short of meeting current demand (Kirschner 2021). Many cities are facing this challenge. The parking deficit rate in downtown Nanjing exceeds 55% (Fig. 3), while the average ratio of cars to parking spaces in urban areas is approximately 1:0.8, far below the ideal standard of 1.1–1.3 parking spaces per vehicle (Nanjing Municipal Urban Management Bureau). And the situation is even more severe in Shanghai. The annual growth rate of car ownership in Shanghai exceeds that of parking spaces. The overall growth rate of car ownership in the past two years is 19.25%, while the growth rate of parking spaces is 11% over the same period (Wu 2023).

2.4 Noise Pollution

The issue of noise pollution stemming from vehicles is a significant concern. On urban expressways, cars generate more noise pollution compared to slow roads (Kohut 2015). This poses adverse impacts on both sides of the road. Moreover, engine noise and horn honking from automobiles are notably prevalent in urban areas with dense road networks, slow-moving traffic, and congestion, such as city centers. In the central urban area of Shanghai, there are currently about 200 km of elevated roads and expressways already built or planned. Meanwhile, 70% of the traffic volume in the central urban area of Shanghai is concentrated on only 20% of the roads and expressways, leading to significant traffic noise in some areas. For example, the average noise level on both sides of the Inner Ring Elevated Road is 75 ~ 77 dB(A), exceeding the prescribed limits of sound environment quality standards, particularly at night, severely affecting residents' sleep environment (Xie and Huang 2018).

3 Novel Transportation—Urban Air Mobility

In response to these challenges, attention has shifted toward the low-altitude domain in search of a novel mode of transportation, thus giving rise to urban air mobility (UAM) which was formally defined in the mid-2010s.

Urban air mobility (UAM), just as its name implies, is a new air transportation system added to the existing public transport system for passengers and cargo in an urban area with a densely populated and built environment. It also is defined as safe and efficient traffic operations for manned and unmanned aircraft to conduct on-demand and scheduled operations including emergency medical evacuations, rescue operations, humanitarian missions, news gathering, ground traffic flow assessment, weather monitoring, package delivery, and passenger transport (Thipphavong et al., n.d.; NASA 2021, Figs. 4 and 5). It connects inner-city locations as well as suburbs and city centers. Broadly, it refers to various concepts related to aircraft-based transportation systems providing services in a metropolitan area (Vascik 2019).

Fig. 4 Bell Nexus envisioned for medical transport missions at a hospital. *Source* NASA

Fig. 5 By combining compressed air foam (CAF) and other extinguishing mediums, a small eVTOL like the one from Esprit can become an effective firefighter. *Source* NASA

3.1 History and Development of UAM

The early concept of UAM began with the idea of "flying cars" in the 1900s, exemplified by inventions such as the auto plane created by Glenn Curtiss in 1917. By the middle of the twentieth century, some air companies started to offer air taxi services in several American cities, enabled by helicopter subsidies (discontinued in 1966) and air mail revenue (Cohen et al. 2021). Between 1964 and 1968, Pan Am offered a connection between the Pan American building in midtown where passengers could check in 40 min prior and JFK's world port; however, it was stopped in 1979 due to an accident that occurred in 1977 resulting in 5 people dead (Jun 11 et al. 2019). In the 1980s, the Trump shuttle offered a connection between Wall Street and LaGuardia airport ended in the 1990s after the acquisition by US Airways (Flanagan, n.d.). Thus far, air taxi services continue to undergo rigorous market testing. In recent years, as advancements in both manned and unmanned aerial technologies continue to reshape the aviation landscape, UAM emerged and captured the public's imagination with its transformative potential, eliciting positive attitudes and interest from citizens (Cohen et al. 2021). The European Union Aviation Safety Agency (EASA) conducted a study on the acceptance of urban air mobility (UAM) among EU citizens, revealing a positive attitude and interest in UAM among EU citizens in 2021 (EASA 2021).

3.2 Aircraft Development

In the early twentieth century, the idea of vertical take-off and landing (VTOL) aircraft emerged and was implemented in the 1940s (Vieira et al. 2019). In the last decade, helicopters were regarded as one type of VTOL that played an important role in air services; however, investors never stopped developing VTOLs. In today's era, UAM relies heavily on electric vertical takeoff and landing aircraft (eVTOL), boasting state-of-the-art technologies that promise passengers a greener, quieter cheaper, and safer flight experience compared to traditional helicopters (Su et al. 2024). In NASA's conceptual model (NASA 2021, Fig. 6) of eVTOLs, costs can be reduced by using electric power and motors, while autonomy helps minimize the expense of flight crews. Additionally, the development of rotors ensures safer and quieter flights compared to traditional helicopters (*Urban Air Mobility*, n.d.)

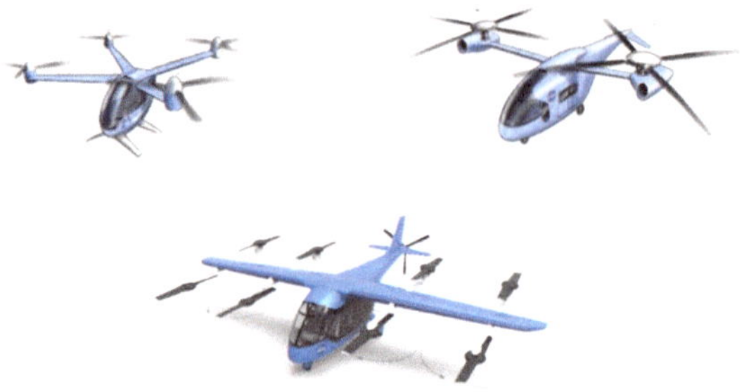

Fig. 6 NASA's UAM concept vehicles. *Source* NASA

3.3 UAM Nowadays

Urban air mobility (UAM) has gained significant traction in recent years, fueled by technological advancements, and increasing interest from both public and private sectors.

The pivotal moment came in 2016 when Uber released its white paper (*VFS—Uber Elevate Press Releases*, n.d.) on UAM, which outlined a comprehensive vision for integrating eVTOLs into urban air transportation. This publication sparked global interest and established UAM as a hot topic worldwide. Following Uber's white paper, NASA and EASA began publishing documents and conducting studies on UAM. In Europe, UAM is anticipated to become a reality within the next 3–5 years, supported by the development of regulatory frameworks and insights from the 2021 UAM study on societal acceptance (*Urban Air Mobility (UAM)*, n.d.).

In the UK, a significant milestone was achieved with the launch of Air One in Coventry in 2022 (*Air One Event*, n.d.), the world's first fully operational Urban-Air Port (Fig. 7) designed for drones and eVTOLs. This facility marks a crucial step in making UAM a practical reality.

Fig. 7 Urban-Air Port. *Source* UNMARKED STUDIO

Currently, UAM is in its initial development phase, allowing for on-demand and scheduled flights. This phase serves as a bridge between the early experimental stages and the future UAM network, which aims to provide seamless point-to-point service.

4 UAM—A Sustainable Transportation Solution to Alleviate Urban Challenges

4.1 Less Time on Trip and Alleviate Congestion

Since last century, the idea of flying taxis has been studied and developed but not received widespread attention; however, growing congestion brings this idea to the public. People are eager to find a transportation method that can save time and alleviate traffic congestion. Will UAM mitigate congestion? Or it is just a utopian idea? UAM discusses the capacity of flying at a low altitude which seems to lead to an excellent solution for traffic congestion (The Future of Air Mobility: Electric Aircraft and Flying Taxis 2021).

Compared to traditional ground transportation, high-speed aircraft provides customers with substantial time savings, swiftly ferrying them from departure points to their destinations. This dynamic catalyzes a growing preference for UAM as a daily transportation option. Such a shift not only promises efficiency but also holds the potential to alleviate traffic congestion. Recently, an eVTOL completed a 20-min journey from Shenzhen Shekou Port to Zhuhai Jiuzhou Port, drastically reducing the one-way ground travel time from 2.5 to 3 h to just 20 min. Set to commence operations in 2026, this route offers unparalleled efficiency compared to other transportation modes while remaining accessible to the public. This advancement is poised to alleviate traffic congestion and reduce travel pressure on the ground route from Shenzhen to Zhuhai (Hu and Zheng 2024).

UAM offers superior long-distance transportation alternatives, simultaneously mitigating traffic congestion and the stress associated with lengthy commutes. However, the first and last kilometers of UAM travel will now be detoured and use ground transportation, which may increase the pressure on a small range of local traffic. Addressing this challenge necessitates the implementation of enhanced accessibility designs for vertiports, thereby alleviating strain on ground transportation networks. Looking ahead, the advent of eVTOL aircraft holds promise for enabling seamless door-to-door flights, offering a viable solution to local traffic pressures in the future (Wang et al. 2023).

4.2 Green Energy Consumption and Emission

Today, one-fifth of global carbon dioxide (CO_2) is produced by traffic; meanwhile, road travel accounts for three-quarters of them. When exploring novel transportation options, it is imperative to assess not only their potential to alleviate current traffic challenges but also their environmental impact, especially greenhouse gas which is not only the culprit of the greenhouse effect but also aggravates the urban heat island effect.

Thanks to the development of eVTOL aircraft, UAM emerges as a sustainable option for journeys exceeding 35 km. An article compared greenhouse gas emissions among air taxis, internal combustion engine vehicles (ICEV), and battery electric vehicles (BEV) across diverse loads and travel distances. Findings indicated that eVTOL aircraft emissions exceeded those of both vehicle types for distances under 35 km due to increased energy consumption and greenhouse gas emissions during takeoff and landing. However, for distances exceeding 100 km with a single pilot, air taxis demonstrated superiority over ICEVs but trailed BEVs. Conversely, when fully loaded (three passengers), eVTOLs outperformed ICEVs and BEVs with an average occupancy of 1.54. These results paint a promising future for sustainable mobility through UAM (Kasliwal et al. 2019).

In real-life scenarios, eVTOLs adhere to direct distances while cars traverse highways, encountering varying road conditions and congestion. UAM seems to present a superior transportation option. An article underscored that eVTOLs surpassed hybrid electric cars in both trip time and CO_2 emissions, the most significant component of greenhouse gases. This was demonstrated through a comparison of the two modes of transportation on the same route from Marina, San Francisco to Downtown San Jose, with the air taxi covering a direct distance of 70 km compared to the car's longer route of 90 km (Donateo et al. 2022).

From this perspective, urban air mobility (UAM) demonstrates significant potential for reducing energy consumption and emissions compared to current automobile mobility, particularly for long-distance commuting. There is reason to believe that with continuous and innovative advancements in eVTOL technology, UAM will play a pivotal role in achieving sustainable mobility objectives.

4.3 Noise

In the report about evaluation on UAM published by EASA, one of the most common concerns from the public is noise, especially when the eVTOLs are taking off and landing. With the ongoing advancements in eVTOL technology, it becomes feasible to ensure that residents on the ground remain undisturbed during their flight. Archer, an aircraft design company, has announced that during the cruise phase of their eVTOLs, operating at approximately 2000 feet with high speed, the noise reaching the ground measures around 45 A-weighted decibels (dBA) (How loud is an eVTOL? n.d.). However, the noise generated during takeoff and landing cannot be ignored. Within a two-kilometer radius of the vertiports, people on the ground can expect noise levels exceeding 65 dBA, with continuous exposure to levels surpassing 45 dBA within a certain range. Hence, the selection of vertiport locations necessitates meticulous consideration (Bauer 2023).

5 Conclusion

When traditional ground transportation systems have significantly enhanced human mobility and urban connectivity, they have also led to challenges such as traffic congestion, environmental pollution, parking shortages, and noise disturbances. In response to these issues, UAM has emerged as a novel transportation solution. UAM, utilizing eVTOL aircraft, promises to alleviate ground traffic congestion, reduce travel time, and provide a greener transportation alternative. Despite noise concerns during takeoff and landing, advancements in eVTOL technology aim to minimize disruptions. UAM offers a sustainable and efficient solution to urban transportation challenges, with the potential to significantly reduce energy consumption and emissions compared to traditional vehicles, particularly for long-distance commutes.

References

Air One Event: Urban-Air Port Ltd. https://www.urbanairport.com/airone (n.d.)

Author: Under Shanghai Inner Ring Elevated Road (2020)

Bauer, M.: Noise footprint assessment at a vertiport for different approach and departure procedures of a tilt-wing air-taxi. SAE Technical Paper 2023-01-1103. https://doi.org/10.4271/2023-01-1103 (2023)

Carlà-Uhink, F.: The impact of roman roads on landscape and space: the case of Republican Italy. In: Horster, M., Hächler, N. (eds.) The Impact of the Roman Empire on Landscapes: Proceedings of the Fourteenth Workshop of the International Network Impact of Empire, Mainz, 12–15 June 2019 (2022)

China Urban Transportation Report 中国城市交通报告 2021 年度: Accessed 20 May 2024. https://jiaotong.baidu.com/cms/reports/traffic/2021/%E7%99%BE%E5%BA%A6%E5%9C%B0%E5%9B%BE2021%E5%B9%B4%E5%BA%A6%E4%B8%AD%E5%9B%BD%E5%9F%8E%E5%B8%82%E4%BA%A4%E9%80%9A%E6%8A%A5%E5%91%8A.pdf (n.d.)

Cohen, A.P., Shaheen, S.A., Farrar, E.M.: Urban air mobility: history, ecosystem, market potential, and challenges. IEEE Trans. Intell. Transp. Syst. **22**(9), 6074–6087 (2021). https://doi.org/10.1109/TITS.2021.3082767

Donateo, T., Ficarella, A., Surdo, L.: Energy consumption and environmental impact of urban air mobility. IOP Conf. Ser. Mater. Sci. Eng. **1226**(1), 012065 (2022). https://doi.org/10.1088/1757-899X/1226/1/012065

Doo, J., Tsairides, M., Pavel, M., Smith, M., Textron, B., Didey, A., Bristol, N., Bennet, U., Hange, C., Bromfield, M., Diller, N., Mooberry, J.: NASA Electric Vertical Takeoff and Landing (eVTOL) Aircraft Technology for Public Services—A NASA Electric Vertical Takeoff and Landing (eVTOL) Aircraft Technology for Public Services—A White Paper NASA Transformative Vertical Flight Working Group 4 (TVF4). https://ntrs.nasa.gov/api/citations/20205000636/downloads/2021-08-20-eVTOL-White-Paper-Final_V48.pdf (2021)

European Union Aviation Safety Agency (EASA): Study on the Societal Acceptance of Urban Air Mobility in Europe. https://www.easa.europa.eu/sites/default/files/dfu/uam-full-report.pdf (2021)

Fitzpatrick, T.: Air and water. In: Climate Change and Poverty: A New Agenda for Developed Nations, 1st edn., pp. 187–210 (2014)

Flanagan, G.: The rise and fall of Donald Trump's $365 million airline. Business Insider. https://www.businessinsider.com/donald-trump-airline-shuttle-transportation-pan-am-eastern-new-york-2019-1 (n.d.)

How Loud Is an eVTOL? Archer.com. https://archer.com/news/how-loud-is-an-evtol (n.d.)

Hu, C., Zheng, X.: The first flight was successful! The "air taxi" is coming. People's Daily. Accessed 9 Apr 2024. https://mp.weixin.qq.com/s/6e4YZHcYGjjgRyLat3fw0Q (2024)

Joshi, M.S., Joshi, V., Lamb, R.: The prisoners' dilemma and city-centre traffic. Oxf. Econ. Pap. **57**(1), 70–89 (2005). http://www.jstor.org/stable/3488836

Jun 11, J.C., Jun 11, 2019 Modified, 2019 Share: In 1977, five were killed in helicopter accident atop midtown's Pan Am building. Gothamist, 11 June 2019. https://gothamist.com/news/in-1977-five-were-killed-in-helicopter-accident-atop-midtowns-pan-am-building

Kasliwal, A., Furbush, N.J., Gawron, J.H., McBride, J.R., Wallington, T.J., De Kleine, R.D., Kim, H.C., Keoleian, G.A.: Role of flying cars in sustainable mobility. Nat. Commun. **10**(1), 1555 (2019). https://doi.org/10.1038/s41467-019-09426-0

Kirschner, F.: Parking and competition for space in urban neighborhoods: residents' perceptions of traffic and parking-related conflicts. J. Transp. Land Use **14**(1), 603–623 (2021). https://www.jstor.org/stable/48646201

Kohut, T.: Noise pollution and the eco-politics of sound: toxicity, nature and culture in the contemporary soundscape. Leonardo Mus. J. **25**, 5–8 (2015). http://www.jstor.org/stable/43832520

Lu, J.: Construction of Shanghai Inner Ring Elevated Road. https://k.sina.cn/article_7517400647_1c0126e4705902hyhf.html?from=photo (1991)

Ng, W.-S., Schipper, L., Chen, Y.: China motorization trends: new directions for crowded cities. J. Transp. Land Use **3**(3), 5–25 (2010). http://www.jstor.org/stable/26189158

Parking issue [online image]. https://news.dahebao.cn/dahe/appcommunity/1759877 (2018)

Samaniego, H., Moses, M.E.: Cities as organisms: allometric scaling of urban road networks. J. Transp. Land Use **1**(1), 21–39 (2008). http://www.jstor.org/stable/26201607

Shanghai Municipal Transportation Commission, Traffic Reports 2021. https://www.shanghai.gov.cn/nw48081/20220615/9a9c2a907ef14b4fa423d4fd7e8a1dd8.html

Su, J., Huang, H., Zhang, H., Wang, Y., Wang, F.-Y.: eVTOL performance analysis: a review from control perspectives. IEEE Trans. Intell. Veh. 1–13 (2024). https://doi.org/10.1109/TIV.2024.3387405

Tallerico, T.: NASA Reference Motor Designs for Electric Vertical Takeoff and Landing Vehicles. https://ntrs.nasa.gov/api/citations/20210017935/downloads/UAM%20Motor%20Design%207_7_2021%20(1).pdf (n.d.)

The Future of Air Mobility: Electric Aircraft and Flying Taxis. https://www.mckinsey.com/~/media/mckinsey/featured%20insights/the%20next%20normal/the-next-normal-the-future-of-air-mobility.pdf (2021)

Thipphavong, D., Apaza, R., Barmore, B., Battiste, V., Burian, B., Dao, Q., Feary, M., Go, S., Goodrich, K., Homola, J., Idris, H., Kopardekar, P., Lachter, J., Neogi, N., Ng, H., Oseguera-Lohr, R., Patterson, M., Verma, S.: Urban air mobility airspace integration concepts and considerations. https://ntrs.nasa.gov/api/citations/20180005218/downloads/20180005218.pdf (n.d.)

UNMARKED STUDIO: Urban-Air Port. https://www.google.com/url?sa=i&url=https%3A%2F%2Fwww.urbanairport.com%2F&psig=AOvVaw2yilEHQfMzEzbWfaj-jfia&ust=1716314331570000&source=images&cd=vfe&opi=89978449&ved=0CBIQjRxqFwoTCKi0g6PnnIYDFQAAAAAdAAAAABAY (2024)

Urban Air Mobility: Rotorcraft.arc.nasa.gov. https://rotorcraft.arc.nasa.gov/Research/Programs/UrbanAirMobility.html (n.d.)

Urban Air Mobility (UAM): EASA. https://www.easa.europa.eu/en/domains/drones-air-mobility/drones-air-mobility-landscape/urban-air-mobility-uam (n.d.)

Vascik, P.D.: Systems Analysis of Urban Air Mobility Operational Scaling (2019)

VFS—Uber Elevate Press Releases: Vtol.org. Accessed 20 May 2024. https://vtol.org/news/uber-elevate-press-releases (n.d.)

Vieira, D.R., Silva, D., Bravo, A.: Electric VTOL aircraft: the future of urban air mobility (background, advantages and challenges). Int. J. Sustain. Aviat. **5**(2), 101 (2019). https://doi.org/10.1504/ijsa.2019.101746

Wang, K., Li, A., Qu, X.: Urban aerial mobility: network structure, transportation benefits, and Sino-US comparison. The Innovation **4**(2), 100393 (2023). https://doi.org/10.1016/j.xinn.2023.100393

Wei, Y.D.: Urban land expansion and structural change in Shanghai, Suzhou and Hangzhou. In: Economic Transition, Development Zones, and Urban Land Expansion in the Yangtze River Delta, China, pp. 17–35. Lincoln Institute of Land Policy (2015). http://www.jstor.org/stable/resrep18481.6

Wu, Y.: Recommendations for optimizing parking in Shanghai. https://www.mjshsw.org.cn/detailpage/jyxc-ec335162-6f8e-47af-954a-5ef4b3b26f5b.html (2023)

Xie, N., Huang, Y.: Research on noise reduction for existing residences along urban elevated roads. Hous. Sci. **38**(9), 7 (2018)

Xu, Y.: The Relationship Between Air Pollution and Car Ownership in Beijing in Recent Years. Harbin Normal University (2014)

Beyond Transit-Oriented Development—Linear or Transit Cities

Halvard Dalheim

Abstract

Managing significant population growth is the central driver for the development of metropolitan strategies for the major capital cities in Australia, particularly Melbourne and Sydney, cities of 5 million people. Outward expansion has been a central policy position for their planning, but a recognition of higher infrastructure costs has shifted the planning focus to urban renewal, specifically, transit-oriented development. The issue of improved accessibility to goods, services and jobs—economic, social and community assets—however, is currently not explicitly considered. This study seeks to understand if accessibility to economic, social and community assets can make a positive contribution to planning for urban renewal. The approach focuses on accessibility to those assets within a walking catchment of train stations on the rail networks of Melbourne and Sydney in the context of accessibility within a '30-min city'. The study findings suggest that the location of transit-oriented development and required infrastructure investments (e.g. schools) is of consequence, if accessibility to economic, social and community assets is to be optimised. Thus, the study highlights that the planning for transit-oriented development and associated infrastructure investments can be reimagined as 'linear transit cities', a conceptual framework to inform the preparation of metropolitan strategies.

Keywords

Transit-oriented development · Urban renewal · Transit corridor · Transit city · Linear city · Urban structure · Planning policy · Urban benefits · Efficiency

H. Dalheim (✉)
School of Architecture, Design and Planning, University of Sydney, Sydney, Australia
e-mail: halvard.dalheim@sydney.edu.au

1 Introduction

Managing population growth is the central driver underpinning the development of metropolitan plans across Australia, with the population of Australia's eight capital cities doubling over the last 50 years (Australian Bureau of Statistics 2022). They continue to grow rapidly, including by 0.5 million in the 12 months to 30 June 2023, the highest on record (Australian Bureau of Statistics 2024). Furthermore, Melbourne's population is projected to grow from 5.1 million people (2023) to 8.0 million people by 2051 (State of Victoria 2023) and Sydney's population is projected to grow from 5.0 million people (2021) to 6.1 million people by 2041 (NSW Government 2024a).

At the same time, there is a recognition across Australia of a significant mismatch between housing supply and demand. For example, in the last decade, housing completions in Sydney peaked in the two years 2017–18 and 2018–19, with around 41,000 and 42,000 net dwelling completions. However, in the year 2022–23 the total number delivered was only 21,100 dwellings (NSW Government 2024b), yet the population growth in the same year was 146,700, suggesting a latent housing demand of around 54,300 dwellings (based on an average dwelling occupancy rate, Australian Bureau of Statistics 2024). Consequently, all states have signed up to a National Housing Accord that aims to deliver 1.2 million dwellings over the next five years, including affordable housing targets (Australian Government 2022).

In managing this growth there is increasing recognition of the relatively higher cost of delivering new greenfield development as opposed to urban renewal. The recent Victorian Housing Statement indicates a fourfold difference (Department of Premier and Cabinet 2023). Multiple studies have also been undertaken on this issue since the work of Newman and Kenworthy (1989) on the impact of urban structure on vehicle travel, such as the assessment of settlement choices

P. Mohammad et al. (eds.), *Geographic Approaches to Climate Change and Mitigation: Urban and Rural Perspectives (Volume 2)*, Advances in Science, Technology & Innovation, https://doi.org/10.1007/978-3-031-92115-5_16

for Sydney (Centre for International Economics 2010) that re-affirms the need to facilitate urban renewal.

In Melbourne, the Victorian Housing Statement seeks to incentivise the provision of more affordable housing, increase housing choice, improve access to community facilities, public spaces and parks, provide homes closer to jobs, hospitals and schools, make precincts about people and places and reduce sprawl. The immediate response focuses on 10 activity centres where 60,000 homes are proposed. In addition, there are policy objectives to building more homes closer to where people currently have the required transport, roads, hospitals and schools and to deliver vital community infrastructure—like footpaths and parks—in a timelier manner. However, other than facilitating housing supply no actions or infrastructure investments are outlined as to how the wider policy outcomes will be achieved.

In recognition of ongoing growth, the Victorian Government has released draft housing targets out to 2051 (State Government of Victoria 2024), that include 1.356 million new dwellings for the established areas (urban renewal). The Victorian Housing Statement (State of Victoria 2023) also indicates the need for a long-term vision for what cities, suburbs and regions may look like in the future.

For Sydney, the NSW Government recently set short term (5 years) housing targets for the established areas, these equate to a total of 189,400 new dwellings (NSW Government 2024). In addition, the NSW Government's Transport Oriented Development Program (2023) aims to deliver 185,800 homes over 15 years in 39 designated station precincts with the aim of providing homes closer to jobs, improving the pattern of urban development by reducing urban sprawl, having more households closer to goods, services, opens space and a vibrant nightlife, managing traffic congestion, promoting a healthy lifestyle and reducing the impacts of climate change. Criteria for selecting the designated precincts, however, only addresses the consideration of stations being within 30 min of a metropolitan centre, housing potential/capacity, utilities and rail infrastructure capacity, policy consistency and local place considerations such as flooding (Department of Planning, Housing and Infrastructure 2024). Development guidance for the program addresses preferred development designs, a need for active street frontages, a response to biodiversity impacts and a freeing up of heritage requirements (NSW Department of Planning, Housing and Infrastructure 2024).

For contextual purposes, it is important to recognise that in the early 2000s there was a renewed emphasis on expanding the passenger rail networks in both aforementioned cities. In Melbourne, one new line has been completed (Sunshine to Wyndham Vale), one is under construction (Melbourne Metro) and one is committed (Suburban Rail Loop). In Sydney, one new line has been completed (Sydney Metro NW), two are under construction (Sydney Metro SW and Sydney Metro Airport) and a further line is committed (Metro West). These investments together with significant growth have raised the issue of an efficient urban form, which is reflected in the current metropolitan plans for each city. In addition, the purpose of these rail projects is in part connected to facilitating residential and commercial development around train stations.

Consequently, for Melbourne and Sydney, while there is a recognised need to manage population growth in the short and long term and actions to address housing supply, with a focus on urban renewal, there is not a clear holistic framework that outlines how the housing initiatives fit within a wider metropolitan framework considering the scale of the long-term task. In that, current metropolitan strategies for both Melbourne and Sydney are principally directed to facilitating growth at individual train stations/centres with separate policy objectives of 20 and 30-min cities for each city, respectively, but there are no actions or commitments as to how the last two objectives will be delivered.

Therefore, it appears that the central policy focus for both Melbourne and Sydney is principally housing supply, in the context of transit-oriented development (TOD) principles, but not whether wider objectives, such as an efficient urban form that maximises both accessibility to goods services and jobs and maximises non-private vehicle trips, will be achieved—a missed opportunity.

This study aims to explore whether a wider network (corridor) view of TODs can provide a greater understanding of the policy objective of accessibility to goods, services and jobs, and equity issue, and thus inform where to target and facilitate residential development opportunities and their required infrastructure investments. More specifically, the study seeks to answer the question—Is the distribution of goods, services and jobs evenly distributed across a city and thus any potential TOD can simply be assessed on local considerations or do differences in distribution suggest location matters when identifying both a TOD and the infrastructure investments to support it, if accessibility is to be optimised?

The objectives of the study are to further understand:

- the benefits and planning principles of transit-oriented development that may inform accessibility to goods, services and jobs.
- the planning considerations for linear cities that may inform a wider networked approach to transit-oriented development.
- the distribution pattern of goods, services and jobs along rail lines within a walkable catchment from train stations and their accessibility from any given train station in the context of a 30-min city.

The anticipated outcomes from these objectives are that:

- the level of goods services and jobs along the rail (transit) corridors will differ for each city, which could then inform the designation of priority corridors for new TODs, and required infrastructure.
- the distribution of goods services and jobs along a single rail (transit) corridor will not be even, which would affirm that location matters and thus inform priority locations for new TODs, and required infrastructure.
- the level of goods services and jobs accessible, within 30 min, from any given station along a rail corridor will differ, which would also affirm that location matters and thus inform priority locations for new TODs, and required infrastructure.
- for sections of any given rail network there will be train stations that have access to the full range of goods, services and jobs expected to be available either locally or across a city, suggesting these are ideal locations for TODs.
- collectively, the outcomes indicate that the planning for transit-oriented development, and the infrastructure to support it should take a rail corridor or 'linear transit city' approach and that such an approach has the potential to be developed into a metropolitan wide urban renewal framework.

For this study goods, services and jobs are categorised into economic and social assets. In that, they can generally be viewed as the infrastructure investments of government (such as schools) or the consequence of policy decisions of governments (such as the designation/facilitation of centres).

This explicit intent to influence accessibility aligns with several of the United Nations Sustainable Development Goals including: SDG3: Ensure healthy lives and promote well-being for all at all ages; SDG10: Reduced inequalities within and among countries; SDG11: Make cities and human settlements inclusive, safe, resilient and sustainable; and SDG13: Take urgent action to combat climate change and its impacts.

The analysis for this study focuses on Melbourne and Sydney. Firstly, due to the combined influence of the scale of each city, currently 5.0 million people, their recognised long-term growth, and commitments by their respective governments to urban renewal initiatives, and therefore, the potential to make a difference in their planning. Secondly, the need for a practical and easily understood planning tool that can better inform how urban renewal is approached when preparing metropolitan strategies, a current gap considered by the author, in the context of his involvement in metropolitan planning for Melbourne and Sydney over the last 15 years, including leading the preparation of the last two plans for Greater Sydney. In terms of wider global application, the central issue of accessibility to goods, services and jobs is more than likely to be relevant to any city. What is not covered in this study is an assessment of any city outside Australia and thus confirmation of wider applicability.

The genesis for this research comes from my involvement in the Victorian Government's working group to plan the Melbourne Metro rail project, where it become evident that along the alignment was a 'string of pearls' that has now been recognised in the metropolitan plan for Melbourne (Department of Environment, Land, Water and Planning 2017). Then following the working group, leading the metropolitan plan for Greater Sydney reinforced a deeper understanding of a linear connections of economic and social assets along a rail corridor—Sydney Metro.

The methodology utilised to assess the presence of goods, services and jobs involved a series of inter-related steps including the identification of rail corridors for analysis, the identification of the goods, services and jobs (economic and social assets) included for assessment, the expansion of social assets into city wide and foundational community assets, the identification of railway station precincts (TODs) to be assessed in the context of a 30-min city, and defining walkability and a 30-min city. The results from the assessments of the rail corridors for both Melbourne and Sydney and the TODs for Sydney were compared using a multi-criteria analysis and the details of each step are outlined in the methodology section.

2 Literature Review

This literature review explores several areas with the aim of understanding that a wider network (corridor) view of accessibility to goods, services and jobs from TODs can provide a greater understanding of where to target and facilitate residential development opportunities and the corresponding required infrastructure. The areas covered include transit-oriented development, linear cities, accessibility and walkability.

Transit-Oriented Development

The concept of transit-oriented development (TOD) emerged in the late 1980s through the work of Peter Calthorpe, with the aim of removing the dependency on private vehicles for travel (Calthorpe 1993). If a networked approach to the planning of TODs is to be considered it is important to recognise the accessibility attributes, both benefits and planning considerations of transit-oriented development.

An understanding of TODs in terms of benefits and planning considerations is well documented, reflecting its extensive use as a tool to manage urban development across the globe. Suggested benefits of TODs that influence accessibility include: a substantial shift towards increased transit use (Crowley et al. 2009); increased residential densities

(Ratner and Goetz 2013); enhanced sense of community, and greater pedestrian and child safety (Brown and Werner 2011); access to attractive, large public open space and public open space with facilities that encourage active use by multiple users are associated with higher levels of walking (Giles-Corti et al. 2005); and promote walking among older adults, by providing good access to shops and services, well-maintained walking facilities, aesthetically appealing places, streets with little traffic and places for social interaction within a neighbourhood (Van Cauwenberg et al. 2012).

Identified considerations for the planning of TODs that influence accessibility and the delivery of economic and social assets include: the development implications of planning controls, the roles of local government and urban design qualities (Mathur and Gatdula 2023; Tsivanidis 2023); the importance of balancing amenity richness and accessibility with density and land use diversity (Zandiatashbar and Laurito 2023); and the potential for TODs to facilitate commercial activities depending on the scale of the centre to its catchment and the complex interactions between a train station and neighbourhood characteristics (Searle et al. 2014; Schuetz 2015).

In Australia, TODs can be viewed in the context of centres' policy which became part of metropolitan plans for Melbourne and Sydney in the 1980s. In Melbourne, 'district centre' policy was introduced to reduce the need for travel and facilitate the development of centres as a focal point for administrative, commercial, retail, cultural and entertainment activities (Melbourne Metropolitan Board of Works 1981). Thirteen centres were originally designated as district centres. Overtime, the policy evolved, whereby the focus is currently on nine 'Metropolitan activity centres', six as per the original policy, with the same intended outcomes but with a greater emphasis on public transport hubs and recognising they are also a focus for high density residential development (Department of Environment, Land, Water and Planning 2017).

In Sydney, centres' policy emerged in 1988 with the designation of two regional centres (CBDs) and 17 sub-regional centres, all on rail lines, two on the Central Coast. The aims were to have retail and office jobs closer to home and provide access to a full range of retail, recreation and community services (Department of Environment and Planning 1988). As with Melbourne, the Sydney centre's policy has evolved. The current focus is on 34 strategic centres, which are linked to the objective of providing residents with access to goods and services within a 30-min radius.

Linear Cities

A networked approach to TODs on rail corridors raises the concept of linear cities. Originally put forward by Arturo Soria y Mata in Madrid in 1882 (Collins 1959), as a tool for shaping urban development, linear cities have been part of urban planning approaches, particularly urban growth corridors, for the last seventy years, with transport efficiency central to its application. Examples include the post war 'Finger Plan' for Copenhagen (Sørensen and Jacob 2019) the 1971 plan for metropolitan Melbourne (Melbourne Metropolitan Board of Works 1971) and the 1968 plan for metropolitan Sydney (State Planning Authority of New South Wales 1968).

Lessons learnt from utilising linear cities as an urban management tool include: linear cities can be used as a tool for connecting more complex structures (Antyufeev and Antyufeeva 2019); linear cities can emerge due to natural and topographic circumstances, as well as reflecting transport options (Antyufeev and Antyufeeva 2019); linear cities should not be seen as an all-encompassing urban form, but more as an element of a wider complex urban form (Doxiadis 1967; Batty 2022); travel times of a transit corridor/linear city need to be faster than that for a private vehicle (Curtis 2008); the mix of activities available within a TOD influence the extent to which other car based trips are required to satisfy the full range of needs of a household including employment (Wolański 2023); attraction of commercial activities in linear corridors can contribute to economic growth (Knowles 2012); and lack of control for managing activity outside a transit corridor can impact on its success (Ryckewaert 2010; Batty 2022).

More recently, Newman et al. (2021) put forward the concept of a 'Transit Activated Corridor'. Their paper outlines several pathways for delivering a transit activated corridor on an existing main road. The approach is based around five principles, the first three principles relate to enabling the transit system, including an emphasis on utilising mid-tier transit options such as trackless trams. The remaining two are urban planning principles and cover urban development and design and walkability and sustainable station precinct design.

The need for transport corridors to include a wide range of activities was recognised in the 1971 metropolitan plan for Melbourne (MMBW 1971), it stated—*The growth corridors must not simply be new dormitory areas where residential settlement will occur, …* (p. 68). The opportunity to deliver on this policy intention has been eroded over time as the green wedges between the urban corridors are progressively turned over to urban development resulting in the tensions as described by Doxiadis (1967) and Batty (2022).

Accessibility and Walkability

If transit-oriented development is to deliver on the accessibility outcomes it proposes then walkability must be central to their planning, as suggested by Calthorpe (1993). Considerations when planning for accessibility and walkability in TODs and linear cities include: there is currently only limited data on the propensity for people to walk to a station and then to a destination (Park et al. 2015); models that test accessibility to public transport (Saghapour et al. 2016); transit networks with headways of less than ten minutes can be used to support interchange transfers (Currie and Loder 2010); changes in accessibility, such as new motorways or railway lines have significant impacts on urban structure overtime, compared to land use change (SGS Economics and Planning 2012); research undertaken in Melbourne suggests that people will walk 400 m to a bus service and 800 m to a train service, in addition, travel distances are influenced by neighbourhoods with gross residential densities exceeding 20 dwellings per hectare, a well-connected street network, and access to nine or more local living destinations (Boulange et al. 2017), low-cost 'feeder' bus systems that complement mass rapid transit by providing 'last-mile' service for passengers' easy access to a system's terminals yield high returns (Tsivanidis 2023), and improved micro-walkability (quality of the pedestrian links) may increases the size of the catchment of a railway station, particularly along major pedestrian corridors (Park et al. 2015).

There are a number of common policy positions across all seven plans for Australia's capital cities including the need to support walkability, the importance of urban renewal being linked to public transport, optimised infrastructure and infrastructure that meets the needs of the community (Dalheim 2023).

3 Methodology

Rail lines (corridors) were assessed based on the published networks for each city including the published timetables, refer Appendices 1 and 2 (Public Transport Victoria 2024; Transport for NSW 2024). In total, nine rail lines were assessed in Melbourne including the Melbourne Metro line, currently under construction and due for opening in 2025. The new line will connect the Sunbury and Pakenham/Cranbourne lines with a new tunnel and stations under the Melbourne CBD. For two rail lines, that divide into two, towards the end of their line, only one arm was assessed in each case—the Lilydale and Cranbourne lines. Also, two small lines—Alamein and Williamstown, were not assessed due to high levels of overlap with other lines. For Sydney, eleven lines were assessed including the new Sydney Metro line. This newly opened line from Tallawong in Sydney's north-west to Sydenham, south of the CBD, is currently being extended to Bankstown, the new section opening in 2025. Several rail lines have spur lines, in these cases only the principal corridor was assessed and the rail line the Blue Mountains were excluded as this area is not classified as being part of urban Sydney.

The economic and social assets (attributes) utilised for the analysis are based on a review of common policy areas of the metropolitan plans for the major state capital cities—*A Metropolis of Three Cities* (Sydney), *Perth and Peel @3.5 million*, *Plan Melbourne*, *ShapingSEQ* (Brisbane) and *The 30-Year Plan for Greater Adelaide* (Dalheim 2023). The review reveals policy consistency in three areas: health and education precincts; major centres including both the CBD and major suburban centres; and access to community facilities and services. Google Maps was utilised for the identification of all the economic, social and foundational community assets. No other source was utilised to identify these assets.

The assessment of major centres is complemented by the presence, or not, of top retail centres in each state, when they are located on a rail line. This in part compensates for the difference in the designation of the major centres as outlined in the relevant government policy for each city, such as the Melbourne policy focuses only on centres in the outer suburbs. Two categories of education facilities are used to reflect the difference between university campuses and single facility tertiary facilities which are grouped with vocational focused Technical and Further Education (TAFE) facilities.

In addition, data was obtained on population and employment levels within 800 m of every station on the eleven Sydney rail corridors. This data was provided by the consultancy firm Mecone, utilising their data tool—Mosaic. Data on the size of retail centres as well as the level of retail at the 20 largest centres in Melbourne was provided by the consultancy firm Urbis from their internal retail database.

Definitions

Having worked in planning for 40 years across four state jurisdictions in Australia, I have concluded that there is no consistent definition of what constitutes community infrastructure. The Queensland Government's Department of Infrastructure, Local Government and Planning (2017) *Community facilities PDA guideline no. 11*, provides a full list of the range of community infrastructure they consider. For this paper, a list of foundational community assets has been developed based on both the Queensland guideline and my experience in leading the planning of new communities on the fringe of metropolitan cities. In creating the list, community assets were considered from two perspectives. Firstly, what infrastructure should be present when a new community is established and secondly, what should be present for a more mature community and collectively these have been named foundational community assets. The list includes supermarkets (including Asian grocery stores), parks, playgrounds, libraries, primary and secondary schools, sports facilities, community centre, medical centres, youth centres, community halls, recreation facilities and pools (not private learn to swim facilities).

The assessment of accessibility to a station within 30 min was limited to centres in Sydney where there is a formal transport oriented development program and the eight state government led initiatives were assessed. The eight TODs together with their the rail line are Bankstown (Sydney Metro), Bays West (Metro West—committed but not built), Bella Vista (Sydney Metro), Crows Nest (Sydney Metro), Homebush (T2 and T3), Hornsby (T1 and T9), Kellyville (Sydney Metro) and Macquarie Park (Sydney Metro). The assessment identified economic and social assets accessible within 30 min from a home within a TOD. Walkability to an asset was measured as being within 800 m walking distance of a railway station, measured by the shortest possible walking route: not a simple straight line distance utilising Google Maps.

A 30-min city reflects the explicit intent of Objective 14 in the current metropolitan plan for Sydney (Greater Sydney Commission 2018) and *Future Transport Strategy* (Transport for NSW 2022). A 30-min journey to work is also recognised by the Marchetti constant (Marchetti 1994). The calculation of a 30-min journey has considered the reality that a train trip requires walking at both ends. To estimate the average walking time for all activities in an 800-m radius, the average walking distance for all residents is 564 m and utilising an average walking speed of 1 min per 100 m, gives 5.64 min walking time, which for simplicity is rounded down to 5 min. Therefore, the rail travel time utilised is 20 min, with no allowance for wait time for a train.

Within the CBD, of both cities, the 800-m station catchments overlap. As this was a consistent error for all rail lines, though it did favour some lines more than others, no attempt was made to address the issue for the analysis for this study. Instead, the analysis considered outcomes with and without the scores for the CBD stations.

A multi-criteria assessment was utilised to assess the relative differences between the presence of economic, social and foundational community assets. Scoring was based on absolute counts for each asset type for each rail corridor or TOD. The tabulated results have been presented in graphical form for this paper.

These results were utilised for a second assessment undertaken for each station on the Sydney rail lines to determine the level of access to all the economic, social and foundational community assets within the context of a 30-min city, translating to a 20-min rail trip, with 5 min walking at each end, including where possible a change to a connecting train.

4 Results

The assessment of accessibility to economic and social assets along each rail corridor in Melbourne and Sydney reveals differences in the number of assets accessible within 800 m of a rail station. One example of the results of this analysis is provided in Appendix 3, noting that 20 such assessments were undertaken. The differences become more pronounced when the assets within the CBD of each city are excluded, refer Figs. 1 and 2.

Within the Melbourne train network, three rail corridors stand out as having high levels of economic and social assets—Melbourne Metro, Werribee to Frankston and Lilydale lines. The Sydney rail corridors were found to have several notable results, the T9 Hornsby to Gordon, T5 Leppington to Blacktown and T1 Western, with a few additional lines having results not far behind. Interestingly, the Sydney rail corridors generally have a higher number of economic and social assets than the Melbourne corridors.

The assessment of the state led TODs for Sydney also reveals differences in the number of economic and social assets accessible within 30 min, refer Fig. 3. One example of the results of this analysis is provided in Appendix 4, noting that eight such assessments were undertaken. Crows Nest is the standout with Macquarie Park, Bella Vista, Homebush and Bay West also performing well, noting that the Bays West assessment is based on a theoretical timetable assessment, as

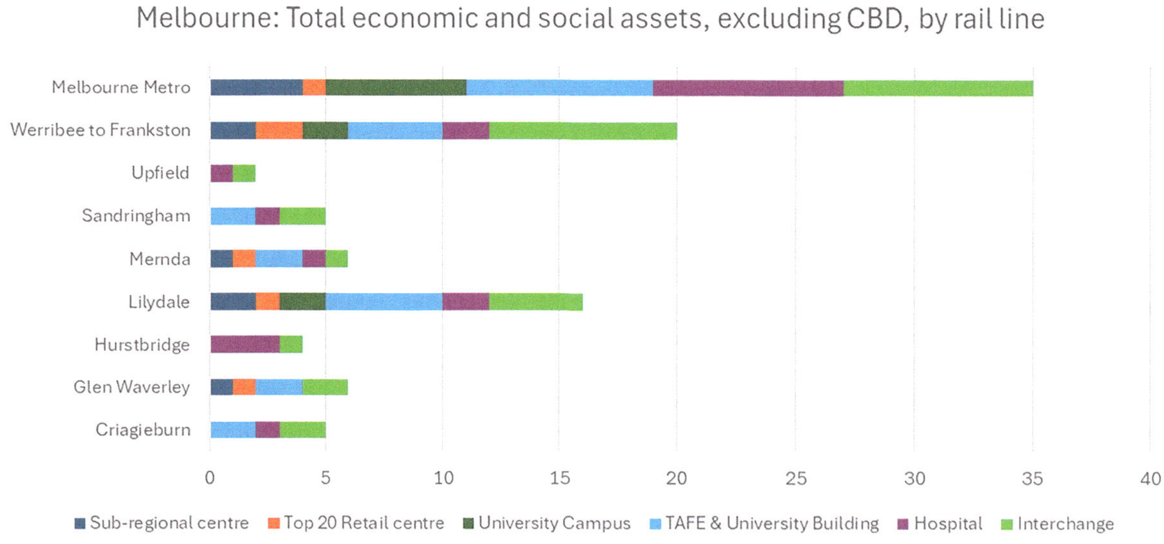

Fig. 1 Melbourne rail corridors relative levels of economic and social assets, excluding the CBD assets. *Source* By author

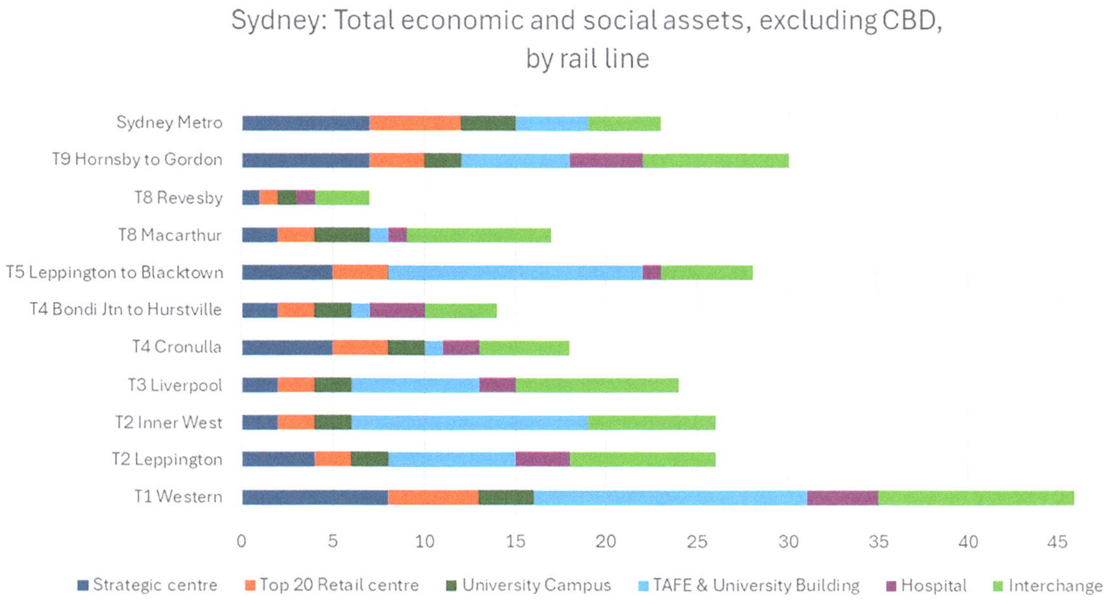

Fig. 2 Sydney rail corridors relative levels of economic and social assets, excluding the CBD assets. *Source* By author

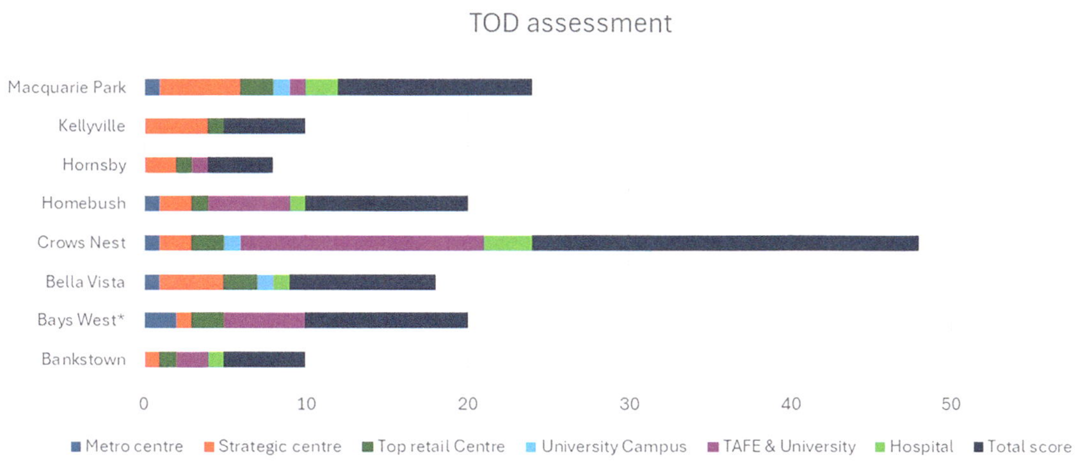

Fig. 3 Sydney state led TODs relative levels of accessibility to economic and social assets. * The Bays West TOD is on a committed yet undeveloped rail line, expected completion in around ten years. *Source* By author

the rail corridor for this TOD although committed, is yet to be developed and is about ten years away from completion.

For the Sydney rail corridors, an assessment was made of the total population and employment within 800 m of each station, with the assessment also considering the influence of removing the values for the CBD locations where densities of employment and population are significantly higher than elsewhere. One example of the results of this analysis is provided in Appendix 5, noting that eleven such assessments were undertaken. The results, refer Fig. 4, show that the five rail lines with the highest population totals, all over 300,000, remain the same when the CBD values are removed and where all population totals are over 200,000. The employment results differ, with five rail lines having employment levels of over 1.0 million, and when the CBD values are removed three different rail lines stand out with employment levels above a quarter of a million.

The assessment of the foundational community assets revealed that their spatial distribution was generally common for individual asset types across rail corridors, refer Figs. 5 and 6. One example of the detailed analysis is provided in Appendix 6, noting that 20 such assessments were undertaken. With the exception of secondary schools and recreation centres, all other asset types were present at stations between 50% and up to nearly 100% for each rail corridor. With reference to the stations that do not have a specific asset, this is outlined in Appendix 7 and discussed in more detail below. Due to the low score for recreation centres across most

Rail line	Population	Employment	Population ex CBD	Employment ex CBD
T1 Western	262,716	880,251	193,866	309,789
T2 Leppington	317,758	1,198,126	205,540	144,735
T2 Inner West	361,551	1,283,984	249,333	230,593
T3 Liverpool	342,683	1,196,077	269,743	239,786
T4 Cronulla	267,149	706,767	197,952	136,878
T4 Bondi Jtn to Hurstville	248,192	702,936	178,995	133,047
T5 Leppington to Blacktown	152,659	161,154	152,659	161,154
T8 Macarthur	231,937	1,196,901	119,719	143,510
T8 Revesby	216,311	1,131,833	104,093	78,442
T9 Hornsby to Gordon	339,509	846,458	270,659	275,996
Sydney Metro	306,087	829,310	236,397	298,045

Fig. 4 Sydney: total population and employment by rail line. *Source* By author

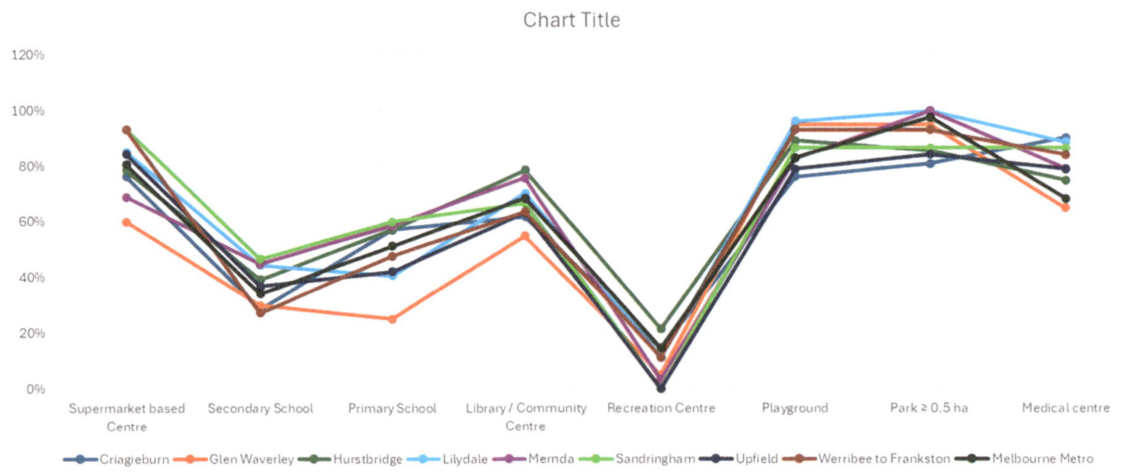

Fig. 5 Melbourne: percent of foundational asset present by rail line. *Source* By author

rail corridors in both cities, they have not been included as an asset type in the second assessment to determine which stations have 30-min accessibility to all economic, social and foundational assets.

Collectively, these results confirm differences in the level of accessibility to economic, social and foundational community assets between and within rail corridors. However, this research is directed to determining whether stations along a rail corridor, or a section of it, can provide access to the full range of economic and social assets that a community would anticipate to be available, within a 30-min travel time. The final assessment examines this for all 186 train stations across the Sydney network, refer Appendix 7.

The assessment reveals that there are 38 train stations that have access to all the economic, social and foundational community assets analysed as well as to the Sydney CBD in terms of 30-min accessibility. There are also 13 train stations that similarly have 30-min accessibility to all economic, social and foundational community assets, excluding the Sydney CBD. When all 51 stations are plotted on the rail network map for Sydney it reveals, for six of the rail lines, linear sections where each station has 30-min accessibility to all the analysed assets, many also have 30-min accessibility to the Sydney CBD, noting some overlap in the Sydney CBD. In summary, the rail line sections that include the 51 stations are as follows (refer Fig. 7).

Fig. 6 Sydney: percent of foundational asset present by rail line. *Source* By author

- Sydney metro line: 18 stations from Bella Vista to Marrickville, the first six without access to the Sydney CBD in terms a 30-min city.
- T1 North Shore line: 8 stations from St Leonards to Central.
- T2 Leppington line: 2 stations, Newtown and Macdonaldtown.
- T4 Cronulla line: 11 stations from Hurstville to Martin Place (excluding Allawah), the first four without access to the Sydney CBD in terms a 30-min city.
- T8 Macarthur line: 6 stations from Sydenham to Town Hall.
- T8 Macarthur line: 5 stations from Mascot to St James.
- T8 Macarthur line: 3 stations, Ingleburn, Minto and Glenfield without access to the Sydney CBD in terms a 30-min city.
- T9 Northern line: 2 stations, Strathfield and Burwood.

In contrast, 19 stations do not have access to either a strategic centre or a TAFE/university facility, nor a hospital or university campus. The remaining 116 stations do not have access to either a university campus, hospital or close proximity to the Sydney CBD or a combination of these, with the most common finding (81 train stations) being the absence of a university campus and not having 30-min access to the Sydney CBD.

As highlighted in Figs. 5 and 6, not all train stations have local foundational community assets and the analysis outlined in Appendix 7 shows that 39 of the 186 train stations do not have a local supermarket (seven with access to all other economic and social assets) and six do not have a local park (two with access to all other economic and social assets). One train station, Mascot, does not have access to a primary school within a 30-min accessibility context and 4 train stations do not have access to a secondary school within a 30-min accessibility context.

5 Discussion and Conclusion

This study has sought to understand whether a wider (corridor) network view of transit-oriented development (TOD) that considers access to goods, services and jobs, as a policy outcome, can inform future decisions about where to locate new TODs and the infrastructure required to support them (e.g. schools). In the first instance, the research reveals that location does matter and, in line with the anticipated outcomes, differences were found in the level of goods, services and jobs—categorised in terms of economic, social and foundational community assets—between rail corridors and on any single rail corridor. When accessibility to these assets was considered in the context of what is accessible in 30 min by public transport and/or walking, differences were found between eight precincts currently designated as TODs in Sydney, where accessibility is defined as being within 800 m from a train station.

Consequently, further analysis revealed bands of train stations, all with access to the full range of economic, social and foundational community assets, on six rail corridors, 51 train stations in total or just over a quarter of all stations. This finding suggests that the planning for TODs, for some parts of the rail network, can be considered as linear cities where public transport and walking are the principal modes of transportation, and when accessibility is added as a public policy outcome. Specifically, noting that a 30-min city is an explicit objective of the current metropolitan plan for Greater Sydney (Greater Sydney Commission 2018) and having public transport networks as a core design element and improving walkability are explicit policy objectives in

Fig. 7 Train stations with full access to all economic, social and foundational community assets—potential TOD locations. *Legend* Yellow dot 30-min accessibility to all assets and the CBD, purple dot only 30-min accessibility to all assets. *Source* Transport for NSW (2024) with annotations by author

most of the current capital city plans in Australia (Dalheim 2023).

The analysis also identified which economic, social and foundational assets (infrastructure investments) are absent (gaps) in terms of a public transport/walkable city. As thresholds for most of these infrastructure investments are understood, the scale of development required to support new invest (induced demand) and thus influence 30-min city outcomes can be calculated and assessed against estimated development potential. This suggests that future infrastructure investments in economic, social and foundational community assets should be a parallel policy, whereby locational principles to locate new or augmented facilities are within walking distance of stations within 20 min by rail, from the source of the demand—a 30-min city outcome.

It is important to also recognise that this research did not consider the capacity for existing infrastructure to accommodate additional demand (development) along a corridor. Also, noting that numerous assets are located just outside the 800-m walkable catchment area and are most likely serving the local community of that area, albeit by a private vehicle. This emphasises the issue raised by Doxiadis (1967) and Batty (2022) that the value of linear cities is eroded when the location for goods, services and jobs are outside the corridor (800-m station catchments).

As identified by Park et al. (2015) there is limited data on the propensity for people to walk to a destination. In *The Next American Metropolis* (1993, p. 56), Calthorpe suggests a 'comfortable walking distance' as 2000 ft (609.6 m). The findings from the literature review suggests that walking trip lengths vary from 400 m (Crowley et al. 2009) to 800 m (Boulange et al. 2017) and while there has been research on issues that influence walking behaviour (Brown and Werner 2011; Giles-Corti et al. 2005; Park et al. 2015; Ryckewaert 2010; Van Cauwenberg et al. 2012) they do not provide data on the average distance travelled. Considering the public policy desire for walkable cities and the walkability expectations for TODs, understanding walkable trips continues to be a major gap in the knowledge base required to inform urban planning. In this study, 800 m was utilised based on the work of Boulange et al. (2017) which focuses on distance travelled to a train station.

It is also prudent to note that rail corridors do not serve the entire community of large cities such as Melbourne and Sydney in terms of walkable catchments, these areas are supported by extensive bus networks. The research of Tsivanidis (2022) on the Bogotá's TransMilenio rapid bus transit network reveals that low-cost feeder buses are effective in providing 'last mile' connections to rapid transit networks. This suggests another area for investigation to further understand 30-min city catchments. In addition, the learnings from the research on Bogotá's rapid bus transit network could be further explored to establish the potential role of rapid bus networks in Melbourne and Sydney being utilised as second tier linear transit cities.

The identification of the economic, social and foundational community assets was based principally on the approach taken in government endorsed strategies. This could be seen as a limitation and further research could be undertaken to determine if there is evidence on the significance of these assets to the community, whether a hierarchy exists that would impact on the method of scoring and their influence on supporting mode shifts to public transport/walking, noting that morning journey to work trips coincide with morning school trips.

The community's accessibility to economic, social and foundational community infrastructure by public transport or walking has been the focus of this study. In doing so, it is recognised that there are wider and well recognised sustainability outcomes from reducing private vehicle trips, which are not covered in this study. Advice to governments on the benefits of addressing accessibility as a public policy outcome can also cover these wider benefits such as, reducing traffic congestion pressures, thus improving the efficiency of the whole transport network and consequently assisting in mitigating against climate change, as well as health benefits resulting from the need to walk when utilising public transport.

This study has focused on accessibility in the context of Australia's two largest cities Melbourne and Sydney both with populations over five million people, with the aim of influencing the urban renewal policies of each city which are currently heavily focused on supporting TOD/centres. Given the nature of these two cities having similar urban structure to many cities in the United States and Canada it may well be the case that the findings here can be translated to North American cities. The principal issue is likely to be the distribution pattern of economic, social and foundational community assets in those cities. A next step could therefore be to study several North American cities and the urban policies that are guiding their growth to affirm the wider applicability of the findings of this study, especially considering that the concept of TODs as a planning framework was developed in the United States.

In conclusion, this study sought to understand whether accessibility to economic, social and foundational community assets within walkable catchments of stations on a city's rail networks can improve the decision-making regarding where to locate new transit-oriented development (TOD) in the context of a 30-min city. The findings suggest that location matters and that parts of rail corridors can be reimagined as 'linear transit cities' where the community has full access to the goods services and jobs, they would anticipate being available in any city. The findings also suggest that a parallel policy relating to the infrastructure investments required to support the growth arising from transit-oriented development (such as schools and parks) should be aligned to support 30-min city outcomes and improve equitable accessibility. Collectively, these conclusions suggest the opportunity to develop a practical framework for managing urban renewal to be developed to inform the preparation of metropolitan strategies.

Funding The undertaking of the analysis for this paper was supported, pro bono, by the provision of data from two sources: **Urbis**, provided data on the scale of retail in centres and at stations (for Melbourne) and **Mecone**, provided data on population and employment within 800 m of rail stations across all stations within Greater Sydney.

Informed Consent Statement Not applicable.

Conflicts of Interest The author declares no conflict of interest.

Appendix 1: Victorian Rail Network Including Melbourne Network—White Area

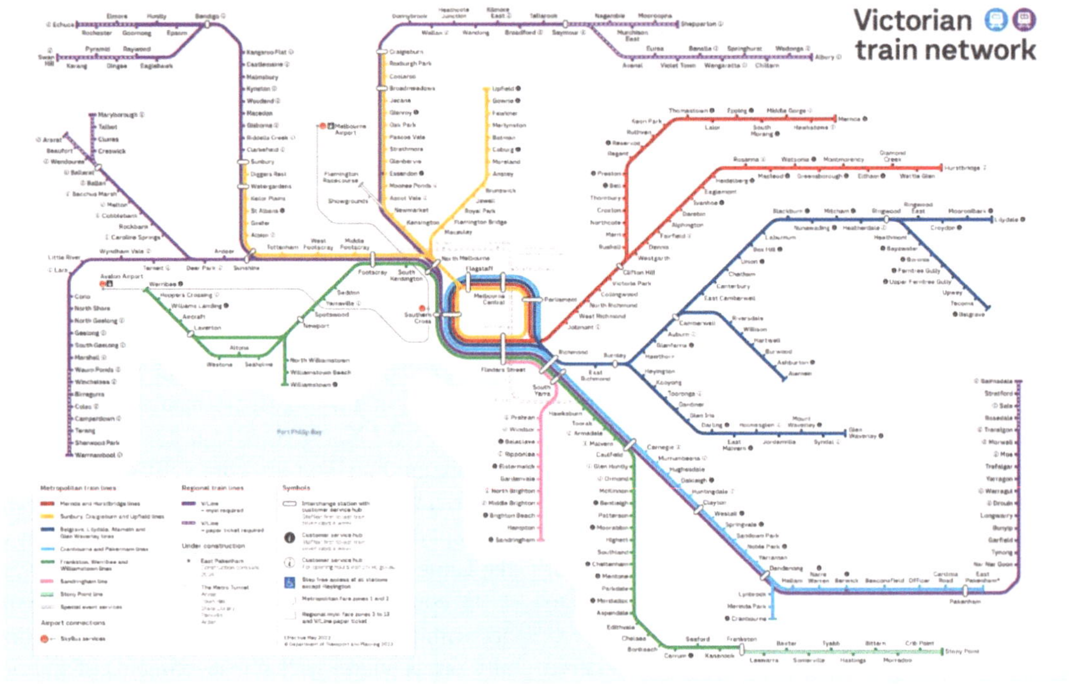

Source Public Transport Victoria (2024)

Appendix 2: Sydney Rail Network

Source Transport for NSW (2024)

Appendix 3: Rail Corridor Analysis: Economic and Social Assets Sydney Example: T9 Hornsby to Gordon

Station	Connect to Sydney CBD	Strategic centre	Top 20 retail centre	University campus	TAFE and university building	Hospital	Interchange
Hornsby		1	1		1		1
Normanhurst							
Thornleigh							
Pennant Hills							
Beecroft							
Cheltenham							
Epping		1			1		1
Eastwood					1		
Denistone						1	
West Ryde							
Meadowbank					1		
Rhodes		1					
Concord West							
North Strathfield							
Strathfield					1		1
Burwood		1					
Redfern				1			1
Central				1			1
Town Hall			1				1
Wynyard							1
Milsons Point							
North Sydney		1					
Waverton							
Wollstonecraft							
St Leonards		1			1	2	
Artarmon							
Chatswood		1	1				1
Roseville							
Lindfield							
Killara						1	
Gordon							

Source By author

Appendix 4: Transit-Oriented Development Analysis: Economic and Social Assets Sydney Example: Crows Nest

TOD	Assessment of accessibility based on 20 min by rail—rankings in brackets						
	Metro centre	Strategic centre	Top retail centre	University campus	TAFE and university	Hospital	Total score
Bankstown	0	1	1	0	2	1	5
Bays West[a]	2	1	2	0	5	0	10
Bella Vista	1	4	2	1	0	1	9
Crows Nest	1	2	2	1	15	3	24
Homebush	1	2	1	0	5	1	10
Hornsby	0	2	1	0	1	0	4
Kellyville	0	4	1	0	0	0	5
Macquarie Park	1	5	2	1	1	2	12

Source By author
[a] Estimate of timetable

Appendix 5: Population and Employment Within 800 m of Station Sydney Example: T9 Hornsby to Gordon

Station	Population	Employment
Hornsby	14,254	8631
Normanhurst	5057	2849
Thornleigh	5449	4195
Pennant Hills	5396	2412
Beecroft	4679	1278
Cheltenham	3066	1648
Epping	14,664	3170
Eastwood	8359	3717
Denistone	3986	3445
West Ryde	9583	4454
Meadowbank	12,975	3822
Rhodes	11,866	9583
Concord West	6292	12,199
North Strathfield	6625	6652
Strathfield	15,965	14,057
Burwood	16,752	11,020
Redfern	19,113	28,913
Central	39,278	97,100
Town Hall	23,525	200,052
Wynyard	6047	273,310
Milsons Point	8421	20,007
North Sydney	12,350	55,632
Waverton	7802	4297
Wollstonecraft	9061	6901
St Leonards	13,202	33,707
Artarmon	9870	2160
Chatswood	17,420	21,446
Roseville	7114	2397
Lindfield	7315	2214
Killara	6828	1110
Gordon	7195	4080
Total	339,509	846,458

Source By author

Appendix 6: Rail Corridor Analysis: Foundational Community Assets Melbourne Example Werribee to Frankston Line

Station	Supermarket based centre	Secondary school	Primary school	Library/ community centre	Recreation centre	Playground	Park ≥ 0.5 ha	Medical centre
Werribee	1			3		1	2	2
Hoppers crossing	1	2				1	1	1
Williams landing	1			1			3	2
Aircraft	1			2		1	1	1
Laverton	1			1		5	3	2
Westona	1	1						
Altona	1	1		1		2	3	
Seaholme	1		1			1	1	
Newport	1		1	2		1	2	3
Spotswood	1		2			4	1	2
Yarraville	1		2	3		2	2	2
Seddon	1	1	2			2	2	2
Footscray	1	2	2	4		4	2	6
South Kensington			1	1		3	2	2
North Melbourne	1					1		1
Southern Cross	1			3		2	3	2
Flinders Street	1			2		2	3	
Richmond	1			1		1	3	2
South Yarra	1	1	1	2		5	7	3
Hawksburn	1		1			5	3	
Toorak	1	1	2	1		2	3	2
Armadale	1		1	1	1	1	3	2
Malvern	1					1	1	4
Caulfield	1			1		1	2	2
Glen Huntly	1	1	2	1		1	3	3
Ormond	1	1	1			1	1	1
McKinnon	1					2	3	2
Bentleigh	1			1		1	2	4
Patterson	1	1				3	2	
Moorabbin	1			1		2		6
Highett	1			2		2	1	8
Southland	1					4	2	4
Cheltenham	1		2	2		2	1	2
Mentone	1	2		1	2	2	2	4
Parkdale	1		1	1			2	1
Mordialloc	1		2	1		3	3	2
Aspendale	1					1	1	
Edithvale	1		1		1	1	2	1
Chelsea	1		1	1	0	2	2	5
Bonbeach						3	1	1
Carrum	1			1	1	3	1	3
Seaford	1		1	2		2	5	6
Kananook			1			3	3	1
Frankston	1	2	3	1		3	3	8

Source By author

Appendix 7: Train station analysis: gaps in accessibility

Station	No university campus	No hospital	No Town Hall station connection in 20 min	No park ≥ 0.5 ha	No supermarket based centre
Allawah					
Arncliffe					
Artarmon					
Ashfield					
Asquith					
Auburn					
Banksia					
Bankstown					
Barangaroo					
Bardwell Park					
Beecroft					
Bella Vista					
Belmore					
Berala					
Berowra					
Beverly Hills					
Bexley North					
Birrong					
Blacktown					
Bondi Junction					
Burwood					
Cabramatta					
Campbelltown					
Campsie					
Canley Vale					
Canterbury					
Caringbah					
Carlton					
Carramar					
Castle Hill					
Casula					
Central					
Chatswood					
Cheltenham					
Cherrybrook					
Chester Hill					
Circular Quay					
Clarendon					
Clyde					
Como					
Concord West					

Station	No university campus	No hospital	No Town Hall station connection in 20 min	No park ≥ 0.5 ha	No supermarket based centre
Cronulla					
Crows Nest					
Croydon					
Denistone					
Doonside					
Dulwich Hill					
East Hills					
East Richmond					
Eastwood					
Edgecliff					
Edmondson Park					
Emu Plains					
Engadine					
Epping					
Erskineville					
Fairfield					
Flemington					
Gadigal					
Glenfield					
Gordon					
Granville					
Green Square					
Guildford					
Gymea					
Harris Park					
Heathcote					
Hills Showground					
Holsworthy					
Homebush					
Hornsby					
Hurlstone Park					
Hurstville					
Ingleburn					
Jannali					
Kellyville					
Killara					
Kings Cross					
Kingsgrove					

Station	No university campus	No hospital	No Town Hall station connection in 20 min	No park ≥ 0.5 ha	No supermarket based centre
Kingswood					
Kirrawee					
Kogarah					
Lakemba					
Leightonfield					
Leppingon					
Leumeah					
Lewisham					
Lidcombe					
Linden					
Lindfield					
Liverpool					
Loftus					
Macarthur					
Macdonaldtown					
Macquarie Fields					
Macquarie Park					
Macquarie University					
Marayong					
Marrickville					
Martin Place					
Mascot					
Meadowbank					
Merrylands					
Milsons Point					
Minto					
Miranda					
Mortdale					
Mount Colah					
Mount Druitt					
Mount Kuring-gai					
Mulgrave					
Museum					
Narwee					
Newtown					
Normanhurst					
North Ryde					

Station	No university campus	No hospital	No Town Hall station connection in 20 min	No park ≥ 0.5 ha	No supermarket based centre
North Strathfield					
North Sydney					
Norwest					
Noth Strathfield					
Oatley					
Olympic Park					
Padstow					
Panania					
Parramatta					
Parramatta North					
Pendle Hill					
Pennant Hills					
Penrith					
Penshurst					
Petersham					
Punchbowl					
Pymble					
Pyrmont					
Quakers Hill					
Redfern					
Regents Park					
Revesby					
Rhodes					
Richmond					
Riverstone					
Riverwood					
Rockdale*					
Rooty Hill					
Roseville					
Rouse Hill					
Schofields					
Sefton					
Seven Hills					
St James					
St Leonards					
St Marys					
St Peters					
Stanmore					
Strathfield					

Station	No university campus	No hospital	No Town Hall station connection in 20 min	No park ≥ 0.5 ha	No supermarket based centre
Summer Hill					
Sutherland					
Sydenham					
Tallawong					
Tempe					
Thornleigh					
Toongabbie					
Town Hall					
Turramurra					
Turrella					
Victoria Cross					
Villawood					
Vineyard					
Wahroonga					
Waitara					
Warrawee					
Warwick Farm					
Waterloo					
Waverton					
Wentworthville					
Werrington					
West Ryde					
Westmead					
Wiley Park					
Windsor					
Wolli Creek					
Wollstonecraft					
Woolooware					
Wynyard					
Yagoona					
Yennora					

Legend

	No access to a strategic centre or a TAFE/university facility
	Full access to all assets and within 20 min of the Sydney CBD
	Full access to all assets but not within 20 min of the Sydney CBD
	No access to a hospital and within 20 min of the Sydney CBD
	No access to a hospital but not within 20 min of the Sydney CBD
	No access to a university campus, hospital and not within 20 min of the Sydney CBD

Source By author

References

Antyufeev, A.V., Antyufeeva, O.A.: Linear cities: controversies, challenges and prospects. IOP Conf. Ser. Mater. Sci. Eng. **687** (2019). https://doi.org/10.1088/1757-899X/687/5/055025

Australian Bureau of Statistics: Population: Census. ABS (2021). https://www.abs.gov.au/statistics/people/population/population-census/latest-release

Australian Bureau of Statistics: 50 Years of Capital City Population Change. ABS (2022). https://www.abs.gov.au/articles/50-years-capital-city-population-change

Australian Bureau of Statistics: Capital City Growth the Highest on Record. ABS (2024). https://www.abs.gov.au/media-centre/media-releases/capital-city-growth-highest-record

Australian Government: National Housing Accord (2022). https://ministers.treasury.gov.au/sites/ministers.treasury.gov.au/files/2022-10/national-housing-accord-2022.pdf

Batty, M.: The linear city: illustrating the logic of spatial equilibrium. Comput. Urban Sci. **2**(8) (2022). https://doi.org/10.1007/s43762-022-00036-z

Boulange, C., Gunn, L., Giles-Corti, B., Mavoa, S., Pettit, C., Badland, H.: Examining associations between urban design attributes and transport mode choice for walking, cycling, public transport and private motor vehicle trips. J. Transp. Health **6**, 155–166 (2017). https://doi.org/10.1016/j.jth.2017.07.007

Brown, B.B., Werner, C.M.: The residents' benefits and concerns before and after a new rail stop: do residents get what they expect? Environ. Behav. **43**(6), 789–806 (2011). https://doi.org/10.1177/0013916510392030

Calthorpe, P.: The Next American Metropolis. Princeton Architectural Press (1993)

Centre for International Economics: The benefits and costs of alternative growth paths for Sydney Economic, social and environmental impacts (2010). https://www.thecie.com.au/publications-archive/costs-and-benefits-of-alternative-growth-paths-for-sydney

Collins, G.R.: Linear planning throughout the world. J. Soc. Archit. Hist. **18**(3), 74–93 (1959). https://doi.org/10.1088/1757-899X/687/5/055025

Crowley, D.F., Shalaby, A.S., Zarei, H.: Access walking distance, transit use, and transit-oriented development in North York City Center, Toronto, Canada. J. Transp. Res. Board **2110**, 96–105 (2009). https://doi.org/10.3141/2110-12

Currie, G., Loader, C.: Bus network planning for transfers and the network effect in Melbourne, Australia. J. Transp. Res. Board. **2145**(1), 8–17 (2010). https://doi.org/10.3141/2145-02

Curtis, C.: Evolution of the transit-oriented development model for low-density cities: a case study of Perth's new railway corridor. Plann. Pract. Res. **23**(3), 285–302 (2008). https://doi.org/10.1080/02697450802423559

Dalheim, H.: Planning Better Cities, a Practical Guide. Palgrave Macmillian (2023)

Department of Environment, Land, Water and Planning: Plan Melbourne 2017–2050. The State of Victoria (2017). https://www.planmelbourne.vic.gov.au/

Department of Infrastructure, Local Government and Planning: ShapingSEQ, South East Queensland Regional Plan 2017. The State of Queensland (2017). https://planning.dsdmip.qld.gov.au/planning/better-planning/state-planning/regional-plans/seqrp

Department of Planning, Housing and Infrastructure: 2022 Sydney Housing Supply Forecast—Methodology: Assumptions and Inputs (2022). https://www.planning.nsw.gov.au/sites/default/files/2023-06/2022-sydney-housing-supply-forecast-methodology-assumptions-and-inputs.pdf

Department of Planning, Housing and Infrastructure: Transport Oriented Development (TOD) Program—Assessment Criteria (2024). https://www.planning.nsw.gov.au/sites/default/files/2024-03/transport-oriented-development-program-assessment-criteria.pdf

Department of Planning, Lands and Heritage: Perth and Peel @ 3.5 Million. Government of Western Australia (2018). https://www.dplh.wa.gov.au/perth-and-peel-@-3-5-million

Department of Planning, Transport and Infrastructure: The 30-Year Plan for Greater Adelaide, 2017 Update. Government of South Australia (2017). https://livingadelaide.sa.gov.au/the_plan

Doxiadis, C.A.: On linear cities. Town Plann. Rev. **38**(1), 35–42 (1967). http://www.jstor.org/stable/40079669

Giles-Corti, B., Broomhall, M.H., Knuiman, M., Catherine Collins, C., Kate Douglas, K., Ng, K., Lange, A., Donovan, R.J.: Increasing walking: how important is distance to, attractiveness, and size of public open space? Am. J. Prev. Med. **28**(2S2), 169–176 (2005). https://doi.org/10.1016/j.amepre.2004.10.018

Greater Sydney Commission: A Metropolis of Three Cities. State of New South Wales (2018). https://www.greater.sydney/metropolis-of-three-cities

Knowles, R.D.: Transit oriented development in Copenhagen, Denmark: from the finger plan to Orestad. J. Transp. Geogr. **22**(5), 251–261 (2012)

Marchetti, C.: Anthropological invariants in travel behavior. Technol. Forecast. Soc. Change **47**(1), 75–88 (1994). https://doi.org/10.1016/0040-1625(94)90041-8

Mathur, S., Gatdula, A.: Review of planning, land use, and zoning barriers to the construction of transit-oriented developments in the United States. Case Stud. Transp. Policy (2023). https://doi.org/10.1016/j.cstp.2023.100988

Melbourne Metropolitan Board of Works: Planning Policies for the Melbourne Metropolitan Region. Melbourne Metropolitan Board of Works (1971). https://www.planning.vic.gov.au/guides-and-resources/strategies-and-initiatives/melbournes-strategic-planning-history/planning-policies-for-metropolitan-melbourne-1971

Melbourne Metropolitan Board of Works: Metropolitan Strategy Implementation. Melbourne Metropolitan Board of Works (1981). https://www.planning.vic.gov.au/guides-and-resources/strategies-and-initiatives/melbournes-strategic-planning-history/metropolitan-strategy-implementation-1981

Metro: Network Maps, 27 Apr 2024. https://www.metrotrains.com.au/corporate/

Newman, P.G., Kenworthy, J.R: Cities and automobile dependence: a sourcebook. Ashgate Publishing Limited (1989)

Newman, P., Davies-Slate, S., Conley, D., Hargroves, K., Mouritz, M.: From TOD to TAC: why and how transport and urban policy needs to shift to regenerating main road corridors with new transit systems. Urban Sci. **5**, 52 (2021). https://doi.org/10.3390/urbansci5030052

New South Wales. Department of Environment and Planning: Sydney into Its Third Century: Metropolitan Strategy for the Sydney Region. Department of Environment and Planning, Sydney (1988). https://archive.org/details/sydney-into-its-third-century-metropolitan-strategy-for-the-sydney-region./page/n3/mode/2up

New South Wales Government: Population Projections, Explore the Data, Metropolitan or Regional (2024a). https://www.planning.nsw.gov.au/research-and-demography/population-projections/explore-the-data

New South Wales Government: Trends in Housing Supply (2024b). https://www.planning.nsw.gov.au/policy-and-legislation/housing/housing-supply-insights/quarterly-insights-monitor-q1/trends-in-housing-supply

NSW Department of Planning, Housing and Infrastructure: Guidance to Transport Oriented Development (2024). https://www.planning.nsw.gov.au/sites/default/files/2024-05/guidance-to-transport-oriented-development.pdf

NSW Government: Transport Oriented Development Program (2023). https://www.planning.nsw.gov.au/sites/default/files/2023-12/transport-oriented-development-program.pdf

NSW Government: Housing Targets (2024). https://www.planning.nsw.gov.au/policy-and-legislation/housing/housing-targets

Park, S., Deakin, E., Jang, K.: Can good walkability expand the size of transit-oriented developments? Transp. Res. Record **2519**, 157–164 (2015). https://doi.org/10.3141/2519-17

Public Transport Victoria: Network Maps, 27 Apr 2024. https://www.ptv.vic.gov.au/more/maps/

Ratner, K.A., Goetz, A.R.: The reshaping of land use and urban form in Denver through transit-oriented development. Cities **30**, 31–46 (2013). https://doi.org/10.1016/j.cities.2012.08.007

Ryckewaert, M.: The ten-year plan for the port of Antwerp (1956–1965): a linear city along the river. Plan. Perspect. **25**(3), 303–322 (2010). https://doi.org/10.1080/02665433.2010.481179

Saghapour, T., Moridpour, S., Thompson, R.G.: Public transport accessibility in metropolitan areas: a new approach incorporatingpopulation density. J. Transp. Geogr. **54**, 273–285 (2016)

Schuetz, J.: Do rail transit stations encourage neighbourhood retail activity? Urban Stud. **52**(14), 2699–2723 (2015). https://doi.org/10.1177/0042098014549128

Searle, G., Darchen, S., Huston, S.: Positive and negative factors for transit oriented development: case studies from Brisbane, Melbourne and Sydney. Urban Policy Res. **32**(4), 437–457 (2014). https://doi.org/10.1080/08111146.2014.931280

SGS Economics and Planning: Long run economic and land use impacts of major infrastructure projects, Unpublished (2012)

Sørensen, E., Jacob, T.: The Copenhagen metropolitan ‘finger plan’: a robust urban planning success based on collaborative governance. In: Hart, P., Compton, M. (eds.) Great Policy Successes, pp. 218–243 (2019). https://doi.org/10.1093/oso/9780198843719.003.0012

State Government of Victoria: Statewide Draft Housing Targets (2024). https://engage.vic.gov.au/project/shape-our-victoria/page/housing-targets-2051

State of Victoria (Department of Premier and Cabinet): Victoria’s Housing Statement, The Decade Ahead | 2024–2034. Victoria Government (2023). https://www.vic.gov.au/housing-statement

State of Victoria Department of Transport and Planning: Victoria in Future 2023, Population and Household, Projections to 2051 (2023). https://www.planning.vic.gov.au/__data/assets/pdf_file/0022/703453/DTP0552-Victori-in-Future-2023-report.PDF

State Planning Authority for New South Wales: Sydney Region Outline Plan 1970–2000 AD. State Planning Authority for New South Wales (1968). https://archives.cityofsydney.nsw.gov.au/nodes/view/1754543

Suburban Rail Loop Authority: Suburban Rail Loop East Project Rational and Benefits (2021). https://srleastees.vic.gov.au/#/rationale/Benefits. Cited 29 Apr 2024

Sun, Z., Scraftonb, D., Allan, A., Somenahalli, S.: A comparison of transit-oriented development in Sendai, Japan and Adelaide. Aust. Plan. Pract. Res. **36**(3), 337–351 (2021). https://doi.org/10.1080/02697459.2020.1859214

The Treasury: Delivering the National Housing Accord. Australian Government (2023). https://treasury.gov.au/housing-policy/accord

Transport for NSW: Directions for On-Street Transit. NSW Government Bus Industry Taskforce Launches White Paper for Better Bus Services | NSW Government (2013)

Transport for NSW: Routes and Timetables, 27 Apr 2024. https://transportnsw.info/routes/train

Transport for NSW: Future transport strategy, state of new south wales (Transport for NSW) (2022)

Tsivanidis, N.: Evaluating the impact of urban transit infrastructure: evidence from Bogota’s Transmilenio. Unpublished manuscript (2022)

Tsivanidis, N.: Evaluating the impact of urban transit infrastructure: evidence from Bogotá’s TransMilenio. University of California, Berkeley (2023). https://static1.squarespace.com/static/55bb98e2e4b0ba843f39599e/t/62882b12d140ce5b51ff6e24/1653091092837/TsivanidisTransMilenio_5.2022.pdf

Van Cauwenberg, J., Van Holle, V., Simons, D., Deridder, R., Clarys, P., Goubert, L., Nasar, J., Salmon, J., De Bourdeaudhuij, I., Deforche, B.: Environmental factors influencing older adults’ walking for transportation: a study using walk-along interviews. Int. J. Behav. Nutr. Phys. Act. **9**(1), 85 (2012). https://doi.org/10.1186/1479-5868-9-85

Van Dyck, D., Deforche, B., Cardon, G., De Bourdeaudhuij, I.: Health & Place 15 Neighbourhood Walkability and its Particular Importance for Adults with a Preference for Passive Transport, pp. 496–504 (2009). https://doi.org/10.1016/j.healthplace.2008.08.010

Van Dyck, D., Deforche, B., Cardon, G, De Bourdeaud’hui, I.: Neighbourhood walkability and its particulalr importance for adults with a preference for passive transport. Health & Place **15**, 496–504 (2009)

Wolański, M.: The potential role of railway stations and public transport nodes in the development of “15-minute cities”. Infrastructures **8**, 141 (2023). https://doi.org/10.3390/infrastructures8100141

World Population Review (2024). https://worldpopulationreview.com/world-cities

Zandiatashbar, A., Laurito, A.: An empirical analysis of the link between built environment and safety in Chicago’s transit station areas. J. Am. Plan. Assoc. **89**(2), 225–239 (2023). https://doi.org/10.1080/01944363.2022.2069590

Analyzing the Logistics Management Performance of Road and Rail Freight Transportation with a Hybrid Approach by Integrating DEA and MCDM Methods

Kadhim H. Abbood and Ferenc Mészáros

Abstract

Analyzing the logistics management performance of road and rail freight transport is essential to improving overall visibility along the supply chain. In this paper, data envelopment analysis (DEA) and the multicriteria decision-making (MCDM) method, using the analytic hierarchy process (AHP), were integrated to assess the logistics management performance of road and rail freight transportation from a sustainability perspective. First, a global review of the integration of DEA and MCDM was conducted to evaluate the sustainability of logistics performance in road and rail freight transportation, covering the period from 2010 to 2023 using the Scopus database. A survey was then conducted, and the AHP technique was used to determine the weight of different criteria, representing all three sustainability parameters, namely environmental, social, and economic, that impact the logistics management performance of the freight transportation process. In addition, the efficiency score for both transport modes was calculated using the DEA tool. Finally, a real-world case of road and rail freight transport in Hungarian industries was considered to assess logistics management performance efficiency. The findings show that companies prioritize economic factors over social and environmental impacts. However, transportation policies support reducing social and environmental impacts for a more sustainable logistics sector. Therefore, decision-makers must work more efficiently to raise awareness of the sustainable logistics performance of the freight transport sector.

Keywords

Logistics · Freight transport · Scopus · Sustainability · Road and rail · Integrating DEA and MCDM methods

1 Introduction

Freight transport logistics involves the planning, organization, management, control, and execution of freight transportation activities across the supply chain. Production and distribution networks depend on efficient, high-quality logistics systems to coordinate the movement of raw materials and finished commodities both within the EU and beyond. It is essentially a commercial activity and an industrial duty (Marinov et al. 2010). Efficient and ecologically friendly logistics systems help communities compete in economic development. As metropolitan areas expand and economic activity increases, logistics facilities seek space and locations just outside city centers and along urban borders. Decisions about land use and facility locations within global logistics networks will shape the landscape, resource utilization, and economic and social geography of urban regions for decades to come (Sgura Viana and Delgado 2019). Urban freight logistics operations impact all three aspects of sustainability: environmental, economic, and social. These operations are influenced by various stakeholders, both governmental and commercial, who may have different goals regarding the three elements of sustainability (Janjevic et al. 2019). Since the mid-1980s, much debate has centered on the relationship between transportation and urban design. Most studies have focused on passenger transportation, with relatively little analysis of the connections

K. H. Abbood (✉) · F. Mészáros
Department of Transport Technology and Economics, Faculty of Transportation Engineering and Vehicle Engineering, Budapest University of Technology and Economics, Budapest, Hungary
e-mail: kadhimhadi.abbood@edu.bme.hu

F. Mészáros
e-mail: meszaros.ferenc@kjk.bme.hu

P. Mohammad et al. (eds.), *Geographic Approaches to Climate Change and Mitigation: Urban and Rural Perspectives (Volume 2)*, Advances in Science, Technology & Innovation, https://doi.org/10.1007/978-3-031-92115-5_17

between urban characteristics and freight transport activities (Allen et al. 2012). Continental Europe boasts a dense network of roads and railways. Inland rivers play an essential role in Europe. Only a few nations in North Western Europe, such as Belgium, Germany, France, and the Netherlands, have extensive infrastructure. The road is the primary form of land-bound freight transport, followed by rail and inland waterways, with market shares of 75, 18, and 7% (Wiegmans et al. 2018).

In this research, we adopted a three-step method. The first step involved conducting a literature review on integrating the data envelopment analysis (DEA) tool with the multicriteria decision-making (MCDM) method to analyze the logistics performance of both rail and road freight transport. Next, a survey was carried out to gather the opinions of experts working in Hungarian logistics companies regarding the most significant criteria affecting the logistics activities of rail and road freight transport modes. The final step was a case study assessing the performance of Hungarian freight transportation logistics.

The study aims to determine the importance of the selected criterion that impacts the performance of the rail and road freight modes along the logistics activities by considering the experts' evaluation. Additionally, the efficiency of this evaluation is examined to show how it is accurate. Therefore, the current study contributes valuable information to the decision-makers to identify the lack of performance in logistics activities, making it apparent where the weaknesses are. In addition, the research tries to fill out the gaps by analyzing the Hungarian logistics sector from 2000 to 2021. The outcome of this study contributes to the advantages of determining the importance of the selected criterion impact on the logistics activities that are done by road and rail.

The paper is constructed as follows: Sect. 2 shows the results of the comprehensive literature. Section 3 explains the conducted method. Sections 4 and 5 demonstrate the study's outcomes and conclusion, respectively.

2 Literature Review

In this section, the Scopus database was utilized to identify the related research that integrated the data development analysis (DEA) tool with the multi-criteria decision-making (MCDM) method to study freight transport activities. The first step involved searching the Scopus database using specific keywords, such as "Integrating DEA and MCDM" AND "freight transport sector" OR "logistics management" OR "supply chain" to search in the Scopus website from 2010 to 2024. The outcomes of this search determined more than five thousand studies in the first phase. The second step was to read the titles of the articles and choose the one that used the DEA and MCDM methods and focused on the freight transportation industry. The results show that approximately 387 articles met the search criteria at this stage. Then, each research chosen from the second phase was read carefully to determine if it utilized the integration of the DEA and MCDM and focused on logistics management activities of the freight transport sector. Therefore, only nineteen articles were selected because they integrated the DEA and MCDM methods to study one or more modes of freight transportation sector or analyze the transportation sector in general. The final step involved classifying the selected articles based on four criteria. The first classification identified the method employed, defining whether the study integrated the DEA and MCDM tools. The second classification explained the country or region addressed in the study, describing the geographical scope of the inquiry. The third classification determined the study's scope, explaining the focused area of the study. Finally, the type of mode included in the study was the fourth classification, indicating which freight transportation sector modes were identified.

The outcomes of the literature analysis show that all the selected articles utilized the integration of the DEA and MCDM methods. These articles also investigated the freight transportation industry in Asia, Africa, and Europe. Fifty percent of these articles analyzed the freight transport sector in Europe, while 35% and 15% focused on Asia and Africa, respectively as shown in Fig. 2. Additionally, the selected studies evaluated various areas within this sector. For instance, Rosić et al. (2017) introduced an MCDM model named PROMETHEE-RS to develop an optimal composite index. Their work also highlighted the evaluation of two prominent methodologies: DEA and TOPSIS. Similarly, Zeydan et al. (2011) utilized a novel technique to improve supplier selection and assessment quality through a two parts. The first stage involves qualitative performance evaluation using fuzzy AHP to determine criterion weights, followed by fuzzy TOPSIS to rank suppliers. The qualitative data were then converted into quantitative variables for use in DEA to create a quality management system audit. Baltazar et al. (2014) integrated the MCDM and DEA to assess the efficiency of the air freight transport mode from 2006 to 2011 under several constraints. Vrtagić et al. (2021) presented an integrated fuzzy model combining MCDM and DEA to evaluate road freight transport sector safety. Keskin and Köksal (2019) utilized the integration of the MCDM and DEA to estimate the efficiency of the air transport mode in Turkey. Finally, Lai et al. (2015) integrated the MCDM and DEA to assess the efficiency of the air transport mode at 24 major airports in Europe. The final classification was based on the type of freight transport mode analyzed as shown in Fig. 1. The results indicated that most

Fig. 1 Literature analysis of the chosen articles by type of transport mode [*source* own editing based on Scopus database (Scopus 2024)]

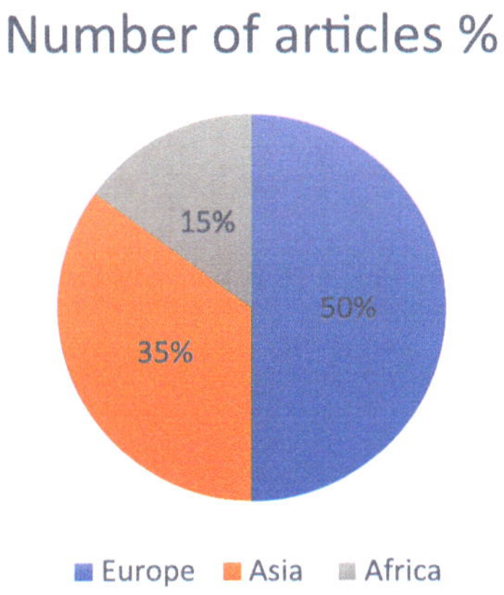

Fig. 2 Literature analysis of the chosen articles by region [*source* own editing based on Scopus database (Scopus 2024)]

articles (45%) focused on the road transport sector. Additionally, 35% of the studies examined air transport, while 10% focused on rail and maritime transport.

After conducting a thorough literature analysis, identifying research gaps is critical for advancing academic understanding and guiding future investigations. The first research gap is that the logistics management activities of the rail freight transport sector have not been efficiently examined. Also, the experts' opinions on which criteria have the most significant impact on the logistics activities for rail and road freight transport modes have not been considered. There is also a lack of analysis on logistics activities in certain countries, such as Hungary. Finally, the existing studies tend to focus on only one or two aspects of sustainability without addressing all three sustainability aspects, which are social, environmental, and economic, in their work. The current study aims to fill these gaps by incorporating expert opinions on criteria representing all three sustainability aspects and ranking their effects based on expert classifications. Furthermore, the study includes Hungarian logistics companies for a more comprehensive analysis.

3 Method

The proposed method addresses some of the research gaps identified in the literature review, as outlined below:

1. Determining criteria that represent the sustainability dimensions, which are environmental, economic, and social, and ranking their importance and effects on the logistics activities of rail and road freight transportation. Thirteen criteria were considered in the current study: speed, goods transported, density, safety, operation costs, CO_2 emissions, energy consumption, technology and equipment, fuel, maintenance and repairs, economy growth, traffic congestion, and infrastructure investment. After that, a survey was conducted to assess the significance of each criterion based on the experts' evaluation. Then, the multi-criteria decision-making using the analytical hierarchy process (MCDM-AHP) method was employed to calculate the weight and rank the selected criteria. The final step involved assessing the efficiency score of the rail and road freight transport modes using a data envelopment analysis tool (DEA). The results of the MCDM-AHP were used as inputs and outputs for the DEA method. After classifying the selected criteria as inputs and outputs, the DEA calculated the efficiency score for the logistics performance of the two modes of freight transport (road and rail) in Hungary.
2. A real case study was conducted to estimate the efficiency score by applying the DEA tool. Also, real data representing these criteria were collected after classifying the selected criteria as inputs and outputs. The data were collected from the Hungarian Central Statistical Office website. Additionally, the efficiency score was calculated using the DEA and compared with the efficiency score estimated in the first phase for the same criteria.

3.1 Integrated MCDM-AHP and DEA

3.1.1 MCDM-AHP

Multi-criteria choice-making (MCDM) involves evaluating options against numerous criteria before making a choice. MCDA is an essential tool in operational research and management science. Several approaches have been created and applied to decision-making in various sectors. Several approaches might help in this process, such as AHP, ANP,

ELECTRE (I, II, III, and IV), PROMETHEE, TOPSIS, Fuzzy, and more (de Oliveira et al. 2023).

The current study utilized an AHP analysis to identify the average weight of the selected criteria, which are classified as inputs and outputs, and to calculate the efficiency score of the logistics activities of the rail and road freight transport industry. The AHP combines many measurements to provide a single set of scores for evaluating decision alternatives (DAs) (Lai et al. 2015). The AHP consists of five key phases (Saaty 2008).

- Specify the decision object.
- Create a multilevel framework by categorizing decision-making variables. The decision's purpose is at the top level, followed by criteria and sub-criteria for comparing DAs and, finally, alternative DAs.
- Construct pairwise decision matrices by comparing the relative relevance of each upper-level criterion to its lower-level equivalent. Let Z be a $n \times n$ pairwise comparison matrix, represented as:

$$Z = \left[a_{fg}\right] = \begin{bmatrix} 1 & a_{12} & \dots a_{1n} \\ 1/a_{12} & 1 & \dots a_{2n} \\ \vdots & \vdots & \ddots \vdots \\ 1/a_{1n} & 1/a_{2n} & \dots 1 \end{bmatrix} \tag{1}$$

where $a_{ff} = 1$ and $a_{fg} = 1/a_{fg}, f, g = 1, 2, \ldots, n$, and $a_{fg} > 0$ (Saaty 2008).

Let $C_1, C_2, \ldots, C_n$ demonstrate the set of criteria and a_{fg} represent a quantified decision on a pair of criteria C_f and C_g. The original 1–9 scale developed by Saaty (1980) is used to compare the significance of one criterion to another. Equal importance, weak importance, essential importance, demonstrated importance, and extreme significance are represented by the values 1, 3, 5, 7, and 9, respectively; the values 2, 4, 6, and 8 are used to find a middle ground between the above values.

The crucial degrees of the decision maker's requirements are determined using the normalization of the geometric mean approach, which is then utilized (Saaty 2008) to calculate the importance degree (Escobar et al. 2004).

If we assume that W_f represents the weight or degree of relevance for the fth criterion, then (Saaty 2008):

$$W_f = \sqrt[n]{\prod_{g-1}^{n} a_{fg}} / \sum_{f=1}^{n} \times \sqrt[n]{\prod_{g-1}^{n} a_{fg} f}, \quad g = 1, 2, \ldots, n \tag{2}$$

where the number of criteria is represented by n (Saaty 2008).

Furthermore, the most significant eigenvalue $\lambda_{\max}$ can be computed using Eqs. (3) and (4) (Saaty 2008):

$$Z \times W_f = \begin{bmatrix} 1 & a_{12} & \dots a_{1n} \\ 1/a_{12} & 1 & \dots a_{2n} \\ \vdots & \vdots & \ddots \vdots \\ 1/a_{1n} & 1/a_{2n} & \dots 1 \end{bmatrix} \times \begin{bmatrix} W_1 \\ W_2 \\ \vdots \\ W_n \end{bmatrix} = \begin{bmatrix} W'_1 \\ W'_2 \\ \vdots \\ W'_n \end{bmatrix} \tag{3}$$

$$\lambda_{max} = \left(\frac{1}{\text{n}}\right) \times \left(\frac{W_1}{W_1} + \frac{W_2}{W_2} + \ldots + \frac{W_n}{W_n}\right) \tag{4}$$

The matrix consistency is evaluated mathematically in the last stage, and any necessary modifications are made to achieve an acceptable consistency. The associated eigenvector is determined by taking the maximum eigenvalue ($\lambda_{\max}$) of the pairwise comparison matrix and calculating it in accordance with the consistency test's premise.

$\lambda_{\max}$ can be used to calculate the Consistency Index (CI), where n is the matrix size (Saaty 1980):

$$\text{CI} = \left(\lambda_{\max} - \frac{n}{n} - 1\right) \tag{5}$$

The consistency ratio (CR) was thus defined by Saaty (1980) as follows:

$$\text{CR} = \text{CI}/\text{CR} \tag{6}$$

where the Saaty scale average value for random matrices is denoted by the random index (RI) (Forman 1990). If CR < 0.1, a matrix is deemed consistent. These five processes are then used to calculate the relative weights of each criterion (Lai et al. 2015).

3.1.2 DEA

A standard DEA model utilizes the productivity ratio of outputs to inputs to quantify the production efficiency of a decision-making unit (Egilmez et al. 2013). DEA uses complex mathematical programming to determine the effectiveness of the rail and road freight transport modes and the weighting of each impact category. The following is the notation for the general input-oriented DEA multiplier model introduced by Charnes et al. (1978).

Objective function (Egilmez et al. 2013):

$$\max_q = \sum_{r=1}^{k} \mu_r y_{ro} \tag{7}$$

Subject to (Egilmez et al. 2013):

$$\sum_{i=1}^{m} v_i x_{io} = 1 \tag{8}$$

$$\sum_{r=1}^{k} \mu_r y_{rj} - \sum_{i=1}^{m} v_i x_{ij} \leq 0, \quad j = 1, \ldots, n \tag{9}$$

$$\mu_r, v_i \geq 0 \tag{10}$$

In this equation, μ_r is the output multiplier, v_i is the input multiplier, o is the decision-making unit (DMU) being evaluated, k is the number of outputs, m is the number of inputs, j is the number of DMUs, y_{rj} is the amount of output produced by DMU j, and x_{ij} is the amount of input used by DMU j. The objective function q calculates the weighted outputs for the evaluated DMU (Egilmez et al. 2013).

3.2 Data Collecting

The data utilized in this research come from two sources. The first set of data was used to calculate the weight of each criterion by applying the AHP tool. The second data set consisted of real data on the logistics activities of the rail and road Hungarian freight transportation industry. These data were utilized as inputs and outputs in the DEA to evaluate the efficiency of the selected criteria. The two data sets are described below:

1. The first data set was obtained by conducting a survey among Hungarian experts working in logistics companies. Twenty experts from fifteen companies, involved in Hungary's rail and road freight transport industry, participated in the survey. The names of the responding companies that are considered in the research are The Cargo-Partner Company, The Hungarian Transportation Research Institute Ltd, HUNGRAIL Magyar Vasúti Egyesület, Harman Professional Solutions, LIEGL & DACHSER Transport and Logistics Ltd, Aspirant European Logistics LIEGL & DACHSER Transport and Logistics Ltd., MAHART CONTAINER CENTER Kft., WTL Logisztikai Kft., John von Neumann University—GAMF Műszaki és Informatikai Kar, RTP CARGO HUNGARIA Kft, and Rail Cargo Hungaria Zrt.,
2. The second data set was extracted from the Hungarian Central Statistical Office website (HCSO 2021). This official website provides data on various sectors, including education, energy, transportation, and more.

4 Results and Discussion

4.1 The Conducted Survey Results

In this section, thirteen subcriteria were considered to evaluate the performance of logistics activities in the rail and road freight transport sector in Hungary, based on the assessments of the Hungarian experts working in logistics companies. The experts were asked to evaluate each criterion for road and rail modes separately, using a scale from 1 to 9. On this scale, 1 represents equal importance, 3 represents somewhat more importance, 5 represents much more importance, 7 represents very much more importance, and 9 represents absolute importance, with 2, 4, and 8 as intermediate values. Tables 1 and 2 present the experts' evaluation of each criterion for both modes.

4.1.1 Calculating the Average Criteria Weighted for Each Criterion by Utilizing Analytical Hierarchy Process (AHP)

After determining the significance of each criterion by the experts, we standardized the data. We estimated how many experts selected the specific AHP scale, 1, 2, 3, ..., 9, for each criterion, as shown in Tables 3 and 4 for both modes. Then, we calculated the average weighted criteria based on the sample's responses for each criterion. Additionally, we used the AHP Online System (AHP-OS) website to estimate the consistency ratio (CR) for each mode separately. Firstly, we input thirteen of the criteria considered in the survey. Then, we manually entered the definition of the criterion manually. Next, we generated a pairwise decision matrix to compare relevant criteria. Each criterion was compared with all others; for example, the first criterion, speed, was compared with goods transported, density, safety, operation costs, CO2 emissions, energy consumption, technology and equipment, fuel, maintenance and repairs, economy growth, traffic congestion, and infrastructure investment based on its average criteria weight. Once this step was completed, the program calculated the priority for each criterion, decision matrix, and consistency ratio (CR). Tables 3 and 4, Figs. 3, 4, 5, and 6 demonstrate the AHP-OS results.

The results indicated that the consistency ratios (CR) for road and rail modes are 7.2% and 3.4%, respectively, both of which are less than the 10% threshold. Therefore, the experts' evaluations are considered reliable for determining the importance of each criterion, and the results are accepted according to Saaty (2008). Given that the expert evaluations are valid,

Table 1 The expert evaluates the criteria for road mode

Criteria	The experts who participated in the survey																			
	1	2	3	4	5	6	7	8	9	10	11	12	13	14	15	16	17	18	19	20
Speed	7	8	6	7	9	8	9	8	7	5	7	6	8	9	8	7	8	6	8	8
Goods transported	4	8	5	3	6	6	3	6	7	4	4	4	4	6	5	6	5	6	5	6
Density	9	8	9	8	9	8	9	8	8	6	7	8	9	8	9	9	9	9	8	6
Safety	9	7	9	9	5	6	2	7	9	3	5	7	7	7	5	1	9	6	7	8
Operation cost	4	8	9	7	7	9	7	9	8	7	7	7	9	9	4	6	7	9	9	9
CO_2 emissions	9	9	9	9	9	6	7	7	8	9	9	9	9	7	8	9	9	9	1	5
Energy consumption	5	7	9	7	9	7	7	7	8	7	8	8	9	8	8	9	8	8	1	5
Technology and equipment	9	8	9	7	5	7	6	6	8	8	8	9	9	6	8	8	8	8	6	7
Fuel	9	8	9	7	9	6	7	6	7	8	9	9	9	8	8	9	8	9	8	5
Maintenance and repairs	8	8	5	6	8	8	6	5	9	7	7	8	9	7	5	7	8	8	6	5
Economy growth	9	8	9	5	5	8	7	6	9	8	8	9	9	9	8	9	7	9	6	6
Traffic congestion	9	8	8	9	7	6	8	6	8	8	9	9	9	6	8	8	7	9	1	4
Infrastructure investment	9	7	9	6	7	7	6	5	8	8	9	8	9	6	4	9	7	8	8	6

Table 2 The expert evaluates the criteria for rail mode

Criteria	The experts who participated in the survey																			
	1	2	3	4	5	6	7	8	9	10	11	12	13	14	15	16	17	18	19	20
Speed	8	8	9	9	6	6	4	8	8	9	8	8	4	9	6	5	4	5	6	7
Goods transported	4	8	8	7	9	7	8	9	9	9	8	8	9	7	8	8	8	8	9	6
Density	9	8	7	9	9	5	5	7	8	9	6	7	7	9	5	6	6	5	6	5
Safety	4	5	4	6	3	5	1	6	9	8	3	6	2	6	9	2	3	8	8	8
Operation cost	4	8	6	5	5	9	4	9	8	9	5	6	5	9	8	8	6	8	1	4
CO_2 emissions	7	7	5	9	6	6	3	7	5	9	5	5	2	6	6	5	2	4	8	5
Energy consumption	8	7	8	8	6	6	3	6	8	9	7	7	8	8	5	7	9	6	8	3
Technology and equipment	7	8	7	3	2	7	6	6	8	8	6	6	7	6	7	3	6	6	8	5
Fuel	2	8	6	4	6	6	3	5	7	8	8	8	9	8	6	7	3	7	8	5
Maintenance and repairs	9	8	6	7	5	8	6	5	9	9	7	6	9	8	6	6	5	9	6	9
Economy growth	7	8	7	6	5	7	5	5	9	8	8	8	9	8	7	8	8	8	8	7
Traffic congestion	9	8	5	7	5	7	2	4	9	7	9	9	9	6	4	1	4	6	6	6
Infrastructure investment	9	8	8	8	5	7	9	8	9	8	9	8	8	7	8	9	5	9	9	4

Table 3 The number of the AHP scales for each criterion of road mode

The selected criterion	AHP scale								
	Equal importance	Somewhat more important	Much more important	Very much more important	Absolutely more important	Intermediate values			
	1	3	5	7	9	2	4	6	8
Speed	0	0	1	5	3	0	0	3	8
Goods transported	0	2	4	1	0	0	5	7	1
Density	0	0	0	1	9	0	0	2	8
Safety	1	1	3	6	5	1	0	2	1
Operation costs	0	0	0	7	8	0	2	1	2
CO_2 emissions	1	0	1	3	12	0	0	1	2
Energy consumption	1	0	2	6	4	0	0	0	7
Technology and equipment	0	0	1	3	4	0	0	4	8
Fuel	0	0	1	3	8	0	0	2	6
Maintenance and repairs	0	0	4	4	2	0	0	3	7
Economy growth	0	0	2	2	8	0	0	3	5
Traffic congestion	1	0	0	2	6	0	1	3	7
Infrastructure investment	0	0	1	4	5	0	1	4	5

Table 4 The number of the AHP scales for each criterion of rail mode

The selected criterion	AHP scale								
	Equal importance	Somewhat more important	Much more important	Very much more important	Absolutely more important	Intermediate values			
	1	3	5	7	9	2	4	6	8
Speed	0	0	2	1	4	0	3	4	6
Goods transported	0	0	0	3	6	0	1	1	9
Density	0	0	5	4	5	0	0	4	2
Safety	1	3	2	0	2	2	2	4	4
Operation costs	1	0	4	0	4	0	3	3	5
CO_2 emissions	0	1	5	3	2	2	1	4	1
Energy consumption	0	2	1	4	2	0	0	4	7
Technology and equipment	0	2	1	5	0	1	0	7	4
Fuel	0	2	2	3	1	1	1	4	6
Maintenance and repairs	0	0	3	2	6	0	0	6	3
Economy growth	0	0	3	5	2	0	0	1	9
Traffic congestion	1	0	2	3	5	1	3	4	1
Infrastructure investment	0	0	2	2	7	0	1	0	8

the ranking of the criteria should be taken into account. Table 5 shows that "goods transported" is the most significant criterion for both road and rail modes, with average weights of 16.1% and 13.9%, respectively. This suggests that decision-makers should prioritize the volume of goods transported to enhance the sector's efficiency. Additionally, "speed" ranks second in road mode, with an average weight of 12.9%. These findings support the need for further investment in building new roads and adopting multimodal transport solutions to increase speed, particularly in rural areas. In contrast, "speed" ranks third in rail mode, with an average weight of 9.4%, indicating that road transport is generally faster than rail for goods transport. Furthermore, "safety" ranks second in rail mode, with an average weight of 10.8%, and fourth in road mode, with an average weight of 7.5%. The remaining criteria are ranked as shown in Table 5.

Table 5 also provides another criteria classification into environmental, economic, and social dimensions. The average weights for the selected criteria were grouped based on their relevance to each dimension. For example, "speed," "traffic congestion," "safety," and "density" represent the social aspects that directly impact customers. "Goods transported," "operating cost," "technology and equipment," "maintenance and repairs," "economic growth," and "infrastructure investment" represent the economic aspects. Finally, "fuel," "CO2 emissions," and "energy consumption" represent the environmental aspects. The results indicate the following values for

	1	2	3	4	5	6	7	8	9	10	11	12	13
1	1	4.00	1.00	2.00	2.00	1.00	2.00	1.00	1.00	2.00	1.00	2.00	2.00
2	0.25	1	4.00	2.00	2.00	4.00	2.00	4.00	4.00	2.00	4.00	2.00	2.00
3	1.00	0.25	1	2.00	2.00	1.00	2.00	1.00	1.00	2.00	1.00	2.00	2.00
4	0.50	0.50	0.50	1	1.00	2.00	1.00	2.00	2.00	1.00	2.00	1.00	1.00
5	0.50	0.50	0.50	1.00	1	2.00	1.00	2.00	2.00	1.00	2.00	1.00	1.00
6	1.00	0.25	1.00	0.50	0.50	1	2.00	1.00	1.00	2.00	1.00	2.00	2.00
7	0.50	0.50	0.50	1.00	1.00	0.50	1	2.00	2.00	1.00	2.00	1.00	1.00
8	1.00	0.25	1.00	0.50	0.50	1.00	0.50	1	1.00	2.00	1.00	2.00	2.00
9	1.00	0.25	1.00	0.50	0.50	1.00	0.50	1.00	1	2.00	1.00	2.00	2.00
10	0.50	0.50	0.50	1.00	1.00	0.50	1.00	0.50	0.50	1	2.00	1.00	1.00
11	1.00	0.25	1.00	0.50	0.50	1.00	0.50	1.00	1.00	0.50	1	2.00	2.00
12	0.50	0.50	0.50	1.00	1.00	0.50	1.00	0.50	0.50	1.00	0.50	1	1.00
13	0.50	0.50	0.50	1.00	1.00	0.50	1.00	0.50	0.50	1.00	0.50	1.00	1

Fig. 3 The pairwise comparison decision matrices of road mode

	1	2	3	4	5	6	7	8	9	10	11	12	13
1	1	2.00	1.00	2.00	1.00	1.00	1.00	1.00	1.00	1.00	1.00	1.00	2.00
2	0.50	1	2.00	4.00	2.00	2.00	2.00	2.00	2.00	2.00	2.00	2.00	1.00
3	1.00	0.50	1	2.00	1.00	1.00	1.00	1.00	1.00	1.00	1.00	1.00	2.00
4	0.50	0.25	0.50	1	2.00	2.00	2.00	2.00	2.00	2.00	2.00	2.00	4.00
5	1.00	0.50	1.00	0.50	1	1.00	1.00	1.00	1.00	1.00	1.00	1.00	2.00
6	1.00	0.50	1.00	0.50	1.00	1	1.00	1.00	1.00	1.00	1.00	1.00	2.00
7	1.00	0.50	1.00	0.50	1.00	1.00	1	1.00	1.00	1.00	1.00	1.00	2.00
8	1.00	0.50	1.00	0.50	1.00	1.00	1.00	1	1.00	1.00	1.00	1.00	2.00
9	1.00	0.50	1.00	0.50	1.00	1.00	1.00	1.00	1	1.00	1.00	1.00	2.00
10	1.00	0.50	1.00	0.50	1.00	1.00	1.00	1.00	1.00	1	1.00	1.00	2.00
11	1.00	0.50	1.00	0.50	1.00	1.00	1.00	1.00	1.00	1.00	1	1.00	2.00
12	1.00	0.50	1.00	0.50	1.00	1.00	1.00	1.00	1.00	1.00	1.00	1	2.00
13	0.50	1.00	0.50	0.25	0.50	0.50	0.50	0.50	0.50	0.50	0.50	0.50	1

Fig. 4 The pairwise comparison decision matrices of rail mode

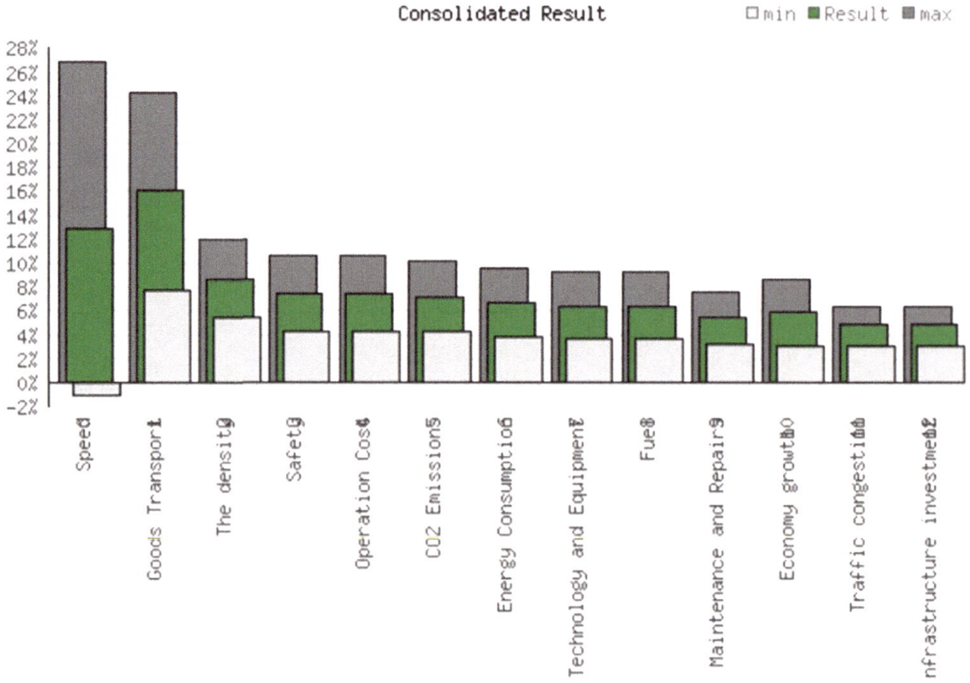

Fig. 5 The consolidated results of road mode

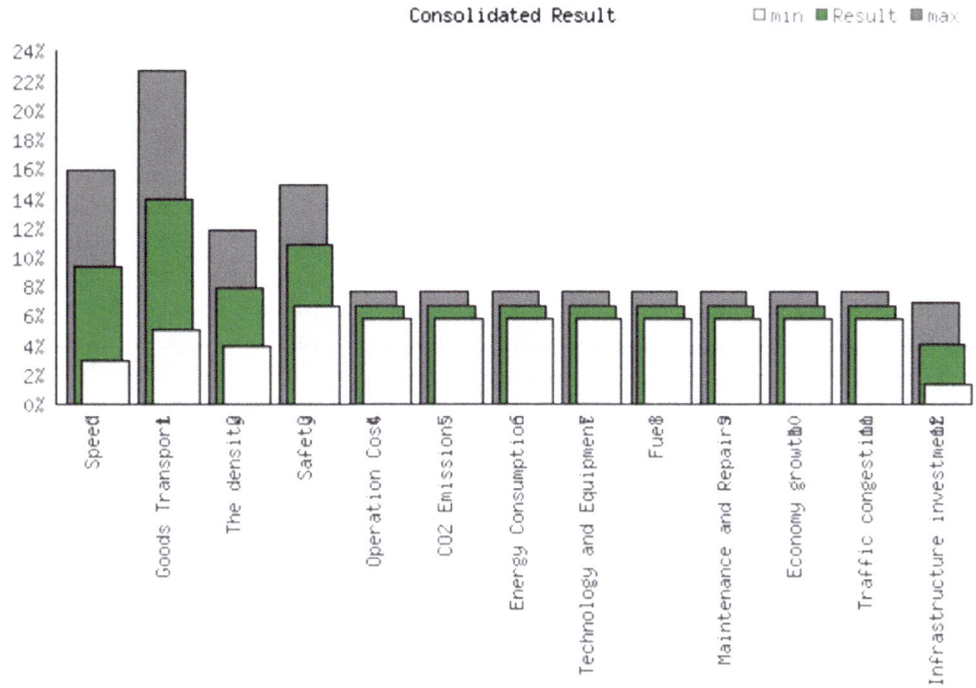

Fig. 6 The consolidated results of rail mode

Table 5 The average criterion weighted for road and rail modes

Road			Rail		
Criterion	Average weighted (%)	Type of criterion	Criterion	Average weighted (%)	Type of criterion
Goods transported	16.1	Economic	Goods transported	13.9	Economic
Speed	12.7	Social	Safety	10.8	Social
Density	8.7	Social	Speed	9.4	Social
Safety	7.5	Social	Density	7.9	Social
Operation cost	7.5	Economic	Operation cost	7.0	Economic
CO_2 emissions	7.1	Environmental	CO_2 emissions	6.7	Environmental
Energy consumption	6.7	Environmental	Energy consumption	6.7	Environmental
Technology and Equipment	6.4	Economic	Technology and equipment	6.7	Economic
Fuel	6.4	Environmental	Fuel	6.7	Environmental
Economy growth	5.9	Economic	Economy growth	6.7	Economic
Maintenance and repairs	5.4	Economic	Maintenance and repairs	6.7	Economic
Traffic congestion	4.8	Social	Traffic congestion	6.7	Social
Infrastructure investment	4.8	Economic	Infrastructure investment	4.1	Economic

Type of criterion	Total average weighted (%)	Type of criterion	Total average weighted (%)
Social	33.7	Social	34.8
Economic	46.1	Economic	45.1
Environmental	20.2	Environmental	20.1

each aspect in both road and rail modes: the economic, social, and environmental aspects of road mode are 46.1%, 33.7%, and 20.2%, respectively, while for rail mode, the values are 45.1%, 34.8%, and 20.1%, respectively. The authors concluded that while companies tend to focus primarily on economic aspects to increase profitability, they should also emphasize improving the social, environmental, and economic dimensions.

4.1.2 The DEA Results

The DEA method was utilized to assess the efficiency score of the logistics performance of the rail and road freight transportation industry. The efficiency score is obtained by dividing the output by the inputs. The closer the efficiency score is to 1, the more efficient the mode is. In this section, the results from the AHP were used as inputs and outputs by the DEA method to identify the efficiency score of the logistics activities of the rail and road freight transport sector. First, inputs and outputs were defined as follows: inputs are speed, goods transported, density, safety, operation costs, energy consumption, technology and equipment, fuel, maintenance and repairs, economy growth, and infrastructure investment, while outputs are CO_2 emissions and traffic congestion. The average weighted criteria for each criterion were used to apply the DEA tool. The results showed that both rail and road modes are efficient, achieving the highest possible efficiency score of 1. These findings support the experts' evaluations and the accuracy of their analysis.

According to the consistency ratio (CR) and the efficiency score, the experts accurately assessed the selected criterion that impacts the logistics performance of road and rail transport modes. The consistency ratio confirmed the acceptability of the average weight assigned to each criterion, and the DEA results reflected the efficiency score for these criteria based on their average weight. As shown in Table 5, the criteria were classified into economic, social, and environmental dimensions, and the average weights for each classification were summed. The results indicated that the economic aspects had the highest value in the evaluation for both road and rail modes, at 46.1% and 45.1%, respectively. Social aspects ranked second for road and rail modes, at 33.7% and 34.8%, respectively. Environmental aspects ranked last, with 20.2% for road and 20.1% for rail. This suggests that companies prioritize economic aspects over social and environmental ones, which presents a significant issue in the transportation sector. This finding is one of the significant contributions of the study. As a result, decision-makers should emphasize improving the environmental and social aspects of logistics activities in both transport modes. Additionally, the experts' evaluations should be considered, and the criteria should be further improved based on their rankings.

4.2 The Hungarian Logistics Activities Results (Case Study)

Hungary is a landlocked Central European country that lies in the Carpathian Basin. Transit movements have historically been concentrated at the intersections of the East–West (from Southeastern Europe/Asia to Western Europe) and North–South ("Amber road" Baltic-Adriatic) Corridors. As a result, transportation has always played an important role in the nation's economic life. Geographic conditions influenced the development of Budapest, Hungary's capital city, with the Carpathian Basin's central locations situated at the border of different terrains (mountains vs. plains) as well as at the crossing points of the region's main river Danube (with strategic fortresses) (Abbood and Mészáros 2023).

The logistics service providers' net turnover of HUF 3400 billion accounts for roughly 5% of the overall net turnover of the national economy. Hungary has 40,000 logistics companies, with a large proportion being small and medium-sized enterprises. The Hungarian logistics sector employs 259 thousand workers, representing 6.5% of the country's total employment, and contributes 6.3% to Hungary's GDP.

According to the World Bank's 2016 Logistics Performance Index (LPI), Hungary ranked 31st globally.

The HUF 32.6 billion in funding provided to the sector has resulted in HUF 81.6 billion in development (TRADE 2019). Goods traffic has much future potential in the Hungarian market regarding transportation logistics. Over the last 15 years, Hungary's road network has undergone tremendous development and investment. Additionally, railway lines are also being modernized. The volume of commodities moved by road constantly expands, while rail transport remains stagnant or shows minimal growth. These improvements aim to facilitate a shift toward more environmentally friendly and energy-efficient modes of transport (TRADE 2019).

The logistics performance of the Hungarian rail and road freight transport industry was assessed using the DEA tool. The DEA method was utilized to identify the efficiency score of this sector by considering criteria that impact its performance. Data were collected from the Hungarian Central Statistical Office website, representing factual information about the selected criteria used in the survey. After collecting the data, the inputs and outputs were identified and used in the DEA method simulation. The results showed that rail and road freight transport modes are efficient from a logistics perspective, with each achieving an efficiency score of 1. This serves as further evidence of the accuracy of the experts' evaluations of the selected criteria. Therefore, decision-makers should consider the priority and ranking of the importance of these criteria based on the experts' assessments.

This study has some limitations, including the challenges in communicating with a large number of experts to gather their opinions, which would provide a more comprehensive understanding of the logistics performance of the freight transport sector. Additionally, there is a lack of data for certain criteria, such as accidents. Existing databases provide general information about fatalities and injuries in road and rail transport, but they do not specify whether the accidents involved freight or passenger transport. As a result, although accidents are a critical social indicator, including this criterion in the study is challenging.

5 Conclusion

The logistics performance activities of the rail and road freight transport industry were assessed in the current study. The study consists of three main parts as a fellow. First, a comprehensive literature review was conducted to examine the use of the MCDM-AHP and DEA methods in assessing the efficiency of the freight transportation sector. Second, an integration of MCDM-AHP and DEA methods was employed to determine the efficiency score of rail and road freight transport from a logistics performance perspective. In this step, a survey was conducted with Hungarian logistics industry experts to evaluate the impact of selected criteria on logistics performance for both rail and road modes. After calculating the average weight for each criterion, the DEA method used these values as inputs and outputs to determine the efficiency score of both modes. Finally, the efficiency scores for logistics performance in the rail and road sectors in Hungary were calculated using actual data with the DEA tool. The results indicated that both modes achieved a perfect efficiency score of 1, reflecting the accuracy of the Hungarian experts' evaluations of the selected criteria. Consequently, decision-makers should focus on improving the criteria based on their rankings. The study faced some limitations, such as the small number of experts participating in the survey and the availability of actual data for certain criteria. The authors intend to extend the research by analyzing all modes of freight transportation and including other countries to gain a broader understanding of logistics performance.

References

Abbood, K., Mészáros, F.: Carbon and energy footprint analysis of Hungarian transportation activities using a multi-region input-output model. Curr. Res. Environ. Sustain. **5** (2023). https://doi.org/10.1016/j.crsust.2023.100208

Allen, J., Browne, M., Cherrett, T.: Investigating relationships between road freight transport, facility location, logistics management and urban form. J. Transp. Geogr.geogr. **24**, 45–57 (2012). https://doi.org/10.1016/j.jtrangeo.2012.06.010

Baltazar, M.E., Jardim, J., Alves, P., Silva, J.: Air transport performance and efficiency: MCDA vs. DEA approaches. Procedia Soc. Behav. Sci. **111**(Lcc), 790–799 (2014). https://doi.org/10.1016/j.sbspro.2014.01.113

Charnes, A., Cooper, W.W., Rhodes, E.: A Data Envelopment Analysis Approach to Evaluation of the Program Follow Through Experiment in US Public School Education, pp. 1–64. Management Sciences Research Group, Graduate School of Industrial Administration, Carnegie-Mellon University (1978)

de Oliveira, M.S., Steffen, V., de Francisco, A.C., Trojan, F.: Integrated data envelopment analysis, multi-criteria decision making, and cluster analysis methods: trends and perspectives. Decision Anal. J. **8**, 100271 (2023). https://doi.org/10.1016/j.dajour.2023.100271

Egilmez, G., Kucukvar, M., Tatari, O.: Sustainability assessment of U.S. manufacturing sectors: an economic input output-based frontier approach. J. Clean. Prod. **53**, 91–102 (2013). https://doi.org/10.1016/j.jclepro.2013.03.037

Escobar, M.T., Aguarón, J., Moreno-Jiménez, J.M.: A note on AHP group consistency for the row geometric mean priorization procedure. Eur. J. Oper. Res.oper. Res. **153**(2), 318–322 (2004). https://doi.org/10.1016/S0377-2217(03)00154-1

Forman, E.H.: Random indices for incomplete pairwise comparison matrices. Eur. J. Oper. Res.oper. Res. **48**(1), 153–155 (1990). https://doi.org/10.1016/0377-2217(90)90072-J

Janjevic, M., Knoppen, D., Winkenbach, M.: Integrated decision-making framework for urban freight logistics policy-making. Transp. Res. Part D: Transp. Environ. **72**, 333–357 (2019). https://doi.org/10.1016/j.trd.2019.05.006

Keskin, B., Köksal, C.D.: A hybrid AHP/DEA-AR model for measuring and comparing the efficiency of airports. Int. J. Product. Perform. Manag.manag. **68**(3), 524–541 (2019). https://doi.org/10.1108/IJPPM-02-2018-0043

Lai, P.L., Potter, A., Beynon, M., Beresford, A.: Evaluating the efficiency performance of airports using an integrated AHP/DEA-AR technique. Transp. Policy **42**, 75–85 (2015). https://doi.org/10.1016/j.tranpol.2015.04.008

Marinov, M., Zunder, T., Islam, D.: Concepts, models and methods for rail freight and logistics performances: an inception paper. In: Proceedings Media of the 12th World Conference on Transport Research, pp. 1–26. Newcastle University (2010)

Office, H.C.S.: Hungarian Transport Performance (2021). Retrieved from https://www.ksh.hu/transport

Rosić, M., Pešić, D., Kukić, D., Antić, B., Božović, M.: Method for selection of optimal road safety composite index with examples from DEA and TOPSIS method. Accid. Anal. Prev.. Anal. Prev. **98**, 277–286 (2017). https://doi.org/10.1016/j.aap.2016.10.007

Saaty, T.L.: The analytic hierarchy process. McGraw Hill, New York. Agric. Econ. Rev. (1980). Retrieved from Saaty, T.L.: The analytic hierarchy process. McGraw Hill, New York. Agric. Econ. Rev. **70**(804), 10-21236 (1980)

Saaty, T.L.: Decision making with the analytic hierarchy process. J. Manuf. Technol. Manag.manag. **26**(6), 791–806 (2008). https://doi.org/10.1108/JMTM-03-2014-0020

Scopus: Integration the data development analysis (DEA) tool with the multi-criteria decision-making (MCDM) method to study freight transport activities (2024)

Sgura Viana, M., Delgado, J.P.M.: City logistics in historic centers: multi-criteria evaluation in GIS for city of Salvador (Bahia—Brazil). Case Stud. Transp. Policy **7**(4), 772–780 (2019). https://doi.org/10.1016/j.cstp.2019.08.004

TRADE, F.I.: Logistics Sector In Hungary. Flanders Investment & Trade Market Survey (2019). Retrieved from https://www.flandersinvestmentandtrade.com/export/sites/trade/files/market_studies/2019-Hungary_Logistics.pdf

Vrtagić, S., Softić, E., Subotić, M., Stević, Ž., Dordevic, M., Ponjavic, M.: Ranking road sections based on MCDM model: new improved fuzzy SWARA (IMF SWARA). Axioms **10**(2) (2021). https://doi.org/10.3390/axioms10020092

Wiegmans, B., Champagne-Gelinas, A., Duchesne, S., Slack, B., Witte, P.: Rail and road freight transport network efficiency of Canada, member states of the EU, and the USA. Res. Transp. Bus. Manag.manag. **28**, 54–65 (2018). https://doi.org/10.1016/j.rtbm.2018.10.004

Zeydan, M., Çolpan, C., Çobanoğlu, C.: A combined methodology for supplier selection and performance evaluation. Expert Syst. Appl. **38**(3), 2741–2751 (2011). https://doi.org/10.1016/j.eswa.2010.08.064

Fatalities in Road Traffic Accidents in Karachi: Lack of Walkability Infrastructure and Role of Motorization

Andaleeb Z. Rizvi

Abstract

The World Health Organization (WHO) has identified that developing countries are more prone to road traffic fatalities, and Pakistan is no exception. WHO says that almost half of the world's annual road traffic fatalities of approximately 1.3 million people are pedestrians, cyclists, and motorcyclists, and more than 90% happen in developing countries. Reasons for this include the increasing use of motorized vehicles, and the absence of pedestrian-friendly policies on national and local levels (Leather et al. in Walkability and Pedestrian Facilities in Asian Cities State and Issues. Metro Manila: Asian Development Bank, 2011). With a focus on building roads, idealizing the car as a status symbol, and a lack of pedestrian-focused planning policy, in Pakistan's largest metropolitan city, pedestrians are increasingly becoming vulnerable on the roads (Hasan and Raza 2015). This paper examines available data on motorization and pedestrian deaths in Karachi as a symptom of poor walkability infrastructure and an absence of pedestrians in the planning policy. The paper also analyzes secondary and primary data to examine fatalities in road traffic accidents (RTAs) in Karachi and the role of walkability infrastructure and motorization in increasing the number of accidents. The study also looks at the existence or lack of engineering interventions to prevent road traffic accidents, based on observations and photographic evidence from the different localities across Karachi. This paper establishes vehicle-centric policies and poor walkability infrastructure as causes of fatalities in RTAs. It considers RTAs preventable with correct planning and design decisions by putting the pedestrian in policy. This study of mortalities caused by RTAs and poor walkability is useful for planners, decision-makers, and road users.

Keywords

Transportation · Motorization · Walkability · Pedestrian safety · Road infrastructure

1 Introduction

Road traffic accidents (RTAs) continue to pose a significant public health risk in Karachi, claiming numerous lives and causing countless injuries each year (Zubair and Kazmi 2013). This conference paper, titled "Fatalities in Road Traffic Accidents in Karachi: Lack of Walkability Infrastructure and Role of Motorization," analyzes the complex connections between the city's walkability infrastructure, and the alarming rate of RTAs. Drawing upon a diverse range of data sources, including official statistics from the Pakistan Bureau of Statistics and Sindh Police Department, and data collected by the Edhi Foundation, this paper aims to shed light on the extent of the problem. Furthermore, it explores the role of rapid motorization in exacerbating these issues and scrutinizes the current walkability infrastructure in Karachi via observation and photographic documentation. This study seeks to provide a comprehensive understanding of the situation, thereby informing potential engineering interventions and policy changes to enhance pedestrian safety and reduce RTAs in Karachi.

This paper is interdisciplinary, drawing upon urban planning, transportation engineering, public health, and policy studies. Its findings may have implications for other cities similar to Karachi in urbanization and motorization patterns. The study uses qualitative and quantitative research methods, including analysis of accident reports, traffic data, infrastructure assessments, and policy reviews.

A. Z. Rizvi (✉)
Department of Architecture and the Built Environment, Nazeer Hussain University, Karachi, Pakistan
e-mail: andaleeb.hussain@nhu.edu.pk

P. Mohammad et al. (eds.), *Geographic Approaches to Climate Change and Mitigation: Urban and Rural Perspectives (Volume 2)*,
Advances in Science, Technology & Innovation, https://doi.org/10.1007/978-3-031-92115-5_18

The primary aim of this conference paper is to investigate the correlation between the lack of walkability infrastructure, increased motorization, and the high rate of fatalities in RTAs in Karachi. To do so, the paper looks at the number of RTAs and fatalities in Karachi. The paper also evaluates the walkability infrastructure in Karachi to identify its shortcomings, while looking at the factor of motorization and its contributions to RTAs.

The paper aims to contribute to the ongoing discourse on urban planning and safety, specifically in the context of developing cities like Karachi. It also aims to influence policymaking toward creating safer, more walkable cities, which put "people before motor vehicles in urban transport policy, and planning, and commit to non-motorized transport" (Miraftab and Kudva 2014). While focusing on Karachi, the paper aims to contribute to the broader discourse on urban safety and sustainable city planning. It seeks to provide insights that could be valuable for urban planners, policymakers, and researchers working in similar contexts.

2 Methodology

This paper employed a mixed methods approach, incorporating both qualitative and quantitative research methods to comprehensively understand Karachi's motorization, road traffic fatalities, road traffic accidents (RTAs), walkability infrastructure, and engineering interventions. By integrating correlational and photographic research methods, the investigation first collected official data on motorization and fatalities published by the Pakistan Bureau of Statistics, as well as provisional data on Karachi obtained from the Sindh Excise and Taxation Department and Sindh Police Department. This data collection aimed to examine the relationship between increased motorization and the high rate of fatalities in RTAs. Furthermore, data on fatalities and injuries compiled by the Edhi Foundation, a private philanthropic organization, was also gathered to assess the frequency and accuracy of reported RTAs in police records.

A comprehensive review of the existing literature on walkability infrastructure and engineering interventions was conducted to gain insights into the current state of knowledge in this field, identifying any gaps that can be addressed.

Additionally, photographic documentation of the available walkability infrastructure and engineering interventions specific to Karachi was carried out with the help of students, capturing visual evidence of the current state of the infrastructure, and its maintenance status, and analyzing potential interventions and user behavior. The photographic evidence was then utilized to validate the findings from the literature review and assess the effectiveness of implemented engineering interventions.

The photographs were taken via smartphones in different parts of the city, especially near busy areas to understand the behavior of pedestrians and other users and to analyze the condition of the available infrastructure. Figure 1a shows the locations on the map of Karachi, where the photographic survey was undertaken.

This methodology offers a systematic approach to collect, analyze, and validate data about RTAs and walkability infrastructure in Karachi. It facilitates a comprehensive understanding of the existing situation and can inform future interventions to improve road safety. Ultimately, this methodology reinforces the argument that better motorization policies and improved walkability infrastructure can potentially mitigate fatalities resulting from RTAs.

3 Results

The following section provides data on motorization, RTAs, reasons for RTAs, and findings of a walkability survey.

3.1 Motorization Data

Almost half of the world's annual road traffic fatalities of approximately 1.3 million people are pedestrians, cyclists,

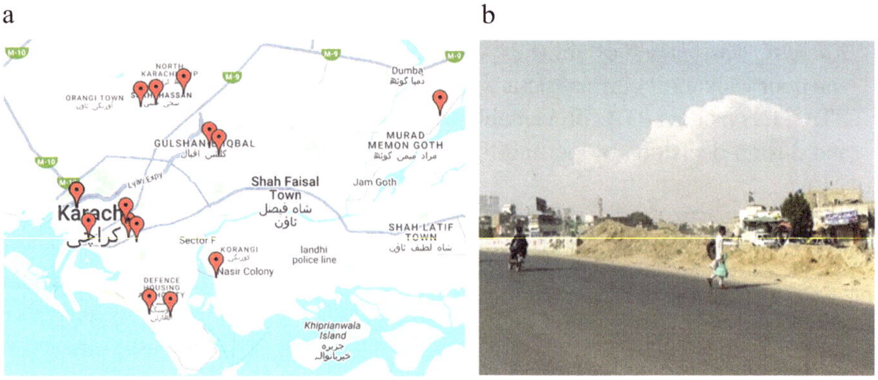

Fig. 1 **a** Map showing locations surveyed in Karachi for photographic documentation of pedestrian infrastructure and interventions. Note that map labels are present in both English and Urdu (image by author, 2024). **b** Pedestrian crosses Rashid Minhas Road without a zebra crossing. There is no footpath due to road works (image by Muhammad Huzaifa, 2023)

Table 1 Data shows the number of registered vehicles across Pakistan between 2009 and 2020 (Pakistan Statistical Year Book 2020, 2022)

Year	Motorcars jeeps and station wagons	Motor cabs/ taxis	Buses /mini buses	Trucks	Motor cycles/ 2 wheels	Motor cycles/ motor rickshaws	Others 3 wheels	Total
2009	1,657,860	106,463	195,163	210,944	3,215,583	167,910	1,005,441	6,559,364
2010	1,726,347	122,882	198,790	216,119	4,305,121	201,827	1,081,916	7,853,002
2011	1,881,560	124,651	202,476	225,075	5,781,953	266,390	1,178,890	9,660,995
2012	2,094,289	143,859	215,374	240,888	7,500,182	323,189	1,270,788	11,788,569
2013	2,281,083	145,234	220,347	247,197	9,169,528	380,579	1,340,963	13,784,931
2014	2,437,735	145,424	224,403	253,574	11,006,421	466,185	1,406,819	15,940,561
2015	2,715,322	167,678	229,290	261,845	13,081,400	559,114	1,487,460	18,502,109
2016	2,933,668	170,759	235,614	269,471	15,230,960	671,403	1,555,975	21,067,850
2017	3,196,542	170,890	242,194	278,120	17,518,365	763,076	1,643,489	23,812,676
2018	3,495,581	171,117	249,198	284,949	19,796,577	843,300	1,725,445	26,566,167
2019	3,703,649	171,179	253,996	288,652	22,001,277	919,020	1,799,789	29,137,562
2020	3,833,616	171,462	255,409	293,460	23,407,865	951,425	1,844,302	30,757,539

and motorcyclists, and more than 90% occur in developing countries (World Health Organization 2009).

Pakistan Bureau of Statistics (PBS) data shows that 5,816 persons were killed in road traffic accidents (RTAs) across Pakistan in 2020–21, up by 6.9% from 2019 to 20 when 5436 people died in RTAs, according to police reports.

A review of data from Pakistan Statistical Year Books 2018–2020 shows that countrywide, registered vehicles jumped 5.6% to 30,757,539 (30.8 million) in 2020 from 29,137,562 (29.1 million) in 2019. Of these. 23,407,865 (23.4 million) were motorcycles (two-wheelers), showing an addition of 1.4 million in 2020 alone. Table 1 gives in detail the growth of vehicles in the country, with a breakdown by vehicle type.

Sindh Excise and Taxation Department: The trend was similar in Karachi, where provisional data from the Sindh Excise and Taxation Department puts the number of registered vehicles at more than four million up till December 31, 2019. This included 2.7 million motorcycles, 1.3 million motor vehicles, 150,000 rickshaws, and 5000 buses.

According to a news report published in Tribune in November 2019, the Sindh Excise and Taxation Department registered 100,000 vehicles in the city annually, a majority of which were privately owned. This annual addition of vehicles has drastically spiked the total number of cars and motorcycles plying on city roads from the previously reported figure of 2.614 million in 2011 (Urban Resource Centre et al. 2015). Of the total number of registered vehicles in 2019, merely 0.001% account for public transport (Table 2). Individually owned vehicles are mostly motorcycles followed by cars, which correlates with the countrywide trend noticed in the PBS data.

3.2 RTAs in Karachi

The now-defunct Road Traffic Injury Research and Prevention Centre collected road traffic incident data from five major trauma centers in Karachi, including Jinnah Postgraduate Medical Centre, Civil Hospital Karachi, Abbasi Shaheed Hospital, Liaquat National Hospital, and Aga Khan University Hospital. The 2007–2014 study showed that out of 9126 accidents, the highest number of fatalities were suffered by two-wheelers (both cyclists and motorcyclists) at 37.2% of 3393 deaths (Shah et al. 2018). It was

Table 2 Provisional data was collected personally from the Sindh Excise and Taxation Department (data collected by author, 2020)

Types of vehicle	2011 (estimated)	2019	% change
Motorcycles/2-wheelers	1.296 million	2.7 million	107
Rickshaws	105,684	150,000	42
Buses/mini-buses	22,313	5000	− 77
Taxis	47,165	–	− 100
Motor vehicles	1.142 million	1.300	14
Total	2.614 million	4.155 million	59

followed by 35.8% or 3300 pedestrian deaths (Shah et al. 2018).

To correlate the data and understand the potential gaps in reporting caused by cases not reported to the police, police data was compared with data shared by the Edhi Foundation.

3.3 Reasons of RTAs

Police records revealed that 91% of road traffic accidents in Karachi occurred due to road user errors (Downing 1991). However, it has also been observed that less than 1% of drivers in Karachi stop for pedestrians on uncontrolled crossings (Downing 1991), while an older study shows that only 20% of pedestrians in the city use designated crossings against 48% in Bangkok and 89% in the UK (Jacobs et al. 1981).

Missing or inappropriate infrastructure also results in RTAs in Karachi. It includes a lack of sidewalks or footpaths, an absence of designated pedestrian crossings, no designated safe zones for waiting, missing or inadequate signage and markings, no dedicated bike lanes, poorly maintained sidewalks, pedestrian bridges that are inaccessible or inconvenient, bus stops that are not pedestrian-friendly, inappropriately timed traffic signals, signal-free corridors, and inconsistent or unclear road markings. Table 5 gives the comparison of injury severity before and after the inauguration of a signal-free corridor in Karachi.

Table 5 presents data on the prevalence of RTAs in Karachi's signal-free corridors. To gain a deeper understanding of the pedestrian safety challenges in these corridors, a photographic survey was conducted at several locations across Karachi (Fig. 1a). The survey revealed several instances of missing pedestrian infrastructure, as exemplified by Fig. 1b.

Poor maintenance and operation of existing pedestrian infrastructure were visible in the photographic survey, which might cause RTAs in different localities. Figures 2a, b, and 3a show the dilapidated state of pedestrian infrastructure.

Limited or minimum utilization of available infrastructure. Observations revealed underutilization of existing pedestrian infrastructure (Fig. 3b). Pedestrians were observed

a b

Fig. 2 **a** This picture shows a littered and poorly designed footpath with a broken curb next to a residential apartment building on Abul-Hasan Isfahani Road in Karachi (image by author, 2021). **b** Example of a missing safety feature on a pedestrian bridge in Karachi. This image shows a missing railing, which can be especially dangerous at night in low visibility (image by Faraz Uddin, 2023)

a b

Fig. 3 **a** Broken road, footpath, and greenbelt show sewage accumulation near Mumtaz Hospital on Sher Shah Suri Road (image by author, 2021). **b** A man walks on the road instead of the footpath on Rashid Minhas Road (image by Muhammad Huzaifa, 2023)

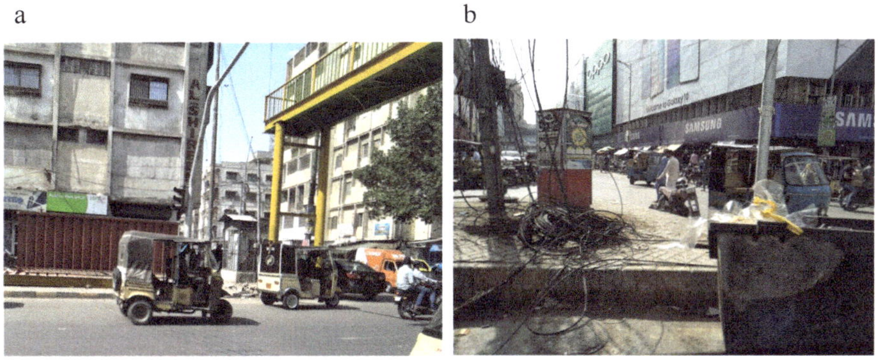

Fig. 4 **a** An under-construction pedestrian bridge on Nishtar Road at the Britto Road intersection (image by author, 2020). **b** A pile of cables, poles, garbage collection trolley, and other paraphernalia can be seen at the Zaibunnisa Street, and Castle Street intersection aisle (image by author, 2020). The picture was taken by the author in 2020

walking on roadways despite available sidewalks, as the surfaces were uneven, broken, or blocked. Pedestrians also cited poor sidewalk conditions (Fig. 3a), lack of crossings, and encroachments (Fig. 4a) as reasons for walking on roadways.

3.4 Walkability Survey

A 2009 study on "walkability" states that 21% of daily trips in Karachi were made on foot, while nearly 66% of commuters who used available modes of public transport walked part of their commute. Every trip begins and ends on foot. Therefore, walking is an integral part of the whole transport system. However, conventional land use and transport planning practices in Asian cities, including Karachi, still pay little attention to walking (Leather et al. 2011).

Observations revealed that occupied footpaths and roadsides used for parking forced pedestrians to walk further from the footpaths and navigate between moving cars.

With occupied footpaths, and roadsides being used for parking rickshaws and other vehicles, people are not only forced to walk further from the footpaths, they also have to often navigate between moving cars while crossing streets and main roads. This adds to the incidences of RTAs.

3.5 Engineering Interventions

As evidenced in the photographic survey of the city's infrastructure, existing engineering interventions are poorly maintained and very limited. The following are observations of the existing engineering interventions or lack thereof.

Walkways are pedestrian linkages that include sidewalks, paths, and highway shoulders (Federal Highway Authority 1998). Walkways in Karachi are poorly maintained, broken, encroached or have missing sections, forcing people to walk on the road and discouraging them from walking entirely if they can afford a car (Figs. 1b, 2a, and 3a). Those who have no choice but to walk, often walk on the road due to a lack of walkways poor conditions of walkways, and encroachments (Fig. 4a), which makes them vulnerable to RTAs. Problems are multiplied for those with disabilities. They either avoid walking or have to depend on others for their survival and basic chores. There is barely a concept of shoulders in Karachi. In case a shoulder exists on some roads, it is used by motorcyclists. Footpaths are walkways intended for use by mostly pedestrians, in Karachi observation reveals that surfaces of a majority of footpaths are either unpaved, encroached, are used for parking and/or garbage collection or are used by the homeless or beggars. There are safety issues as well for women (Raza 2016).

Intersections should be designed and built to encourage pedestrian usage (Federal Highway Authority 1998). Walkways provide mobility for pedestrians, but they need to ultimately go across roads and streets at intersections, which brings people and vehicles together. Observations in Karachi reveal intersections are one of the most challenging sections for pedestrians. Intersections are unsafe and limit mobility severely due to speed, lack of crosswalks, and ill attitudes of drivers. There seems to be a lack of attention toward intersections in Karachi (Fig. 4b).

Crosswalks are well-defined to encourage pedestrians to cross at reasonable locations and to increase motorists' awareness of pedestrians (Federal Highway Authority 1998). As already pointed out, in Karachi maintenance of crosswalks is close to zero, and locations where some crosswalks might be visible are mostly in cantonments or other posh localities like the Defense Housing Authority (DHA) (Fig. 5).

Curb ramps enable people on wheelchairs and those using walkers and sticks easier transition from walkways to roads and streets (Federal Highway Authority 1998). Lack of ramps excludes people with disabilities and deprives them of equal opportunity to work, play, participate in public activities and move freely in urban spaces (Doussard 2024). In some places where ramps are provided, they are poorly designed and not up to standard to enable independent movement for those with disabilities.

Fig. 5 This photograph is an example of a faded crosswalk in the DHA area of Karachi. Though a pedestrian is using the crosswalk, the vehicle did not stop for the pedestrian's right of way (image by Hamza Khan, 2023)

Curb extensions and curb radii shorten crossing distances and reduce pedestrians' exposure to motor vehicles. Pedestrians are more vulnerable to accidents when they are crossing wide streets (not all wide streets have pedestrian bridges to enable crossing or barriers to stop pedestrians from crossing in Karachi). Parked cars also hinder visibility, and pedestrians can often not see oncoming motorcycles and vehicles, which results in accidents. "Also, when streets intersect at an acute or obtuse angle or have a large curb radius, motorists can make turns at relatively high speeds" (Federal Highway Authority 1998), an issue observed in Karachi.

Signing and marking are also missing in Karachi overall to alert motorists of pedestrian activity, and pedestrians of any oncoming dangers, the photographic survey and observations showed.

Pedestrian amenities such as landscaping, rest areas, benches, lit walkways, buffer areas, handrails, and waste containers are available in limited areas. This lack of amenities discourages walking and also affects the overall well-being of citizens.

Arterial streets (signal-free corridors) increase the risk of accidents for pedestrians, especially in the absence of appropriate infrastructure (Kumar et al. 2010). Motorists resort to speeding on arterial streets, which increases the risk of RTAs (Lateef 2010) (Fig. 1b).

Bridges also create many barriers for pedestrians if they do not have appropriate overpasses and underpasses.

Traffic calming is perhaps one of the most abundantly used methods in Karachi in the form of speed bumps. However, these speed bumps are often not up to standards, lack reflectors, and become the reason for the accumulation of rain and sewage water on the streets due to poor sanitation infrastructure, leading to potholes and general disrepair. Speed bumps though are made on a mostly self-help basis near schools, hospitals, shopping areas, mosques, imambargahs, and other locations where higher pedestrian activity is expected. These also become a hindrance for those with disabilities as adjacent inclusive infrastructure is non-existent.

Maintenance is the most neglected aspect of infrastructure in Karachi. Broken pavements, and sidewalks (Figs. 1b, 2a, and 3a) create safety and access problems for pedestrians.

Educational interventions and awareness campaigns are limited. Most of the traffic awareness campaigns focus on fastening seatbelts or wearing helmets. A comprehensive educational program is missing when it comes to awareness about pedestrian rights of way. Motorists in Karachi are not obligated to yield or stop for pedestrians. A "pedestrian program" is missing. Ideas regarding "trip length reduction" and information on "walking route maps" are also missing. Walking events are however held in certain exclusive circles.

Construction zones exclude pedestrians and they are made to walk on uneven surfaces that are unsafe, and poorly lit (Fig. 1b). Construction zones often encroach on sidewalks or crosswalks, and pedestrians end up making unsafe detours, which make them vulnerable to oncoming traffic on busy streets.

Land use development is increasingly making Karachi anarchic, ugly and a divided city—it is however "loved and glorified by its comparatively well-to-do citizens, public intellectuals and those in search of an identity" as per architect Arif Hasan's article Ugly Karachi in Dawn newspaper, published on November 6, 2016 (Hasan 2016).

Law enforcement agencies fail to enforce traffic laws related to pedestrians and vehicles in Karachi. This is also a reason for the increased number of crashes and resultant injuries and fatalities, as well as a lower number of walkers. Motor vehicle speeds are also not controlled. According to WHO's Report on Road Safety, Pakistan does have a national speed limit law, but there is no good speed limit law, while the maximum urban speed limit is 90 km/h, the rural speed limit is 110 km/h and the motorway speed limit is 130 km/h. The report says that though local authorities can modify limits, the enforcement is average.

4 Discussion

This section interprets the results and explores their significance. It discusses the findings that show what increasing motorization means in the context of this research on RTAs and walkability infrastructure in Karachi. It begins with a discussion of the national trends, and how those national trends align with this research or how such trends might influence Karachi's situation specifically.

4.1 Motorization Data

Global Statistics and Trends: Pedestrians, cyclists, and motorcyclists account for nearly half of all RTA fatalities worldwide. This trend is mirrored in Pakistan, where rapid motorization, coupled with the lack of pedestrian-centric policies at both national and local levels, exacerbates the situation (Nazir 2016; Leather et al. 2011).

Vehicle Ownership in Pakistan: Tables 1 and 2 illustrate the escalating number of privately owned vehicles in Pakistan, with a specific focus on Karachi. The data reveals a stark contrast between the burgeoning number of private vehicles and the scant number of buses in the city. This disparity exacerbates the transportation challenges faced by the populace, compelling them to invest in private vehicles.

The Role of Motorcycles: Motorcycles, in particular, are a popular choice due to their affordability and maneuverability in traffic congestion. However, motorcycles offer less protection compared to cars, rendering riders more susceptible to injuries in collisions. This surge in motorcycle usage could be a significant factor in the escalating RTAs.

Additional Contributing Factors: A multitude of factors could be contributing to the rise in RTAs, in addition to the increase in motor vehicles. These include infrastructure deficiencies, sub-standard road conditions, insufficient signage, absence of dedicated lanes for motorcycles, aggressive driving behavior, speeding, and a lack of awareness among vulnerable road users. Furthermore, inadequate enforcement of traffic laws could be perpetuating unsafe driving practices.

4.2 RTAs in Karachi

Establishment of the Road Traffic Injury Research and Prevention Centre: In response to an alarming number of RTAs in 2007, the Road Traffic Injury Research and Prevention Centre was established (Lateef 2010). This joint project of Jinnah Postgraduate Medical Centre, Aga Khan University Hospital, and NED University of Engineering and Technology was approved by the Federal Ministry of Health, Government of Pakistan, and initiated operations in January 2007.

Findings of the Centre: The Centre's data was consistent with global RTA statistics, indicating the highest fatalities among two-wheelers (cyclists and motorcyclists), followed by pedestrians. Although the Centre is no longer operational, the trends it studied persist, as evidenced by official data from the Sindh Police Department (Table 3). This data shows that from 2018 to 2022, the highest numbers of fatalities were suffered by those on two-wheelers, followed by pedestrians.

Limitations in the Police Data: The police data only includes cases where a First Information Report (FIR) was filed, and therefore does not portray the absolute numbers of deaths and injuries due to RTAs. Table 4 compares data collected by the Edhi Foundation (a philanthropic ambulance service) and the Sindh Police Department. It's important to note that both the police and Edhi data only relate to RTAs and do not include railway or other accidents. For a more complete picture of RTAs, data from all ambulance service providers, trauma centers, and police stations should be compiled with specific details, such as geographical reference information, types of vehicles involved, trauma details, gender, age.

Classification of Accidents: Police record fatal accidents as those where death occurred either on the spot or later

Table 3 Provisional data was collected personally from the Sindh Police Department (DIG Traffic Office) (data collected by the author, 2024)

Category	2022		2021		2020		2019		2018	
	Killed	Injured	Killed	Injured	Killed	Injured	Killed	Injured	Killed	Injured
Pedestrians	38	15	50	30	47	32	42	39	51	21
Paddle cyclists			1		1		2	1	2	1
Motorcyclists	137	91	153	100	124	87	152	116	146	96
Other road users	18	53	20	87	17	56	25	87	18	61
Total	193	159	224	217	189	175	221	243	217	179

Table 4 Comparison between official police data versus Edhi data collected directly from the Edhi Foundation Office at Tower, Karachi[1] (data collected by the author, 2024)

Year	Police FIR data				Edhi ambulance data	
	Accident		Persons			
	Fatal	Non-fatal	Killed	Injured	Killed	Injured
2020	169	54	189	175	679	9111
2021	190	69	224	218	1034	9460
2022	191	57	193	159	781	8943
Total	550	180	606	552	2494	27,514

Table 5 Comparison of injury severity before and after the inauguration of Signal-free corridor I (Shahrah-e-Faisal to SITE Industrial Area) in Karachi (Kumar et al. 2010)

Accident severity	Five months before the inauguration	Five months after the inauguration
Minor	154	377
Serious	45	52
Fatal	7	9
Total	206	438

on, whereas non-fatal accidents refer to injuries to a person, persons, or property. Police data is added to the reports generated by the Pakistan Bureau of Statistics, which produces the yearly statistical books. Therefore, important information related to RTAs is missing in the final official data produced by the government institute.

4.3 Reasons of RTAs

A majority of RTAs over some time have been blamed on human error, including pedestrian errors, car driver judgment errors, failure to use protective measures, poor road design, and excessive speed (Lateef 2010). However, the findings from Table 5 highlight a high prevalence of RTAs in Karachi's signal-free corridors. Our photographic survey results (Figs. 1b, 3a and b) further corroborate this concern, showing a lack of designated pedestrian crossings, and proper walking areas in many locations.

Missing or inappropriate infrastructure: Since 2006, Karachi's developers have focused on signal-free corridors, flyovers, additional lanes on roads, and road widening exercises to enable smooth vehicular flow at the cost of those without private vehicles. While the strategy has failed to resolve the problematic traffic flow in Karachi, it has quadrupled the problems pedestrians face. This brings us to the phenomenon of pedestrian bridges for crossing thoroughfares and signal-free corridors (Raza 2016). Pedestrian bridges were installed so people cross roads safely without creating conflict. However, these bridges do not take into account children, the elderly, and the differently abled. Distances and time have also not been taken into consideration in the design of these bridges, which can go up to 3 km in some cases. This has increased the crossing times for pedestrians, and they seek alternative means like cutting barriers placed in the center of the signal-free corridors to get across. This too has added to fatalities and injuries caused by RTAs in case of a collision with an oncoming vehicle (Kumar et al. 2010).

Poor maintenance and operation of existing pedestrian infrastructure: Our photographic survey (Fig. 1a) revealed several instances of poor maintenance and lack of pedestrian infrastructure. This is exemplified by Fig. 3b. This figure shows a busy road undergoing construction, yet there is no designated pedestrian crossing or proper footpath available. This situation creates a dangerous environment for pedestrians who are forced to navigate traffic or walk on uneven surfaces.

Limited or minimum utilization of available infrastructure: The underutilization of existing pedestrian infrastructure is a significant concern, as observed in Karachi. Despite the availability of sidewalks, pedestrians prefer to walk on roadways (Fig. 3b). This preference is primarily due to the poor conditions of the sidewalks, which are often uneven, broken, or blocked. The lack of crossings and encroachments further exacerbate the issue, forcing pedestrians to choose roadways over sidewalks for their safety and convenience. This behavior indicates a gap between the provided infrastructure and the pedestrian's actual needs and expectations. It is crucial to understand that infrastructure, no matter how well-designed, will only be effective if it is utilized. Therefore, the design and maintenance of pedestrian infrastructure should be user-centric, taking into account the needs and preferences of the pedestrians. Moreover, the presence of encroachments on sidewalks highlights the need for stricter regulations and enforcement to ensure that pedestrian spaces are not compromised.

[1]Edhi Foundation's Information Bureau Chief Azeem Khan provided the data.

4.4 Walkability Survey

The ADB's walkability survey conducted in Karachi reveals insightful patterns in pedestrian behavior and infrastructure utilization. The survey shows that walkability parameters across different areas in the city vary significantly.

Residential areas scored the highest overall walkability rating of 59.11, indicating better pedestrian infrastructure and safety measures (Table 6). This could be attributed to the lower traffic volumes and slower speeds typically found in residential areas. However, the relatively low score for disability infrastructure (20) across all areas, including residential, suggests a lack of consideration for inclusivity in infrastructure planning.

In contrast, public transport terminals, which are critical nodes in the transport network, scored the lowest overall rating of 35.56 (Table 6). This is concerning as these areas typically see high pedestrian volumes. The low scores for obstructions (30), amenities (26.67), and motorist behavior (26.67) suggest that these areas are not pedestrian-friendly, potentially discouraging walking and impacting the overall efficiency of the transport system.

Commercial and educational areas scored moderately on the walkability index, with ratings of 49.63 and 47.78, respectively (Table 6). However, the low scores for disability infrastructure and obstructions indicate room for improvement.

The observed use of footpaths and roadsides for parking, which forces pedestrians to navigate between moving cars, likely contributes to the low walkability scores in certain areas and increases the incidences of RTAs. This highlights the need for better regulation of parking and dedicated pedestrian infrastructure.

In conclusion, while walking is a significant component of the transport system in Karachi, the walkability survey results suggest that pedestrian infrastructure needs to be improved, particularly in areas with high pedestrian volumes such as public transport terminals.

Discussion on Implementing Pedestrian Improvements: Cities worldwide are recognizing the importance of pedestrian improvements to make walking safer and more appealing. Karachi too needs to develop a non-motorized transport environment by enhancing sidewalks, road crossings, bikeways, and more. This strategy encourages people to opt for walking for everyday trips, thereby reducing traffic congestion and pedestrian-vehicle accidents.

However, in Karachi, the focus has been primarily on pedestrian bridges, neglecting other crucial infrastructure. As the trend of car and motorcycle ownership increases, promoting pedestrian-oriented developments could help reduce vehicular movements that would otherwise be generated if one had to drive everywhere.

This strategy requires promoting pedestrian-oriented developments where people can walk between multiple uses and to public transit stops. It also necessitates a well-designed pedestrian system with sidewalks that are safe for all types of users, including elders, children, and the disabled.

Multiple neighborhoods can be linked into one urban center where services are coordinated and lined by a pedestrian system. For this purpose, the presence of the pedestrian in policy is necessary. The government's focus seems to be on "going upwards in the form of pedestrian bridges" to enable walking. This method is faulty and needs to be completely revised.

Table 6 Data was quantified from an ADB Sustainable Development Working Paper, in which 272 people were interviewed in Karachi (Leather et al. 2011)

Surveyed walkability parameters	Residential areas	Educational areas	Commercial areas	Public transport terminals
Walkability rating	59.1	47.78	49.63	35.56
Security from crime	60	56.67	53.33	46.67
Obstructions	52	43.33	53.33	30
Disability infrastructure	20	30	20	23.33
Amenities	52	43.33	50	26.67
Motorist behavior	64	50	63.33	26.67
Grade crossing safety	72	50	43.33	36.67
Availability of crossings	76	43.33	60	53.33
Availability of walking paths	72	56.67	53.33	36.67
Walking path modal conflict	64	56.67	50	40

The following is an assessment of engineering, educational, and enforcement mechanisms that could help improve the pedestrian environment in Karachi.

5 Conclusion

This paper presents evidence on vehicle-centric policies and poor walkability infrastructure as causes of fatalities in RTAs in Karachi based on both primary and secondary data. It considers RTAs preventable if correct planning and design decisions are taken by the policy and decision-makers by putting the pedestrian in the policy. The assessment of the relationship between fatalities in RTAs and poor walkability can be useful for planners, decision-makers, and road users in improving the safety of pedestrians as well as other victims of RTAs. This can help detect issues and their causes, and plan accordingly to increase the awareness of pedestrian rights and the importance of walking for well-being. A detailed evaluation of factors that contribute to traffic accidents, including spatial characteristics, vehicle-oriented planning policies, poor engineering and educational interventions, as well as poor law enforcement, are key to designing and implementing effective measures to reduce the rising number of pedestrian fatalities. Findings of this study can be useful in informing urban policies and interventions to improve the safety and quality of pedestrian areas, promoting a non-motorized mode of transport and, consequently, active lifestyles. In particular, this paper proposes that a pedestrian-friendly approach should be adopted in urban planning and design. This paper states that RTAs are preventable via appropriate planning and design solutions.

6 Way Forward

This author suggests some key recommendations that are necessary to mitigate the impacts of rapid motorization in Karachi and to improve the walkability infrastructure.

6.1 Cohesive Data Collection Mechanism

One major need is for a comprehensive and cohesive data collection mechanism for RTAs, as massive discrepancies exist in the data on RTAs collected by the Sindh Police Department versus philanthropic organizations, hospitals, and other entities and institutions. A collaborative approach will give a clearer picture of the fatalities and injuries caused by RTAs. For a more holistic understanding of RTAs, it is recommended to compile data from various sources, including all ambulance service providers, trauma centers, and police stations. This data should be detailed and include the following specifics:

Geographical Reference Information: The exact location of the accident can provide insights into accident-prone areas and help in planning preventive measures.

Types of Vehicles Involved: Information on whether the accident involved two-wheelers, cars, trucks, or pedestrians can help identify the most vulnerable road users.

Trauma Details: Information on the nature and severity of injuries can guide medical response and treatment protocols.

Gender and Age: Demographic data can reveal patterns and trends in accident victims, aiding in targeted safety campaigns.

Time of Accident: Identifying peak accident times can help in scheduling increased traffic monitoring and control.

Weather Conditions: Understanding the impact of weather on accident rates can inform about necessary precautions during adverse conditions.

Road Conditions: Information on road quality and infrastructure can help identify areas needing improvement.

Safety Equipment Usage: Data on the use of safety equipment like seat belts, helmets, etc., at the time of the accident can emphasize the importance of these measures.

6.2 Putting People in the Policy

Instead of focusing only on vehicles, vehicle safety, and awareness campaigns, the attitudes, limitations, and aspirations of people should also be understood when designing policies as well as traffic-related infrastructure. In the case of pedestrian bridges, it has been documented that people do not use the bridges if they are far apart or are too high for comfort. It shows a need to take input from the public when putting up such investments in the city that have the potential to go to waste. A breakdown of this approach could be:

Understanding User Attitudes: Discussing the importance of understanding the attitudes of people toward traffic rules and infrastructure.

Addressing User Limitations: Highlighting the need to consider physical, economic, and other limitations of users when designing policies and infrastructure.

Incorporating User Aspirations: Exploring how the aspirations and preferences of people can be incorporated into policy-making.

Pedestrian Bridges: Analyzing the usage of pedestrian bridges and the factors influencing how people use or avoid pedestrian bridges because of height, poor maintenance, lighting, security, and other issues.

Public Input in Infrastructure Investments: Public hearings should be carried out for every infrastructure investment in accessible spaces to discuss the importance of infrastructure investments and to avoid potential wastage of resources.

6.3 Long-Term Planning

Instead of short-term goals, there is a need to come up with longer-term planning policies based on cohesive data collection mechanisms for appropriate projections in an expanding city like Karachi.

6.4 Focusing on Non-Motorized Transport and Public Transport

While the National Transport Policy, 2018 talks about prioritizing non-motorized transport and public transport, it still considers private transport complimentary and does not include a discussion on the ride-hailing apps, which have become an integral part of the urban transportation network. Therefore, there is a need for the policy to assess the impacts of ride-hailing apps on the transportation network as well as the impact of such services on the incidence rates of RTAs.

References

Doussard, C.G.: Universal design for the workplace: ethical considerations regarding the inclusion of workers with disabilities. J. Bus. Ethics (2024)

Downing, A.B.-I.: Road Safety in Developing Countries: An Overview. PTRC Education and Research Services, London (1991)

Federal Highway Administration: Implementing Pedestrian Improvements at the Local Level. U.S. Department of Transportation (1998)

Hasan, A.: Ugly Karachi. Dawn Newspaper, Karachi (2016)

Hasan, A., Raza, M.: Responding to the transport crisis in Karachi. IIED Urban Res. Cent. (2015)

Jacobs, G., Sayer, I., Downing, A.: A Preliminary Study of Road User Behaviour in Developing Countries. Transport and Road Research Laboratory (1981)

Kumar, R., Ali, M.S., Ahmed, A.: An appraisal of signal free corridor in Karachi via empirical study of road accidents and pedestrian movement concerning road crossing. In: Third International Symposium on Infrastructure Engineering in Developing Countries, Karachi (2010)

Lateef, M.: Estimation of fatalities due to road traffic crashes in Karachi, Pakistan, using capture-recapture method. Asia-Pac. J. Public Health 332–341. https://doi.org/10.1177/1010539509356808

Leather, J., Fabian, H., Gota, S., Mejia, A.: Walkability and Pedestrian Facilities in Asian Cities State and Issues. Asian Development Bank, Metro Manila (2011)

Miraftab, F., Kudva, N.: Cities of the Global South Reader, 1st edn. Taylor & Francis, London (2014)

Nazir, E.N.: Road safety challenges in Pakistan: an overview. J. Transp. Secur.secur. **9**, 161–174 (2016)

Pakistan Bureau of Statistics: Pakistan Statistical Year Book 2020. Islamabad (2022)

Raza, M.: Exploring Karachi's Transport System Problems: A Diversity of Stakeholder Perspectives. International Institute for Environment and Development (2016)

Shah, S., Ahmad, N., Ha, A.: Pedestrians' exposure to road traffic crashes in urban environment: a case study of Peshawar, Pakistan. J. Pak. Med. Assoc. 615–623 (2018)

World Health Organization: Global Status Report on Road Safety: Time for Action (2009)

Zubair, S., Kazmi, S.: Spatial framework for the assessment of road traffic accidents in Karachi. J. Basic Appl. Sci. **9**(67), 525–532 (2013)

Policies and Governance for Sustainable Urban Development

Science in the City: Implementation of Evidence-Based Green Policies in Urban Environments

Davide Testa and Francesco Berni

Abstract

Recently, city science has been emerging as an autonomous field of research and making itself useful to the community through increasingly close collaboration between universities and local authorities, as evidenced by the creation of a European network of City Science Offices. On the other side, citizen science has been proven to be a powerful approach to addressing complex challenges by engaging ordinary people in scientific research and data collection and harnessing their collective intelligence. Both approaches can be useful for a wide range of urban challenges, but especially the second one emerged predominantly in the fields of ecology, biodiversity, and climate change. The aim of this paper is to demonstrate the use of these two approaches not only for scientific research in itself but also for the preservation of well-being and a healthy environment in the urban context, especially in medium-sized cities. The implications of adopting the approach proposed by this paper imply first of all that cities should equip themselves with research offices and dedicated technology infrastructure. Furthermore, local authorities should start developing policies based on the knowledge and data obtained through these tools.

Keywords

City science · Citizen science · Sustainability · Evidence-based policy making · Public policy

1 Introduction

Recently, the need for ecologic transition has heavily affected urban settings and consequently forced cities to revise traditional land-use governance strategies, opening up to active citizen input and moving toward more scientific and data-driven, evidence-based government strategies (Cavallo Perin and Racca 2019; Beccarello and Di Foggia 2022).

In Italy, as well as in other countries, there's a noticeable problem with a lack of crucial expertise within local governments. Various factors contribute to this, making it difficult to attract skilled individuals who are willing to work for local authorities. This issue is compound by the fact that the average age of the workers in these entities is 50.7 years, significantly higher than the average of 44.2 years recorded in 2001.[1] Additionally, small- and medium-sized cities often lack the necessary data and information required for developing policies that are grounded in data, his digital and technological gap is especially prevalent in these areas (World Economic Forum 2022). This raises the question: how can local governments bridge this gap between research and policy design? One potential solution lies in tapping into the expertise of both citizens and academia, combining civic engagement and scientific knowledge to support better policy making efforts (Iaione 2021; Testa et al. 2022).

☆Paragraphs n. 1, 4, 5 are attributed to Francesco Berni and n. 2, 3, 6 are attributed to Davide Testa.

D. Testa (✉)
LUISS Guido Carli, Rome, Italy
e-mail: dtesta@luiss.it

D. Testa · F. Berni
City Science Office, Reggio Emilia, Italy
e-mail: francesco.berni@comune.re.it

[1] Forum PA Data Insight, Lavoro pubblico 2023, Maggio 2023.

P. Mohammad et al. (eds.), *Geographic Approaches to Climate Change and Mitigation: Urban and Rural Perspectives (Volume 2)*,
Advances in Science, Technology & Innovation, https://doi.org/10.1007/978-3-031-92115-5_19

To tackle these issues, various initiatives have emerged at the European level: spanning from the regulatory measures[2] to funding mechanisms, such as the Horizon Europe package, succeeding the Horizon 2020 strategy. This occurs in a context where cities play an increasingly prominent role in European funding and development strategies,[3] particularly empathizing environmental concerns. Examples include the strategy for the 100 Climate-Neutral and Smart Cities by 2030, or the Global Covenant of Mayors for Climate & Energy.

In particular, this paper focuses on medium-sized cities in Italy[4] as they represent the typical territorial dimension of the country. These cities, alongside inland areas and small towns, face significant challenges due to the research-policy gap besides, a structural lack of skills and an overall limited attractiveness for companies and investors, despite possessing considerable innovative capacity. Establishing collaborative relationships between research institutions and local governments, may offer a solution enabling the achievement of ambitious goals through multi-actor collaborations as discussed by Berni et al. (2022) and Iaione (2022), and further developed in this paper.

The concept of combining administration and scientific expertise, akin to already established practices—for companies that have long relied on research and innovation departments—prompts both academics and practitioners to question the possible developments in the public sector. Moreover, it is reasonable to assume that a highly scientific relationship between local governments and academia could leverage the analysis of data produced by the urban environment and analyzed through high-level skills for meaningful insights.

Recognizing the importance of data and engaging with city residents leads to the adoption of citizen science; meanwhile, the pursuit of data-driven policies leads to the emergence of city science inside the public administration.

2 A Scientific Approach to Tackle the Biggest Challenges Facing Cities: Reggio Emilia as a Case Study

Given the aforementioned challenges, an important question arises: how can efficient public policies be developed to tackle major challenges such as ecologic transition, ensuring a high standard of reliability and effectiveness, especially in Italian medium-sized cities? This paper addresses the question by examining the empirical case study of Reggio Emilia, Italy.

To answer the paper's question, first of all, existing research could assess the importance of developing data-driven policies in a local government looking at how they can significantly increase the well-being of the citizenry. "Urban data" (Dughiero et al. 2021): a subset of "big data" focusing at improving city services and welfare, encompasses various types such as "sensor data on urban infrastructure and moving objects, user data on society and humans, governmental administration data, customer and transaction record data, and arts and humanities data" (Pan et al. 2016).

The case study of Reggio Emilia highlights two synergistic approaches, both rooted in the open innovation paradigm (Chesbrough 2003): (a) City science, based on the relationship between the academia and the local government[5]; (b) Citizen science, a 360-degree approach to scientific research, which has proven particularly effective in urban and climate challenges and still needs to be properly developed (Liu et al. 2023).

In combining these two dimensions, the municipality of Reggio Emilia, in collaboration with university actors, developed a multidisciplinary research team composed of a mixture of full researchers, PhD students and public servants, which is called City Science Office (Berni et al. 2022). Additionally, Reggio Emilia is a member of the European network of City Science Offices, coordinated by the city of Amsterdam under the City Science Initiative, a Horizon 2020 program promoted by the EU Commission's Joint Research Centre (Nevejan 2020).

In this framework, citizen science represents a specific approach used to support decision-making processes, and in particular to develop data-oriented policies in the environmental and ecologic field (Prandi 2022; Berni et al. 2023). An example of that approach is the "Reggio City of Trails" project, launched by the municipality of Reggio Emilia in 2020. It is a project aiming at the regeneration of urban environmental commons—i.e., urban pathways—through an active collaboration between the local government, associations and citizens, through participatory workshops carried out in neighborhoods. The new urban footpaths

[2] For instance, the Communication "Twinning the green and digital transitions in the new geopolitical context", which is part of the EU Green Deal, the new Data governance act (Regulation EU 2022/868), and the previous Open Data Directive (Directive EU 2019/1024).

[3] We can think, for example, of Urban Innovative Actions, or the URBACT IV Cooperation Programme (2021–2027). At the national level, moreover, it is worth mentioning the Smarter Italy program, It is a program developed in concert by the Ministry of Business and Made in Italy, the Ministry of University and Research and the Department for Digital Transformation, and implemented by the Digital Italy Agency.

[4] We can consider medium-sized cities, almost in the EU framework, those between 50,000 and 250,000 inhabitants (Andreani et al. 2013). Reggio Emilia, our case study, has around 170,000 inhabitants.

[5] City science, as a tool, is mostly a purely institutional one, although there is no clear definition of it (Duminy and Parnell 2020).

were designed with a genuine citizen science approach, co-projecting the trails—that increased from 85 to 246 km in total length—through designing walks carried together with the people.

Furthermore, another relevant project for the fight against urban pollution and climate change, in Reggio Emilia, is the Urban Forestation, launched in 2019 and based on a previous study conducted by the University of Venice (IUAV), that produced a mapping of urban heat islands. This project, therefore, was based on a typical and effective city science approach, in support of a data-oriented strategy dedicated to the containment of that issue through a massive forestation project. Citizens, moreover, as well as some private companies acting in accordance with a social responsibility perspective, participated actively in the implementation of the urban forestation project.

Those projects are showing how public administrations deal with limited resources, such as the deep "research-policy gap," which consists in a widespread lack of skills in reading and interpreting data, the failure to collect and analyses properly the specific datasets useful for a purpose and the frequent unavailability of environmental data. This evidently causes heavy difficulties in developing data-oriented policies, particularly on those crucial issues where it is strongly needed, such as ecologic transition (Dingelstad et al. 2022; Keller et al. 2017).

The study therefore aims to propose a solution to these drawbacks, helping local government pursue challenges of the utmost importance, such as climate, through increased knowledge, based on the analysis of urban data. The proposed solution passes first of all through the consolidation of a "city science" that local government would have available, among its own staff or even making use of experts from the academy, as in the case of the City Science Office of Reggio Emilia (Berni 2021). Moreover, adopting a collaborative approach even in data collection and analysis, of which citizens (in a broad sense: inhabitants, tourists, students, workers) should become protagonists: and this is what citizen science is all about.

Finally, the aim of this paper is also to present a model that can be exported to other local governments of approximately the same dimensions, leading to the sharing of best practices and further improvement of innovative policies through comparison among those cities.

3 Literature Relevant to the Research Question: An Analysis Conducted Through Four Search Queries

For the purposes of this paper, is it possible to identify four relevant topics on which to research studies and academic literature: open innovation, city science, citizen science and data-oriented policies.

Open innovation (European Commission 2016), first of all, is a classical topic of urban studies[6] related to urban policies: the key assumption is that the development of local government in the direction of widespread welfare and optimization of urban services cannot rely only on in-house resources, such as the employees of the public administration (Shibasaki et al. 2020). The local government, therefore, can open its doors to the expertise of civic innovators (Chesbrough 2003; Mirowski 2018) drawn from the broad citizenry and from the studies of the academics in a variety of fields, such as architecture, law, geography, sociology, economics and political science.

This approach leads to city science, which implies a comprehensive consideration of the urban context, supported both by theoretical and empirical research. While the first can count on various collective books and papers produced by research groups focused on the city, and in particular on the collaborative experiences that are on the rise in many local governments (*ex multis*: Iaione et al. 2019; Pizzolato et al. 2022), the second is often developed around case studies and the concrete issues of a city. The academia and the local governments, therefore, can benefit each other, bringing science to the administration and case studies to the academia.

In this regard, every aspect of city functioning, and citizen relations is taken into consideration, including housing, transportation, climate change, public health, social, and economic inequalities (Iaione and De Nictolis 2017). Urban studies, in this perspective, are the theoretical prerequisite in order to build innovative urban policies based on city science, seeking to narrow the research-policy gap to zero. Noteworthy, as mentioned earlier, the direct involvement of citizens and the use of expertise and data gathered from the public is another key component of open innovation.

About citizen participation in urban applied research, and thus citizen science as a research method for the development of innovative policies, first of all some methodological literature can be found (Veeckman et al. 2021), including papers about ICT, digitalization and computer science: among them, for instance, feel really interesting some researches about

---

[6] Urban studies, in general, are a broad and interdisciplinary field of research that is expanding in relation to the rediscovered importance of cities and city networks in a highly globalized and interconnected world (Pizzolato 2022). The real milestones of urban studies can be traced back to papers like Lefebvre's work "The Right to the City" (Lefebvre 1996) and Saskia Sassen's "Cities in a World Economy" (Sassen 1994).

gamification, which is a technique of participation in scientific research aimed at encouraging the inclusion of people otherwise more difficult to engage, like the youth, through transforming the collection of datasets and information functional to the research question into a game (Cardoso et al. 2019).

The theme of civic participation through digital technologies (De Marchi et al. 2001), is emerging in the work of many research groups, which are involved in citizen science and are producing some interesting products (Berti Suman and Alblas 2023; Hultquist et al. 2022; Schade et al. 2020), like the book "Citizen Science," published by UCL in 2018 (Hecker et al. 2018). It is also possible to rely on almost other two trustworthy sources of knowledge on the topic, in particular if looking for case studies to compare: the work of private associations—the European Citizen Science Association,[7] the recently established Citizen Science Italia ETS[8] —and, above all, the work of the Joint Research Centre of the European Commission, which recognizes the strategic importance of citizen science and brings together the most relevant collected works[9] (Maccani et al. 2020).

Based on the existing literature, Citizen science concentrates most of its potential in the widespread collection of urban data and, for this reason, is primarily useful for the development of data-oriented innovative policies (Figueiredo Nascimento et al. 2016; Schade et al. 2021; Criscuolo et al. 2023). The importance of data-oriented policies for a sustainable development of cities obviously (Irwin 1995) lies, again, in overcoming the research-policy gap that prevents making the most efficient choices and developing policies based on a holistic knowledge of all urban features (Pan et al. 2016; Dughiero et al. 2021; Cavallo Perin and Racca 2019; Ruijer et al. 2023). The availability and ability to interpret data, coupled with a collaborative strategy that ensures the democratic and collective approach to policy making, seems thus to be capable of ensuring better policies for all citizens.

4 Methodology

The methodology followed in this study, as shown in Fig. 1, is twofold: it combines empirical methods, including case studies and a "research to action" approach, and theoretical frameworks.

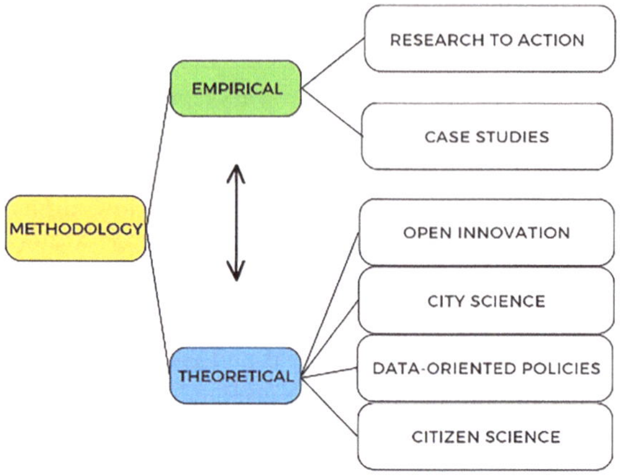

Fig. 1 Methodology. Display of the methodology adopted, partly theoretical—divided into four areas of investigation—and partly empirical, developed in the field. Figure by D. Testa

The empirical aspect of the study is closely tied to the very nature of the City Science Office of Reggio Emilia, the organizational unit established in the Municipality of Reggio Emilia aimed at supporting the development of innovative urban policies through field research (Berni et al. 2022), to which the authors belong. It is worth noting that empirical research is becoming increasingly important, not only in the academic circles but also in disciplinary fields where traditionally it has not been developed as much, such as in legal research (Petersen and Chatziathanasiou 2021)—as well as among local authorities, which are interested in taking advantage of the best practices developed elsewhere.

Case studies, therefore, are pivotal to the "research to action" approach. To the aim of this paper, the main case study investigated is represented by the recent environmentally significant policies pursued by the municipality of Reggio Emilia with specific reference to two recent experiences, it also invites comparison with other cities engaged in city science and citizen science.

The "research to action" method (Somekh 1995) is gaining traction particularly within the European innovation system, with initiatives like Horizon 2020 and Horizon Europe funding, which—like the City Science Initiative—involve universities and cities in the realization of joint projects, in which the city is the testbed, the innovator, and the academy provides the scientific support and conceptualizes the outcomes with a view to future replicability. In the Reggio Emilia case study, the "research to action" approach is exemplified by the collaboration between civil servants, municipal managers, and researchers from the City Science Office (Berni et al. 2022).

On the other hand, case studies and the empirical research approach still need a strong theoretical basis for each of the previously mentioned pillars: for this reason, a review of the literature concerning the development of innovative policies

[7] European Citizen Science Association, https://www.ecsa.ngo/.

[8] Citizen Science Italia ETS, https://www.museonaturalemaremma.it/csi/.

[9] EU Commission JRC, Publications, https://joint-research-centre.ec.europa.eu/scientific-activities-z/citizen-science-eu-policies/publications_en.

at the local level—as developed before—is needed. Specifically this paper delves into the impact of digital technologies on citizen science opportunities and to the benefits carried by a data-oriented and evidence-based administrative model for the quality of life of citizens and for the city as a whole. This has been accomplished in particular by investigating the literature on the four topics examined in the previous section.

Reggio Emilia represents the effort to build a sustainable, research-based development of a medium-sized city and it supports us investigating how citizens can contribute to this model of urban development, thus democratizing research and bridging it beyond academia's confines.

5 Results of Research and Analysis of the Case Study

Cities today face monumental challenges, particularly the imperative of ecological transition, which demands local authorities the need to seek innovative ways to construct effective, efficient and collaborative public policies. These are wicked problems (Rittel and Webber 1973) that require to be addressed with innovative tools and strong professional competencies.

In Italy, one pressing issue is the progressive reduction in public administration personnel largely due to austerity measures initiated in response to the economic crisis of the past decade. This has resulted in an aging workforce and a lack of professional updating. Additionally, there is a structural lack of professional profiles dedicated to data analysis and interpretation, aiming to the development of evidence-based policies. Sometimes, the lack of proper data collection poses obstacles impossible to overcome to any attempt to improve urban policies in a scientific manner.

These challenges are particularly daunting for medium-sized cities, a significant category in the Italian territory, as they must deal both with a lack of economic and professional resources, especially in comparison with the major cities and the few metropolitan cities in the Italian Country.

The case study of Reggio Emilia demonstrates that, through an approach based on a structured collaboration between the local government, the university and the community, it is possible to develop evidence-based policies and integrated projects that can contribute to improve the quality of life in the urban environment. That city, in fact, gave birth to a real process of administrative innovation by establishing the City Science Office within the local government, integrating multiple external and internal professional skills, and including citizens in urban projects, through a citizen science approach.

6 Conclusion and Key Elements for Future Scenarios

Finding on the Reggio Emilia case study and on the theoretical framework adopted through this paper, several key points emerge, which are also related to the potential replicability of the city science and citizen science strategies in different contexts, and, according to Fig. 2, may be summarized in three main dimensions:

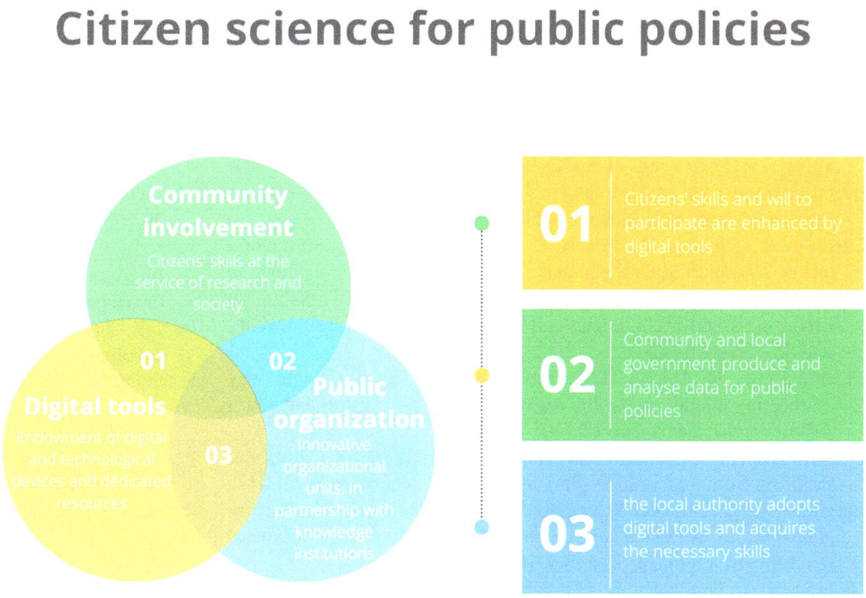

Fig. 2 The three pillars of citizen science for public policies: the figure particularly highlights the mutual intersection between the public actor, citizens and digital tools. Figure by D. Testa

(a) Public organization: the local authority's ability to establish innovative organizational units, open to external and especially academic experts, and the capacity of the local government to assimilate the new strategies emerging from this city science approach.
(b) Community involvement: the presence of a local community aiming to be actively involved in public choices, along with a local authority capable of developing programs and activating adequate tools capable of enabling the best skills of citizens.
(c) Digital tools and costs: the endowment of digital and technological devices to support research and data-oriented policies, such as digital platforms, can be crucial for citizen science and city science practices. Connected to that, another key element is the availability of economic resources dedicated to support the costs of city science and citizen science initiatives, ensuring the continuity and the improvement of these efforts over time.

References

Andreani, V., Giorgio, C., Marinuzzi, G.: The Italy of medium-sized cities [L'Italia delle città medie]. I Comuni. Quaderni di analisi ANCI-IFEL, p. 4 (2013)

Beccarello, M., Di Foggia, G.: Sustainable development goals data-driven local policy: focus on SDG 11 and SDG 12. Admin. Sci. **12** (2022)

Berni, F.: The neighbourhood architect. Innovating the public authority in the field of collaborative regeneration of urban spaces [L'Architetto di Quartiere. Innovare l'ente pubblico nel campo della rigenerazione collaborativa degli spazi urbani]. In Atti della XXIII Conferenza Nazionale SIU—Società italiana degli Urbanisti (2021)

Berni, F., De Franco, L., Levi, N.: The City Science Office of Reggio Emilia: research and innovation paths in the energy and social fields [Il City Science Office di Reggio Emilia: percorsi di ricerca e innovazione in campo energetico e sociale], Diritto e Società, p. 4 (2022)

Berni, F., Santangelo, G., Testa, D.: Open innovation processes in Reggio Emilia: science, politics and active citizenship [Processi di innovazione aperta a Reggio Emilia: scienza, politica e cittadinanza attiva]. In: Proceedings of the XXV SIU National Conference (2023)

Berti Suman, A., Alblas, E.: Exploring citizen science over time: sensing, technology and the law. Sustainability 5 (2023)

Cardoso, B., Ribeiro, M., Prandi, C., Nunes, N.: When gamification meets sustainability: a pervasive approach to foster sustainable mobility in Madeira. In: Proceedings of the 1st ACM Workshop on Emerging Smart Technologies and Infrastructures for Smart Mobility and Sustainability (2019)

Cavallo Perin, R., Racca, G.M.: Smart cities for an intelligent way of meeting social needs. In: Auby, J.-B. (ed.) Le Futur Du Droit Administratif. Lexisnexis (2019)

Chesbrough, H.: Open Innovation: The New Imperative for Creating and Profiting from Technology. Harvard Business School Press (2003)

Criscuolo, L., L'Astorina, A., van der Wal, R., Colucci Gray, L.: Recent contributions of citizen science on sustainability policies: a critical review. Curr. Opin. Environ. Sci. Health **31** (2023)

De Marchi, B., Funtowicz, S., Guimarães-Pereira, A.: From the right to be informed to the right to participate: responding to the evolution of European legislation with ICT. Int. J. Environ. Pollut. **15**(1), 1–21 (2001)

Dingelstad, J., Borst, R.T., Meijer, A.: Hybrid data competencies for municipal civil servants: an empirical analysis of the required competencies for data-driven decision-making. Public Pers. Manag. **4**, 458–490 (2022)

Dughiero, F., Michieli, A., Spiller, E., Testa, D.: Governing with urban big data in the smart city environment: an Italian perspective. Ius Publicum Netw. Rev. **1** (2021)

Duminy, J., Parnell, S.: City science: a chaotic concept—and an enduring imperative. Plan. Theory Pract. **4**, 648–655 (2020)

European Commission: Open Innovation, Open Science, Open to the World—A Vision for Europe. https://ec.europa.eu/digital-single-market/en/news/open-innovation-open-science-open-world-vision-europe

Figueiredo Nascimento, S., Cuccillato, E., Schade, S., Guimarães Pereira, A.: Citizen Engagement in Science and Policy-Making. Publication Office of the European Union (2016)

Hecker, S., Haklay, M., Bowser, A., Makuch, Z., Vogel, J., Bonn, A.: Citizen Science. Innovation in Open Science, Society and Policy. UCL Press (2018)

Hultquist, C., De Sherbinin, A., Bowser, A., Schade, S.: Editorial: Open citizen science data and methods. Front. Clim. **4** (2022)

Keller, S., Lancaster, V., Shipp, S.: Building capacity for data-driven governance: creating a new foundation for democracy. Stat. Public Policy **1** (2017)

Iaione, C.: City, science and innovation: the right to science for the city as a cornerstone of a new urban governance oriented towards sustainable development and intergenerational responsibility [Città, scienza e innovazione: il diritto alla scienza per la città come pietra angolare di una nuova governance urbana orientata allo sviluppo sostenibile e alla responsabilità intergenerazionale]. Munus **3** (2021)

Iaione, C.: Urban sustainable development and innovation partnerships. Italian J. Public Law **14**(2), 521–605 (2022)

Iaione, C., De Nictolis, E.: Urban pooling. Fordham Urban Law J. **44**(3), 665–702 (2017)

Iaione, C., De Nictolis, E., Foster, S.: The Co-Cities Open Book for Just and Inclusive Cities (2019). https://commoning.city/the-co-cities-open-book/

Irwin, A.: Citizen Science: A Study of People, Expertise and Sustainable Development (Environment and Society). Routledge (1995)

Lefebvre, H.: The right to the city. In: Kofman, E., Lebas, E. (eds.) Writings on Cities. Wiley-Blackwell (1996)

Liu, F., Zhang, C., Zhang, Y., Liu, H.: A data-driven approach for the measurement and improvement of regional industrial ecological efficiency for carbon peaking and carbon neutralization. Environ. Sci. Pollut. Res. **30** (2023)

Maccani, G., Goossensen, M., Righi, V., Creus, J., Balestrini, M.: Scaling up Citizen Science. Publications Office of the European Union (2020)

Mirowski, P.: The future(s) of open science. Soc. Stud. Sci. **48**(2), 171–203 (2018)

Nevejan, C.: City Science for urban challenges. Pilot assessment and future potential of the City Science Initiative 2019–2020, report for the European Commission JRC (2020)

Pan, Y., Tian, Y., Liu, X., Gu, D., Hua, G.: Urban big data and the development of city intelligence. Engineering **2** (2016)

Petersen, N., Chatziathanasiou, K.: Empirical research in comparative constitutional law: the cool kid on the block or all smoke and mirrors? I-CON **5**, 1810–1834 (2021)

Pizzolato, F.: Floating cities, networks and territory: transnational constitutionalism between functionalism and democratic instance [Floating cities, reti e territorio: il costituzionalismo transnazionale tra funzionalismo e istanza democratica]. In: Pizzolato, F., Rivosecchi, G., Scalone, A. (eds.) La città oltre lo Stato. Giappichelli (2022)

Pizzolato, F., Rivosecchi, G. Scalone, A.: The City Beyond the State [La città oltre lo Stato]. Giappichelli (2022)

Prandi, C. (2022). Citizen Science: State of the Art and Opportunities in the Regional Context [Citizen science: stato dell'arte e opportunità nel contesto regionale]. Citizer Science report 1.0

Rittel, H., Webber, M.: Dilemmas in a general theory of planning. Policy Sci. **4**, 155–169 (1973)

Ruijer, E., Porumbescu, G., Porter, R., Piotrowski, S.: Social equity in the data era: a systematic literature review of data-driven public service research. Public Admin. Rev. **2**, 316–332 (2023)

Sassen, S.: Cities in a World Economy. Pine Press (1994)

Schade, S., Pelacho, M., van Noordwijk, T., Vohland, K., Hecker, S., Manzoni, M.: Citizen science and policy. In: Vohland, K., et al. (eds.) The Science of Citizen Science. Springer (2021)

Schade, S., Tsinaraki, C., Manzoni, M., Berti Suman, A., Spinelli, F.A., Mitton, I., Kotsev, A., Delipetrev, B., Fullerton, K.T.: Activity Report on Citizen Science—Discoveries from a Five Year Journey. Publications Office of the European Union (2020)

Shibasaki, R., Hori, S., Kawamura, S., Tani, S.: Integrating urban data with urban services. In: Hitachi-Utokyo Laboratory (ed.) Society 5.0. A People-Centric Super-Smart Society. Springer (2020)

Somekh, B.: The Contribution of Action Research to Development in Social Endeavours: a position paper on action research methodology. Br. Educ. Res. J. **21**(3), 339–355 (1995)

Testa, D., Berni, F., Santangelo, G.: From energy community to climate city contract: towards climate-neutral neighbourhoods in Reggio Emilia [Dalla comunità energetica al climate city contract: verso quartieri a impatto climatico zero a Reggio Emilia]. Diritto e Società **4** (2022)

Veeckman, C., Van Herck, B., Carpentier, M., Van Laer, J., Sterken, M.: Citizen Science Roadmap for Local Government. A Hands-on Manual for Citizen Science, by and for Cities and Towns. SCIVIL, Leuven, Belgium (2021)

World Economic Forum: Shaping the Future of Small and Medium-Sized Cities: A Framework for Digital Transformation. Insight Report (2022)

Fostering Green Consensus: A World Café Approach to Participatory Democracy and Climate Action in Local Communities

Umberto Lavorata

Abstract

This paper explores the potential of the World Café approach to induce a transformation in individual preferences and to facilitate democratic dialogue and action on climate change at the micro-level. In an era marked by widespread skepticism toward institutions on local, national, supranational levels, the implementation of deliberative and participatory models of democracy has emerged as a promising response to address pressing global issues, including climate change and local environmental concerns. The Intergovernmental Panel on Climate Change (IPCC) underscores the significance of deliberative policy-making processes in formulating adaptive strategies, emphasizing the role of local communities in combatting the climate crisis. In particular, this research assesses the capacity of a World Café initiative, conducted in four local communities across Emilia-Romagna and Calabria regions, to stimulate shifts in individual preferences and foster a *green consensus* among participants. A total of 58 individuals participated in these quasi-experiments. Using an action-research method characterized by features of the qualitative research, our study explores the following three key themes of discussion: (1) Separate Waste Collection; (2) Green Urban Areas; (3) Renewable Energy. The results provide insights into the current state of citizens' participation in grassroots environmental issues at a micro-level and offer innovative solutions that emerged from the dialogues, which stakeholders may consider for implementation. Each World Café session empowered participants to establish networks, fostering collective intelligence. This study contributes to the ongoing discourse on the role of local communities in shaping environmentally sustainable policies and actions, aligning with the goals of the 2030 Agenda, particularly SDGs 7, 11, 12, and 13, by promoting community engagement and consensus building for a sustainable future.

Keywords

Participatory democracy · Citizens' participation · Climate action · World café · Sustainable development

1 Introduction

In the context of escalating climate crisis, it is imperative to comprehend the mechanisms through which informal discussions can alter individual and collective perspectives. This research aims to ascertain whether the World Café format can effectively bridge diverse viewpoints and cultivate a shared understanding that motivates action. Furthermore, urban and rural communities frequently confront disparate environmental challenges and possess varying degrees of access to resources. To this end, this research seeks to identify strategies for inclusively engaging both demographics, leveraging their distinctive insights and experiences to co-create solutions.

One of the goals of this paper is to demonstrate that, although urban contexts are more dynamic by definition, there are many communities, such as those in Italian rural/inner areas, willing to make an effective contribution through active participation (Moro and Sorice 2022) to address the climate challenge starting from a bottom-up perspective. Furthermore, this paper intends to provide a comprehensive overview of the research, abstaining from delving into the intricacies of each quasi-experiment and localized context.

U. Lavorata (✉)
Department of Political and Social Sciences, University of Bologna, Bologna, Italy
e-mail: umberto.lavorata2@unibo.it

P. Mohammad et al. (eds.), *Geographic Approaches to Climate Change and Mitigation: Urban and Rural Perspectives (Volume 2)*,
Advances in Science, Technology & Innovation, https://doi.org/10.1007/978-3-031-92115-5_20

Climate change is one of the most pressing challenges of our time, with significant impact on the planet and human societies (IPCC 2023; UNFCCC 2015; World Bank 2023). Climate action is nevertheless urgently needed: climate resilience requires an integrated view with adaptation and mitigation, in order to advance a real sustainable development (IPCC 2023).

As a result, it is important to take a localized approach to climate action, taking into account the specific needs and circumstances of each community when developing and implementing strategies to address climate change.

However, the United Nations Human Settlements Program (UN-Habitat) pays special attention to the urban–rural linkages, intended as an interdependent dynamic, considering the continuous flow of people, goods and environmental services between rural, peri-urban and urban areas (UN-Habitat 2019). While urban areas are crucial arenas in terms of adaptation and mitigation to climate change, the role played by rural/inner areas should not be underestimated. Indeed, at a European level, the importance of rural areas toward a green transition is also emphasized by EU COM (2021) 345 final, mentioning four areas for action, each representing a key factor that will influence the development of rural areas by 2040, i.e., (1) stronger (empowered communities, access to services, social innovation); (2) connected (digital connectivity, transport links, and new mobilities); (3) prosperous (diversification of economic activities, sustainable food production); (4) resilient (resilience to climate change, environmental resilience, social resilience).

Deliberative processes (e.g., citizens' juries, deliberative polls®, etc.) and citizens' participation in local democratic life could play a pivotal role in achieving the climate neutrality goals, set by European Union, as well as by municipalities and regional governments. Nevertheless, all deliberative processes require many days or years of interactions, as well as financial resources (e.g., for the Bologna's citizens' jury in 2006, each juror received a gift voucher—Giannetti et al. 2007).

The objective of our research was to make a contribution to the strengthening and analysis of the concept of participation within specific contexts, with a particular focus on climate change challenge. The objective was to ascertain whether fieldworks conducted in the World Café format (Brown and Isaacs 2005; Löhr et al. 2020; Steier et al. 2015) could effect a change in individual preferences, as would occur in a genuine deliberative process (Fishkin and Luskin 2005). The change of individual preferences, as well as other factors (e.g., satisfaction levels, knowledge acquisition, potential impacts of the quasi-experiments), was verified using pre- and post-event questionnaires, recalling the deliberative poll® (Fishkin 2018) and the World Café experiment by Alunni-Menichini et al. (2023).

We therefore used action-research method (Egmose 2019; Elden and Levin 1991; Ventura and Shahar 2022; Waardenburg et al. 2020), characterized by typical features of qualitative approach (Corbetta 2015). Action research is able to build transformative processes of the individual-environment system, by employing a method focused on empowerment through collective learning (Mannarini 2009).

Given that the needs and citizens' participation differ from one local context to another, it was essential to consider specific issues that would be relevant and, at the same time, open in comparison with the areas considered, also in order to structure the World Cafés as spaces for social learning and concerted action (Collins 2014), making such initiatives attractive for residents.

However, the World Café sessions were arranged in collaboration with some local stakeholders, such as associations (i.e., Cittadinanzattiva, ARCI), a local municipality (i.e., Municipality of Gazzola), and a local ethological donkey farm in Santa Caterina (i.e., Avamposto Agricolo Autonomo), in order to avoid or at least limit the impression of the "outsider researcher" (Kerstetter 2012).

The decision to examine two distinct Italian regions, Emilia-Romagna in the north and Calabria in the south, is based on the premise that socio-historical differences in civic participation exist between these areas. As Putnam (1993) posited, the economic and social circumstances in southern Italy fostered a culture of individualism rather than collective efforts. In contrast, other regions, such as Emilia-Romagna, faced similar challenges, which led to the emergence of cooperative organizations. Putnam's research commences with the observation that in the 1970s, a uniform institutional model was implemented across all Italian regions. He identifies a significant correlation between institutional effectiveness and the presence of a distinctive local political culture, designated as "civicness" or civic culture (Putnam 1993). This culture is distinguished by a pervasive political engagement among citizens, reinforced by a robust interpersonal trust and a tradition of cooperation. As Almagisti (2016) observes, the term "social capital" was initially employed by Putnam (1993) to examine political phenomena, particularly the institutional performance of Italian regions. According to Putnam, a society is considered to possess a strong civic culture when it exhibits a high level of social capital.

This paper is structured as follows: Sect. 1 discusses the literature review about deliberative democracy and the change in individual preferences, incorporating perspectives from various scholars as well as insights from grey literature (e.g., IPCC); Sect. 2 explores the World Café approach, the case selection and the Café's themes, the methods used for stakeholders' and participants' recruitment, the structure of the World Cafés, synthesizing the results of discussions, and presenting the findings from the pre- and post-event questionnaires; Sect. 3 investigates strengths and weaknesses of our fieldworks, putting in relation the results from pre- and post-event questionnaires. The conclusions highlight the efficacy of the World Café approach in rural and urban contexts, considering the possible avenues for future research.

1.1 Deliberative Democracy and the Change in Individual Preferences: Literature Review

Deliberative democracy is an innovative model of democracy in which the influence of citizens is not merely based on a decision, but it is based on shaping a decision through debate (Moro 2009). Thus, deliberative democracy could be considered as a deep analysis of processes that lead to a decision, concerning several issues (e.g., climate change and local environmental concerns). In its Sixth Assessment Report (IPCC 2022), for instance, the Intergovernmental Panel on Climate Change emphasizes the importance of civil society's engagement for developing and implementing mitigation strategies.

However, over the years the attention of scholars on deliberative democracy and, in general, on participatory forms of democracy, has significantly grown (e.g., Helbing et al. 2023; Talan et al. 2023; Willis et al. 2022). Within this context, while the "vulnerable" liberal conception of democracy could be seen as the aggregation of individual preferences, the "less vulnerable" deliberative democracy could be intended as a process of open discussion leading to an agreed judgment on policy (Miller 1992).

While deliberative theory offers a model of democracy as a forum through which judgments and preferences are formed and altered through reasoned dialogue between free and equal citizens, against the picture of democracy as a procedure for aggregating and effectively meeting the given preferences of individuals (O'Neill 2002), "inclusivity" constitutes a keyword of the whole concept of deliberative democracy.

Additionally, deliberative processes have the potential to benefit participants by increasing their knowledge and understanding of complex issues (Celaya 2019) that, in turn, could motivate participants to re-evaluate their policy preferences (Fishkin and Luskin 2005). Nevertheless, participation represents a fundamental prerequisite for the initiation of deliberation.

As highlighted by Bobbio (2010), one of the more intriguing aspects of deliberative processes is that each participant arrives at the dialogic forum with his or her own opinion on the topic under debate. These starting positions are based on preferences, ideas about the state of the world, and cause-and-effect linkages. The point is that the nature of these initial positions might vary substantially between contexts and situations. They can be more or less distinct, firm, or pliable. Not all participants start the deliberative process with the same level of structure or conviction. And this first component is likely to influence the subsequent phase.

In other words, participation and deliberation seek to transcend the mere aggregation of preferences, striving to generate collective knowledge and prioritizing the well-being of the community as a whole. Within this framework, Elster's statements are eloquent about the ideal role of politics, advocating for a shift in focus: "[...] the central concern of politics should be the *transformation of preferences* rather than their aggregation. On this view the core of the political process is the public and rational discussion about the common good, not the isolated act of voting according to private preferences. [...]" (Elster 2016, pp. 34–35).

By prioritizing transformation over aggregation, we can empower citizens to critically examine their preferences and work together toward solutions that benefit everyone, for instance from the environmental point of view.

However, deliberative poll® constitutes a practice of deliberative democracy that employs a pre- and a post-event questionnaire to assess the impact of the process on individual preferences (Fishkin 2018): a deliberative poll® typically gathers a cohort of 100–200 randomly selected participants, who initially respond to a questionnaire about a specific policy issue; then, those participants, organized into small groups, are engaged in discussions, at the presence of experts and facilitators; at the end of the process, another questionnaire assesses the transformation or persistence of initial preferences (Floridia 2017).

Nevertheless, deliberative polls® are essentially "mini-publics" (Fung 2007), deliberative arenas bringing together randomly selected citizens, conceived as a representative microcosm, a representative sample of the society (Floridia 2017): the idea is that the group (i.e., mini-public), after learning about the issues, becomes a miniature representation of the public's potential views if everyone had access to the same information, allowing them to contribute in a well-informed deliberation (Floridia 2017). According to Floridia (2017), the deliberative poll® was designed to bring together three key aspects of democracy, i.e., political equality, nontyranny, and deliberation, stating that deliberative poll® can be considered as an "operational translation of a theoretical model of democratic deliberation" (Floridia 2017, p. 149, translated by the author).

Beyond such features, the dialogical aspect represents a key concept of deliberative democracy: for instance, in his analysis of the Australian Citizens' Parliament, Dryzek (2009) asserted that "deliberation is a particular kind of communication that ideally induces reflection about preferences, beliefs, and values in non-coercive fashion, and that connects particular interests to more general principles" (p. 3).

According to Dryzek (2009), a key virtue of this approach is its emphasis on reciprocity, whereby communication occurs within a framework that is both understandable to and acceptable to those who do not necessarily share the initial point of view. This distinguishes deliberation from the adversarial style of debate, where the goal is not to win but to understand. Deliberation permits the possibility of changing one's position when presented with new evidence or arguments.

However, such concepts refer to deliberative democracy practices that, as already pointed out, require a significant investment of resources, particularly in terms of money and time. Nevertheless, this understanding is crucial for analyzing the framework of the quasi-experiments presented in the following sections, allowing us to consider key elements of deliberative democracy (such as the dialogical aspect) and connecting them with specific practices, such as deliberative polls®, from an operational point of view.

Our quasi-experiments do not fully meet the criteria of deliberative democracy due to the lack of random selection of participants, and the absence of experts. There are two main reasons for this: (1) financial constraints (random selection of participants and the inclusion of experts are both time-consuming and expensive processes, as demonstrated, for instance, by the above-mentioned Bologna's citizens' jury by Giannetti et al. (2007), where a company in charge did a random selection of participants) and (2) the nature of fieldworks (designed to explore general environmental issues, without a particular policy issue, that would be emerged or not during the Cafés).

Therefore, our fieldworks may be more accurately considered participatory quasi-experiments which could offer a valuable contribution through the following three aspects, fostering a community-driven approach to research and potential deliberation: (1) deeper local context (each fieldwork situates the research within the specific context of participants' experiences), (2) unforeseen issues (the fieldworks have the potential to uncover unforeseen local issues that might not have been considered in the initial design of each quasi-experiment), and (3) foundation for deliberation (each fieldwork could serve as a springboard for future deliberative processes).

2 Combining the World Café and the Change in Individual Preferences: Insights from Our Research

The World Café is a dialogical and collaborative approach, aimed at creating a network of exchanges between people with different viewpoints and information on a given topic (Alunni-Menichini et al. 2023; Brown and Isaacs 2005; Steier et al. 2015). As highlighted on the *Participedia* platform, there are nearly 400 participatory methods. It is noteworthy that the World Café approach is a widely utilized method, including in the context of addressing climate change at the local level. For example, it was employed in the development of the Flood Risk Management Plan of the Po Basin in Italy in 2019 (Partecipedia 2019).

Considering, for example, the first World Café experience in 1995, the co-founder Juanita Brown stated that "the Café process had somehow enabled us to become more aware of our 'collective knowing' at increasing levels of scale, as the conversations cross-pollinated through the evolving rounds of conversation" (Steier et al. 2015, pp. 212–213). The World Café is able to foster a participatory environment, aligning with the core principles of the action-research method: "action research is a way of knowing rooted in engagement. This engagement derives from diverse forms of relationships, including relationships between people, and between people and places and things. What is clear is that the culture of inquiry that is action research, whatever form it takes, requires attention to the relationships that allow it to unfold" (Steier et al. 2015, p. 211).

Additionally, although it is a flexible approach which can be tailored to fit a wide range of contexts and topics (Alunni-Menichini et al. 2023), seven design principles must be taken into account to arrange a successful World Café session, considered for our quasi-experiments as well: (1) Set the context ("clarify the purpose and broad parameters within which the dialogue will unfold", Brown and Isaacs 2005, p. 40); (2) Create a hospitable space ("ensure the welcoming environment and psychological safety that nurtures personal comfort and mutual respect", p. 40); (3) Explore questions that matter ("focus collective attention on powerful questions that attract collaborative engagement", p. 40); (4) Encourage everyone's contribution ("enliven the relationship between the "me" and the "we" by inviting full participation and mutual giving", p. 40), keeping in mind that "the purpose of the Café is not to criticize, but to contribute" (Brown and Isaacs 2005, p. 100); (5) Cross-pollinate and connect diverse perspectives, as shown by the several arrows within Fig. 1 ("use the living-system dynamics of emergence through intentionally increasing the diversity and density of connections among perspectives while retaining a common focus on core questions", Brown and Isaacs 2005, p. 40); (6) Listen together for patterns, insights, and deeper questions ("focus shared attention in ways that nurture coherence of thought without losing individual contributions", p. 40); (7) Harvest and share collective discoveries ("make collective knowledge and insight visible and actionable", p. 40).

Each fieldwork commenced with the delineation of these fundamental principles that are indispensable for the optimal outcome of the World Café.

Figure 1 is crucial to understanding the underlying workings of the World Café. It presents a dichotomy between

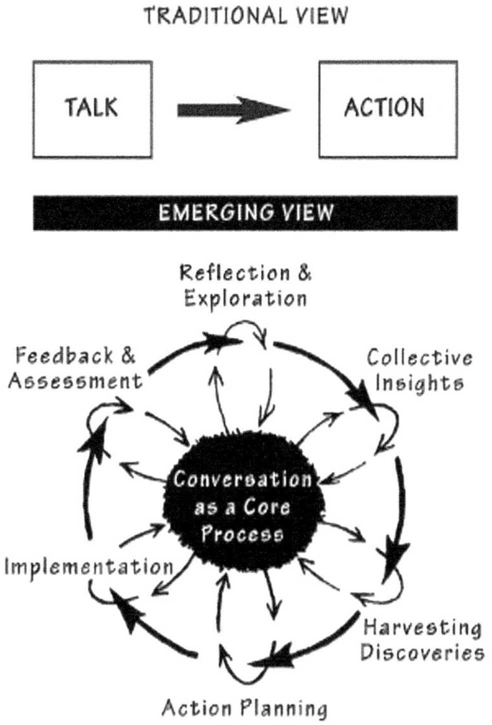

Fig. 1 Traditional view versus emerging view in the World Café. *Source* Brown and Isaacs 2005, p. 37

a linear "traditional view" of the transition from "talk" to "action," and an "emerging view" that presents a more complex, cyclical process. In the "traditional view," communication is conceived as a unidirectional process, moving from talk to action in a direct line, and implying a straightforward progression. In contrast, the "emerging view" portrays conversation as a fundamental, continuous process situated at the core of an ongoing cycle, underscoring the significance of dialogue in each stage. The cycle commences with the phase designated as "Reflection and Exploration," wherein nascent ideas and potential avenues of inquiry are contemplated. Subsequently, participants proceed to the "Collective Insights" phase, wherein shared understandings are developed through group discussion. These insights culminate in the "Harvesting Discoveries" phase, wherein valuable ideas and pivotal findings are identified and organized. The subsequent phase, "Action Planning," entails the organization of these discoveries into a logical plan for action. Subsequently, the plan is implemented, which is followed by the phase of implementation. The final stage of the cycle is "Feedback and Assessment," during which the results of the implemented actions are evaluated, generating reflections that inform future iterations of the process. The "emerging view" proposes a recursive model in which each phase informs the subsequent one. Throughout this process, conversation remains a crucial element to facilitate adaptation and continuous improvement.

Within such a framework, our Cafés were designed with a dual purpose, as mentioned earlier: (1) assess if a change in individual preferences occurs through a World Café discussion, building a *green consensus* and (2) create a favorable context for democratic dialogue about climate change in order to foster the emergence of innovative solutions around the specific topics at a local level.

Therefore, we opted for an innovative research approach, considering the principles of the World Café (Brown and Isaacs 2005; Löhr et al. 2020; Steier et al. 2015) together with a key feature of deliberative processes, i.e., the change in individual preferences (Bobbio 2010; Dryzek 2009; Dryzek and Niemeyer 2010; Fishkin and Luskin 2005; Fishkin 2018; Floridia 2017), using a qualitative design (Corbetta 2015). Such a combination allowed us to focus on three elements: (1) the individual/participant, (2) the inclusion of some local stakeholders, and (3) the (potential) impacts of the World Café on participants.

However, as already argued, one element missing from our quasi-experiments was the random selection of participants which is typical of the so-called mini-publics. Thus, the *open door* method was chosen, conscious of the biases that could inevitably have manifested (e.g., participation of *already active* citizens and *already interested* in the topics). Nevertheless, as noted by Bobbio (2019), it is also worth noting that such biases can, in certain circumstances, have positive effects on participants engaged in the process. Individuals with a high level of interest in a topic tend to be more informed and engaged, which can increase attention and concern for the issues being discussed. This heightened interest can, in some cases, help offset social inequalities in participation, especially if the topic at hand is relevant to disadvantaged groups. In summary, while the *open door* method has the potential for bias, it can also encourage more active and informed participation, contributing to a richer discussion.

Finally, the World Café format seemed to emerge as a particularly pertinent approach for collecting rich qualitative data: the participants' perspectives demonstrated how our quasi-experiments provided them with a valuable opportunity for exchange and interaction, capable of fostering the emergence and circulation of ideas useful for developing resilient communities to the challenge of climate change.

2.1 Case Selection and the Café's Topics

As stated in the introduction, the selection of Emilia-Romagna and Calabria regions, in the north and south Italy respectively, might be motivated by a desire to explore the contrasting socio-historical factors influencing civic engagement in these two regions. According to Putnam (1993), while economic and social conditions in southern Italy fostered individualism ("vertical bonds"), regions like Emilia-Romagna responded to similar challenges by developing cooperative structures ("horizontal bonds").

While the historical and cultural distinctions between northern and southern Italy are beyond the scope of this article, it is nonetheless important to outline the specific research areas and the methods employed for participants' and stakeholders' engagement.

However, the selection of the areas examined in this research should not be viewed solely through a political lens; rather, it should be considered from a comprehensive perspective that encompasses sociology, geography, and politics. Indeed, participatory and deliberative democracy is inextricably linked with the concept of territory, where collective actions are of paramount importance in terms of its organization and transformation (De Salvo 2021). As will be detailed in the subsequent section concerning fieldwork activities, a question pertaining to the post-event questionnaire is dedicated to the concept of "sense of community." This question will inquire as to whether the World Café has increased the sense of community among participants. As posited by De Salvo (2021), rural areas are capable of fostering a "sense of community" that is discernible in participatory practices, actions of active territoriality, and the experimentation of shared cultural and social experiences.

In this vein, the research focused on the following four areas:

- Bologna, in Emilia-Romagna.
- Gazzola, in the Province of Piacenza (Emilia-Romagna).
- Cosenza, in Calabria.
- Santa Caterina dello Ionio, in the Province of Catanzaro (Calabria).

Therefore, the research considered two urban areas (Bologna and Cosenza), and two rural/inner areas (Gazzola and Santa Caterina). A nuanced understanding of the territorial and geographical complexities within these regions has significantly informed the design of our research, guiding our exploration of the diverse contexts and their attendant challenges and opportunities: Gazzola, for instance, is a Municipality in the Trebbia River Park (it. Parco Fluviale Regionale del Trebbia), a protected area encompassing the confluence of the Piacenza hills and the lower Po Valley. The Trebbia River Park, established in 2009, covers a total area of 4031 hectares and comprises eight municipalities, including Gazzola (Emilia-Romagna Region 2019). Also, the Municipality of Santa Caterina is located within the context of a natural park, i.e., Parco Naturale Regionale delle Serre, which, encompassing a mosaic of woodlands, forests, and rural fields, the area boasts a rich biodiversity, evident in its diverse flora and fauna (Serre Regional Park 2024). Furthermore, Santa Caterina is a territory characterized by hills, mountains and coastline, where a "collective archeology" project is underway (De Matteis and Dolce 2021), focusing primarily on fostering spaces for social interaction and sharing.

On the other hand, considering the Municipality of Bologna, it is noteworthy that Bologna is among the nine Italian cities participating in the EU mission "Climate-Neutral and Smart Cities by 2030" (European Commission 2024).

Additionally, the four municipalities (Bologna, Gazzola, Cosenza, and Santa Caterina) are signatories to the Covenant of Mayors for Climate and Energy, an initiative launched by the European Commission intended as a platform to mobilize and empower local governments in achieving the European Union's ambitious climate goals. Signatories to the Covenant commit to implementing actions that contribute to the EU's target of a 40% reduction in greenhouse gas emissions by 2030. Additionally, they endorse a comprehensive strategy that addresses both climate change mitigation and adaptation measures, and energy poverty (European Commission 2024).

A significant element in comprehending the territorial context of the selected areas is the extent of citizen involvement in local participatory processes. The map provided by the Observatory on Participation of the Emilia-Romagna Region is notable for its level of detail and indicates the degree of participation in participatory processes at the local Italian level. The categories of participation include empowerment, participatory design, information, consultation, and a fifth category, "not calculated." With regard to the "type of territorial impact," it is evident that the municipal level accounts for the greatest number of participatory processes within the Emilia-Romagna region (1137 processes out of 1518) (Observatory on Participation of the Emilia-Romagna Region 2024).

Conversely, with respect to the Calabria region, the level of participation is not quantified, while the total number of mapped participatory processes is limited to only 8 (Observatory on Participation of the Emilia-Romagna Region 2024). Additionally, it is also noteworthy that the number of municipalities in Emilia-Romagna involved in participatory processes is considerably higher than in Calabria, with a total of 330 municipalities participating in such processes in Emilia-Romagna, as opposed to 137 in Calabria (Observatory on Participation of the Emilia-Romagna Region 2024). Nevertheless, it would be beneficial to ascertain whether and to what extent participatory processes (either ongoing or concluded) exist in the specific areas under investigation, with due consideration of the data provided by the Observatory on Participation of the Emilia-Romagna Region (2024):

- With regard to the Municipality of Bologna, the number of concluded participatory processes is 166, while 19 are currently ongoing. The primary thematic areas include welfare, territory, and the environment.

- The Municipality of Gazzola (PC) records 78 participatory processes, of which 68 have concluded. The primary thematic area is the environment.
- With regard to the City of Cosenza, no data are available concerning regional and local administrations' participatory processes.
- Concerning Santa Caterina dello Ionio (CZ), no data are available.

From these data, a significant discrepancy emerges between the considered areas. The urban area of Emilia-Romagna, specifically Bologna, and the inner area, specifically Gazzola, appear to be particularly active in terms of participation. In contrast, the data from Calabrian areas is cause for concern, with few ongoing and/or concluded participatory processes, which are mostly organized at a top-down level.

However, concerning our Cafés, the selection of local contexts was informed by a confluence of these factors, while the specific themes explored in each fieldwork were shaped by distinct considerations. A common topic across all four fieldworks was the focus on separate waste collection, aligned with Sustainable Development Goal (SDG) 12 ("Ensure sustainable consumption and production patterns"). This issue manifested as a social problem particularly within the urban areas of Bologna and Cosenza, as evidenced by reporting in local and national newspapers.

The second topic discussed within the two urban contexts was green areas, related to the SDG 11 ("Make cities and human settlements inclusive, safe, resilient, and sustainable"), and aligned with the already mentioned EU mission.

Considering the two rural/inner contexts, in addition to the focus on separate waste collection, the broader theme of renewable energy was explored, aligned with SDG 7 ("Ensure access to affordable, reliable, sustainable and modern energy for all"). While all SDGs provide a general framework within which our research was situated, our quasi-experiments were conducted with a strong focus on SDG 13 ("Take urgent action to combat climate change and its impacts"), which is also closely linked to the action-research method.

2.2 The Methodological Approach: Stakeholders' and Participants' Recruitment

Prior to the commencement of the quasi-experiments conducted for the author's Ph.D. research, the author had already established a formal collaboration with Cittadinanzattiva Emilia-Romagna (Cittadinanzattiva ER), an organization which promotes citizens' activism in the areas of protection of rights, care of common goods, and support for vulnerable people. It was within the offices of Cittadinanzattiva ER, in the Bologna city center, that the idea of organizing a participatory quasi-experiment in the World Café format emerged, inspired also by the "Community PRO" project (carried out by Cittadinanzattiva in several Italian local communities).

Crucially, Cittadinanzattiva ER offered immediate and multifaceted support to facilitate the fieldworks conducted in the Emilia-Romagna region. Its support encompassed several key elements: venue provision, generously providing access to its offices in Bologna, offering a dedicated space to conduct the first Café; social media amplification, actively promoting the initiative through their social media channels on platforms such as Facebook and Instagram, expanding the project's reach and attracting potential participants. Cittadinanzattiva ER played a pivotal role in disseminating information by designing and distributing printed posters in key locations throughout the city, including study halls and public venues.

While our collaboration with Cittadinanzattiva ER predated the empirical phase of the research and proved instrumental in promoting our fieldwork in Bologna, identifying the other of local stakeholders necessitated further investigation:

- In the case of Gazzola, given the absence of Cittadinanzattiva in the Province of Piacenza, in order to facilitate the fieldwork, the identification of alternative stakeholders was required. To achieve this, we held a series of meetings (both in Bologna and remotely) with the local city councilor responsible for cultural and educational policies. These meetings served two key purposes: (1) selecting the optimal location for the fieldwork (i.e., the Municipal Library) and (2) promoting the initiative among residents. Furthermore, the event was disseminated through two key communication channels: (1) the official website of Cittadinanzattiva ER, expanding awareness beyond the immediate Gazzola community and (2) the official pages of the Gazzola Municipal Library, ensuring residents were informed. It is important to recognize that the decision to conduct a participatory quasi-experiment in a library is not a trivial one. Indeed, libraries may be considered a stronghold of participatory democracy and "hubs" for deliberative democracy (Kranich 2010), providing citizens with the opportunity to promote innovation and change through their empowerment, thereby creating integrated solutions that contribute to the construction of an increasingly sustainable future.
- Before Santa Caterina's fieldwork, we conducted a series of online meetings with Avamposto Agricolo Autonomo's owners to determine the optimal location for the Café. Initially, the plan involved hosting the World Café session outdoors within the AAA fields. However, due to unforeseen weather conditions, a contingency location was

chosen (i.e., a private terrace). AAA not only played a crucial role in selecting the venue but also actively promoted the event to the residents of Santa Caterina. The promotional campaign involved the distribution of paper flyers in strategic locations throughout the village, such as local bars, restaurants, and other public spaces. Nevertheless, the participation/absence of participants was probably influenced by the fact that our fieldwork coincided with a protest against the construction of an offshore wind park in the Ionian Sea.

- Recruiting local stakeholders in Cosenza presented a significant challenge, as evidenced by the limited response rate: only one out of the four organizations contacted, ARCI Cosenza expressed interest in collaborating on the project (ARCI Cosenza was carrying out the project "CoGreen Community-Based Action for a Green Transition," co-funded by the European Commission, and related to local environmental issues, at the neighborhood level). They readily agreed to host our participatory quasi-experiment in their offices and undertook the initiative's promotion through a multifaceted campaign. This campaign leveraged ARCI's social media presence on platforms such as Facebook and Instagram, alongside traditional methods like paper flyers and newsletter distribution. Furthermore, ARCI Cosenza expressed a strong desire to continue the collaboration for a larger-scale participatory and deliberative process in the future.

The disparity in the simplicity of engaging specific stakeholders is predominantly attributable to the unavoidable contrasts between smaller contexts (such as rural areas) and larger ones (such as urban areas). Moreover, the difficulties encountered in the process of gathering stakeholders using the World Café approach have been elucidated by Löhr et al. (2020). Consequently, while in Emilia-Romagna we initiated the research with Cittadinanzattiva networks, in Cosenza we faced greater challenges due to the absence of a preliminary relationship between our team and potential stakeholders.

Concerning rural areas, the presence of local institutional representatives among the participants can be attributed to the closer, more interconnected community dynamics and the stronger, often personal ties between residents and local authorities (Denters and Klausen 2018). In contrast, in urban settings, the scale and complexity of the social environment, as well as the often more formalized and segmented interactions between stakeholders, likely contributed to the absence of institutional representatives.

It is noteworthy that a dedicated booking page was created for each fieldwork on the Eventbrite® platform. The booking page served a dual purpose: it provided potential participants with key information about the initiative and facilitated the monitoring of participation levels through online bookings, allowing for verification of whether the minimum number of participants (i.e., 12 participants for a single World Café—Löhr et al. 2020) was reached. Moreover, after his/her registration, each (potential) participant received by email some informative material and easily readable brochures about the topics of the Café sessions, with the aim of providing them with additional information.

Nevertheless, the quasi-experiments conducted in Bologna, Gazzola, Cosenza, and Santa Caterina demonstrate both similarities and differences across various dimensions of participants' recruitment. It is important to note that the number of participants is comparable across the various contexts, with 14 participants in Bologna, 17 in Cosenza, 14 in Gazzola, and 13 in Santa Caterina. This suggests that, despite differing local contexts, a comparable level of participant engagement was attained.

A notable discrepancy, however, emerged in the recruitment strategies used to attract participants. In Bologna and Gazzola, participants primarily registered through a dedicated online portal on Eventbrite®. In contrast, engagement with the online platform was minimal in Cosenza and Santa Caterina. This discrepancy stemmed not from the quasi-experimental design itself but rather from varying levels of responsiveness to the recruitment process. Furthermore, although each fieldwork began with an introductory session to present the topics and the World Café rules, the prior dissemination of informational materials via email led to uneven levels of participant preparedness. This variation likely arose because some participants joined sessions last-minute without preregistration, while others were unable to review the materials in advance due to scheduling constraints. In this framework, the effectiveness of the World Café as a participatory approach becomes evident. By facilitating the *cross-fertilization* of knowledge, the Café creates an informal environment where individuals with diverse expertise can interact and gain new insights. Consequently, the Café helps mitigate potential selection bias, which may occur when the *open door* primarily attracts those already highly interested in the topics at hand (Bobbio 2019).

On the other hand, it is important to highlight that the documentation of each fieldwork was easily accessible through the Heroots website: such platform functions as a comprehensive database for sustainability-themed events across Italy, featuring an interactive map for user-friendly navigation. During the event upload process, meticulous attention was paid to identifying key stakeholders involved (e.g., Cittadinanzattiva, Arci, Avamposto Agricolo Autonomo). Additionally, a direct link to the event description was included for convenient reservation, and the alignment of these activities with the Sustainable Development Goals (SDGs) was clearly articulated.

Finally, it is important to emphasize that each initiative was designed and conceived with a balance between academic rigor and participant comfort, aligning with the principles of action-research and the World Café approach, which advocate for an informal environment. However, meticulous planning ensured all details were addressed: projectors were provided in every seat to facilitate presentations, including the introductory speech, questions to participants, final harvest, and sharing of collective discoveries. Additionally, all necessary World Café tools were procured (pens, notepads, Post-it® notes, placemats, talking sticks, posters), as suggested by the Practical Guidelines of the World Café (Brown and the World Café Community 2002). Furthermore, refreshments were provided in the form of a sustainable buffet, featuring locally sourced food, minimal use of disposable plastics, and fair trade-certified coffee and tea.

Nevertheless, as previously highlighted, when mentioning the *open door* method, but also referring to Eventbrite®, it is essential emphasizing once again that our participatory quasi-experiments were open to all citizens of the selected areas, so that the diversity of perspectives was ensured. In other words, our project was based on the assumption that participation, not merely reduced to statistics or understood as the episodic and intermittent involvement of voting (Moro and Sorice 2022), is in itself something positive within a local community. Thus, our research adhered with the growing recognition that participatory and/or deliberative processes, particularly at a micro-level, should not be *closed* but rather *open* to all citizens. In an era where scientists and scholars are urged to engage with ordinary citizens to make the understanding of climate change more accessible (e.g., Brulle et al. 2012), the idea of making deliberative processes closed (and limiting them to randomly selected citizens) may not always be desirable, particularly within more undersized, and often underestimated contexts such as Italian rural and inner areas.

2.3 The World Café Structure

The quasi-experiments were organized following the above-mentioned seven design rules of the World Café (Brown and Isaacs 2005), also drawing on some empirical examples (e.g., Alunni-Menichini et al. 2023). To facilitate each conversation table, we designated a consistent host who remained the same across all conversation rounds. This approach was intended to foster discursive continuity, ensuring that key themes and ideas could evolve coherently as participants shared and built upon each other's insights. Additionally, each host maintained a dedicated notebook, capturing notes and reflections that documented the flow and substance of discussions. This practice enabled the application of the seventh principle/design rule of the World Café, which emphasizes the importance of making collective discoveries visible and actionable ("making collective knowledge and intuition visible and usable"—Brown and Isaacs 2005, p. 40). By systematically documenting these insights, the hosts helped transform individual contributions into a shared repository of knowledge that participants could reference and leverage, thus enriching the overall conversational experience and reinforcing the collaborative learning process.

Participants were invited to sit at the table of their choice when they arrived at the Café, while we randomly chose the hosts. It is important to emphasize that the host was not *neutral* in the conversations, as was the case, for example, in the World Café by Alunni-Menichini et al. (2023), but he/she interacted with other participants in a meaningful way. The main distinction between our fieldworks and Alunni-Menichini's Café lied in the topics selected for the session (i.e., environmental issues in our case, and substance users experiencing homelessness in Montreal, Canada, in Alunni-Menichini's research): the host within Alunni-Menichini's Café was defined as "neutral," with a crucial role in involving vulnerable people in the conversation process. However, within the Café organized in Canada, the hosts were part of the research group with experience in conducting focus groups. The added value of our quasi-experiments can be seen in the fact that the hosts were randomly chosen among the participants, ensuring the informality of the initiative. Such an informality was further enhanced by the choice of locations, while in the Canadian study the Café was held inside the university.

The World Café structure is illustrated in Fig. 2. Another aspect to consider is that, for logistical reasons, not all four quasi-experiments were conducted in the same time slot: for example, in the case of Gazzola, the Café was held after dinner, from 8 to 10 pm, while in Bologna, Cosenza, and Santa Caterina, the fieldwork took place in the afternoon (from 5 pm in Cosenza and Santa Caterina, and from 6:30 pm in Bologna) and lasted three hours, including breaks and rotations. Each round was followed by a rotation and began with a concise summary provided by the host, in order to stimulate deeper discussions. At the end of the World Café session, we collated and synthesized hosts' notes into a slide presentation, shared with all participants, facilitating a final round of feedback, and encouraging participants to offer additional comments on the findings.

As showed in Fig. 2, the World Café was structured in two rounds of conversation, with the second round focused on green urban areas in Bologna and Cosenza, and renewable energy in the two rural/inner contexts, Gazzola and Santa Caterina.

However, it is worth noting that the first fieldwork conducted in Bologna served as a trial run. In that case, the general question on *sustainability* was included in the first round, while a general question on climate change was asked together with the opening question on the *conversation*. In subsequent fieldworks, the general question on climate change (i.e., "If we talk about climate change, what are we

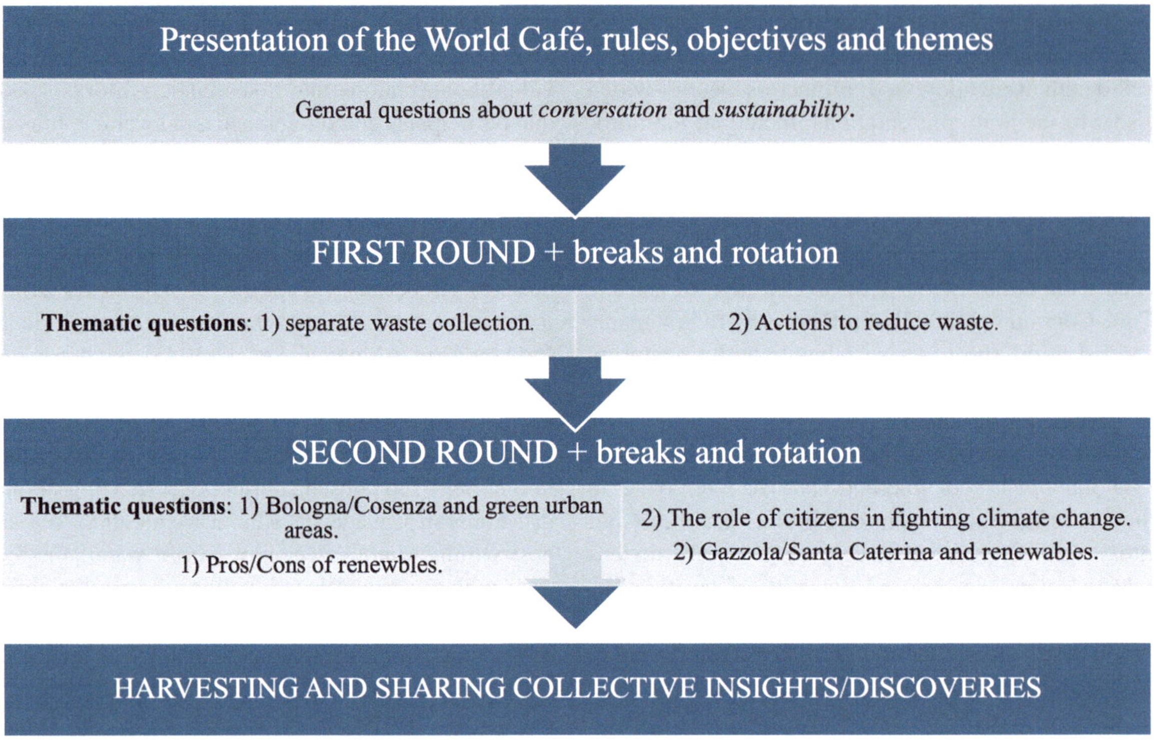

Fig. 2 Final World Café structure (*source* By author)

seeing/facing as members of this local community? What will happen if immediate action is not taken?") was instead included in the slides for a possible third round of conversation or, if the minimum number of participants was not reached, to potentially transform the World Café into a focus group.

In summary, the results of conversations demonstrated the World Café format's capacity to foster a space for sharing ideas and perspectives on complex issues, such as those explored in each session:

- Concerning Bologna's fieldwork, participants highlighted the importance of *reuse* over *recycle*, outlining three key areas for improvement, i.e., prioritizing reuse, promoting appliance repair services, and minimizing waste generation at the source. The conversation emphasized the value of restoring worth to waste and the significance of reusing organic waste at a local level. However, participants also recognized separate waste collection as essential for achieving local environmental sustainability.

  Considering the second topic, i.e., green urban areas, the discussion acknowledged the benefits of green areas for residents, biodiversity, and social interaction. Proposals for increasing greenery in the historic center included utilizing climbing plants, maximizing space for trees, restricting traffic, and fostering the creation of urban gardens. Notably, all participants stressed the crucial role of citizens and community-driven initiatives in achieving these goals.
- Gazzola's participants offered different perspectives from both the individual (e.g., consume less) and the collective (e.g., separate waste collection in the workplace) action point of view. Participants' enthusiastic discussions in proposing practical measures, such as expanding recycling bin availability or advocating for door-to-door collection, underscored the participants' dedication to environmental stewardship.

  Concerning the second topic, i.e., renewable energy, participants focused on the critical roles of education and affordability in facilitating the adoption of renewable energy sources. Additionally, participants underscored the importance of sustainable transportation as a core strategy for reducing $CO2$ emissions. Public lighting upgrades were recognized, and the conversation emphasized the need for greater investment in alternative energy options for the transportation sector.
- Santa Caterina's participants suggested solutions for improving separate waste collection, such as reusing organic waste locally (like Bologna's Café), highlighting the importance of promoting plastic-free campaigns (particularly during the summer), and suggesting actions which may be addressed through specific policies, e.g., public drinking water to reduce plastic water bottles.

In the second round about renewables, multiple perspectives emerged, mentioning environmental impacts, local energy autonomy, and effects of wind turbines on regional landscape.

- In Cosenza, the first round on separate waste collection was focused on shared bins for separate collection, strengthening existing recycling zones (called "ecological islands"), and encouraging individual responsibility through mindful shopping and reducing disposables. The discussion also emphasized redesigning collection systems to better suit local needs and demographics.

  In the second round of discussion on green urban areas, participants proposed possible actions to fight climate change at a local and global level, highlighting the role of green urban areas in terms of mitigation and adaptation to climate change, and recognizing the role of green urban areas in terms of psychological well-being.

An aspect which needs to be emphasized is that none of the participants, both in Gazzola and Santa Caterina, knew about the EU Covenant of Mayors for Climate and Energy. Nevertheless, compared to Gazzola's case (the PAESC, i.e., the *Plan of Action for Sustainable Energy and Climate*, was fully accessible online), we haven't found any paperwork online about a PAESC for Santa Caterina.

Overall, all participants actively contributed to the discussions, sharing personal and collective experiences. They unanimously emphasized the critical need for collective action to address climate change from a participatory perspective.

2.3.1 The World Café Participants

A total of 58 individuals participated in the four World Café sessions, representing a diverse range of backgrounds and professions. Of these participants, 56 (96%) completed the pre-event questionnaire. A total of 51 participants provided information about their profession in the dedicated section of the pre-event questionnaire. The group included a range of professionals, students, and individuals from various walks of life, highlighting a spectrum of expertise and viewpoints: the participants comprised 12 students, who brought fresh ideas and perspectives, and one lawyer, potentially offering insights into legal frameworks relevant to the session topics. Additionally, one medical doctor was present, along with one communications officer, 6 employees from various sectors, and 2 volunteers, illustrating a strong representation from both professional and civic-oriented backgrounds. Specialized professionals were also present: one concrete technologist, one coachbuilder, one housewife, one business consultant, one university professor, one metalworker, one steelworker, one farmer, one head of artisan laboratory for vegetable preserves, one massage therapist, one artist/illustrator, one biologist/chemist, 2 architects (one of which was official in the Ministry of Culture), 3 teachers, one biologist, one journalist, one geologist, one engineer/teacher, one project owner. 4 retired persons, 2 unemployed persons further contributed to the diversity. The contributions of these participants were informed by a foundation of everyday experiences and practical knowledge, which proved to be a valuable addition to the discussions.

The diversity of participants facilitated a robust exchange of perspectives, thereby enhancing the scope of insights obtained during the sessions.

Each fieldwork began with the participants completing the pre-event questionnaire, which was divided into two sections. This questionnaire was divided into two main sections. The first part gathered sociodemographic information, including participants' gender, age group, education level, employment status, and any political or associative affiliations. The second part aimed to delve into their expectations and motivations, providing a preliminary look into what they hoped to gain or contribute to the fieldwork.

Among the participants who responded to the pre-event questionnaire ($n = 56/58$), the majority were men (55%; $n = 31/56$), and the most present age group was 26–33 years (21 out of 56 participants).

Additionally, a substantial proportion (64%) held a university degree. Concerning the question "Are you an active member of political parties/associations?," 27 participants (48%) did not declare themselves active members of political parties and/or associations. This detail provided insight into the level of civic participation exhibited by the group, suggesting the presence of both engaged and less affiliated within the participants.

The questionnaire also explored participants' past experiences with participatory and deliberative processes. To better understand their familiarity with such processes, participants were asked:

- "Have you ever taken part, even in the past, in participatory-deliberative processes?."
- "Specifically, have you already participated, even in the past, in community meetings (e.g., citizens' assemblies) on local and/or global environmental issues?."

For the first question, participants had four multiple-choice options to reflect the range of formats they might have encountered: deliberative polls, participatory budgeting, citizens' juries, other. The results showed that 5 respondents participated in deliberative polls, 4 in participatory budgeting, one in citizens' juries, and 16 chose the "other" option, suggesting some exposure to additional forms of engagement not explicitly listed.

On the other hand, 13 respondents participated in community meetings (e.g., citizens' assemblies) on local and/or global environmental issues. Overall, the questionnaire results sketched a portrait of a relatively young, educated, and diverse group, some well-versed in participatory processes, bringing valuable experience, while others were newcomers to the world of civic engagement, bringing fresh perspectives and enthusiasm.

Lastly, the second section of the pre-event questionnaire provides insight into the motivations and expectations driving people to participate. Participants' answers painted a vivid picture of what they hoped to achieve, revealing a range of aspirations that spoke to both personal growth and community impact.

Upon being queried as to their expectations, participants proffered a plethora of responses. Some sought to deepen their understanding, such as one respondent who expressed a desire "to improve my awareness and knowledge on an issue that directly affects us every day." Others hoped to enrich their perspectives by engaging with new ideas, as one participant put it, "listening to new ideas and opinions from other citizens." This curiosity and openness reflected a shared recognition that the inclusion of diverse viewpoints could enhance collective problem-solving.

For many, the event represented an opportunity to grow as citizens, even in small ways. One participant mentioned an aspiration to "improve my being a citizen, in my own small way," while others emphasized the desire to learn practical ways to make a difference locally, stating goals like "learn more about the topics and discover how to make a local contribution."

Several responses demonstrated a forward-looking hope for change. For example, some participants spoke of "improving for the future" and "developing new ideas," reflecting an ambition to foster innovation and resilience in their communities. The theme of creating bonds also emerged, with one participant aiming to "forge territorial bonds," illustrating the role of connection and cooperation in addressing shared challenges.

Environmental awareness featured prominently in many of their answers, underscoring a collective concern for ecological well-being. Responses like "create awareness about the environment in which we live" and the hope that "a concrete and definitive action to improve the environment can be found" revealed the participants' strong commitment to sustainable action. Others articulated a desire for a "greater and better understanding of which actions can be implemented to support the ecological transition," signifying their interest in practical, actionable insights that could guide both personal and collective efforts.

These limited examples of answers showcase the diverse motivations which drove people to participate at the World Café. A common thread was the desire to deepen understanding, cultivate dialogue, and contribute to positive societal or environmental change. Participants expressed a thirst for knowledge, a dedication to civic participation, and a drive to explore avenues for personal and collective growth. Ultimately, these responses revealed a commitment to continuous learning, collaborative action, and proactive problem-solving in the face of pressing environmental challenges.

2.3.2 Participants' Perspectives: Insights from the Post-event Questionnaire and Effects of the World Café

One week after the World Café, a second questionnaire, structured on the model by Alunni-Menichini et al. (2023), was sent to all participants, in order to evaluate:

- Satisfaction levels (overall rating of the activity, using a Likert scale from 1—very dissatisfied to 5—very satisfied), through also open-ended questions such as "What did/didn't you like in this laboratory?", and "You have any suggestions for the next (eventual) World Café?."
- Knowledge acquisition ("The event represented a learning opportunity for you?" – "yes/no/I don't know").
- (Potential) impacts of the laboratory: "In your opinion, the World Café was able to…"; "You believe that the World Café has positively contributed to…"; "Considering the debate, do you think you will have to change/will you change anything in your lifestyle? If yes, what?" (open-ended question); "Does the effectiveness of mitigation strategies (to reduce the levels of pollution in the atmosphere) and adaptation (to implement strategies to limit the effects) to climate change also depend on the involvement of citizens (as in the World Café format)?," using Likert scale (from 1 – strongly disagree – to 5 – strongly agree); "Has the initiative, in general, increased your sense of community?"; "Considering the aspects of awareness, risk perception, and commitment to action, obviously related to climate change, you believe that your point of view has changed?," using Likert scale (from 1—not at all to 5—extremely).

While 56 participants (96%; $n = 56/58$) responded to the pre-event questionnaire, the post-event questionnaire was completed by 53 participants (91%; $n = 53/58$), after multiple reminders. It is important to note that not all 53 respondents answered all of the questions, as the questionnaire did not have any mandatory questions beyond those specifically designed to assess changes in preferences on the specific topics. Indeed, the question about overall satisfaction was answered by 52 respondents out of 53 post-event respondents: 27 participants (52%; $n = 27/52$) were "very satisfied" (indicating 5 points), 23 participants (44%; $n = 23/52$) indicated 4 points, while only 2 participants (4%; $n = 2/52$) indicated 3 points.

At the same time, the participatory quasi-experiment received many positive comments. For example: "I really enjoyed discussing with unknown people about common issues that concern all of us. Mutual respect was very nice, speaking in rounds, which allowed the exchange of ideas, opinions, debate, but without creating misunderstanding"; "[I appreciated] Meeting the neighbors with whom you never interact"; "[I liked] the possibility of discussing/reasoning together among participants, according to a non-hierarchical method"; "[I appreciated] the topic covered, the engagement, being a participant, and having learned useful notions for our planet."

However, according to 50 out of 53 respondents (94%), the World Café represented a learning opportunity, as also demonstrated by further comments: e.g., "[I appreciated] The free exchange of views/opinions, and the newly learned knowledge."

Nevertheless, as already highlighted, while Bologna's quasi-experiment was conceived as a trial run, in the following three quasi-experiments (i.e., in Gazzola, Santa Caterina, and Cosenza), questions were added to the pre- and post-event questionnaires to understand whether the quasi-experiment itself was capable of causing a transformation in individual preferences regarding the specific topics of each Café. To this end, the following questions were added:

- "Which of the following actions do you believe is required to enhance the standard of the environment in which you live?," reproposed as "Based on what was discussed during the World Café, in your opinion, what is required to enhance the standard of the environment in which you live?", in the post-event questionnaire. The four options of answer were: "more landfill sites"; "more incinerators"; "more materials recovery facilities (e.g., for organic waste, plastics, etc.)"; "other."
- "What is necessary in order to take action against climate change, also valorizing the territory?," reproposed as "After the laboratory, do you believe that reacting to climate change requires [:]," in the post-event questionnaire. The four options of answer, for rural/inner areas, were: "Invest more in renewable energy sources"; "Invest in energy from fossil fuels"; "Invest in nuclear energy"; "other." The four options of answer, for Cosenza's fieldwork, were: "create more green areas"; "create more infrastructures, also at the expense of green areas"; "create more infrastructures, well-integrated with urban green"; "other."

In the pre-event phase, 34 out of 42 participants (81%) expressed the view that the establishment of "more materials recovery facilities (e.g., for organic waste, plastics, etc.)" was a fundamental measure to enhance the standard of the local environment. On the other side, 7 out of 42 participants (17%) selected the option "other," while 1 participant (2%) checked the box "more landfill sites". Such a view was largely echoed in the post-event phase where 33 out of 39 participants (85%) concurred with the same perspective on "more materials recovery facilities (e.g., for organic waste, plastics, etc.)." It is important to acknowledge, however, that post-event questionnaires remained incomplete, as three questionnaires are pending.

Nevertheless, various comments were collected in the section dedicated to alternative perspectives, by selecting the box "other," in the post-event questionnaire. For example: "More respect for the environment by citizens, more controls by the administrations, more efficiency of services"; "Generalized consciousness-raising"; "It is crucial to encourage environmental education activities and limit the production of waste upstream"; "It should be implemented creative ways of virtuous recycling"; "Social services and 'soft' infrastructures based on collaborative practices between public and private are needed."

Concerning the question about the other Café's topic (i.e., renewable energy), 19 out of 26 participants (73%) (the pre-event respondents were 26 within the two inner/rural areas) believed that it was necessary to "Invest more in renewable energy sources"; 2 participants (8%) believed that it was necessary to "Invest in nuclear energy," while 5 participants (19%) checked the box "other." In the post-event questionnaire, taking into account the three missing responses, 16 out of 23 respondents (70%) confirmed that it was necessary to "Invest more in renewable energy sources," 1 participant (4%) believed that it was necessary to "Invest in energy from fossil fuels," while 6 participants (26%) checked the box "other," stating that "It is important to re-evaluate and reuse the already generated power," "We should invest in energy communities," "Energy should be saved," "It is necessary to consume less," "Responsibilization is necessary," "Specific attention should be required not only to the production of primary energy but also to the circular recovery of energy, to improve the quality of life, starting from food resources."

For the Cosenza's second topic (i.e., green urban areas), in the pre-event phase, according to 11 out of 16 participants (69%), in order to take action against climate change, also valorizing the territory, it was important to "create more green areas"; 4 participants checked the box "create more infrastructures, well-integrated with urban green," and 1 participant checked the box "other," without specifying. In the post-event questionnaire, 9 participants out of 16 participants (56%) believed that reacting to climate change required the creation of "more green areas," while 5 participants (31%) thought that it was necessary to "create more infrastructures, well-integrated with urban green." On the other side, 2 participants (12%) checked the box "other," specifying: (1) "Even more than green areas, the marine ecosystem deserves to be respected, being a crucial oxygen source for the planet,"

(2) "Working together on activities focused on awareness/information."

However, the World Café had a noticeable impact on individual preferences. Although we do not have the exact number of participants who experienced a shift in preference (as the questionnaires were anonymous), we observed 13 responses that differed from those in the pre-event questionnaire concerning specific topics. This uncertainty arises because the questionnaire included two distinct questions, allowing participants to adjust their preferences on one or both topics.

Furthermore, several comments were added in the post-event questionnaire, as answers to the open-ended question "Considering the debate, do you think you will have to change/will you change anything in your lifestyle? If yes, what?." For example: "Increased focus on resource conservation and more attention in recycling"; "I seek more alternative solutions, by evaluating the experiences of others"; "Yes, by installing photovoltaic at home"; "I will propose to more people to come and compost at my earthworm farm in the countryside"; "[Yes,] my participation in sustainable activities"; "Yes, I will take part in the activities promoted by local associations (e.g., plastic-free events), in order to protect local territory and environment"; "I will pay more attention to separate waste collection; concerning meeting places with friends, I will prefer green areas, organizing, where necessary, waste collection days (together with the people I met at the World Café)."

However, in the post-event questionnaire, according to 35 out of 53 participants (66%), the World Café was effectively able to encourage citizens' participation in democratic life, promoting active participation of citizens (43%; $n = 23/53$). Additionally, according to 45 out of 53 participants (85%), the World Café was able to increase their "sense of community."

In the post-event questionnaire, the questions about "awareness," "risk perception," and "commitment to action," were included (i.e., "Considering the aspects of awareness, risk perception, and commitment to action, obviously related to climate change, you believe that your point of view has changed?"):

- Concerning "awareness," 15 out of 53 participants (28%) indicated 5 points (i.e., extremely changed point of view).
- About "risk perception," 17 out of 53 participants (32%) indicated 5 points (i.e., extremely changed point of view).
- The aspect of "commitment to action" was indicated with 5 points (i.e., extremely changed point of view) by 28 out of 53 participants (53%).

The results demonstrate a multifaceted representation across the three dimensions of awareness, risk perception, and commitment to action. Participants exhibited varying degrees of awareness, with a spectrum of responses encompassing minimal and substantial changes. Similarly, diverse attitudes were observed in risk perception and commitment to action, underscoring the intricate nuances of participants' perspectives.

Finally, 38 out of 53 participants (72%) were "strongly agree" on the fact that the effectiveness of mitigation and adaptation strategies to climate change also depends on the involvement of citizens, as in the World Café format.

The results provide insight into the extent to which the findings align with the initial research objectives. The findings demonstrate the effectiveness of the World Café as a participatory instrument in fostering relationships among participants and stakeholders, resulting in some modifications in individual preferences and perspectives on individual lifestyle modifications.

3 Discussion and Lessons Learned

The results presented above demonstrate how they are aligned with the initial research objectives. However, the change in individual preferences is typically observed in deliberative processes (e.g., Dryzek and Niemeyer 2010; Fishkin and Luskin 2005) and, more specifically, in reference to deliberative polls®, where this aspect is verified through a pre- and a post-event questionnaire (Floridia 2017). Nevertheless, the deliberative poll® is a mini-public, based on random selection of participants (Bobbio 2019; Fishkin 2018; Fung 2007; Sintomer 2018).

Thus, concerning our quasi-experiments, we note two limits: (1) the absence of participants' random selection and (2) the absence of deliberation at the end of each fieldwork. The two limits are both related, and the absence of a representative sample of the population leads to the fact that statistical discussions about the findings are precluded.

However, the contribution of this approach is that: (1) it fostered a change in individual preferences, stimulating participants' engagement; (2) it created a space for open dialogue and learning, with a high level of satisfaction among participants; (3) it went beyond raising awareness, with possible concrete actions, both on a collective and individual level; (4) it was aligned with action-research method, promoting collective learning and a commitment to action on sustainability issues; (5) it served as a springboard for deliberations by fostering dialogues and bridging the gap between talk and action.

In fact, the World Café approach (Brown and Isaacs 2005; Löhr et al. 2020; Steier et al. 2015) is not closely related to deliberative model of democracy, although in some cases the World Café was used for deliberative democratic evaluation (e.g., Alunni-Menichini et al. 2023).

Our Cafés could be conceived as jumping-off points for real deliberative processes. Nonetheless, the findings from conversations and questionnaires indicate that *open door* method (Bobbio 2019) should not be ruled out a priori when organizing a participatory laboratory, but it can be a valuable tool for participatory and potentially deliberative processes. In fact, for instance, almost half of the total participants (48%; $n = 27/56$) did not declare themselves active members of political parties and/or associations. Furthermore, another bias which may be present using the *open door* could be related to participants' time availability (e.g., retired persons may represent the majority of participants) (Bobbio 2019): instead, findings from our quasi-experiments indicate that not only retirees or students were present, but individuals holding various occupational positions were also present (e.g., teachers, employees, architects, a geologist, a medical doctor, etc.). Moreover, in some countries, the *open door* method is preferred for citizens' participation. In Portugal, for instance, Article 2 of the national Constitution has promoted a culture of participatory democracy based on *open door* processes that allow citizens to participate at any step (Allegretti 2021). As a result, Portugal has no tradition of mini-publics or deliberative processes, and concepts like *sortition* and *random selection* have not been important to political plans for citizens' participation (Allegretti 2021). On the other hand, attempts to use random selection have been hampered by procedural obstacles from the Privacy Protection Authority (Allegretti 2021).

Within our fieldworks participants believed that the engagement of civil society (also through approach such as the World Café) is an essential element to addressing climate crisis at a local and global scale, in line with IPCC (2022, 2023).

On the other hand, the findings about whether participants had ever engaged in participatory-deliberative processes, even those unrelated to environmental issues, may reflect diverse levels of citizens' participation and awareness in those areas and regions, presenting a different picture compared to the one suggested by Putnam (1993), with the "vertical" culture in the south and the "horizontal" civic culture in the north.

Nevertheless, concerning the principal aim related to the change in individual preferences, specifically considering the three quasi-experiments (in Gazzola, Santa Caterina, and Cosenza) where further and specific questions were present in the questionnaires, we can derive the following conclusions:

- In the pre-event questionnaire, 34 out of 42 participants (81%) expressed the view that establishing more materials recovery facilities (e.g., for organic waste, plastics, etc.) was a fundamental measure to improve the local environment. In the post-event phase, 33 out of 39 participants (85%), taking into account that three responses to the post-event questionnaires are pending, shared the same perspective. Nevertheless, preferences remained consistent, mirroring the initial phase, although initially, in Cosenza, 11 out of 16 participants shared the same perspective, whereas, in the post-event phase, they were 13 out of 16. Thus, it constituted a slight shift in preferences.
- The framework underwent modifications when examining issues related to other discussion topics, i.e., renewable energy and green urban areas. A meaningful transformation occurred in the case of Santa Caterina: 8 out of 13 participants emphasized the importance of investing in renewables, while 5 held alternative perspectives. After the event, the distribution changed: 6 participants advocated for increased investment in renewables, 1 participant favored fossil fuels, and 6 expressed alternative viewpoints. It suggested that the World Café indeed influenced the emergence of diverse perspectives within the sustainability framework, regardless of the preference for fossil fuels. Concerning Cosenza's fieldwork and the discussion on green urban areas, preferences experienced shifts as well: in the pre-event phase, 11 out of 16 participants believed in the necessity of creating more green spaces, while in the post-event, this number decreased to 9 out of 16. Additionally, the number of participants advocating for infrastructures well-integrated with urban green increased from 4 to 5. Furthermore, while 1 participant held an alternative perspective initially, the post-event results indicated that 2 participants shared an alternative viewpoint. Conversely, concerning renewable energy (in Gazzola), preferences did not undergo meaningful changes. For instance, the supporters of nuclear energy ($n = 2/13$, in Gazzola) were absent in the post-event results. Due to the lack of three responses, it is uncertain whether these supporters altered their preferences; however, if they have, the World Café may have served as a catalyst for such changes.

Such a general overview illustrates how the World Café approach could foster a change in individual preferences. However, it is crucial to recognize that the positive effects of the World Café extend beyond the aspects considered thus far. In fact, results from the post-event questionnaire, as already mentioned, indicated the 51% of participants ($n = 27/53$) as "very satisfied," with several positive comments about the fieldworks. Moreover, a high portion of participants (66%; $n = 35/53$) found the fieldworks to be effective in promoting citizens' participation in democratic life, representing a learning opportunity according to almost all participants (94%; $n = 50/53$). Furthermore, our quasi-experiments appeared to have elicited changes in behaviors, both from the collective (e.g., "I will propose to more people to come and compost at my earthworm farm in the countryside") and the individual point of view (e.g., "I will use the car less"; "[I will install] photovoltaic at home").

These results align with the action-research method which has been employed to address sustainability and environmental issues at a local level (Collins 2014; Egmose 2019), using a collective learning orientation (Mannarini 2009). The three aspects of "awareness," "risk perception," and particularly "commitment to action," retrieved from the Climate Perception Index (CPI) of Meta's Data for Good platform, are likewise consistent with the method and approach employed.

As also demonstrated by our quasi-experiments, the World Café is a useful tool to bridge dialogue with tangible action, promoting the intersection of different perspectives through learning empowerment and transformation, and co-generative dialogues, fostered by the cross-pollination of ideas (Brown and Isaacs 2005; Steier et al. 2015). In this respect, the World Café constitutes a potential tool able to provide rich qualitative data (Alunni-Menichini et al. 2023), also considering its flexible nature.

Finally, the combination of the World Café and the change in individual preferences is compelling, and it can be used as a catalyst for local deliberative processes, since the Café is centered on dialogical process, unavoidable condition for successful participatory and deliberative laboratories. The World Café offers a valuable tool for stakeholders to utilize in organizing a more expansive participatory and deliberative democratic process, beginning with the issues that emerged from our fieldworks.

The stakeholders involved have the potential to act as advocates for the issues that emerged from the fieldworks, which could result in significant impacts at both the regional and national levels. This is particularly true given that organizations like ARCI and Cittadinanzattiva operate on a national scale. Moreover, these stakeholders could facilitate further participatory laboratories, engaging both existing participants and newcomers. Such an approach would enable a more profound examination of the specific challenges, facilitating meaningful discourse that would ultimately inform targeted actions to address the issues and concerns raised.

4 Conclusions

The World Café has proven to be a useful collaborative method for addressing environmental issues, thanks to its accessibility and the short duration of each fieldwork, which likely encouraged participation. The approach was successful in promoting the exchange of perspectives and ideas specific to each local context, while treating the broad issue of climate change considering both the local and global implications.

Our findings suggest that the World Café is a valuable participatory approach, both in urban and rural/inner contexts, able to potentially improve collaboration between residents, and between residents and other stakeholders (e.g., local associations, municipalities in rural/inner areas). Moreover, it could act as a real engine of change at a local level in terms of environmental and social sustainability, creating a pleasant and democratic space for exchange, with an inter- and trans-disciplinary lens.

The innovative aspect of our quasi-experiments lies in addressing highly topical issues within specific contexts such as Italian rural/inner areas, using the World Café as a potential tool for a transformation in individual preferences. Precisely in the inner areas (i.e., Gazzola and Santa Caterina), the fieldworks were organized in collaboration with local municipalities (e.g., Gazzola Municipal Library) and appreciated by some local councilors both in Gazzola and Santa Caterina, while in the two urban areas, figures of the municipal administration were perceived as distant, as emphasized by some participants during the Cafés. However, according to Löhr et al. (2020), one of the weaknesses of the World Café approach is the challenge of connecting with stakeholders: in our research, involving stakeholders from the third sector (such as Cittadinanzattiva and ARCI) proved to be successful, while engaging local administrators was more straightforward in rural/inner contexts, but notably absent in urban areas. Furthermore, our Café sessions were included by Cittadinanzattiva ER into the Social Impact Assessment of Activities (2022), underscoring the broader influence of our activities within the organizational context of Cittadinanzattiva ER.

In conclusion, although our quasi-experiments present several positive points in reference to the applicability within local contexts, some limits should be taken into account: first of all, a single meeting, although it created social bonds between stakeholders and participants, and between residents, giving a rough idea of citizens' participation, is probably not enough to create effective and long-lasting changes. On the other hand, a continuous interaction between citizens and stakeholders would be required to avoid and/or limit the academic nature of the participatory and/or deliberative laboratories which could be intended and implemented as *permanent territorial structures*.

However, by conducting World Café discussions within two distinct Italian regions (i.e., Emilia-Romagna and Calabria), we were able to draw comparative insights into the approach's effectiveness across different contexts, thereby further enriching our understanding of the method's adaptability and potential for broader application. Such comparative research enabled the identification of similarities and differences in citizens' participation, discussion dynamics, and outcomes.

Thus, future research could investigate a number of pivotal areas. One area of focus could be an assessment of the impact of the World Café approach on diverse participant types. This could be achieved by examining how the approach affects individuals with varying levels of climate change knowledge and different political or associative affiliations. Furthermore, studies could examine the long-term impact of the World Café by following participants over time to ascertain whether their attitudes and behaviors undergo a change as a result of engaging in a participatory or deliberative setting. This could potentially lead to the establishment of a genuine "community of practice" (Boffi et al. 2016).

Another area of interest is the development and testing of new World Café formats, such as hybrid sessions that integrate both online and in-person participation with various technologies. Finally, adapting the World Café format to align with the deliberative democracy model, while maintaining the *open door* method, may be particularly valuable in smaller or rural areas. These adjustments could enhance citizens' participation and address local challenges collaboratively, reflecting Egmose's (2019) vision of addressing issues "*with* people."

By addressing these challenges and adapting the World Café to the distinctive requirements of rural contexts, future research can investigate the potential of this approach to facilitate more inclusive and effective deliberative practices in underrepresented territories. Our findings could serve as a foundation for broader research on deliberative processes, particularly with regard to complex environmental issues that were raised in the course of the Café discussions. For example, in the context of offshore wind farm construction in the Ionian Sea, a deliberative framework grounded in the principles of energy justice (Sovacool and Dworkin 2014) and energy democracy (Burke and Stephens 2017; Szulecki 2018) could prove particularly advantageous. As posited by Sovacool and Dworkin (2014), "energy justice ensures [...] that communities are meaningfully informed and represented in energy decisions" (p. 5). A participatory and deliberative process could therefore facilitate the participation of all stakeholders, including residents, policymakers, industry representatives, and environmental organizations, in a thoughtful and inclusive dialogue. Such an approach would facilitate a comprehensive evaluation of the wind farm project's impacts, ensuring that all perspectives are taken into account and that decision-making is balanced across economic, social, and environmental dimensions. This is in accordance with the recommendations set forth by Sovacool et al. (2017), which include criteria such as "availability," "affordability," "due process," "transparency and accountability," "sustainability," "intergenerational and intragenerational equity," "responsibility," "resistance" and "intersectionality."

Ultimately, the outcomes of these Cafés hold immense value for a range of stakeholders, including organizations like Cittadinanzattiva, ARCI, and local municipalities, by offering a wealth of innovative ideas derived from our participatory quasi-experiments. These discussions could act as a bridge, transforming bottom-up insights into practical strategies and closing the gap between theoretical dialogue and real-world action. The incorporation of local perspectives into the participatory framework is not merely advantageous; it is imperative for the development of sustainable and equitable climate solutions.

Acknowledgements The author gratefully acknowledges FSE REACT-EU and MUR for financial support of the Ph.D. fellowship PON Research and Innovation 2014-2020 (D.M. 1061/2021), Innovation and Green, XXXVII Cycle in Future Earth, Climate Change and Societal Challenges. Title of the project: *La political economy della neutralità climatica. Costruire il "consenso verde" con nuovi modelli di democrazia deliberativa e partecipativa* (Supervisor: Prof. Rosa Mulè).

References

Books

Almagisti, M. (2016). Una democrazia possibile: politica e territorio nell'Italia contemporanea. Carocci

Boffi, M., Colleoni, M., Rossetti, M.: La percezione della qualità dell'ambiente in Italia: Un'analisi socio-territoriale. In: Castrignanò, M., Landi, A. (eds.) La città e le sfide ambientali globali, pp. 57–73. FrancoAngeli (2016)

Brown, J., Isaacs, D.: The World Café: Shaping Our Futures Through Conversations That Matter. Berrett-Koehler Publishers, Inc. (2005)

Collins, K.: Designing social learning systems for integrating social sciences into policy processes: some experiences of water managing. In: Manfredo, M.J., Vaske, J.J., Rechkemmer, A., Duke, E.A. (eds.) Understanding Society and Natural Resources: Forging New Strands of Integration Across the Social Sciences, pp. 229–251. Springer (2014). https://doi.org/10.1007/978-94-017-8959-2_11

Corbetta, P.: La ricerca sociale: metodologia e tecniche: I. I paradigmi di riferimento. Società editrice il Mulino (2015)

De Matteis, S., Dolce, R.: Geografie collettive: una prospettiva rurale. In: De Matteis, S., Maida, A., Riverso, R. (eds.) Santa Caterina dello Ionio. Ambiente, stratificazioni culturali, paesaggi rurali, pp. 191–201. Rubbettino editore srl (2021)

De Salvo, P.: Legami territoriali, spazio rurale e cittadinanza attiva. In: De Salvo, P., Burini, C., Pizzi, M. (eds.) Territorialità e partecipazione civica: Teoria e casi, pp. 17–30. FrancoAngeli (2021)

Dryzek, J.S., Niemeyer, S.: Pluralism and meta-consensus. In: Dryzek, J.S. (ed.) Foundations and Frontiers of Deliberative Governance, pp. 85–116. Oxford University Press (2010). https://doi.org/10.1093/acprof:oso/9780199562947.003.0005

Egmose, J.: Action Research for Sustainability: Social Imagination Between Citizens and Scientists. Routledge (2019)

Elden, M., Levin, M.: Cogenerative learning: bringing participation into action research. In: Whyte, W.F. (ed.) Participatory Action Research, pp. 127–142. SAGE Publications, Inc. (1991). https://doi.org/10.4135/9781412985383

Elster, J.: Sour Grapes: Studies in the Subversion of Rationality. Cambridge University Press (2016)

Fishkin, J.S.: Democracy When the People Are Thinking. Oxford University Press (2018)

Floridia, A.: Un'idea deliberativa della democrazia: genealogia e principi. Società editrice il Mulino (2017)

Fung, A.: Minipublics: deliberative designs and their consequences. In: Rosenberg, S.W. (ed.) Deliberation, Participation and Democracy,

pp. 159–183. Palgrave Macmillan, UK (2007). https://doi.org/10.1057/9780230591080_8
Mannarini, T.: La cittadinanza attiva: Psicologia sociale della partecipazione pubblica. Società editrice il Mulino (2009)
Moro, G.: Cittadini in Europa: l'attivismo civico e l'esperimento democratico comunitario. Carocci (2009)
Moro, G., Sorice, M.: Partecipazione democratica: dialogando di sogni e libertà. Castelvecchi (2022)
Putnam, R.D.: Making Democracy Work: Civic Traditions in Modern Italy. Princeton University Press (1993)
Sovacool, B.K., Dworkin, M.H.: Global Energy Justice: Problems, Principles, and Practices. Cambridge University Press (2014)
Steier, F., Brown, J., Silva, F.: The World Café in action research settings. In: Bradbury, H. (ed.) The SAGE Handbook of Action Research, pp. 211–219. SAGE Publications Ltd. (2015). https://doi.org/10.4135/9781473921290
Ventura, J., Shahar, D.: Socially Oriented Action Research in Design Practice. SAGE Publications, Ltd. (2022).

Journal Articles

Allegretti, G.: Common patterns in coping with under-representation in participatory processes: evidence from a mutual learning space for Portuguese Local Authorities (LAs). Innovation: Eur. J. Soc. Sci. Res. **34**(5), 729–765 (2021). https://doi.org/10.1080/13511610.2021.1997573
Alunni-Menichini, K., Bertrand, K., Brousselle, A.: Combining the World Café and the deliberative democratic evaluation: a win-win strategy. J. Participatory Res. Methods **4**(1), 1–23 (2023). https://doi.org/10.35844/001c.66191
Bobbio, L.: Types of deliberation. J. Public Deliberat. **6**(2), 1–24 (2010). https://doi.org/10.16997/jdd.105
Bobbio, L.: Designing effective public participation. Policy Soc. **38**(1), 41–57 (2019). https://doi.org/10.1080/14494035.2018.1511193
Brulle, R.J., Carmichael, J., Jenkins, J.C.: Shifting public opinion on climate change: an empirical assessment of factors influencing concern over climate change in the U.S., 2002–2010. Clim. Change **114**(2), 169–188 (2012). https://doi.org/10.1007/s10584-012-0403-y
Burke, M.J., Stephens, J.C.: Energy democracy: goals and policy instruments for sociotechnical transitions. Energy Res. Soc. Sci. **33**, 35–48 (2017). https://doi.org/10.1016/j.erss.2017.09.024
Dryzek, J.S.: The Australian citizens' parliament: a world first. J. Public Deliberat. **5**(1), 1–7 (2009). https://doi.org/10.16997/jdd.73
Fishkin, J.S., Luskin, R.C.: Experimenting with a democratic ideal: deliberative polling and public opinion. Acta Politica **40**(3), 284–298 (2005). https://doi.org/10.1057/palgrave.ap.5500121
Giannetti, D., Lewanski, R., Bobbio, L., Romano, I., Crosby, N.: Cinque risposte a Lyn Carson. Rivista Italiana di Politiche Pubbliche **2**, 143–162 (2007). https://doi.org/10.1483/25117
Helbing, D., Mahajan, S., Fricker, R. H., Musso, A., Hausladen, C. I., Carissimo, C., Carpentras, D., Stockinger, E., Sanchez-Vaquerizo, J.A., Yang, J.C., Ballandies, M.C., Korecki, M., Dubey, R.K., Pournaras, E.: Democracy by design: Perspectives for digitally assisted, participatory upgrades of society. J. Comput. Sci. **71**, 102061 (2023). https://doi.org/10.1016/j.jocs.2023.102061
Kerstetter, K.: Insider, outsider, or somewhere between: the impact of researchers' identities on the community-based research process. J. Rural Soc. Sci. **27**(2), 99–117 (2012)
Kranich, N.C.: Academic libraries as hubs for deliberative democracy. J. Deliberat. Democracy **6**(1) (2010). https://doi.org/10.16997/jdd.102
Löhr, K., Weinhardt, M., Sieber, S.: The "World Café" as a participatory method for collecting qualitative data. Int J Qual Methods **19**, 1609406920916976 (2020). https://doi.org/10.1177/1609406920916976
Miller, D.: Deliberative democracy and social choice. Polit. Stud. **40**(1 Suppl.), 54–67. https://doi.org/10.1111/j.1467-9248.1992.tb01812.x
O'Neill, J.: The rhetoric of deliberation: some problems in Kantian theories of deliberative democracy. Res. Publica. **8**(3), 249–268 (2002). https://doi.org/10.1023/A:1020899224058
Sintomer, Y.: Deliberative polls and the systemic democratization of democracy. Good Soc. **27**(1–2), 155–164 (2018). https://doi.org/10.5325/goodsociety.27.1-2.0155
Sovacool, B.K., Burke, M., Baker, L., Kotikalapudi, C.K., Wlokas, H.: New frontiers and conceptual frameworks for energy justice. Energy Policy **105**, 677–691 (2017). https://doi.org/10.1016/j.enpol.2017.03.005
Szulecki, K.: Conceptualizing energy democracy. Environ. Polit. **27**(1), 21–41 (2018). https://doi.org/10.1080/09644016.2017.1387294
Talan, A., Rao, A., Sharma, G.D., Apostu, S.A., Abbas, S.: Transition towards clean energy consumption in G7: can financial sector, ICT and democracy help? Resour. Policy **82**, 103447 (2023). https://doi.org/10.1016/j.resourpol.2023.103447
Waardenburg, M., Groenleer, M., De Jong, J.: Designing environments for experimentation, learning and innovation in public policy and governance. Policy Polit. **48**(1), 67–87 (2020). https://doi.org/10.1332/030557319X15586040837640
Willis, R., Curato, N., Smith, G.: Deliberative democracy and the climate crisis. Wiley Interdiscip. Rev.: Clim. Change **13**(2), e759 (2022). https://doi.org/10.1002/wcc.759

Legislation & Cases

Communication from the Commission to the European Parliament, the Council, the European Economic and Social Committee and the Committee of the Regions. A long-term Vision for the EU's Rural Areas—towards stronger, connected, resilient and prosperous rural areas by 2040 (2021) (COM/2021/345 final)

Reports & Grey Literature

Brown, J., the World Café Community: Il World Café (2002). https://www.theworldcafe.com/wp-content/uploads/2015/07/WorldCafeGuidaPractica.pdf
Emilia-Romagna Region: Observatory on Participation (2024). https://www.osservatoriopartecipazione.it/italia/mappa
Emilia-Romagna Region: Trebbia Regional Park (2019). https://ambiente.regione.emilia-romagna.it/it/parchi-natura2000/aree-protette/parchi/treb
European Commission: Covenant of Mayors for Climate & Energy (2024). https://eu-mayors.ec.europa.eu/en/home
European Commission, Directorate-General for Research and Innovation: EU Missions, 100 Climate-Neutral and Smart Cities: Cities on a Journey to Climate Neutrality. Publications Office of the European Union (2024). https://doi.org/10.2777/169604
IPCC: Summary for policymakers [P.R. Shukla, J. Skea, A. Reisinger, R. Slade, R. Fradera, M. Pathak, A. Al Khourdajie, M. Belkacemi, R. van Diemen, A. Hasija, G. Lisboa, S. Luz, J. Malley, D. McCollum, S. Some, P. Vyas, (eds.)]. In: Shukla, P.R., Skea, J., Slade, R., Al Khourdajie, A., van Diemen, R., McCollum, D., Pathak, M., Some, S., Vyas, P., Fradera, R., Belkacemi, M., Hasija, A., Lisboa, G., Luz, S., Malley, J. (eds.) Climate Change 2022: Mitigation of Climate Change. Contribution of Working Group III to the Sixth Assessment Report of the Intergovernmental Panel on Climate Change.

Cambridge University Press, Cambridge, UK and New York, NY, USA (2022). https://doi.org/10.1017/9781009157926.001

IPCC: Summary for policymakers. In: Core Writing Team, Lee, H., Romero, J. (eds.) Climate Change 2023: Synthesis Report. Contribution of Working Groups I, II and III to the Sixth Assessment Report of the Intergovernmental Panel on Climate Change, pp. 1–34. IPCC, Geneva, Switzerland (2023). https://doi.org/10.59327/IPCC/AR6-9789291691647.001

Partecipedia: Flood Risk Management Plan of the Po Basin (2019). https://participedia.net/case/piano-di-gestione-del-rischio-alluvioni-del-bacino-del-po-flood-risk-management-plan-of-the-po-river-basin

Serre Regional Park: Parco Naturale Regionale delle Serre (2024). https://www.parks.it/parco.serre/Epar.php

UN, Department of Economic and Social Affairs, Sustainable Development (2023). https://sdgs.un.org/goals

UNFCCC: The Paris Agreement. What is the Paris Agreement? (2015). https://unfccc.int/process-and-meetings/the-paris-agreement

UN-Habitat: The Urban-Rural Linkages: Guiding Principles and Framework for Action to Advance Integrated Territorial Development (2019). https://unhabitat.org/sites/default/files/2020/03/url-gp-1.pdf

World Bank: Climate Change (2023). https://www.worldbank.org/en/topic/climatechange/overview

Contribution to Conference Proceedings

Denters, B., Klausen, J.: Governance in Close Proximity to the People? Contemporary Local Government and Politics [Abstract]. ECPR General Conference, Hamburg (2018). https://ecpr.eu/Events/Event/SectionDetails/733

Thesis

Celaya, C.M.: The Efficacy of Deliberative Democracy [Doctoral dissertation, Harvard University]. Harvard University's DASH repository (2019). https://dash.harvard.edu/bitstream/handle/1/42029773/CELAYA-DISSERTATION-2019.pdf?sequence=1&isAllowed=yhttps://dash.harvard.edu/bitstream/handle/1/42029773/CELAYA-DISSERTATION-2019.pdf?sequence=1&isAllowed=y

Investigating 50 Years of Temperature Change in Africa and Implementing Nature-Based Solutions as a Climate Change Adaptation Tool

Laila El Ghazouani, Carey Duncan, Tamsin Faragher, and Hassan Radoine

Abstract

Temperature rise is a significant aspect of climate change primarily due to anthropogenic activities. At the same time, rising temperatures contribute to broader shifts associated with climate change. This phenomenon poses major environmental, health, and socio-economic challenges, especially in the context of rapid urban growth. African cities, in particular, are at the forefront, confronting substantial temperature increases and their negative effects. Consequently, this paper aims to (1) illustrate the changes in air temperatures across African countries over the past five decades (1973–2023) and (2) advocate for the use of nature-based solutions (NBS) to mitigate urban temperatures. This would be done initially by protecting and restoring ecosystems, and subsequently fostering the creation of new ones. The findings suggest that climate change in Africa encompasses both warming and cooling trends. Furthermore, NBS can serve as effective instruments for addressing temperature-related hazards and enhancing climate resilience in Africa.

L. El Ghazouani (✉) · H. Radoine
Citinnov-Mohammed VI Polytechnic University, Rabat, Morocco
e-mail: laila.elghazouani@um6p.ma

H. Radoine
e-mail: hassan.radoine@um6p.ma

C. Duncan
International Federation of Landscape Architects (IFLA) Africa, Versailles, France
e-mail: carey@careyduncan-design.com

T. Faragher
African Centre for Cities-University of Cape Town, Cape Town, South Africa
e-mail: tamsinfaragher@gmail.com

Keywords

Nature-based solutions · Climate change · Mitigation · Adaptation · Landscape · Temperature rise

1 Introduction

1.1 The Global Context Along with the Consequences of Rising Temperatures

According to the Sixth Assessment Report of the IPCC, the global temperature has increased significantly since industrialization, underlining the urgency to mitigate climate change impacts. The world is already experiencing the effects of climate change, giving way to a wide range of new and unusual phenomena, including rising temperatures, ecosystem disruption, rising sea levels, and unpredictable floods and droughts (Bedair et al. 2023). It is evident that higher temperatures are detrimental to public health and human well-being, worsening air quality (El Ghazouani et al. 2019), leading to heat stress (Lee 2022), and exacerbating respiratory and cardiovascular disorders (Achebak 2024). In a global analysis of heat waves and human mortality, temperatures of only 20 °C were lethal (i.e. were associated with excess mortality) when relative humidity was high (80%), whereas at low relative humidity (20%), the lethal temperature increased to 30 °C (Mora et al. 2017).

1.2 Africa's Vulnerability and the Need to Mitigate Climate Change

Significant rises in temperatures in Africa over recent decades have been highlighted in previous studies. Temperatures are increasing significantly, and Urban Heat Island (UHI) effects are very pronounced, notably due to increasing urbanization

P. Mohammad et al. (eds.), *Geographic Approaches to Climate Change and Mitigation: Urban and Rural Perspectives (Volume 2)*, Advances in Science, Technology & Innovation, https://doi.org/10.1007/978-3-031-92115-5_21

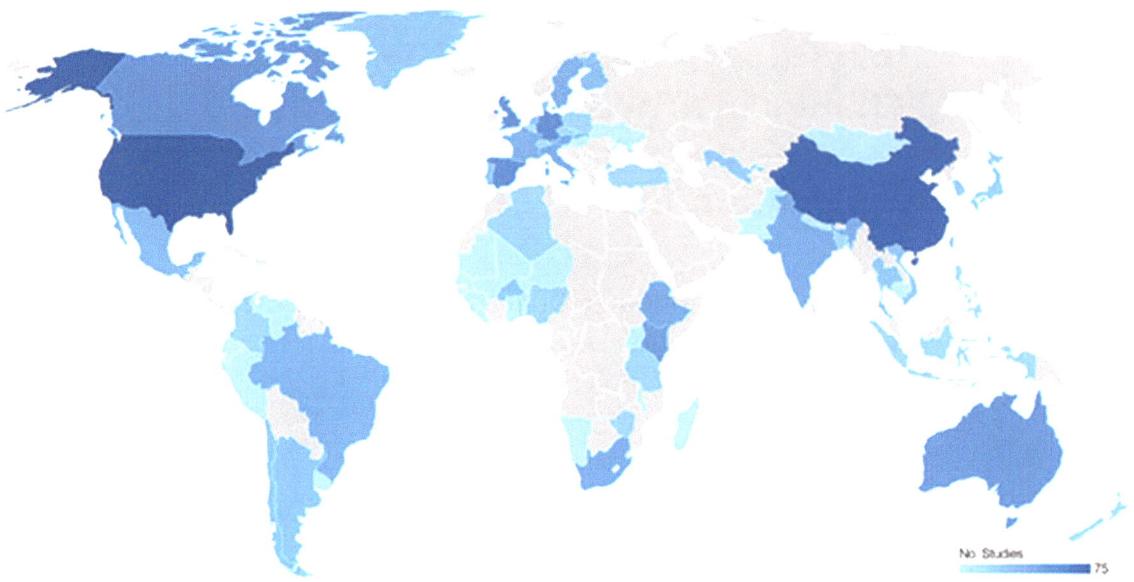

Fig. 1 Global distribution of studies examining the effectiveness of nature-based interventions to address climate impacts, Chausson et al. (2020)

(El Ghazouani 2021), a lack of vegetation (Li 2021), and rising anthropogenic activities. Alarmingly, climate modelling indicates that Africa is the most exposed to climate change hazards, including heat. The situation is exacerbated by the urban sprawl of African cities where urban populations are expanding at an average rate of 3.2% per year leading to increased emissions (Ruocco et al. 2015) and where Africa's young, growing populations' vulnerability is compounded by high levels of poverty (Fotso-Nguemo et al. 2022). Furthermore, many African cities suffer from inadequate infrastructure, making them more susceptible to climate impacts such as floods and heat waves (Lwasa 2018). Given these alarming figures, it is urgent to take action to mitigate climate change and promote adaptation to ensure the ongoing liveability of African cities.

1.3 Introduction to Nature-Based Solutions

Over the last few years, the concept of nature-based solutions (NBS) has gained popularity as a way to adapt to and engage against climate change (Babi Almenar 2021). Strategies based on a hybrid model that combines 'grey' hard infrastructures, 'soft' knowledge and information systems, and green measures (i.e. NBS) promote transformative adaptation and mitigation to increase the overall sustainability of urban systems (EEA 2016). According to the International Union for Conservation of Nature (IUCN), many NBS actions should explore place-specific approaches to tackle environmental challenges by reintroducing natural assets, leveraging nature to safeguard a stable and biodiverse future (Chausson 2020).

1.4 Advantages of NBS Implementation

Nature-based solutions have numerous advantages for addressing environmental challenges. Green (planted) roofs and rain gardens can increase water retention and reduce stormwater runoff helping to prevent urban flooding in intense rain events (Haghighatafshar et al. 2018). NBS can also contribute to reduce air temperature through shading and evapotranspiration, limiting the negative impacts of frequent and intense heat waves (Zardo 2017). Recent evidence found that parks were 1 °C cooler on average than non-park areas, presumably because of high evaporative cooling, low heat storage, and shade provided by trees (Hobbie and Grimmi 2020). Most of the time, these benefits come with additional urban health co-benefits in terms of ecosystem service supply, including recreation, aesthetic value, air quality, even social cohesion, and economic opportunities (Cortinovis et al. 2022).

Interestingly, some studies on reducing outdoor urban temperatures established that white roofs and those painted with reflective finishes ('grey' solution) were more effective at cooling than green roofs (Kabisch 2017; Brousse et al. 2024). This seems to suggest that planting is not the only way to tackle warm temperatures in urban environments, which is advantageous for cities like Cape Town where trees struggle to thrive due to the natural conditions and the extremely dense urban environments.

1.5 Significance of Studying NBS in Africa

In the African context, the study by Chausson et al. (2020) illustrates the limited number of studies on the effectiveness of Nature-Based Solutions (NBS) across the continent (Fig. 1). This lack of research poses a significant barrier to the implementation of NBS and to securing financial support from donors, who require evidence to justify their investments. Addressing this gap is critically urgent, emphasizing the need to prioritize scientific efforts in Africa to evaluate the effectiveness of NBS.

This current paper attempts to provide an in-depth examination of the change in air temperature in Africa over the past 50 years and advocates for the implementation of NBS as an adaptation tool. After reviewing existing literature in this introductory section, we will describe the data sources used and the methodologies adopted and suggest various interpretations to explain the observed temperature changes. Finally, and by providing three case studies as examples, we will advocate the use of NBS for climate adaptation as part of urban and climate resilience.

2 Data and Methods

2.1 Extracting Air Temperature Data

This study set out to map temperature differences in Africa over 50 years and draw comparative analyses in different contexts. Air temperature data were taken from the Copernicus network on the ERA5 platform (Copernicus.eu). ERA5 is an Earth Observation product of the Copernicus program, providing estimates of several atmospheric, terrestrial, and oceanic climate variables at a resolution ranging from 30 to 9 km. ERA5 data are produced by combining measurements, observations, and simulation inputs using data assimilation techniques (Copernicus.eu). The use of ERA5 in scientific work has been tested and validated by many studies such as Sabater (2019) and Wang et al. (2020) who all agree that the ERA5 dataset is crucial for understanding various climate phenomena including climate features. Additionally, it has been used extensively in climate studies conducted in Morocco (Khomsi et al. 2020; Kaissi et al. 2024).

Data extraction involved entering the desired mean annual air temperature and the X and Y coordinates of the entire African continent. The data presented in.NetCFF format was then imported into ArcGIS Pro, converted from Kelvin to degrees Celsius, and represented as annual mean temperature maps using the 'Zonal statistical by region' function.

2.2 Mapping Deviation and Comparative Assessments

After separately mapping the mean annual temperatures for 1953 and 2023, the 'Raster calculator' option on ArcGIS Pro was used to calculate the temperature difference over the 50 years. Presented alongside these maps are the vegetation cover map displaying major land use in Africa, and the country location map offering a clear view of temperature correlation with significant agglomerations.

2.3 Framework for Results Analysis

A systematic approach was developed to analyse the findings comprehensively. A structured template featuring specific criteria provided a uniform method for changing temperature trends in Africa over the last 50 years ensuring consistently high-quality interpretations. This is accomplished by analysing, assessing and crossing the interplay of the following factors: geography, topography, city size, biodiversity, and vegetation cover.

3 Results and Discussion

3.1 Interpretation of Temperature Differences: Both Increase and a Decrease

Upon examining the temperature variation map (Fig. 2), we noted some unexpected results that shed light on the complexity of climate change. Temperature changes varied between minus 5.30 °C (cooling) to 8.9 °C (warming). Both trends pose significant challenges for human life and biodiversity and are part of a long list of important issues facing the African continent. It is worth noting that these outcomes are influenced by broader climatic patterns such as precipitation and humidity, set against a backdrop of demographic growth, rural flight, and rapid urbanization.

Correlating these results with available literature reveals similar trends. The Intergovernmental Panel on Climate Change (IPCC) highlighted significant warming trends in air temperatures across Africa, projecting a rise of 3–6 °C by the end of the twenty-first century (IPCC 2021). In a related study, Chen (2022) reported a consistent increase in air temperatures across Africa with notable trends in both maximum and minimum temperatures. This reveals that warming patterns vary significantly by geographic latitude with some regions experiencing more intense heat waves. However, the cooling observed in Africa was not widely reported in the literature.

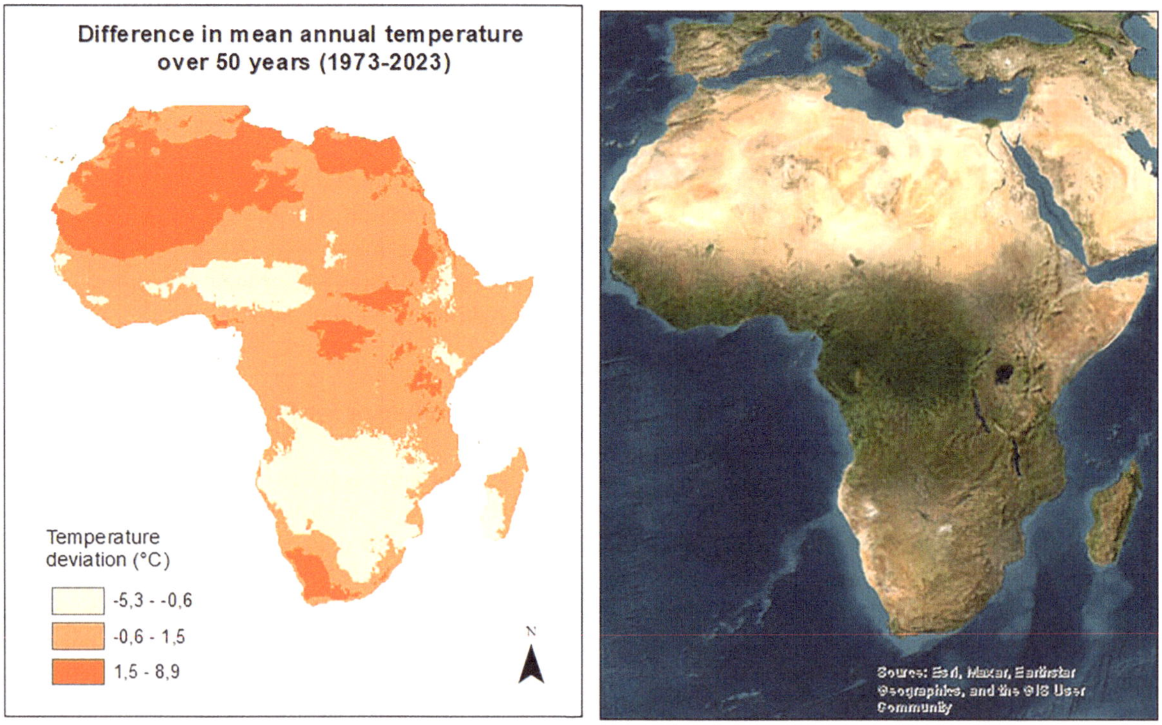

Fig. 2 Visual comparison of temperature variation over 50 years and current vegetation cover in Africa. *Source* By authors

3.2 Interpretation of Temperature Distribution Considering Both Urban and Landscape Criteria

Our initial analysis, based on the systematic approach outlined in this paper, focused on interpreting the results in relation to plant cover. This angle gave us an understanding of the role of vegetation in climate interactions and temperature changes. By scrutinizing this relationship, we aimed to provide a foundational understanding of how temperatures and plant cover correlate and how African landscapes interact with temperatures.

3.2.1 The Status Quo: The Sahara and the Kalahari Desert

The Sahara region is of particular interest when discussing climate change on the African continent. This region warmed by up to 8.9 °C as shown in Fig. 2. Such a large temperature change is likely to exacerbate evaporation rates contributing to severe droughts. In response to climate change, the African Strategy on Climate Change (ASCC) explores the potential for integrated solutions and improved governance to enhance climate resilience (Dan-Woniowei et al. 2022).

On reviewing the temperature variation map, it is relevant to note that the Kalahari Desert has not warmed as much as the Sahara due to differences in geographical, climatic, and ecological characteristics. Notably, despite constantly adapting to changing conditions, vegetation in the Kalahari (grasses, shrubs, and sparse tree cover) has maintained a generally stable ecosystem condition over time (Mayaud et al. 2017). Also, the proximity of the Kalahari to the Atlantic Ocean greatly may have to influence its air temperature due to ocean currents and moisture-laden winds.

3.2.2 Equatorial Forests and Temperature Rise

Africa's equatorial alignment has yielded two distinct climatic regimes, each reacting differently to climate change. The Sahel region grapples with severe droughts, while equatorial forests face threats from changing rainfall patterns and human deforestation. Interestingly, the Congo Basin in the equatorial region has experienced a cooling of up to minus 0.6 °C (Fig. 2), a trend linked to heavy rainfall and the presence of large wilderness areas. This temperature decrease has the potential to disrupt the balance of tropical ecosystems causing rainforests to experience stress, leading to a significant reduction in their role in CO_2 sequestration (Huntingford et al. 2013). The correlation between large wilderness forested areas and the cooling of air temperatures seems to suggest that nature is playing a vital role in regulating temperatures. Any threat to the forest cover, whether it be caused by human activity such as logging or changes in the rainfall patterns which call into question forest health and extent, could have serious effects on temperatures. This trend was explained by (Todd 2004) revealing that central Equatorial Africa experiences rainfall anomalies that are driven by large-scale atmospheric circulation patterns, particularly influenced by sea surface temperatures in the tropical North Atlantic.

3.2.3 Temperature Assessment According to Major Cities in Africa

Finally, as the African continent develops and the prominence of its cities grows, the thermal responses of these urban areas, especially regarding Urban Heat Islands, Urban Health, and Urban Resilience, warrant attention. The World Bank projects that the urban population in Africa will reach 60% by 2050, up from the current 43%, emphasizing the need to tackle temperature rise challenges to ensure sustainable resilience.

This study reveals significant temperature variation in Africa's metropolises and major cities including Cape Town, Lagos, Casablanca, and Cairo (Fig. 3). However, Johannesburg, South Africa displays a moderate temperature change attributed to the dense tree cover across the city. Originally a rolling grassland devoid of trees aside from those along small rivers, modern Johannesburg now boasts the largest man-made forest in the world, due to the profusion of exotic trees significantly impacting the city's temperature dynamic (Turton et al. 2006).

In addition to these climatic imbalances, these cities are grappling with a rapidly growing population. Cairo, with over 20 million residents, is Africa's most densely populated city. This growth is accompanied by striking economic disparities, the proliferation of informal settlements, and an increase in vulnerable communities.

3.3 Implementing Nature-Based Solutions to Address Heat Challenges: Three Case Studies

3.3.1 NBS for Climate Change Mitigation

Given these alarming facts, an appropriate strategy to mitigate fluctuating climate changes (ranging from colder to hotter temperatures) would be to both maintain and protect existing planted areas and launch new initiatives to increase urban greening. One such initiative is the Great Green Wall project in the Sahel region. It is a game-changing African-led initiative that aims to restore the continent's degraded landscapes and transform millions of lives in the Sahel (Mirzabaev 2022). It's also particularly interesting to see how all the climate models assess the positive effect this project could have.

An important question is understanding the mechanisms through which these green interventions combat climate change. In their article in issue 07 of the African Journal of Landscape Architecture (AJLA), Breed and Du Plessis investigate this issue. One of the answers lies in the cooling function that vegetation cover and water bodies can give through transpiration and evaporation, respectively, by increasing the

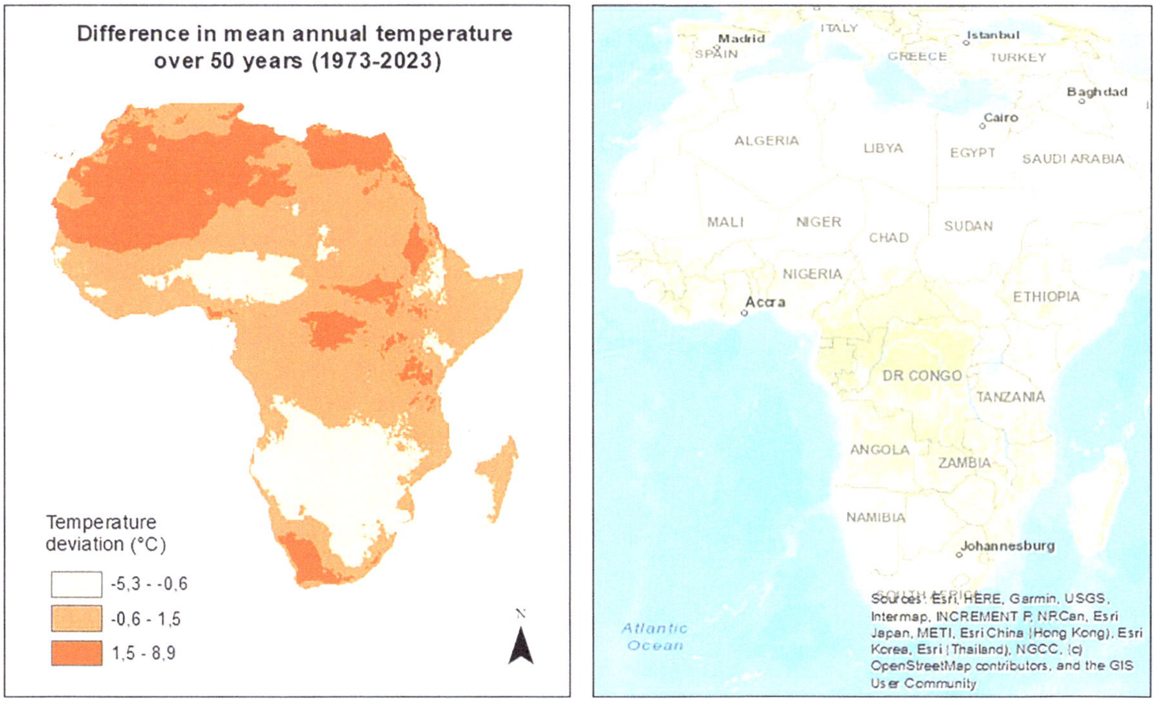

Fig. 3 Visual comparison of temperature variation over 50 years and country map in Africa. *Source* By authors

net latent heat flux and reducing the net sensible heat flux in urban areas (Breed 2024). The author also point to the shade provided by vegetation that reduces the net radiant heat incident to the surrounding urban surface and which also influences air temperatures.

3.3.2 Assessment of Three Case Studies of NBS Implementation

The Msimbazi River Basin City Park Project in Tanzania exemplifies the use of large scale NBS to reduce temperatures and combat frequent flooding (Hoefsloot, 2024). The planned City Park would principally feature unpaved surfaces, with 4000 predominantly native trees and 80,000 square meters of forested areas intended to minimize the urban heat island effect and improve air quality and micro-climate on a regional level. To add to these positive outcomes, the planners also included facilities designed to encourage a sense of community belonging, such as sports fields, areas to practise urban farming, an area for festivals and events and some 15 000 new homes in the flood-resilient parts of the project. In this way, and with the inclusion of several pedestrian bridges to improve mobility across the river, the project is set to not only solve physical and climate phenomena but also to cater for the health and wellbeing of the population.

A second notable case study is a project in Paterson Park, Johannesburg situated in a relatively dense residential area. Landscape architects transformed a crumbling concrete culvert into an attractive natural stream using NBS. Not only has this rehabilitation served to cool the area, but it also resulted in ecological, social and technical benefits such as improved rainwater filtration and increased water retention capability as shown in Fig. 4 (Breed et al. 2023). The creation of new wetland habitats and the integration of walkways, play spaces and bridges bring life—human, fauna, and flora—and art into the park.

Fig. 4 Transformation into an attractive, natural drainage line capable of retaining four times the initial capacity of rainwater and bringing associated ecological and environmental solutions with it, Paterson Park, Johannesburg (Johan Barnard, no date)

The third case to consider is Rwanda and its forestry policy. Could it be that Rwanda is in the category with the most moderate temperature changes (− 0.6–1.5 °C) shown in Fig. 3. The vast forested areas in Rwanda have been under pressure from high population growth resulting in the conversion of wooded areas into agricultural land on the one hand, and from general degradation because of the advanced age of productive trees on the other (Karangwa 2016). The 2004 *National Forest Policy* made great strides in restoring dying forests by promoting sustainable management, involving stakeholders and promoting agroforestry on agricultural land, thereby enhancing ecosystem services. Increased targets for forest cover by 2020 in conjunction with Rwanda's *Economic Development and Poverty Reduction Strategy* (EDPRS), which set targets for reducing wood energy consumption, have underpinned the increased commitment to restore 2 million hectares of degraded land resulting in the fulfilment of several interconnected goals including improving ecosystem quality and resilience, thus by providing opportunities for rural livelihoods and securing energy and water supply through low carbon economic development (Karangwa, 2016).

3.3.3 Governance and Socio-economic Gains of Implementing NBS

NBS and governance are intricately linked, as the successful implementation and scaling of NBS require strong governance frameworks. The study of Karangwa (2016) points out some key lessons learned illustrating the multiplying effect that nature-based solutions can have on the landscape. They underline the importance of having champions to garner support, of being contextually specific and shaping programmes within a broader national context, highlighting the importance of collaboration and coordination between different government agencies.

Advocating for NBS can contribute to social ties establishing a sense of community (Kazmierczak 2013) contrary to the use of 'grey' infrastructure that has marked the urban fabric in recent decades. NBS also create work opportunities on a continent with a high unemployment rate. These green jobs can be developed in landscaping, urban forestry and water treatment to support sustainable development and urban biodiversity.

The motivation for using NBS and providing a network of green spaces therefore includes an economic component. Ngulani and Shackleton (2019) researched the temperature differentials in Bulawayo, Zimbabwe using fortnightly readings in winter and summer, between 12 urban green spaces and the surrounding built-up urban fabric at increasing distances from each green space. They reported a mean difference in temperature between green spaces and neighbouring built-up areas of 3.6 °C, ranging from 0.9 °C to 6.1 °C in summer and from 0.9 °C to 5.7 °C in winter. The cooling effects extended up to 1.7 km beyond the centre of green spaces. They also found that the magnitude of cooling increased slightly with increasing green space size which provides another argument to support the importance of implementing NBS at a large scale. Finally, the authors were able to translate the cooling into a cost on energy saving of approximately US$532/household/year which is a considerable amount in the context of a weak economy and low incomes in Zimbabwe (Ngulani and Shackleton 2019).

4 Conclusion

The present research aimed to examine the variation in air temperatures across Africa in the last 50 years according to a framework combining urban and landscape criteria. One of the findings to emerge from this study is that some areas have warmed by up to + 8.9 °C, while others have cooled down by − 5 °C.

These shifts were interpreted in broader contexts, such as land cover, where broadly speaking, the Sahara region warmed, and equatorial forests cooled. Africa's major cities, in their drive towards metropolitan status, stand out for their warming because of extensive imperviousness and the concentration of anthropogenic activities. The scientific community have confirmed that we are already experiencing the consequences where ecosystem equilibria have been disrupted and the vulnerability of populations is at risk.

We see an interconnectedness in the results of NBS in dealing with climate change—as a solution, they aid in adaptation towards resilience regarding several phenomena: flooding, drought, temperature change... but also provide opportunities for recreation and social cohesion as well as economic advantages. NBS that comprise isolated, localized actions have little effect; however large-scale actions, or many coordinated ones on a very local scale, can have a combined positive effect on the environment.

More research is needed to compare other weather data with the temperature changes we observed over five decades. Looking at other climate variables such as rainfall and humidity may show interesting correlations, and a deeper look at the type of vegetation could reveal further support for our hypothesis. Another direction for research would be to simulate the impact of NBS on the climate and the expected cooling, providing the necessary projections in terms of climate moderation, financial commitment, and health benefits in a wider climate-resilient context. In this regard, it is interesting to encourage urban development and landscape architectural projects to take localized specific temperature readings before and after implementation.

References

Achebak, H., Rey, G., Lloyd, S.J., Quijal-Zamorano, M., Méndez-Turrubiates, R.F., Ballester, J.: Ambient temperature and risk of cardiovascular and respiratory adverse health outcomes: a nationwide cross-sectional study from Spain. Eur. J. Prev. Cardiol. **31**(9), 1080–1089 (2024). https://doi.org/10.1093/eurjpc/zwae021

Babi Almenar, J., Elliot, T., Rugani, B., Philippe, B., Navarrete Gutierrez, T., Sonnemann, G., Geneletti, D.: Nexus between nature-based solutions, ecosystem services and urban challenges. Land Use Policy **100**, 104898 (2021). https://doi.org/10.1016/j.landusepol.2020.104898

Bedair, H., Alghariani, M.S., Omar, E., et al.: Global warming status in the African continent: sources, challenges, policies, and future direction. Int. J. Environ. Res. **17**, 45 (2023). https://doi.org/10.1007/s41742-023-00534-w

Breed, C.A., Du Plessis, T., Engemann, K., Pauleit, S., Pasgaard, M.: Moving green infrastructure planning from theory to practice in sub-Saharan African cities requires collaborative operationalization. Urban For. Urban Green. **89**(open in a new window) (11(open in a new window)), 128085 (2023). https://doi.org/10.1016/j.ufug.2023.128085

Brousse, O., Simpson, C., Zonato, A.,Martilli, A., Taylor, J., Davies, M., Heaviside, C.: Cool roofs could be most effective at reducing outdoor urban temperatures in London (United Kingdom) compared with other roof top and vegetation interventions: a meso scale urban climate modeling study. Geophys. Res. Lett. **51**, e2024GL109634 (2024). https://doi.org/10.1029/2024GL109634

Chausson, A., Turner, C.B., Seddon, D., Chabaneix, N., Girardin, C.A.J., Key, I., Smith, A.C., Woroniecki, S., Seddon, N.: Mapping the effectiveness of nature-based solutions for climate change adaptation Glob. Change Biol. **26**, 6134–6155 (2020)

Cortinovis, C., Olsson, P., Boke-Olén, N., Hedlund, K.: Scaling up nature-based solutions for climate-change adaptation: Potential and benefits in three European cities. Urban For. Urban Green. **67**, 127450 (2022). https://doi.org/10.1016/j.ufug.2021.127450

Dan-Woniowei D., Ikporukpo I., Samuel B.: Climate change governance and ecosystem sustainability in sub-Saharan Africa. Adv.Soc. Sci. Res. J. (2022). https://doi.org/10.14738/assrj.98.12791

EEA: Urban adaptation to climate change in Europe 2016, EEA report no. 12 (2016). https://doi.org/10.2800/021466

El Ghazouani, L., Bounoua, L., Nigro, J., Mansour, M., Radoine, H., Souidi, H.: Combining satellite data and spatial analysis to assess the UHI amplitude and structure within urban areas: the case of Moroccan cities. Urban Sci. **5**(3), 67 (2021). https://doi.org/10.3390/urbansci5030067

Fotso-Nguemo, T.C., Vondou, D.A., Diallo, I., Diedhiou, A., Weber, T., Tanessong, R.S., Nghonda, J.P., Yepdo, Z.D.: Potential impact of 1.5, 2 and 3 °C global warming levels on heat and discomfort indices changes over Central Africa. Sci. Total Environ. **804**, 150099 (2022)

El Ghazouani L, Ouhechou A, Khomsi K, Radoine H.: The interaction between urban heat island and air quality in Casablanca. In: Conference paper, 7th International Renewable and Sustainable Energy Conference (IRSEC) (2019). https://doi.org/10.1109/IRSEC48032.2019.9078170Fie

Haghighatafshar, S., Nordlöf, B., Roldin, M., Gustafsson, L.G., la Cour Jansen, J., Jönsson, K.: Efficiency of blue-green stormwater retrofits for flood mitigation conclusions drawn from a case study in Malmö Sweden. J. Environ. Manage. **207**, 60–69 (2018). https://doi.org/10.1016/j.jenvman.2017.11.018

Hobbie, S.E., Grimm, N.B.: Nature-based approaches to managing climate change impacts in cities. Philos. Trans. R. Soc. b: Biol. Sci. **375**(1794), 20190124 (2020). https://doi.org/10.1098/rstb.2019.0124

Hoefsloot, J.: Msimbazi River Basin City park project. Afr. J. Landsc. Arch. (AJLA) 7 (2024)

Huntingford, C., Zelazowski, P., Galbraith, D., et al.: Simulated resilience of tropical rainforests to CO_2-induced climate change. Nature Geosci. **6**, 268–273 (2013). https://doi.org/10.1038/ngeo1741

Intergovernmental Panel on Climate Change (IPCC). IPCC_AR6_WGI: Climate change. Phys. Sci. Basis 202 (2021)

IUCN https://iucn.org/our-work/nature-based-solutions, Consulted on 10 Nov 2024

Kabisch, N., Korn, H., Stadler, J., Bonn, A. (eds.): Nature-based Solutions to Climate Change Adaptation in Urban Areas: Linkages between Science, Policy and Practice. Springer International Publishing (2017). https://doi.org/10.1007/978-3-319-56091-5

Kaissi, O., Belaqziz, S., Kharrou, M.H., Erraki, S., El Hachimi, C., Amazirh, A., Chehbouni, A.: Advanced learning models for estimating the spatio-temporal variability of reference evapotranspiration using in-situ and ERA5-Land reanalysis data. Model. Earth Syst. Environ. **10**(2), 1915–1939 (2024)

Khomsi, K., Najmi, H., Chelhaoui, Y., Souhaili, Z.: The contribution of large-scale atmospheric patterns to pollution with PM10 : the new Saharan Oscillation index. Aerosol Air Qual. Res. (2020). https://doi.org/10.1002/essoar.10502133.1

Lee S.Y., Lung, S.C.C., Chiu, P.G., Wang W.C., Tsai, I.C., Lin T.H.: Northern hemisphere urban heat stress and associated labor hour hazard from ERA5 reanalysis. Int. J. Environ. Res. Public Health (2022). https://doi.org/10.3390/ijerph19138163

Martin, T.C., Richard, W.: Climate variability in central equatorial Africa: influence from the Atlantic sector. Geophys. Res. Lett. (2004). https://doi.org/10.1029/2004GL020975

Mayaud, J.R., Bailey, R.M., Wiggs, G.F.: Modelled responses of the Kalahari Desert to 21st century climate and land use change. Sci. Rep. **7**, 3887 (2017). https://doi.org/10.1038/s41598-017-04341-0

Mirzabaev, A., Sacande, M., Motlagh, F., et al.: Economic efficiency and targeting of the African Great Green Wall. Nat. Sustain. **5**, 17–25 (2022). https://doi.org/10.1038/s41893-021-00801-8

Mora, C., et al.: Global risk of deadly heat. Nat. Clim. Change **7**, 501 (2017). https://doi.org/10.1038/nclimate3322

Ngulani, T., Shackleton, C.M.: The degree, extent and value of air temperature amelioration by urban green spaces in Bulawayo. Zimb. S. Afr. Geogr. J. **102**(3) (2019). https://www.tandfonline.com/journals/rsag20Volume. https://doi.org/10.1080/03736245.2019.1685405

Ruocco, A., Paolo, G., Guy, W. (2015). 3 Urbanisation and climate change in Africa: setting the scene. https://doi.org/10.1007/978-3-319-03982-4_1

Turton, A., Schultz, C., Buckle, H., Kgomongoe, M., Malungani, T., Drackner, M.: Gold, scorched earth and water: the hydropolitics of Johannesburg. Int. J. Water Resour. Dev. **22**(2), 313–335 (2006). https://doi.org/10.1080/07900620600649827

Wang, Y., Feng, J., Luo, M., Wang, J., Qiu, Y.: Uncertainties in simulating central Asia: sensitivity to physical parameterizations using weather research and forecasting model. Int. J. Climatol. **40**(14), 5813–5828 (2020)

Zardo, L., Geneletti, D., Pérez-Soba, M., van Eupen, M.: Estimating the cooling capacity of green infrastructures to support urban planning. Ecosyst. Serv. **26**, 225–235 (2017). https://doi.org/10.1016/j.ecoser.2017.06.016

The Adoption of the URBiNAT Methodology for the Co-creation of a NBS Park in Nova Gorica and Porto

Marco Acri and Gonçalo Canto Moniz

Abstract

The Horizon 2020 project named URBiNAT, (The URBiNAT project has received funding from the European Union's Horizon 2020 research and innovation programme under grant agreement No 776783—www.urbinat.eu) *Healthy Corridors as Drivers of Social Housing Neighbourhoods for the Co-creation of Social, Environmental, and Marketable NBS*, has targeted the regeneration of social housing districts by combining nature-based solutions, NBS, through a distinctly participatory approach. The project, focussing on the way to recreate places in urban alienated areas, integrated the traditional NBS with socio-economic and solidarity economy practices, recalling the ancient need of humans to gather in groups and construct social bonds to better defend against wilderness and danger. The generation of urban corridors, not merely green but healthy, to connect social housing neighbourhoods to other more central or wealthy urban areas relies on the idea that corridors are not solely shortcuts but also places of proximity, integrating nature, culture, education, and sport, where social bonds can be generated and eventually nourished. The project included eight cities from Europe and Iran, which developed different concepts of healthy corridors based on their specificities, objectives, and participatory dynamics. This paper will dive into two case studies: the pilot city of Porto, which was acting as frontrunner in the project by co-creating a new park in the Campanhã neighbourhood, and the city of Nova Gorica, which acted as follower city and co-developed with locals a NBS park to partly reconnect with the Italian city of Gorizia at the margin of the famous border. Both cities resulted in providing regenerated areas where resilience is expressed both by means of enhanced natural resources and biodiversity, and in terms of social cohesion and sharing, demonstrating how citizen participation may lead to the transformation of spaces into places.

Keywords

Co-creation · Healthy corridors · Nature-based solutions · Participatory approach · Citizens' validation · Living labs

1 Introduction

Urban structures in Europe underwent a dramatic change in the twentieth century, mainly due to the need to accommodate the growing population that was migrating to cities in search of industrial, commercial, and service jobs. Urban planners attempted to address the housing crisis by designing neighbourhoods that expanded cities outward, creating suburban zones characterised by high-density residential complexes. They were initially influenced by the Garden City movement and later by well-known figures like Le Corbusier and his Athens Charter. Suburbanisation was pushed into European cities in large measure after World War II. Housing estates built between the 1950s and the 1970s were designed with the neighbourhood units' model in mind. These estates, which were created by states, local governments, and cooperatives, were frequently found on the outskirts of cities and featured big, densely populated structures meant to serve as working-class housing. Originally intended to be planned, structured communities, these "oases of order" frequently became isolated urban fabrics cut off from the larger city. Even though

M. Acri (✉)
Graduate School, University of Nova Gorica, Nova Gorica, Slovenia
e-mail: marco.acri@ung.si

G. C. Moniz
Department of Architecture, Centre of Social Studies, University of Coimbra, Coimbra, Portugal
e-mail: gmoniz@ces.uc.pt

P. Mohammad et al. (eds.), *Geographic Approaches to Climate Change and Mitigation: Urban and Rural Perspectives (Volume 2)*, Advances in Science, Technology & Innovation, https://doi.org/10.1007/978-3-031-92115-5_22

each home in these neighbourhoods usually had a sizable green space, the overall urban design of these neighbourhoods often produced areas that felt disconnected from the city centre. Even though there were many, the public areas were occasionally just vacant lots between structures that lacked the necessary infrastructure to comply with contemporary laws and standards of living. These neighbourhoods' road networks frequently ended in *cul-de-sacs* or had few access points, which made it difficult to integrate these enclaves with the larger urban fabric. Many of these social housing neighbourhoods suffered from stigma over time as impoverished places with a focus on social homogeneity and financial difficulties. When these communities struggled with problems like poor public amenities, high unemployment, and restricted access to urban centres, the initial hope that these housing projects would provide a bright future faded. The deterioration of these formerly auspicious urban extensions brought attention to the necessity of all-encompassing urban regeneration approaches. Modern urban planning has placed a greater emphasis on the rehabilitation of these social housing neighbourhoods because of its recognition of their potential for transformation. In order to create more integrated and liveable environments, there is an increasing emphasis on revitalising public spaces in addition to building renovations. This entails developing community involvement in the revitalisation, boosting connectivity with the city, and upgrading infrastructure. Modern challenges call for rethinking public space, involving citizens in its co-creation and management. According to the Public Space Charter (UNHABITAT and Biennial Public Space 2013), participatory processes are crucial for inclusive urban spaces.

This paper is meant to introduce an innovative approach to social housing neighbour regeneration focussed on distinctly participatory processes to co-create healthy corridors, namely spaces for social interrelations, individual and collective well-being, and environmental resilience. Such an approach was introduced by the URBiNAT project.

In 2017, after a long collaborative preparation that lasted more than one year with an increasing collaborative commitment of nearly thirty European and non-European partners, the Horizon 2020 project URBINAT was approved and launched. The primary goals of URBiNAT were the integration and revitalisation of underutilised or disconnected city districts, mostly characterised by social housing urbanism. In fact, the project's focus was on the regeneration of public spaces and on the way residents can contribute to such regeneration by co-creating new spatial and social bonds within and between different neighbourhoods. As a material and immaterial nature-based solution (NBS) that takes into account the full physical, mental, and social well-being of its citizens, URBiNAT aimed at co-creating so-called "Healthy Corridors."

2 Conceptual Framework

Beyond the conceptual frameworks for the regeneration of the urban natural environment (Klaus and Kiehl 2021; Pedersen Zari et al. 2022) and of the historic urban landscape (Bandarin et al. 2015; Bandarin and Oers 2012; Dobričić and Acri 2021), there is a need to introduce new spatial and economic models to boost the social cohesion and trust that may enable the effects of cultural and natural urban preservation. In fact, what has been lost in the twentieth century was more the use of the public space as described by Nolli in its plan of Rome, the same public space that generates social bonds, also through frictions. Where can we learn about the social and public space? Arguing that public spaces should be open, democratic, and inclusive (Kinkaid 2020; Lefebvre et al. 2013), forms to achieve this *optimum* should be envisaged.

Inclusiveness is partly tackled by the Habitus concept (Bourdieu 1985), which refers to how social structures shape individual behaviours and perceptions. In the context of public spaces, *habitus* can influence how different social groups interact with and perceive these areas. Public spaces that cater primarily to certain demographics may inadvertently exclude others, reinforcing social inequalities. Bourdieu's framework encourages urban planners to consider the diverse social backgrounds and practices of city inhabitants, ensuring that public spaces are designed to be inclusive and welcoming to all. This is opening up interesting scenarios on the too often and commonly mentioned "community identity," which is confused with the logical territorial identity, the same that nourish the HUL, Historic Urban Landscape concept, and the *Genius Loci.* The human experience of space is indeed focussing on how individuals perceive and interact with their environments. This concept of "topophilia," anticipating the dynamic by which a space turns into place (Tuan 2011), explores the emotional connections people form with places, highlighting the importance of creating public spaces that resonate with their users on a personal and emotional level.

Inclusiveness and openness to different social groups recall the "Right to the City" as a powerful idea that advocates for the democratisation of urban spaces (Lefebvre et al. 1996). Lefebvre argues that all inhabitants, not just the privileged few, should have a say in how cities are designed and managed. This concept underscores the importance of participatory planning processes that include marginalised communities, ensuring that public spaces reflect the needs and desires of all urban residents. Moreover, Lefebvre's critique of the commodification of urban space calls for a reimagining of cities as spaces for social interaction and community building rather than mere economic transactions. The debate on inclusiveness was interestingly matched by the one of "urban ethnography" to understand the complexities of public spaces in contemporary cities (Delgado Ruiz 1999). In this work, the author examines how power dynamics, social conflicts, and cultural expressions manifest in urban settings and highlights the importance of recognising and addressing the diverse and often conflicting needs of urban populations, ensuring that public spaces are sites of inclusion rather than exclusion. A concept stressed by Ali Madanipour (Madanipour 2003), who examined the dynamics of public and private spaces within urban environments, elaborating how the interplay between these spaces influences social inclusion and exclusion, highlighting the impact of urban design on community interaction and accessibility. Madanipour advocated the need for thoughtful urban planning to create inclusive public spaces that foster social cohesion and equitable access for all urban residents.

To contribute to inclusiveness, physical spaces are though fundamental in fostering social interaction and community cohesion (Sennet 1992, 2006). Sennett argues that well-designed public spaces can facilitate encounters between diverse groups, promoting social integration and reducing isolation. His advocacy for "open" cities, characterised by flexible, adaptive spaces that encourage interaction, aligns with the goals of creating inclusive public spaces that are accessible and welcoming to all. A discourse that was further developed by adding the ethical and political dimensions of urban planning, advocating for a humanistic approach to urbanism where the well-being of inhabitants takes precedence over economic imperatives (Paquot 2007). Paquot's vision includes the creation of public spaces that are not only inclusive but also promote a high quality of urban life, with a focus on sustainability and environmental stewardship. It is a matter of producing the space but also of representing it to be closer to normal users. Indeed, spatial representations shape our understanding and organisation of the world, and the traditional cartographic views often simplify and distort the complexity of urban spaces, asking for more nuanced and inclusive representations that reflect the lived realities of urban dwellers (Farinelli 2003). Thus, the production of space asks for appropriate governance and policy capable of shaping urban environments (Bourdin 2005), given that comprehensive urban policies that integrate social, economic, and environmental considerations to create inclusive and sustainable cities question that confirm the need of being inclusive not merely in the results but also in the process, as the European governance concept advocates (EU 2001). A governance that recognises not only the need to involve vertically and horizontally different stakeholders, with a stress on citizens, but also the individual capacity and intrinsic legacy that each citizen holds, so clear in the landscape definition given by the Council of Europe (CoE 2000). Again, citizens as actors, as new owners of the space in a new form of "appropriation," based on use and care instead of ownership, embracing the emerging idea of commons and claiming new practices of management (Ostrom 2012).

Nevertheless, modern cities main focus is resilience, which may be achieved by matching circularity (Foster 2019; Gravagnuolo et al. 2019; Moniz et al. 2023) and renaturing or greening (Mahmoud et al. 2022). The former asks mostly for practices, new approaches, and attitudes; the latter is more based on urban physical transformations, ranging from the application of extensive nature-based solutions, such as parks, green roofs, and façades, and for the creation of green belts or corridors, which should enable the increase of biodiversity by generating preferential "highways" of human and non-human migrations. Nevertheless, we have learnt that corridors are not only speedy shortcuts for fast and dislocations but can often transform into spaces, we can say here *passages* (Benjamin et al. 2000), where other uses are possible, such as strolling, resting, and interacting, which are even advisable. So, if corridors are not merely shortcuts, becoming spaces of interaction and proximity, can we imagine (green) corridors that become places by merging the circular, green and human dimensions, where the human component is taking the lead? Can we create corridors by combining solutions of nature, sustainable economy, and socialisation?

In this general framework, we can identify 2 main gaps to be investigated, tackled by means of two case studies from the URBiNAT project:

- Integration of Inclusivity and Public Space Transformation. While the text highlights the need for inclusivity in public spaces, a research gap exists in developing innovative spatial and economic models that combine cultural preservation, natural regeneration, and active social inclusion. Specifically, how can public spaces be designed to balance diverse community identities, address socio-economic inequalities, and foster meaningful interactions across different groups?
- Reimagining Green Corridors as Multidimensional Spaces. There is limited exploration of green corridors as interactive spaces that combine ecological benefits with human-centric activities. A gap exists in investigating how corridors can transition from being functional pathways to vibrant "places" that integrate nature-based solutions, sustainable economies, and community-building practices while prioritising the human experience.

Both gaps highlight opportunities for interdisciplinary approaches combining urban planning, sociology, ecology, and governance to address modern challenges in urban design and management.

3 The URBiNAT Methodology

3.1 The Healthy Corridor Concept

How did the URBiNAT project combine these different needs and pushes to regenerate social housing districts?

The core of the project has been the definition of healthy corridors as a possible approach to positively changing the situation of underserved neighbourhoods. The concept of "healthy corridors" is borrowing the idea of "green corridors," adding the social dimension. It can be defined as a linear network of places that is planned, implemented, and managed with a variety of interconnected uses in mind, such as ecological, educational, recreational, cultural, or aesthetic. Healthy Corridors aims at enhancing the neighbourhood's overall health by providing ingredients for the quality of life that are typical of a communitarian social life. Therefore, green spaces can be more than just green if the local communities that will use them are involved from the beginning to co-create solutions to address their social and environmental challenges. In this perspective, a healthy corridor is a material and immaterial public area that serves as a walkway as well as a venue for social, cultural, and educational activities that connects and links neighbourhoods; it is a shortcut as well as a "passage" where strolling, observing, resting, and chatting are done without time obligations or comfort restrictions. In addition to fostering mobility between various spaces and communities, it can occupy public and private urban plots or commons. It can also improve living conditions by fostering a safe and healthy environment, reducing the segregation effect that currently exists through social and urban cohesion, and boosting urban resilience to climate change.

3.2 Co-diagnostics, Co-design, Co-implementation Through Validation

The co-creation process is organised in four stages: co-diagnostics, the co-design, the co-implementation, and the co-monitoring—that engage the local community in the urban regeneration of their territory by activating a living lab (Moniz et al. 2022; Steen and van Bueren 2017). These laboratories are located in the real context, and the citizens progressively take control of the process, being equal actors in the decision-making and validation of the urban plan.

The design of such healthy corridors, though, asks for objective evidence of the needs in the city and the objective priority to be given to a certain neighbourhood, as well as significant statistical evidence of the subjective needs in the selected neighbourhoods. In this respect, even if a policymaker and urban planner perspective would have been sufficient to identify local priorities, a set of so-called co-diagnostics activities have been implemented, both looking at first the available information at the urban and local level, such as socio-economic, territorial, and social data, and, in a second step, the use and perception of the site by citizens and daily visitors. This means gathering objective data from traditional sources to have an objective state of the art of the city and its neighbourhoods or parishes, to be further investigated through more citizens' science-dictated approaches based on citizens direct or indirect feedbacks (Bagnasco et al. 2024; Ferilli et al. 2021; Giglione et al. 2024).

If the first step permits confirming the focus on one neighbourhood instead of others, the second step gathers information from locals that would not be possible through traditional assessment tools. The local data gathering was performed through different methodologies, such as (mentioning some):

- Territorial mapping, performed through on-site observation, is meant to highlight impediments and potentialities in the neighbourhood;
- Behavioural mapping, performed on-site by observing the use made by citizens and visitors of the area in different moments of the day and of the week (Fig. 1);
- Photovoice, performed in any location and done by enabling users and citizens to express their perceptions of the site by confronting individual photos and images;
- Walkthrough, performed by strolling in groups in the area with open debates about feelings, perceptions, wills, needs, etc.
- Focus groups are performed in any location, opening debates on emerging issues to be solved, ranging from basic needs, such as a lack of benches or unsafety, to dreams, such as renaturing, building infrastructure and facilities, etc.
- Cultural mapping involves interviewing citizens and users about the available cultural and social resources of the area, such as calendars of events, local associations, local private initiatives, and cultural heritage and assets.

The adoption of methodologies to get in contact with citizens and users also enables a more detailed mapping of the local stakeholders and the construction of trust among individuals, both towards the authority that performs the investigation and towards other members of the community, opening up a sense of mutual understanding. Such core steps offer the basis for a subsequent and successful co-design phase. Before the co-design was through, URBiNAT introduced a fundamental moment in the process, "validation." The validation happens when, after the gathering of data from existing sources and from observation and sharing with citizens, the planner submits its interpretation to local stakeholders, mainly citizens, to get their approval on the correctness of the diagnostics. Validation is the moment when citizens see the final picture of their contribution, agree or disagree on some points, and, once positively done, become co-responsible for the process. Validation is a fundamental moment in all the stages

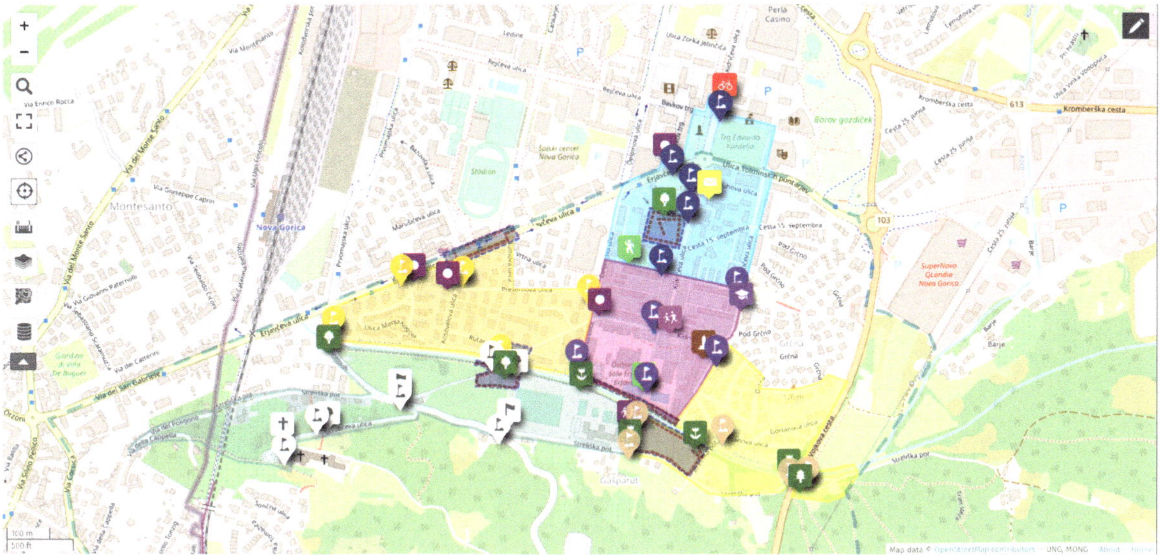

Fig. 1 Working map for the behavioural and territorial mapping in Nova Gorica, produced during the co-assessment phase in Nova Gorica. URBiNAT Loal Diagnostics in Nova Gorica, courtesy of Municipality of Nova Gorica, author Maja Debevec

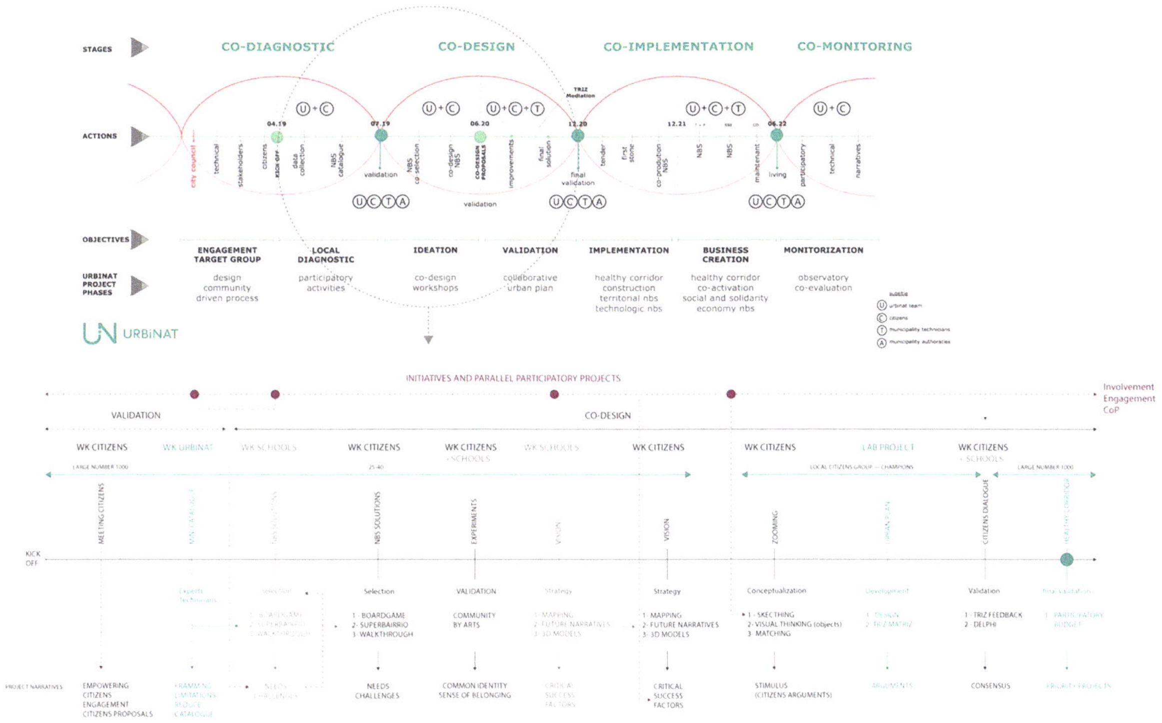

Fig. 2 Co-implementation steps, in URBiNAT co-creation methodology. *Source* URBiNAT (Design: GUDA)

of the URBiNAT co-creation methodology, namely in the transition of each stage (Fig. 2).

The co-diagnostics is considered also the stage for the creation of the Living Lab, a tangible and intangible space of regenerative ideas for a neighbourhood, where citizens are empowered and take the lead of the transformation accompanied by thematic experts such as planners and technicians of the local authorities (Nunes et al. 2021).

3.3 The Co-design of the Healthy Corridor and the Combination of NBS

Once the site is understood in its strengths and weaknesses, healthy corridors are built upon by adopting a strategy of integration of NBS chosen by the living lab from the URBiNAT NBS catalogue and from other sources, including brand new creations that may fit better with the improvement of the

Fig. 3 The Campanha area (South Below, North Above) as considered by the URBiNAT project. In this figure, the main existing accessibility axes are highlighted as important factors for the need of transformation through proximity spaces, as the healthy corridor. *Source* URBiNAT—Biopolis (taken from URBiNAT D.2.1)

area. The URBiNAT catalogue consists of four different typologies of NBS, namely territorial, technological, participatory, and solidarity economy ones, for a total of more than forty solutions. If the territorial and technological NBS are normally present in the common imaginary, the added value of URBiNAT was to consider participation and a social and solidarity economy. NBS is part of the "natural" practices of humans to aggregate and organise to survive and, often in the past, get defended from nature and its wilderness. The first ones are products and infrastructures, while the second ones are processes and services. (Conserva et al. 2021).

4 The Cases of Porto and Nova Gorica

Porto and Nova Gorica are two completely different cities. The former is a million-person urban agglomerate with a dense economy, urbanisation stress, and population density, while the latter is a little green city that is looking for a new future in collaboration with its neighbouring city named Gorizia, located in another country, Italy, but with no spatial discontinuity (only political and administrative).

4.1 The Front Runner City of Porto

Portugal's second biggest city is Porto. Like most European cities, it expanded rapidly in the second half of the twentieth century, taking up enormous land areas with social housing projects that were not always accompanied by well-thought-out infrastructure planning. These projects frequently featured the best architectural designs. Despite the fact that these extended neighbourhoods now resemble a part of the historic city due to their frequent proximity to the city's centre and other tourist attractions, this phenomenon has the effect of segregating many of these neighbourhoods.

The Campanhã parish's specific study area (Fig. 3) was successfully defined by the URBiNAT project with the local community, allowing for a thorough analysis of the region and dialogue with local actors. The purpose of this was to find public spaces where NBSs could be installed if there was a need for them. During the co-diagnostic phase, all the project methodologies were used to collect the data. When approaching the area, to have confirmation about the need to give it a priority within the large urban sprawl of Porto, existing public areas with high social impact, primary services, facilities, and green spaces were also taken into consideration. This helped to prevent investment overlaps and foster strategic synergies that will increase the project's impact in the area. To improve local living conditions in a more sustainable way, Porto city's Campanhã parish has been named a priority area for urban regeneration.

This area faces severe socio-economic challenges that are exacerbated by its remoteness, in addition to other problems like poor urban conditions, limited accessibility, safety worries, and security concerns. In addition to the various socio-economic and environmental obstacles that resulted in the designation of the intervention area as crucial for the execution of the Healthy Corridor, it is imperative to draw attention to the serious connectivity problems that arise primarily from topographical features (the Campanhã valley) and physical barriers (highway infrastructure). This 4 ha area's exact borders were drawn through a deductive exercise. It is situated at the intersection of several key facilities and services (e.g. Oriental City Park, Corujeira Square, Campanhã Cemetery, local schools, sport facilities, and Oliveira Urban Garden) and three municipal housing neighbourhoods for social rental (Cerco do Porto, Falcão, and Lagarteiro), but being at an intersection, it suffers from physical barriers, i.e. the infrastructure as the highway cut, and social backwardness due to the economic divide and the presence of criminality.

By the time the co-diagnostic was over, enough framework data about the city, the parish, and the study area had been gathered to enable the identification of a new work scale—the project's intervention area—when it was combined and synthesised (Table 1).

The flexibility of the developed activities and the applied methodology, in addition to the close proximity that the new URBiNAT living lab offered, allowed the participants to transform the above-listed needs into new ideas and proposals that were inspired by the URBiNAT catalogue. This permitted them to think creatively and innovatively about how to rethink this area while also rethinking and reshaping the local economy, culture, and learning environment. At this point, thanks also to the design ability of the local team and task force, i.e. the landscape design solutions proposed by the University of Porto, the affected area was assigned to, and the New NBS for Porto was defined. The impact of the Healthy Corridor—which is made up of both material and immaterial solutions—was amplified because many new NBS were assigned to other strategic areas within the study area that were not part of the intervention area.

Following the proposal of various New NBS, the URBiNAT technical team, the municipal technicians, and political representatives carefully examined them in terms of their viability from both a technical and financial standpoint.

Following this systematic analysis, all submissions fell into three primary groups:

- Category 1: proposals that are compatible with the objectives of the project and the municipal strategy and that have been found to be technically and financially feasible within URBiNAT.
- Category 2: which is to be channelled to the respective municipal departments to be integrated into ongoing and/or future projects.
- Category 3: proposals that could not be further developed or implemented within the financial framework of URBiNAT but were pertinent for the territory and/or aligned with the municipal strategy for the future.

As a result, nearly 38 NBS solutions were proposed and almost fully implemented, divided into four new main categories, namely: (1) public space and nature; (2) Socio-economic and solidarity practices; (3) culture and sports; (4) education and environment, enabled by an important morphological transformation meant to improve the resilience of the area and mobility (Fig. 4).

The redevelopment of the local school by creating rest areas nearby, the revitalisation of the community gardens with linked social care, the preservation of existing heritage assets discovered during the excavations, the transformation of the old farm into a public space, the stabilisation of the steep banks, the preparation of the ground for an open terraced green theatre, and the provision of new social services, the majority of which are led by the community, such as the community market, the cultural platform known as Campanh'UP, the community movable kitchen, and the outdoor leisure and sport practices, such as yoga lessons, should be listed among the most pertinent interventions.

The park was completed and opened in February 2024, witnessing an immediate social appropriation of the space with a constant increase of users from the neighbourhood and the entire city. The establishment of this park showed how an original idea of a corridor, meant to mainly improve the connectivity opportunities for a neighbourhood, could become a local activator, attracting use and care by means of social awareness and co-governance. Activities in the park are continuing with the lead of citizens and the newly established local associations.

4.2 The Follower City of Nova Gorica

Nova Gorica was founded in 1947 and was born out of the new Yugoslavia–Italy border established after WWII. This division disrupted a bilingual community, creating socio-economic tensions. While Nova Gorica flourished with its industries and casinos, Gorizia became a hub for diplomacy and defence. Architect Edvard Ravnikar designed Nova Gorica with an emphasis on public parks and minimal private land. After 1991, the region's socio-economic landscape shifted significantly, with the 2008 economic crisis exacerbating the situation. In response, the cities began collaborating through the EGTC, established in 2010, to improve health services and develop biking routes. In December 2020, their joint efforts were recognised when they were named the Future EU Culture Capital 2025. Applying the URBiNAT approach in Nova Gorica presents unique challenges compared to other cities in the partnership. Nova

Table 1 Selected list of feedbacks and desire by locals on the potential transformation of the selected area, based on the results of the local diagnostics.

About the green areas and the sport facilities	• Connections and mobility should be improved to enable shorter walks to commerce and avoid infrastructural barriers; • Space multifunctionality was needed to provide facilities to the different residents of the neighbourhood, elders included; • Security by presence (residents and passersby), given the socio-economic difficulties of some areas of the neighbourhood; • Liveability as an extended concept, namely providing safety, health, and pleasure to stay; • More active inhabitants, social life, and especially well-being, boosting socio-economic opportunities; • Open-air physical activity and leisure, initially absent
About the conditions of the paths (especially pedestrian)	• Mobility and connections (explored with the walkthroughs), enhancing the "quantity" of space dedicated to pedestrians; • Local safety, namely making these paths safer (lighting, impediments, surfaces, etc.) • Mitigate urban fragmentation, by improving the connections and the shortcuts; • Novel pedestrian options to shorten the walking distance between pertinent locations
About increasing the visibility of all spaces in the area, including the improvement of lighting	• Local safety for all • Connections and mobility for all • Criminal activities reduction
About the quality of the resting and leisure areas	• Space multifunctionality, given the basic residential destination of the area; • Security by presence (residents and passersby), namely building a community monitoring; • Liveability, in terms of good comfort and relaxation; • Inclusiveness of the public space, open to anyone with no discrimination; • Open-air physical activity and leisure facility provision
About the socialisation boosting—erasing the discriminations	• Community empowerment, giving strength to local champions and voices; • Sense of belonging related to the territorial identity of the area (proudness of being in a historic social housing district of Porto); • Identity and uniqueness (re)-creation, namely helping boost the cohesion of the community members; • Inclusiveness of the public space, enabling all to contribute and participate; • Improving the connectivity with existing and new social initiatives;
Improve the literacy towards environmental protection	• Providing opportunities for environmental education, i.e. showing the benefits of nature in urban contexts and personal life
About economic opportunities generation	• Activate inclusion through solidarity activities • Local economy re-invention, offering alternative commercial opportunities (considering the distance from existing ones) • Favouring the entrepreneurship of local products and services; • Again, boosting the connection with existing initiatives and potential new ones;

Source URBiNAT local plan Porto
It is clearly visible that accessibility and safety in the area, more than green and biodiversity are the perceived needs, confirming the expected baseline of the municipality and the research team. This list permits urban planner and landscape architects to propose a landscape and space transformation considering these inputs and organising spaces for local social activities to be performed independently

Gorica is a small city with 13,000 inhabitants, and its municipality, along with Gorizia, has a combined population of approximately 67,450. Unlike other cities, Nova Gorica does not face severe issues like criminality and social injustices. However, there is a pressing need to reconnect the people and revive the area's historical ties. The initial idea of the URBiNAT corridor in Nova Gorica and Gorizia aimed to act as a socio-cultural catalyst, bridging communities on both sides of the border and revitalising underused cultural and environmental spots. The idea to focus on the Koren stream as the corridor's centrepiece was significant not only for its centrality for the potential low-impact commuting communities of both cities but also for the presence of water that, despite having been neglected for several years, could serve as an attractive element to bring people together.

The area selected for redevelopment in Nova Gorica (Fig. 5) was initially chosen through a detailed analysis of border cities and their dynamics, confirming initial perceptions of its potential. Supported by a task force, various actors assessed the site, aiming to convert a neglected urban

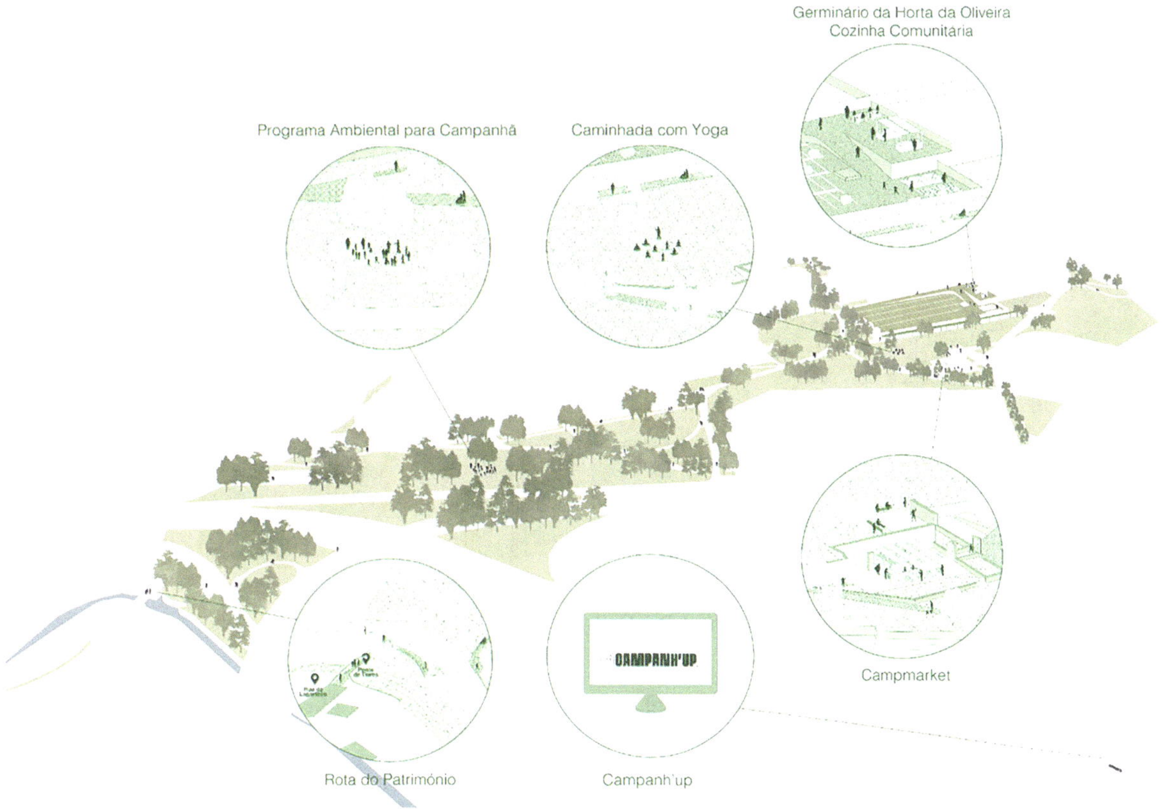

Fig. 4 Conceptual plan of the "spatialisation" of the park as a cluster of NBS. The area is transformed through traditional landscape architecture approaches, integrated by social functions. *Source* URBiNAT-CES/UC (Design Vitório Leite)

space into a nature-based co-living area benefiting from the Koren stream. Local Diagnostics showed citizens' interest in enhancing green spaces, leisure, and sports areas. The selected site, a green part of the city, has seen little change over the past 70 years and includes a former sewage canal and historic paths. The plan focusses on connecting cultural spots through green paths, highlighting the Koren stream, to improve living quality by integrating cultural and natural assets. The area is characterised by hilly terrain, the Koren stream, and underwater channels from Kostanjevica Hill and Panovec forest, supporting rich vegetation and diverse wildlife, including boars, salamanders, and deer. To the north, it is bordered by a socialist-era residential neighbourhood and private villas in Grčina, approaching the Italian border. Key northern sites include the Living Lab in the X-Centre, the Municipality, Europe Square, and the train station, central to the Culture Capital Initiative. The southern boundary is defined by Kostanjevica Hill and Panovec forest, with pathways connecting to Nova Gorica and Gorizia. The eastern limit is marked by the Koren stream's retention basin and Panovec Park, while the western boundary meets the Italian border and the former customs zone. The area, rich in natural and cultural assets, is underused due to a lack of services. Selection criteria for the intervention area include the presence of water, underused cultural assets, the border's role in the Culture Capital 2025 initiative, potential for new pedestrian paths, enhancement of sports facilities, and increased biodiversity.

The site was selected when thinking about the pilot to be used in the project, looking at the possible enhancement of the Koren stream as a unique central water element of the city, trying to replicate the excellent work done in Ljubljana with the Ljubljanica River. After a thorough territorial and socio-economic analysis of Nova Gorica and Gorizia, with the aim to reinforce the crossborder mechanisms already in place in the cities, the area was assessed by adopting key methods utilised to perform a local diagnostics on site (Ferilli et al. 2021), beyond the assessment of the existing data, include walkthrough and photovoice activities, facilitating collaboration among researchers, municipal officials, citizens, and local institution representatives. This collaborative effort aims to enhance the area's quality in terms of leisure, health, and ecosystem services while also shifting the population's mindset away from car dependency towards

Fig. 5 Map with a focus on the core area, displaying two focal points for the cultural and participatory future of the city (Taken from URBiNAT D.2.7). Orientation is South below and North Above. *Source* URBiNAT—UNG

sustainable urban development. The main findings of the local co-diagnostics highlighted the will of citizens to have a more dynamic space, namely an area that people do not merely cross but use for different purposes throughout the day and in all seasons. Citizens asked for better contact with the stream and the existing waters in the area, having leisure and sports facilities for all, especially children and youngsters, and also considering the presence of two primary schools and the secondary school dormitory (Table 2).

Based on the feedback from all the stakeholders, the local task force worked on a co-design process. A long co-creation experiential process lasting the full summer 2022 focussed on engaging families and children through on-site activities like Mobilet (a wooden, temporary kiosk to provide basic services such as kids monitoring, drinks, games, etc.), bike rides, crossborder meetings and performances, as well as in-room workshops at schools and student houses. These activities demonstrated a high level of attention to green environments and resilience, as well as an active participation in sharing the conversion of the space into something new, still green but more open to interaction and proximity (Dobricic and Acri 2021). The results reflect a community committed to enhancing their urban environment, driven by a shared vision of sustainability and resilience. The integration of children's perspectives and active participation underscores the project's inclusive approach, fostering a collective commitment to long-term urban transformation (Acri et al. 2024). The design of the combination of all these data, based on a collaborative passive and active contribution by citizens, led to the ideation of a park, named NBS Koren Park, seen as a condensation of nature-based solutions, either inspired by the URBiNAT catalogue or by other existing good experiences, including those already implemented by the city of Nova Gorica. The park has been designed to have three main interventions, namely a water theatre to be connected to the EU Culture Capital 2025, a terraced standings football open field, and a phytoremediated swimming pool, collecting and filtering the waters from the hill (Fig. 6).

The NBS park gathers, to integrate the main territorial interventions, numerous other NBS as per URBiNAT categories, namely social, such as adopting a tree, community art workshops, and educational spaces, but also solidarity solutions, such as solidarity markets, flea markets, farmer networks, complementary currency, etc. The quality of this plan, which was replicated in other areas of the city thanks to the boost of the EU Culture Capital, promoted Nova Gorica to the role of Frontrunner in a new EU-funded Horizon Europe project named GreenInCities,[1] which has absorbed many other NBS proposed in the NBS Koren Park.

The ongoing co-creation process of the Healthy Corridor in Nova Gorica integrates the city's strategic development plans, requiring official approval. This process adheres to the five steps defined by URBiNAT: Communication, Information, Consultation, Participation, and Co-Building.

[1] www.greenincities.eu

Table 2 Selected list of feedbacks and desire by locals on the potential transformation of the selected area, based on the results of the local diagnostics

About having more and better green facilities	• A greater variety of green, such as fruit trees, herbs, and more varied vegetation; • More green spaces, trees, avenues, parks, and natural shade • Better care for the existing vegetation • Maintaining tall trees and overgrown grass is important • Setting up animal shelters • Improving riverbed greening for diverse vegetation that will increase biodiversity • Creating grassy, sandy relaxation areas
Concerning enhancing area upkeep	• Cleaning, organising, and maintaining the Koren stream and water to make it more accessible to people • Including all of the pipelines and inlets that enter the Koren • Building bridges again using more organic materials like glass and wood • Hanging signs to prevent free movement of dogs, to allow kids to play safely • Upkeep of the walkway • Multiple waste collection containers, including ones for pets • Handling issues with rodents (especially rats at the moment along the stream banks); • Outlawing the feeding of untamed animals • Moreover, a thought that "Koren might be the city's jewel!"
About areas for socialising	• Offering a local pub or hangout • More outdoor areas for gathering with friends and relaxing in the great outdoors • Setting up a location for gatherings for those with limited mobility • Considering locations for cross-generational gatherings • Having a public washbasin/sink available • There is not enough urban equipment (lights, trashcans, benches with tables, etc.), and the equipment that is there is not well maintained • An increase in water features • Improved lighting in public areas
About the walking paths and the water course access	• Setting up a nature-related educational learning path that is (circular) and includes educational boards • Recently constructed nature trails • Several river crossings made possible by piers, bridges, stepping-stones, etc • Increase and improve access to water • The bed has to be restored and the stream controlled • Storm water runoff collection nets • Resolving issues with wastewater • Better upkeep and sanitation of the dry hold
About recreational and sporting opportunities:	• New sports attractions, such as skate parks and dirt parks • Exercise outdoors • A football pitch in the public with "artificial" grass • To allow outside guests to use the football pitch that is currently on-site and not publicly accessible • Setting up a court for beach volleyball • Child playgrounds that are maintained and open to the public • Improving the layout of the paths for jogging and walking
Regarding the area's viability and traffic	• Wheelchair and pram accessibility is unrestricted • Transportation infrastructure modernisation and regulation • Securer links and the removal of hazards from the current sports facilities connecting Koren, the Panovec trim track, and Kostanjevica • A paved path, well lit and safe, allowing youngsters to reach the bus and train stations far from the dormitory • Boost pedestrian security • On the new route along the Koren, there are confusing and insufficient traffic signals • Students who ride scooters and skateboards to school would prefer more paved roads • More soft grass areas are what pedestrians' desire

Source by the author based on URBiNAT, Nova Gorica Urban Plan

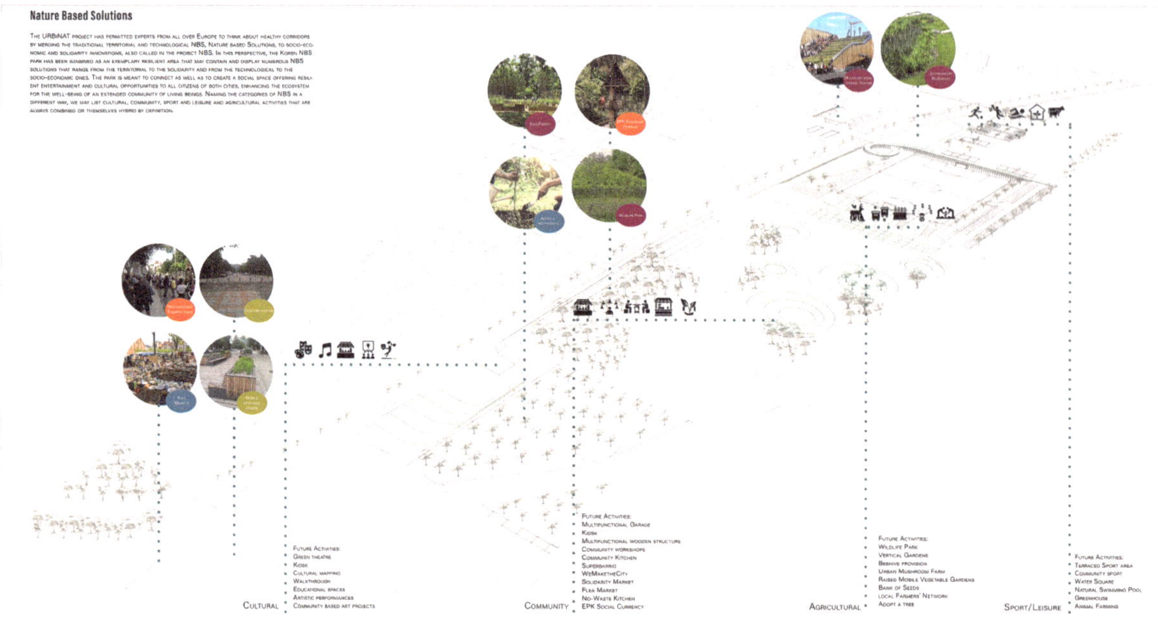

Fig. 6 An axonometric view of the park showing the main territorial interventions, such as the water theatre, the phytoremediated swimming pool and the football field with terraces green standings, combined with other territorial works, such as a terraced orchard, terraced doline, etc. This picture represents only the main transformations, especially physical, happening in the site, that will benefit of an event and gatherings all over the year, such as community lunches, flea markets, sport activities, cultural talks, etc.. ©URBiNAT-UNG (Design Saša Dobričić)

5 Conclusion

Among the eight urban plans, Porto as the frontrunner and Nova Gorica as the follower explored the healthy corridor as a park with a network of paths and green areas that provide an ecosystem service where social activities can enhance nature quality and contribute to the citizens' well-being. Although the two urban plans use the same strategy, the two intervention areas are very different: in Porto, a terrain vague with a slope topography, marked with several informal paths; in Nova Gorica, a flatter area structured by and focussed on a stream with several places for physical activities, but also with previous cultivation spots. Both places are at the limits of the urban areas, connecting to other cities and other territories, taking advantage of the idea of creating a park that is not meant to be only for leisure but to be a destination, a place instead of a space, for the everyday lives of people that live, work, and study in the area.

On the one hand, a common issue is the fact that in both cities, the urban plan was finally designed by landscape architects, both academic partners of the municipalities (Biopolis and UNG), with the knowledge that clusters of NBS work better because they provide multiple and complementary impacts. On the other hand, the long co-creation process with co-implementation allowed Porto to develop the immaterial solutions that needed time to experiment and consolidate, results that were only partly achieved in the Slovenian city due to the role of the follower and the consequent lack of direct funding for implementation. Nevertheless, considering that Nova Gorica integrated the healthy corridor in the Capital of Culture 2025 initiative and in a new Horizon Europe project, it is expected to achieve the same level of development with the immaterial NBS, such as the complementary currency for which UNG is working jointly with local and international partners.

In terms of NBS, both plans co-created solutions to manage the water, as a retention basin in Porto and as a water stream in Nova Gorica, to bring culture to the public space with an open-air amphitheatre in both cities, to promote a social economy approach with social markets and community gardens, and to engage directly children, given the presence of local schools and other services for youngsters, to enhance their opportunity for leadership.

The potential for urban regeneration in social housing districts, once peripheral but today central, is immense. URBiNAT has demonstrated that, using the concept of a corridor as a place, making space instead of a shortcut, and adopting nature as a leitmotiv to enable social participation, awareness, and care, transformations are possible with little financial commitment. Investing in local communities and permitting them to act and contribute to all phases of redevelopment is a long-term investment with a social return. By leveraging the unique properties of these neighbourhoods, such as their green spaces and strong community networks, urban planners can create more sustainable and inclusive environments that do not purely impact the resilience of the city but also the capacity of individuals in small communities to play a role. This involves strengthening connections to the city, adopting innovative partnership schemes, and ensuring that regeneration efforts are driven by the interests and aspirations of the residents and by their care and commitment, as the common good theory has suggested.

Incorporating green spaces into urban environments is a crucial aspect of creating inclusive cities. Green cities not only provide environmental benefits, such as improved air quality and biodiversity, but also offer social and psychological benefits. Accessible green spaces can serve as communal gathering points, fostering social cohesion and providing residents with opportunities for recreation and relaxation. Ensuring that green spaces are equitably distributed across urban areas is essential to prevent the reinforcement of existing social inequalities. The URBiNAT methodology did show not only to work but to be strongly advisable as common practice in the cities involved (as stated by the Mayor of Porto in the final conference). In this respect, the URBiNAT project lasted 6 years and permitted investigations that are out of the traditional cycles in urban planning. In particular, it benefitted from important funding to experiment with several participatory practices that are usually not possible in urban dynamics. Nevertheless, it showed cities the importance of involving citizens' in all processes to make them stewards of their neighbourhood and permit them to construct alternative forms of managing their commons. This means that if the preparation process was longer and more expensive than usual, the effects of having committed citizens would be able to continue shaping their places beyond political cycles. The important achievement of URBiNAT was indeed to include social and solidarity solutions in the set of NBS, giving to citizens fertile soil for the growth of their social bonds. In this perspective and looking at the results of the URBiNAT cities, in particular in this case those of Porto and Nova Gorica, we can say that the methodology was successful, not only for the quality of the transformation of the space, be it a park or a corridor, but for the consolidated understanding that citizens want to be enabled to commit and contribute to the change, which is an important policy recommendation and something that the EU commission and the evaluators confirmed several times.

Acknowledgements This paper is the result of a collaborative work that involved numerous partners and individuals in the URBiNAT project. We share with all of them the theoretical framework and methodological results. Special thanks to all those who contributed to the co-design of the NBS parks in Porto and Nova Gorica, in particular both cities and their Living Labs.

Funding This research was funded under the framework of Horizon 2020 research project URBiNAT: Healthy corridors as drivers of social housing neighbourhoods for the co-creation of social, environmental, and marketable NBS. The URBiNAT project has received funding from the European Union's Horizon 2020 research and innovation programme under Grant Agreement No 776783.

References

Acri, M., Dobričić, S., Kobe, A.: Healthy corridor urban plan Nova Gorica. In: Moniz, G.C. (ed.) URBINAT—D2.7 Healthy CORRIDORS URBAN PLANS FOR FOLLOWER Cities. University of Coimbra, CES (2024)

Bagnasco, A.M., Ferilli, G., Acri, M., Lameiras, J.M.: Participatory approach, mixed methods, and urban sustainable development in the Horizon 2020 URBiNAT project. In: Canto Moniz, G., Bechet, B., Lameiras, J.M., Acri, M., Nunes, N., Ferreira, I., Tasheva-Petrova, M., Andersson, I., Caitana, B., Ferilli, G. (eds.), Proceedings of the International Conference on Nature for an Inclusive and Innovative Urban Regeneration (NATiURB 2022) vol. 24, pp. 19–28. Atlantis Press International BV (2024). https://doi.org/10.2991/978-94-6463-469-3_3

Bandarin, F., van Oers, R.: The Historic Urban Landscape: Managing Heritage in an Urban Century. Wiley Blackwell (2012)

Bandarin, F., van Oers, R., Bandarin, F. (eds): Reconnecting the City: The Historic Urban Landscape Approach and the Future of Urban Heritage. John Wiley & Sons Inc. (2015)

Benjamin, W., Tiedemann, R., Benjamin, W.: I 'passages' di Parigi. Einaudi (2000)

Bourdieu, P.: The social space and the genesis of groups. Theory Soc. **14**(6), 723–744 (1985)

Bourdin, A.: Urbanisme: Des modèles en mutation. Presses Universitaires de France (2005)

CoE: European Landscape Convention. Council of Europe (2000). https://www.coe.int/en/web/landscape/text-of-the-european-landscape-convention

Conserva, A., Farinea, C., Villodres, R.: New NBS Co_Creation of URBiNAT NBS (Live) Catalogue and Toolkit for Healthy Corridor (Deliverable Del4.1, p. 212). IAAC (2021)

Delgado Ruiz, M. (1999). *El animal público: Hacia una antropología de los espacios urbanos*. Editorial Anagrama.

Dobričić, S., Acri, M. (eds): Creative Cities: Which (Historic) Urban Landscape. Mimesis (2017)

Dobricic, S., Acri, M.: Still life—Natura Morta: the landscapes of proximity. Ri-Vista. Res. Landsc. Arch. **19**(1), 64–83 (2021). https://doi.org/10.36253/rv-10298

EU: Eu White paper on Governance. EU Commission (2001). https://eur-lex.europa.eu/EN/legal-content/summary/white-paper-on-governance.html

Farinelli, F.: Geografia: Un'introduzione ai modelli del mondo. Giulio Einaudi editore (2003)

Ferilli, G., Acri, M., Dobričić, S., Boon, L., Kobe, A., Dodangeh, A., Ben Chaabane, N.Z., Prisca Ohler, L.: URBiNAT local diagnostic report for follower cities (2.6; p. 308). IULM (2021)

Foster, G.: Planning the circular city: focus on buildings' environmental impact. Bulletin of the Calza Bini Center, pp. 117–123 (2019). https://doi.org/10.6092/2284-4732/7063

Giglione, F., Maffei, S.P., Acri, M., Polo, E.: Participatory planning for a widespread and resilient cross-border park. In: Canto Moniz, G., Bechet, B., Lameiras, J.M., Acri, M., Nunes, N., Ferreira, I., Tasheva-Petrova, M., Andersson, I., Caitana, B., Ferilli, G. (Eds.), Proceedings of the International Conference on Nature for an Inclusive and Innovative Urban Regeneration (NATiURB 2022) vol. 24, pp. 255–266. Atlantis Press International BV (2024). https://doi.org/10.2991/978-94-6463-469-3_21

Gravagnuolo, A., Angrisano, M., Fusco Girard, L.: Circular economy strategies in eight historic port cities: criteria and indicators towards a circular city assessment framework. Sustainability **11**(13), 3512 (2019). https://doi.org/10.3390/su11133512

Kinkaid, E.: Re-encountering Lefebvre: toward a critical phenomenology of social space (2020). https://doi.org/10.1177/0263775819854765

Klaus, V.H., Kiehl, K.: A conceptual framework for urban ecological restoration and rehabilitation. Basic Appl. Ecol. **52**, 82–94 (2021). https://doi.org/10.1016/j.baae.2021.02.010

Lefebvre, H., Kofman, E., Lebas, E.: Writings on Cities. Blackwell Publishers (1996)

Lefebvre, H., Nicholson-Smith, D., Lefebvre, H.: The Production of Space (33. print). Blackwell Publishing (2013).

Madanipour, A.: Public and Private Spaces of the City (0 ed.). Routledge (2003). https://doi.org/10.4324/9780203402856

Mahmoud, I.H., Morello, E., Oliveira, F.L. de, Geneletti, D. (eds): Nature-Based Solutions for Sustainable Urban Planning: Greening Cities, Shaping Cities. Springer (2022).

Moniz, G.C., Acri, M., Mateus, A., Leonor, S., Ferreira, I.: Co-creation of healthy corridors: a participatory approach to the circular city. In: Fusco Girard, L., Nijkamp, P., Nocca F. (eds), Reconnecting the City with Nature and History: Towards Circular Regeneration Strategies. Franco Angeli (2023)

Moniz, G.C., Andersson, I., Hilding-Hamann, K.E., Mateus, A., Nunes, N.: Inclusive urban regeneration with citizens and stakeholders: from living labs to the URBiNAT CoP. In: Mahmoud, I.H., Morello, E., Lemes de Oliveira, F., Geneletti, D. (eds), Nature-Based Solutions for Sustainable Urban Planning, pp. 105–146. Springer International Publishing (2022). https://doi.org/10.1007/978-3-030-89525-9_5

Nunes, N., Björner, E., Hilding-Hamann, K.E.: Guidelines for citizen engagement and the co-creation of nature-based solutions: living knowledge in the URBiNAT project. Sustainability **13**(23), 13378 (2021). https://doi.org/10.3390/su132313378

Ostrom, E.: The future of the commons: beyond market failure and government regulation. Institute of Economic Affairs (2012)

Paquot, T.: The Art of Urbanism. de la Villette (2007)

Pedersen Zari, M., MacKinnon, M., Varshney, K., Bakshi, N.: Regenerative living cities and the urban climate–biodiversity–wellbeing nexus. Nat. Clim. Chang. **12**(7), 601–604 (2022). https://doi.org/10.1038/s41558-022-01390-w

Sennet, R.: The Fall of the Public Man. W. W. Norton (1992)

Sennet, R.: The Open City—Essays | Urban Age (2006). https://urbanage.lsecities.net/essays/the-open-city

Steen, K., van Bueren, E.: The defining characteristics of urban living labs. Technol. Innov. Manag. Rev. **7**(7) (2017)

Tuan, Y.-F.: Space and Place: The Perspective of Experience (7. print). Universe of Minnesota Press (2011)

UNHABITAT & Biennial Public Space: Charter of Public Space. UN (2013). https://habnet.unhabitat.org/sites/default/files/documents/Charter%20of%20Public%20Space.pdf

Mapping of Sea-Level Rise Impacts on the Cultural Heritage of the Apulian Gulf of Manfredonia and Ionian Coast

Arianna Lobascio, Custode S. Fioriello, Giovanni Scardino, Giovanni Scicchitano, and Marina Zingaro

Abstract

One of the most severe consequences of climate change concerns the progressive sea-level rise. Nevertheless, the potential impact on the cultural heritage, especially the Apulian one, is nowadays widely underestimated due to the slow progress of the phenomenon. The studies that map the assets at risk due to sea-level rise are more sporadic and, in any case, present only at the international level. Knowing imminent risks and predicting future scenarios is essential to conserving and protecting our heritage effectively. In order to evaluate which cultural assets and archaeological landscapes in Apulia will be subjected to marine flooding or submersion by 2050 and 2100, the following research reviews and synthesizes the scientific literature regarding the coastal dynamics in Apulia, focussing on two of the most threatened Apulian coasts, namely the Gulf of Manfredonia and the Ionian arc between Ginosa and Taranto. A GIS mapping of the vulnerable coastal assets has been produced by superimposing the available marine flooding projections for Apulia with the institutional heritage census databases. Various data sources were employed, including flood maps created by the Earth Sciences Department of the University of Bari Aldo Moro and territorial information systems about regional and national cultural assets and archaeological landscapes, such as 'CartApulia', 'Vincoli in Rete', and 'Carta del Rischio'. The study's results constitute a preliminary step towards defining a methodological standard applicable in other contexts to acquire a broader assessment of endangered Italian heritage, which is necessary for communal risk management planning. Finally, publishing the obtained maps could contribute to effective risk communication, increasing the communities' awareness of climate change impacts, not only on the environment but also on the cultural heritage.

Keywords

Sea-level rise · Marine flooding · Coastal heritage · Cultural heritage · GIS mapping · Risk map

1 Introduction

The rising sea levels caused by climate change pose a severe threat to cultural heritage sites around the globe (Abdelnour 2020; Marzeion and Levermann 2014; Reimann et al. 2018). Low-lying coastal areas are particularly vulnerable to coastal flooding, especially when exacerbated by other concurrent factors such as shoreline erosion and subsidence (Karegar et al. 2017), all of which jeopardize the preservation of cultural assets that hold immense historical, social, and economic value. The region of Apulia, located in southeastern Italy, is a region of exceptional historical significance, with its extensive yet vulnerable coastline home to numerous cultural heritage sites (Auriemma and Volpe 2020; Campanile and Esposito 2016). Despite its high exposure to the effects of sea-level rise ('SLR' from now on), worsened by coastal erosion and subsidence (Scardino et al. 2020,

A. Lobascio (✉)
PASAP_Med, DIRIUM, University of Bari Aldo Moro, Bari, Italy
e-mail: arianna.lobascio@uniba.it

C. S. Fioriello
DIRIUM, University of Bari Aldo Moro, Bari, Italy
e-mail: custode.fioriello@uniba.it

G. Scardino · G. Scicchitano · M. Zingaro
DiSTeGEO, University of Bari Aldo Moro, Bari, Italy
e-mail: giovanni.scardino@uniba.it

G. Scicchitano
e-mail: giovanni.scicchitano@uniba.it

M. Zingaro
e-mail: marina.zingaro@uniba.it

P. Mohammad et al. (eds.), *Geographic Approaches to Climate Change and Mitigation: Urban and Rural Perspectives (Volume 2)*, Advances in Science, Technology & Innovation, https://doi.org/10.1007/978-3-031-92115-5_23

2022), research addressing the threat to Apulian heritage remains limited.

In the Apulian context, recent research has emphasized the need for coastal risk evaluations (Parete et al. 2024). However, these efforts do not address the vulnerability of cultural heritage sites specifically. This study aims to fill this gap by integrating geomorphological data and cultural heritage censuses by means of a straightforward geospatial analysis to assess the vulnerability of Apulia's coastal heritage to SLR, with a particular focus on the Gulf of Manfredonia.

The Gulf of Manfredonia, characterized by high rates of subsidence and a low-lying plain (Scardino et al. 2022), and the Gulf of Taranto, with its high rates of erosion (Antonioli et al. 2017, 2020; Lambeck et al. 2011; Petrillo 2007; Scardino et al. 2020), represent ideal case studies for understanding the interplay between inundation processes and cultural heritage risk. This paper will serve as a preliminary study for an extensive research project aimed at assessing the vulnerability of the region's coastal heritage to sea-level rise. Unlike previous Italian studies, which emphasize individual renowned sites such as the Venice lagoon (Fletcher and Spencer 2005; Giupponi 2022; Schlumberger et al. 2022; Trovò 2023), this research adopts a regional-scale approach that is neither site-specific nor over-generalized and approximate, as in national or continental overviews. It integrates local geomorphological and cultural data with a replicable geospatial analysis framework.

As looking forward to future conservation challenges, it is crucial to evaluate the extent to which cultural and landscape heritage is at risk of inundation due to sea-level rise. Identifying looming risks and predicting the *ex-post* scenarios are fundamental requirements to protect cultural and landscape assets. This knowledge serves as the groundwork for developing targeted management strategies for climate-related emergencies, even the less obvious ones, while also raising public awareness about climate change and the urgent need to protect cultural heritage. Heading over this direction, this study aims to bridge the gap between environmental risk assessments and cultural heritage management in Apulia, providing a foundational framework for future studies and policymaking in the face of escalating climate challenges.

2 Problem Statement and Study Aim

The potential impact of SLR on Apulian sites is currently being overlooked, despite SLR being expected to heavily impact the Apulian territory and the cultural assets diffusely distributed across the region, as previously described. Apulia's coastline, rich in archaeological and historical sites, faces increasing risks from SLR-driven inundation, shoreline erosion, and subsidence. However, there is a lack of projections to determine how many sites could be submerged or significantly affected in the coming decades, and which will be sooner affected by the SLR consequences. This gap in understanding undermines the efforts to develop effective conservation strategies for the region.

We decided to apply these research questions to two study areas, the Gulf of Manfredonia (Adriatic coast stretching between Manfredonia and Barletta) mainly and the Gulf of Taranto (Ionian coast stretching between Castellaneta and Taranto) partly. These study areas were chosen due to their cultural significance, exposure to SLR risks, and availability of local flood projection maps (Scardino et al. 2020, 2022). Both areas are coastal plains that show evidence of significant shoreline retreat and erosion caused by various natural and anthropogenic factors. Additionally, the Gulf of Manfredonia is characterized by high rates of land subsidence, making it particularly fragile to any change in its coastal dynamic and to the SLR phenomenon. Moreover, these areas are home to a heritage of immense cultural interest, with extensive and valuable evidence of the continuity of human presence throughout the millennia, making them culturally vulnerable.

This study employs two temporal benchmarks, 2050 and 2100, to evaluate the short- and medium-term impacts of SLR on cultural heritage. These benchmarks align with available SLR projection data and enable a comparative analysis of near-future and more distant scenarios.

This study aims to address the gap in knowledge about SLR-induced risks to Apulian cultural heritage through replicable GIS-based geospatial analysis to identify, quantify, and map the sites within the at-risk areas that are expected to be exposed to sea waters due to marine ingression by 2050 and 2100. By integrating predicted SLR scenarios and cultural heritage datasets and geoprocessing them in a GIS environment, this research seeks to provide a clear assessment map of Apulia's cultural assets susceptible to SLR threat in the near future. The results will help address the current research gap, serving as a foundation for future conservation planning, climate resilience strategies, and public awareness efforts.

3 Review of the Literature

National and international scientific research investigating the effects of climate change on cultural heritage has begun in the twenty-first century, focussing mainly on extreme phenomena (Orr et al. 2021; Sesana et al. 2021). The most frequently investigated factors are also the most macroscopic and rapid-onset ones, such as intense rainfall, hydrogeological instability, thermo-hygrometric stress, solar radiation, and anthropic pollutants are commonly taken into consideration as causes of the degradation of cultural heritage, both at a national (Castellari et al. 2014; Gaddi et al. 2019; Trigila et al. 2021) and international level (ARCH 2019–2022; Bonazza et al. 2020). For example, in Italy, the Territorial Information System of the Cultural Heritage Risk Map (Accardo et al. 2005), designed by the Central Institute for Restoration, evaluates the present damage risk to cultural heritage due to pollution, hydrogeological instability, seismic danger, and current coastal dynamics. The latter—heritage vulnerability to present coastal processes—has also been taken into consideration by other authors (Mattei et al. 2019, 2020). While these studies offer critical insights, they primarily focus on immediate and macroscopic risks, leaving the gradual impacts of SLR underexplored.

Globally, SLR has garnered increasing attention as a long-term threat to coastal heritage. National and international research has highlighted the looming nature of this problem, elaborating forecasts on the future transformation of the coastlines (Marsico et al. 2017; Pörtner et al. 2019; Sannino et al. 2022; SaveMedCoasts 2017–2023). Numerous efforts have also been made to develop defence strategies (Leone et al. 2021; LIFE20 1992–2021; START 2015–2018) and modelling techniques (Bosello and De Cian 2013). However, studies specifically mapping cultural assets at risk from SLR remain sporadic, often presenting broad, non-specific national overviews (Jones et al. 2023; Li et al. 2022) or even continental and global perspectives (Marzeion and Levermann 2014; Vousdoukas et al. 2022). Among all, UNESCO has identified 49 world heritage sites along Mediterranean coasts that could be submerged in the future due to coastal erosion (Reimann et al. 2018). The regional-specific studies incorporating geomorphological peculiarities and localized factors are more commonly conducted outside Italy. Notable examples include heritage mapping along the coasts of Portugal (Valente Monteiro et al. 2023), Türkiye and Greece (Zengin 2023), Poland (Ciski and Rzasa 2021), Spain (Garrote et al. 2020), and various regions of the United States (Elliott and Williams 2019; Howey 2020; Reeder-Myers 2015).

Despite growing global attention, studies addressing the regional impacts of SLR on cultural heritage in Italy remain limited. Previous Italian research mainly centres on isolated, renowned case studies (Anzidei et al. 2017, 2019), like the famous and sensational case of the Venice Lagoon (Fletcher and Spencer 2005; Giupponi 2022; Schlumberger et al. 2022; Trovò 2023). While valuable, these studies lack a regional-scale perspective. Apulia, in particular, lacks comprehensive studies combining SLR projections with cultural heritage mapping, despite alarming evidence of erosion and flooding vulnerability. Among Apulian coasts, the most susceptible to erosion and flooding are Lesina and the areas between Manfredonia and Barletta (Scardino et al. 2022), between Brindisi and Otranto (Antonioli et al. 2020; Mastronuzzi and Sansò 2014), and between Ginosa and Taranto (Antonioli et al. 2017, 2020; Lambeck et al. 2011; Petrillo 2007; Scardino et al. 2020). However, these assessments primarily focus on geomorphological processes without integrating cultural heritage inventories.

Even though not always recent or up-to-date, Italian research regarding the census of underwater archaeological heritage is notably solid (Archeomar 2004–2006; Cossa 2012), as is the experimentation regarding the conservation of submerged sites (Petriaggi 2004; Petriaggi and Mancinelli 2004; Ricci et al. 2015). Nevertheless, there is a lack of projection on how many and which coastal assets could become the underwater heritage of the future.

This study builds upon this growing body of research by leveraging geospatial analysis techniques and considering geomorphological factors to bridge the gap between SLR projections and cultural heritage mapping in Apulia and offers a replicable regional-scale methodology for assessing the SLR's potential impact on local cultural assets.

4 Methodology

To achieve the objective of identifying, quantifying, and mapping cultural assets at risk from SLR in the Gulf of Manfredonia and the Gulf of Taranto, this study employs a GIS-based methodology integrating existing and open-access data: local SLR projections and regional cultural heritage geodatabases. The analysis was conducted using the Free and Open Source QGIS software, applying geoprocessing tools to produce SLR-risk maps for cultural heritage by 2050 and 2100.

Data Sources SLR scenario projections in the form of flood maps were sourced from prior studies for the Gulf of Manfredonia (Scardino et al. 2022) and the Gulf of Taranto (Scardino et al. 2020). The former shows the relative SLR projections for 2050, 2100, and 2150, relative to the SSP1-2.6 scenario (IPCC 2021 best-case projections with low greenhouse-gas emission) and SSP5-8.5 scenario (IPCC 2021 worst-case projections with high greenhouse gas emission). The latter shows the relative SLR projections for 2050 and 2010, relative to the AR 5 RCP 2.6 scenario (lower greenhouse gas emission) and the RCP 8.5 scenario (higher greenhouse gas emission). Both SLR models are obtained by combining the effects caused by vertical land movements and horizontal shoreline movements, to indicate which areas are prone to flooding and water-over-land submersion more accurately. For consistency and focus on high-risk scenarios, only the worst-case scenario (SSP5-8.5 and RCP 8.5) projections for 2050 and 2100 were considered here. The relative polygons (vectors) were acquired from the research's authors and imported into a GIS environment using the Free and Open Source QGIS software.

To identify the cultural heritage vulnerable to SLR, two territorial information systems were used as data sources, 'CartApulia' and 'Vincoli in Rete'. 'CartApulia' (Regione Puglia, P.O. 2007–2013—Linea 4.2) is a regional geodatabase collecting all the known Apulian cultural heritage, categorized into three categories—archaeological heritage, architectural heritage, and sites of cultural interest. On the other hand, 'Vincoli in Rete' (Istituto Superiore per la Conservazione ed il Restauro—MiBACT) is a national geodatabase collecting the cultural heritage under the jurisdiction of the Italian 'Ministero della Cultura', categorized into four categories—archaeological heritage, architectural heritage, parks and gardens, and historic centres. Both Territorial Information Systems' datasets were downloaded, imported into the aforementioned QGIS project as points (vectors, each one representing and georeferencing an individual cultural asset), and merged. To eliminate redundancies related to the same assets signalled in both datasets, overlapping entries were compared and amended.

Geospatial Analysis After collecting, uploading, and preparing all the necessary data in the GIS environment, simple geoprocessing methods were applied as follows.

1. Superimposition: The cultural heritage site map was layered onto the modelled SLR scenario map for 2050 and 2100.
2. Extraction by Location: Using QGIS's 'extract by location' tool, punctual cultural assets located within the polygonal projected at-risk areas were identified. This process, repeated for both study areas and both 2050 and 2100 scenarios, allowed for the quantification, identification, and mapping of assets expected to be inundated in the different scenarios and territories considered.
3. Output Maps: These separate analyses resulted in four final SLR-risk maps:

- Gulf of Manfredonia cultural heritage at risk by 2050.
- Gulf of Manfredonia cultural heritage at risk by 2100.
- Gulf of Taranto cultural heritage at risk by 2050.
- Gulf of Taranto cultural heritage at risk by 2100.

This methodology is subject to uncertainties inherent in SLR projections and potential gaps or inaccuracies in cultural heritage databases. Additionally, the decision to focus on worst-case scenarios ensures a conservative approach but may not reflect intermediate outcomes. By leveraging a GIS-based workflow, this methodology provides a replicable framework for assessing SLR's impact on regional cultural heritage. The resulting maps serve as critical tools for conservation planning and prioritizing future research.

5 Results

The analysis reveals significant differences in the vulnerability of cultural heritage between the Gulf of Manfredonia and the Gulf of Taranto under future SLR scenarios. The findings are summarized below.

Gulf of Manfredonia Figures 1 and 2 present the submersion predictions for the heritage in the Gulf of Manfredonia area for 2050 and 2100, respectively. Known cultural heritage sites in the area will be impacted in both SLR scenarios and increasingly over time. Two assets are expected to be inundated by 2050 and a total of 11 assets are expected to be inundated by 2100. As can be imagined, the 2100 scenario is projected to damage the greatest number of sites in comparison with the 2050 scenario due to the gradual progression of the phenomenon.

Table 1 presents an overview of the impacted heritage of the Gulf of Manfredonia, categorized by chronology and typology. Chronologically, the affected sites include examples from the Neolithic period to the Modern era. Typologically, they include six archaeological sites, four architectural heritage sites, and one historical centre. The Roman *villa* in Località San Vito, Cerignola (FG), and the Neolithic and Bronze settlement in Località Vasche di Napoleone, Trinitapoli (BAT), appear to be the specific sites most vulnerable to inundation because they would likely be impacted in approximately 25 years from now.

Gulf of Taranto. In contrast, the analysis for the Gulf of Taranto area did not produce results; apparently, known cultural heritage sites in the area will be impacted in none of the two SLR scenarios, since they are not located within the projected at-risk areas. Despite being prone to shoreline erosion, the immediate risks related to marine ingression appear minimal. The absence of at-risk sites can be attributed to the spatial distribution of known cultural assets and the coastal geomorphology of the region. Furthermore, it should not be underestimated that this is an area with still partly underexplored archaeological potential (Fioriello et al. 2023).

Comparison and Implications. The results show that SLR scenarios are likely to affect the Gulf of Manfredonia more compared to the Ionian coast sites, due to the different geomorphology of the area. The threat to the cultural heritage in the Gulf of Manfredonia appears to be more urgent, possibly because it is exacerbated by the combination of SLR's effects with the territory's geomorphology and the high subsidence rates here recorded, which amplify the marine ingression and shoreline erosion (Scardino et al. 2022). On the other hand, although research shows that the Gulf of Taranto is prone to shoreline erosion, the coastal heritage in the area will suffer minimal immediate SLR-related risks. These findings highlight the need for prioritized conservation efforts and adaptive

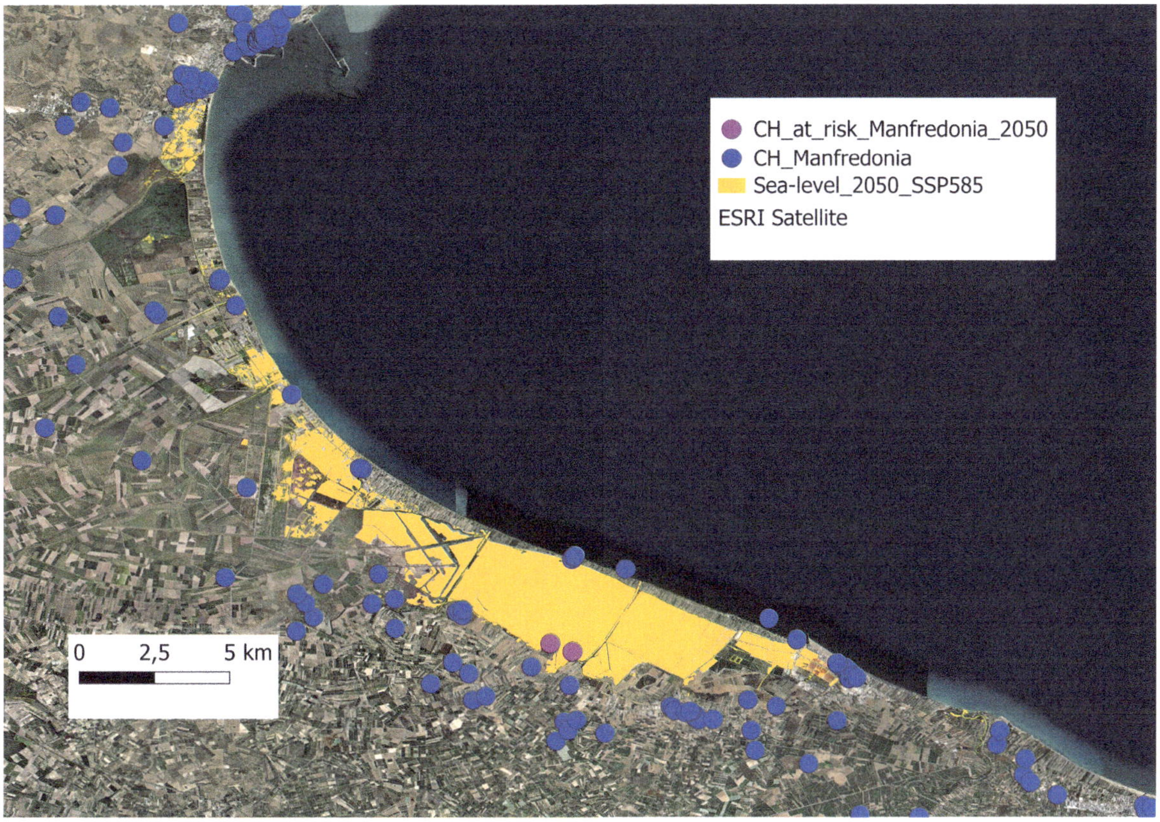

Fig. 1 The purple dots represent the heritage expected to be submerged in the Gulf of Manfredonia area by 2050

management strategies in the Gulf of Manfredonia to mitigate the effects of SLR. The absence of immediate threats in the Gulf of Taranto offers an opportunity to focus on proactive preservation measures in more vulnerable areas, while still warranting long-term monitoring given the potential for future geomorphological changes.

6 Discussion

SLR poses a particularly looming threat to coastal cultural heritage that can be highlighted by means of simple GIS-based geoprocessing tools and geospatial analysis. This study provides the first prediction of cultural assets at risk due to SLR along two coastlines of Apulia, successfully fulfilling its objective of identifying, quantifying, and mapping heritage sites expected to be inundated by 2050 and 2100 in two areas of Apulia. The results form a foundational assessment of cultural heritage vulnerability to SLR in the region and demonstrate the utility of GIS-based methods in addressing this risk.

The findings indicate a stark contrast between the two study areas. The Gulf of Manfredonia emerges as the most vulnerable area, with multiple assets projected to face inundation by 2050 and 2100. This underscores the urgency for targeted conservation strategies. In contrast, the Gulf of Taranto's cultural assets display greater resilience, highlighting the importance of localized assessments to properly allocate resources and direct conservation efforts.

While the number of coastal sites at risk of being endangered may appear relatively low on a regional scale, the actual loss of coastal heritage may be underestimated. Future assessments should consider additional factors that may further amplify the risk, such as scenarios for 2150, worsening predictions on climate change, the potential impact of storm surges and other extreme marine events, continued anthropogenic pressure, and possible undiscovered archaeological sites (Fioriello et al. 2023) requiring preventive archaeological evaluations. Moreover, many assets are located close to the limits of the defined risk area and are therefore excluded from the current geospatial analysis. Their proximity to the risk areas suggests that, even not directly submerged, they may face secondary impacts, such as restricted accessibility or inundation by 2150, therefore necessitating buffer analysis. Finally, verifying the precision of regional geodatabase coordinates is essential, as inaccuracies could affect the reliability of these projections.

It is important to acknowledge that each asset in the study areas represents a unique expression of local history, contributing to the region's cultural identity. Their loss or

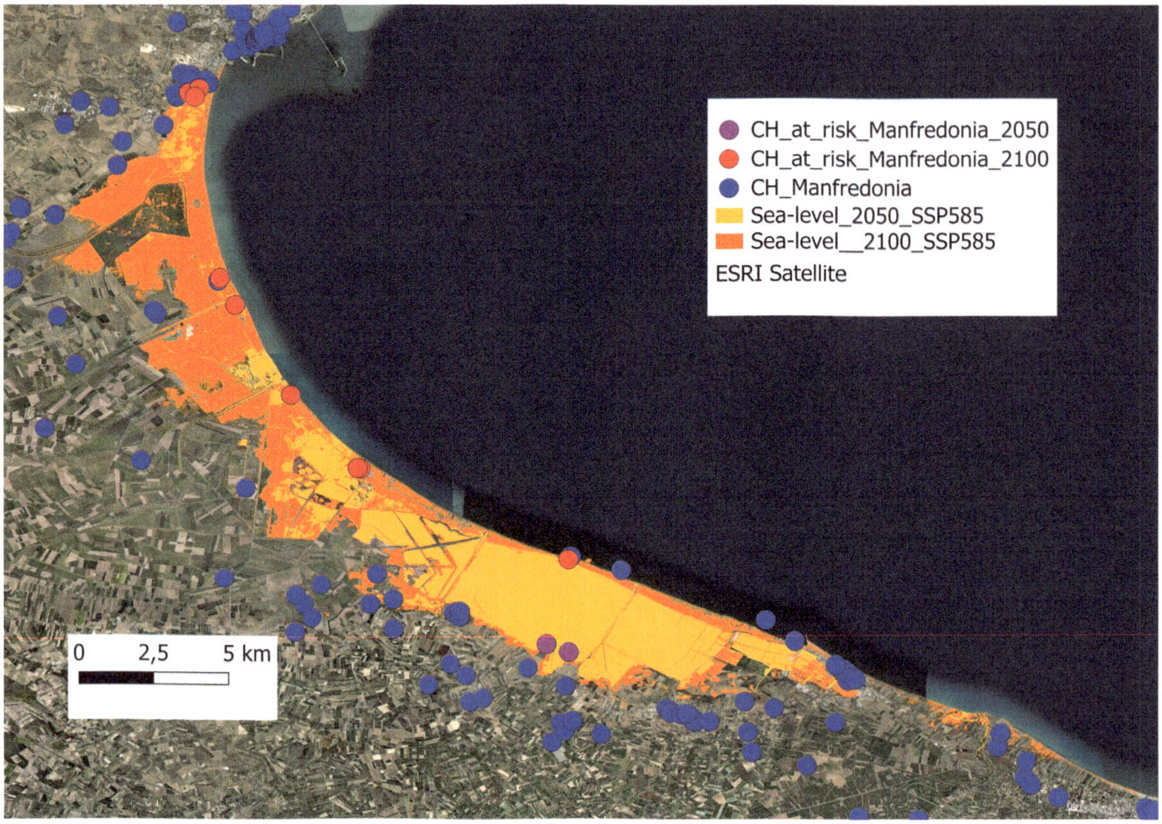

Fig. 2 The red dots represent the heritage expected to be submerged in the Gulf of Manfredonia area by 2100, along with the purple ones

Table 1 List and classification of the impacted heritage of the Gulf of Manfredonia

Name	Location	Age	Typology	Inundation
Roman *villa*	Località San Vito, Cerignola (FG)	Roman	Archaeological	By 2050
Settlement	Località Vasche di Napoleone, Trinitapoli (BAT)	Neolithic and Bronze	Archaeological	By 2050
Spring water collection basin	Siponto, Manfredonia (FG)	Roman	Archaeological	By 2100
Port facilities	Siponto, Manfredonia (FG)	Roman	Archaeological	By 2100
Settlement	Siponto, Manfredonia (FG)	Neolithic to Contemporary	Archaeological	By 2100
'Don Benedetto' tower	Scalo dei Saraceni, Manfredonia (FG)	Modern and Contemporary	Architectural	By 2100
Roman road	Località Sciale Carlone, Manfredonia (FG)	Roman	Archaeological	By 2100
'Rivoli' tower	Zapponeta (FG)	Modern and Contemporary	Architectural	By 2100
Roadman's house 'ex ANAS'	Zapponeta (FG)	Contemporary	Architectural	By 2100
Historic centre of Zapponeta	Zapponeta (FG)	Modern and Contemporary	Historic centre	By 2100
'San Michele' country church	Margherita di Savoia (BAT)	Modern	Architectural	By 2100

damage would represent a disaster in itself and would also lead to an alteration of the landscape, withholding the perception of all heritage values, tangible and intangible. Given the scale and complexity of the challenge, it is unlikely that all heritage sites can be preserved. Therefore, the focus must shift towards establishing priorities, aiming to minimize risk rather than entirely eliminate loss.

In doing so, this research can serve as a foundation for informed decision-making and preventive measure planning by stakeholders, guiding the prioritization and efforts to safeguard Apulia's invaluable cultural heritage from the escalating threats posed by SLR. While an initial step towards more accurate assessments, a study as such offers valuable insights through an integrated, replicable, and scalable methodology, applicable to other regions with available SLR scenarios' data. This straightforward yet effective approach shifts the focus from individual, well-known sites to a broader regional-scale perspective, while still considering the local geomorphological peculiarities.

Addressing the challenges posed by SLR will require improved archaeological surveys and multidisciplinary collaboration. For this reason, we intend to expand our research on Apulian coastal heritage vulnerability to have a comprehensive overview of the expected risks and possible responses to SLR, ranging from risk mitigation to risk adaptation and management.

7 Conclusion

Prior to this study, knowledge about the vulnerability of Apulia's cultural heritage to SLR was notably lacking, despite the well-documented risks of SLR for the region's coastlines. This research achieves its aim of filling a knowledge gap regarding SLR impacts on Apulian heritage by utilizing GIS-based geospatial analysis, successfully identifying, quantifying, and mapping, for the first time, the cultural assets in the Gulf of Manfredonia that are projected to be impacted by SLR. Using a simple yet effective methodology, the study identified 11 at-risk assets, with two expected to be inundated by 2050 and an additional nine by 2100. In contrast, no assets within the Gulf of Taranto were found to be at immediate risk under the current SLR scenarios. A more comprehensive risk assessment in the future could find an even more extended impact.

These results provide actionable insights and a replicable methodology for regional cultural heritage risk assessments. Although an initial overview, it offers an opportunity to raise public awareness on the topic and constitutes a preliminary guidance to start an informed conservation planning in response to the ongoing SLR phenomenon in Apulia, that is both severe and often underestimated when compared to other Italian regions. This research, together with the suggested effort, could contribute to protecting Apulia's invaluable coastal heritage in the face of escalating challenges posed by climate change and SLR.

Acknowledgements We are thankful to Gianluca Martinez for his contribution to improving the presented data. The authors also gratefully acknowledge the anonymous referees for their accurate revisions.

Competing Interests The authors have no conflicts of interest to declare that are relevant to the content of this paper.

References

Abdelnour, M.: The impact of environmental changes on coastal archaeological sites in the Mediterranean. Marit. Archaeol. Grad. Symp. 2020 (2020). https://doi.org/10.33583/mags2020.10

Accardo, G., Cacace, C., Rinaldi, R.: Il Sistema Informativo Territoriale della Carta del Rischio. In: ARKOS. Scienza e Restauro dell'Architettura, VI, pp. 43–52. Nardini Editore (2005)

Antonioli, F., Anzidei, M., Amorosi, A., Lo Presti, V., Mastronuzzi, G., Deiana, D.F.G., Fontana, A., Fontolan, G., Lisco, S., Marsico, A., Moretti, M., Orrù, P.E., Sannino, G.M., Serpelloni, E., Vecchio, A.: Sea-level rise and potential drowning of the Italian coastal plains: flooding risk scenarios for 2100. Quatern. Sci. Rev. **158**, 29–43 (2017). https://doi.org/10.1016/j.quascirev.2016.12.021

Antonioli, F., De Falco, G., Lo Presti, V.M., Scardino, G., Anzidei, M., Bonaldo, D., Carniel, S., Leoni, G., Furlani, S., Marsico, A., Petitta, M., Randazzo, G., Scicchitano, G., Mastronuzzi, G.: Relative sea-level rise and potential submersion risk for 2100 on 16 coastal plains of the Mediterranean sea. Water **12**(8), 2173 (2020). https://doi.org/10.3390/w12082173

Anzidei, M., Bosman, A., Carluccio, R., Casalbore, D., Caracciolo, F.D., Esposito, A., Nicolosi, I., Pietrantonio, G., Vecchio, A., Carmisciano, C., Chiappini, M., Latino Chiocci, F., Muccini, F., Sepe, V.: Flooding scenarios due to land subsidence and sea-level rise: a case study for Lipari Island (Italy). Terra Nova **29**(1), 44–51 (2017). https://doi.org/10.1111/ter.12246

Anzidei, M., Doumaz, F., Vecchio, A., Serpelloni, E., Pizzimenti, L., Civico, R., Greco, M., Martino, G., Enei, F.: Sea level rise scenario for 2100 A.D. in the Heritage site of Pyrgi (Santa Severa, Italy). Mar. Sci. Eng. **8**(2), 64 (2019). https://doi.org/10.3390/jmse8020064

ARCH: ARCH—saving cultural heritage. Retrieved 06 June 2023, (2019–2022) from https://savingculturalheritage.eu/about/project

Archeomar: Archeomar virtual research. Retrieved 06 June 2023, (2004–2006) from https://archeomar.d4science.org/

Auriemma, R., Volpe, G.: La Puglia e il mare. Un viaggio nei paesaggi costieri e subacquei. In: Sbonias, K., Kapetanios, A. (ed.), Corfu's Coastal Environments, Cultural Heritage & Local Communities, pp. 227–246. Ionio Panepistimio Epitropi Erevnon (2020)

Bonazza, A., De Nuntiis, P., Sardella, A., Ciantelli, C., Palazzi, E.: ProteCHt2save: valutazione del rischio e protezione sostenibile del patrimonio in ambienti in trasformazione. In Alfonsina, R., Della Giovampaola, I. (ed.), Monitoraggio e manutenzione delle aree archeologiche: cambiamenti climatici, dissesto idrogeologico, degrado chimico-ambientale, pp. 141–147. L'Erma di Bretschneider (2020)

Bosello, F., De Cian, E.: Climate change, sea level rise, and coastal disasters. a review of modeling practices. Energy Econ. **46**, 593–605 (2013). https://doi.org/10.1016/j.eneco.2013.09.002

Campanile, D., Esposito, C.: Itinerari turistici tra le torri costiere di Puglia. Catalogo della mostra (Monopoli, Castello Carlo V, 9–24 ottobre 2015–Giovinazzo, Vedetta del Mediterraneo, 29 ottobre-8 novembre 2015). Mario Adda Editore (2016)

Castellari, S., Venturini, S., Giordano, F., Ballarin Denti, A., Bigano, A., Bindi, M., Bindi M., Bosello F., Carrera L., Chiriacò M.V., Danovaro R., Desiato F., Filpa A., Fusani S., Gatto M., Gaudioso D., Giovanardi O., Giupponi C., Gualdi S., Guzzetti F., Lapi M., Luise A., Marino G., Mysiak J., Montanari A., Pasella D., Pierantonelli L., Ricchiuti A., Rudari R., Sabbioni C., Sciortino M., Sinisi L., Valentini R., Viaroli P., Vurro M., Zavatarelli M.: Elementi per una Strategia Nazionale di Adattamento ai Cambiamenti Climatici. Ministero dell'Ambiente e della Tutela del Territorio e del Mare (2014)

Ciski, M., Rząsa, K.: Threats to cultural heritage caused by the global sea level rise as a result of the global warming. Water **13**(18), 2577 (2021). https://doi.org/10.3390/w13182577

Cossa, A.: La Carta Archeologica Subacquea della Puglia meridionale. Il sistema informativo territoriale in rete. L'archeologo subacqueo, anno XVIII, **3**(54), 3–4 (2012)

Elliott, P., Williams, H.: Evaluating sea-level rise hazards on coastal archaeological sites, Trinity Bay, Texas. J. Island Coast. Archaeol. **16**(2–4), 591–609 (2019). https://doi.org/10.1080/15564894.2019.1701149

Fioriello, C.S., Moro, A., Piccenna, S., Stasolla, V.: Archeologia dei paesaggi tra Puglia e Basilicata: la bassa valle del Bradano. In: Fioretti, G., Campobasso, C. (eds.), Il patrimonio culturale pugliese. Ricerche, applicazioni e best practices. Atti del II congresso Beni Culturali in Puglia, Bari, 28–30 settembre 2022, pp. 117–122. Fondazione Pasquale Battista (2023)

Fletcher, C.A., Spencer, T. (ed.): Flooding and Environmental Challenges for Venice and its Lagoon: State of Knowledge. Cambridge University Press (2005)

Gaddi, R., Cusano, M., Fornasier, M.F., Bonanni, P., Astorri, F., Ricchiuti, A., Iadanza, C., Spizzichino, D., Trigila, A.: Le attività di ISPRA per la tutela dei beni culturali: la partecipazione al progetto Artek. ISPRA, Rapporti 310 (2019)

Garrote, J., Díez-Herrero, A., Escudero, C., García, I.: A framework proposal for regional-scale flood-risk assessment of cultural heritage sites and application to the Castile and León Region (Central Spain). Water **12**(2), 329 (2020). https://doi.org/10.3390/w12020329

Giupponi, C.: Venezia e i cambiamenti climatici: Quale futuro per la città e la sua laguna? Rizzoli (2022)

Howey, M.C.: Harnessing remote sensing derived sea level rise models to assess cultural heritage vulnerability: a case study from the northwest Atlantic ocean. Sustainability **12**, 9429 (2020). https://doi.org/10.3390/su12229429

Jones, B.D., Dickson, M.E., Ford, M., Hikuroa, D., Ryan, E.J.: Aotearoa New Zealand's coastal archaeological heritage: a geostatistical overview of threatened sites. J. Island Coast. Archaeol. **19**(4), 657–677 (2023). https://doi.org/10.1080/15564894.2023.2207493

Karegar, M.A., Dixon, T.H., Malservisi, R., Kusche, J., Engelhart, S.E.: Nuisance flooding and relative sea-level rise: the importance of present-day land motion. Sci. Rep. **7**(1), 11197 (2017). https://doi.org/10.1038/s41598-017-11544-y

Lambeck, K., Antonioli, F., Anzidei, M., Ferranti, L., Leoni, G., Scicchitano, G., Silenzi, S.: Sea level change along the Italian coast during the Holocene and projections for the future. Quatern. Int. **232**, 250–257 (2011). https://doi.org/10.1016/j.quaint.2010.04.026

Leone, E., Kobayashi, N., Francone, A., De Bartolo, S., Strafella, D., D'Alessandro, F., Tomasicchio, G.R.: Use of Nanosilica for increasing dune erosion resistance during a sea storm. Marine Sci. Eng. **9**(6), 620 (2021). https://doi.org/10.3390/jmse9060620

Li, Y., Jia, X., Liu, Z., Zhao, L., Sheng, P., Storozum, J.M.: The potential impact of rising sea levels on China's coastal cultural heritage: a GIS risk assessment. Antiquity 2022, **96**, 406–421 (2022). https://doi.org/10.15184/aqy.2022.1

LIFE20, C.: LIFE+ A_GreeNet. Retrieved 06 June 2023, (1992–2021) from https://www.lifeagreenet.eu/site/

Marsico, A., Lisco, S., Lo Presti, V., Antonioli, F., Amorosi, A., Anzidei, M., Deiana, G., De Falco, G., Fontana, A., Fontolan, G., Moretti, M., Orrù, P.E., Serpelloni, E., Sannino, G., Vecchio, A., Mastronuzzi, G.: Flooding scenario for four Italian coastal plains using three relative sea level rise models. J. Maps **13**(2), 961–967 (2017). https://doi.org/10.1080/17445647.2017.1415989

Marzeion, B., Levermann, A.: Loss of cultural world heritage and currently inhabited places to sea-level rise. Environ. Res. Lett. **9**(3), 34001 (2014). https://doi.org/10.1088/1748-9326/9/3/034001

Mastronuzzi, G., Sansò, P.: Torre Fiumicelli (Otranto) e l'evoluzione storica del litorale adriatico salentino. Studi Costieri **22**, 159–170 (2014)

Mattei, G., Rizzo, A., Anfuso, G., Aucelli, P.P.C., Gracia, J.: A tool for evaluating the archaeological heritage vulnerability to coastal processes: the case study of Naples Gulf (southern Italy). Ocean Coast. Manag. **179**, 104876 (2019). https://doi.org/10.1016/j.ocecoaman.2019.104876

Mattei, G., Rizzo, A., Anfuso, G., Aucelli, P.P.C., Gracia, J.: Enhancing the protection of archaeological sites as an integrated coastal management strategy: the case of the Posillipo Hill (Naples, Italy). Rendiconti Lincei. Scienze Fisiche e Naturali **31**, 139–152 (2020). https://doi.org/10.1007/s12210-019-00867-9

Orr, S.A., Richards, J., Fatorić, S.: Climate change and cultural heritage: a systematic literature review (2016–2020). Hist. Environ.: Policy Pract. **12**(3–4), 434–477 (2021). https://doi.org/10.1080/17567505.2021.1957264

Parete, G., Bruno, M.F., Calabrese, P., Carlucci, R., Chiarulli, M., D'Onghia, G., Fiore, A., Fratino, U., Longo, C., Longo, F., Scorrano, S., Gentile, F.: Climate impacts and adaptation strategies for coastal erosion, aquaculture, and tourism along the Adriatic side of Apulia region. Front. Clim. **6**, 1378253 (2024). https://doi.org/10.3389/fclim.2024.1378253

Petriaggi, R.: Restauro subacqueo delle strutture sommerse della peschiera romana di Torre Astura. In: Ghini, G. (ed.), Lazio e Sabina 2. Atti del Convegno Secondo Incontro di Studi sul Lazio e la Sabina, Roma, 7–8 maggio 2003, pp. 273–276. De Luca Editori d'Arte (2004)

Petriaggi, R., Mancinelli, R.: An experimental conservation treatment on the mosaic floor and perimeter walls of room n. 1 of the so-called Villa con ingresso a protiro. The underwater archaeological park of Baia (Naples). Archeol. Marit. Mediterr.: Int. J. Underw. Archaeol. **1**, 109–126 (2004)

Petrillo, A.F.: Aree costiere: attuali e future criticità. Geologi e Territorio. Periodico dell'Ordine dei Geologi della Puglia, **3/4**, 117–130 (2007)

Pörtner, H.-O., Roberts, D., Masson-Delmotte, V., Zhai, P., Tignor, M., Poloczanska, E., Mintenbeck, K., Alegría, A., Nicolai, M., Okem, A., Petzold, J., Rama, J., Weyer, N.: IPCC. The Ocean and Cryosphere in a Changing Climate: Special Report of the Intergovernmental Panel on Climate Change. Cambridge University Press (2019). https://doi.org/10.1017/9781009157964

Reeder-Myers, L.A.: Cultural heritage at risk in the Twenty-First century: a vulnerability assessment of coastal archaeological sites in the United States. J. Island Coast. Archaeol. **10**(3), 436–445 (2015). https://doi.org/10.1080/15564894.2015.1008074

Reimann, L., Vafeidis, A.T., Brown, S., Hinkel, J., Tol, R.S.J.: Mediterranean UNESCO world heritage at risk from coastal flooding and erosion due to sea-level rise. Nat. Commun. **9**(1), 4161 (2018). https://doi.org/10.1038/s41467-018-06645-9

Ricci, S., Antonelli, F., Sacco Perasso, C., Davidde Petriaggi, B.: Indagini quali-quantitative della colonizzazione biologica di geotessuti utilizzati per la protezione in situ di pavimenti musivi sommersi (Poster presentation). II convengo tematico di biologia e biotecnologie per i beni culturali, Biologia e Archeobiologia: dalla Conoscenza alla Conservazione Preventiva, Palermo (2015). http://bit.ly/4apvcl1

Sannino, G., Carillo, A., Iacono, R., Napolitano, E., Palma, M., Pisacane, G., Struglia, M.V.: Modelling present and future climate in the

Mediterranean sea: a focus on sea-level change. Clim. Dyn. **59**(1–2), 357–391 (2022). https://doi.org/10.1007/s00382-021-06132-w

SaveMedCoasts: SaveMedCoasts1–2. Retrieved 06 June 2023, (2017–2023) from https://www.savemedcoasts.eu/; https://www.savemedcoasts2.eu/index.php/en

Scardino, G., Sabatier, F., Scicchitano, G., Piscitelli, A., Milella, M., Vecchio, A., Anzidei, M., Mastronuzzi, G.: Sea-level rise and shoreline changes along an open sandy coast: case study of Gulf of Taranto, Italy. Water **12**(5), 1414 (2020). https://doi.org/10.3390/w12051414

Scardino, G., Anzidei, M., Petio, P., Serpelloni, E., De Santis, V., Rizzo, A., Liso, S.I., Zingaro, M., Capolongo, D., Vecchio, A., Refice, A., Scicchitano, G.: The impact of future sea-level rise on low-lying subsiding coasts: a case study of Tavoliere Delle Puglie (Southern Italy). Remote. Sens. **14**(19), 4936 (2022). https://doi.org/10.3390/rs14194936

Schlumberger, J., Ferrarin, C., Jonkman, S.N., Diaz Loaiza, M.A., Antonini, A., Fatorić, S.: Developing a framework for the assessment of current and future flood risk in Venice, Italy. Nat. Hazard. **22**(7), 2381–2400 (2022). https://doi.org/10.5194/nhess-22-2381-2022

Sesana, E., Gagnon, A.S., Ciantelli, C., Cassar, J.A., Hughes, J.J.: Climate change impacts on cultural heritage: a literature review. Wires Clim. Change **12**, 710 (2021). https://doi.org/10.1002/wcc.710

START: START—SisTemi di rApid mapping e contRollo del Territorio costiero e marino. Retrieved 06 June 2023, (2015–2018) from https://www.cmcc.it/projects/start-sistemi-di-rapid-mapping-e-controllo-del-territorio-costiero-e-marino

Trigila, A., Iadanza, C., Lastoria, B., Bussettini, M., Barbano, A.: Dissesto idrogeologico in Italia: pericolosità e indicatori di rischio—Edizione 2021. ISPRA, Rapporti 356 (2021)

Trovò, F.: Gli effetti dei cambiamenti climatici sul patrimonio culturale di Venezia: proposta di un metodo per la definizione di misure di contrasto. Materiali e strutture: problemi di conservazione, **24**(2), 1121–2373 (2023)

Valente Monteiro, C., Barreiros Proença, S., Dal Cin, F.: Rovine costiere Il Patrimonio culturale vulnerabile all'innalzamento del livello medio del mare. In: Bagnoli, L. (ed), La difesa dei beni culturali negli scenari di crisi: Strategie di salvaguardia e tutela del patrimonio culturale, pp. 69–80. Anteferma (2023). https://doi.org/10.57623/979-12-5953-092-9

Vousdoukas, M.I., Clarke, J., Ranasinghe, R., Reimann, L., Khalaf, N., Minh Duong, T., Ouweneel, B., Sabour, S., Iles, C.E., Trisos, C.H., Feyen, L., Mentaschi, L., Simpson, N.P.: African heritage sites threatened as sea-level rise accelerates. Nat. Clim. Chang. **12**, 256–262 (2022). https://doi.org/10.1038/s41558-022-01280-1

Zengin, E.: Inundation risk assessment of Eastern Mediterranean Coastal archaeological and historical sites of Türkiye and Greece. Environ. Monit. Assess. **195**, 968 (2023). https://doi.org/10.1007/s10661-023-11549-3

Identifying Key Obstacles to Solving the Problem of Rehabilitating Belgrade's Cultural Heritage

Biljana Arandelovic

Abstract

Over the last 200 years, Belgrade has been in a very turbulent and dynamic era, with a rich cultural heritage that reflects its historical development. Unfortunately, a large part of the cultural heritage has not survived multiple destructions throughout its history, especially in the twentieth century, during which Belgrade was destroyed and rebuilt several times due to war. The study analyses the current state of Belgrade's rich and unique cultural heritage buildings and provides an overview of the most endangered sites. It also aims to identify the key obstacles to solving the problem of rehabilitating Belgrade's cultural heritage, as well as to determine the extent to which climate change contributes to these obstacles. The results will indicate whether human factors, such as management, lack of adequate financial resources, or failure to comprehend the actual reality of climate change, pose a larger threat to Belgrade's cultural heritage than climate change directly.

Keywords

Cultural heritage buildings · Sustainable heritage rehabilitation · Cultural heritage management · Belgrade · Contemporary cities

1 Introduction

1.1 The Importance of Smart Cultural Heritage Management for Sustainable Cities—An Introduction

Smart cultural heritage management plays a significant role in the development of sustainable cities. Decision-making through cultural management is essential for the rehabilitation of cultural heritage objects in line with modern technology (Clark 2021; Arandelovic 2018; Yates and De Oliveira 2016; Ferretti et al. 2014). This, in turn, leads to the further rehabilitation and improvement of cultural heritage objects (Arandelovic 2018; Arandelovic and Musil 2023). Investments in cultural heritage (Medda and Lipparini 2021), continuous rehabilitation, and adaptive reuse strategies contribute to more sustainable and socially resilient cities (Almeida et al. 2018). Cultural heritage is the key to maintaining an urban consciousness of the past while creating a sustainable future (Tweed and Sutherland 2007). The objective of smart city management is to identify and address these challenges, in order to create a more prosperous future for all (Petti et al. 2020; Pereira Roders and Van Oers 2011).

Rehabilitating cultural property is more important now than ever, and in the future, we can expect that finding the most effective rehabilitation solutions will be a priority for all city administrations, with city management expected to set and then follow standards in the coming years. There are numerous published research studies on the rehabilitation, reconstruction, and adaptation of cultural heritage objects, which encompass global case analyses and proposed solutions, and which showcase the latest technological advancements for addressing the challenges that arise in these processes (Arfa et al. 2022). It is crucial to find a balance between the sustainability of public cultural heritage sites and other key priorities for the development of contemporary

B. Arandelovic (✉)
Faculty of Civil Engineering and Architecture, University of Niš, Niš, Serbia
e-mail: biljana.arandelovic@gaf.ni.ac.rs

P. Mohammad et al. (eds.), *Geographic Approaches to Climate Change and Mitigation: Urban and Rural Perspectives (Volume 2)*, Advances in Science, Technology & Innovation, https://doi.org/10.1007/978-3-031-92115-5_24

cities. One of the key priorities for cities today is to coordinate the requirements for protecting the heritage assets that are most at risk from the direct effects of climate change, through the development of a plan and action strategy. It is crucial to recognize the potential adverse consequences of overlooking cultural heritage in favor of other aspects, such as investor-driven urbanism or accelerated programs that are detrimental to the urban economy. It is essential to view cultural heritage as a pivotal aspect of urban development and to approach it with a balanced perspective.

Ensuring a sustainable future is a top priority for practically every community. This goal can be achieved through the use of up-to-date development strategies, appropriate tools, expertise and information transfer (Guzmán et al. 2017). Cities that own public cultural heritage properties have an advantage over other cities, and they are encouraged to invest in the dissemination of relevant know-how in the sustainable management of these properties (Masini and Soldovieri 2017), such as maintenance, rehabilitation, reuse, conservation, and marketing, in order to ensure that the heritage is protected and enhanced in a sustainable manner. Sustainability-related studies are often centered around the technological perspective, such as studies on environmental balance, the management of product chains, or cleaner manufacturing systems.

It could be reasoned that cultural heritage buildings have a greater ability to adapt to urban challenges due to their sustainable nature. Accordingly, it is important to create the conditions for heritage properties to be sustainable in terms of their ability to communicate with the urban environment, in keeping with contemporary tendencies. The city authorities have a responsibility to reconcile contemporary needs and preserve the historical significance of the buildings.

1.2 Culture Heritage Buildings and Climate Change

The impacts of climate change are manifold and will manifest at various levels including the global economy, and the environmental impact. Additionally, they will impact human health. For sustainable heritage management and the regular rehabilitation of both indoor and outdoor spaces, it is crucial to assess the potential risks of climate change on cultural heritage objects and adapt to all possible rescue and recovery scenarios (Leissner et al. 2015). Climate change can be classified as one of the biggest global risks that can potentially lead to accelerated deterioration, or even the disappearance, of a large part of the world's cultural heritage (Bertolin 2019; Brimblecombe et al. 2011; Gandini et al. 2017). Listed heritage buildings, along with the valuable cultural assets they hold, provide priceless cultural and artistic testimonies of our past, which must be appropriately safeguarded for generations to come. Over the last few decades, concern regarding climate change has grown, with it becoming critically relevant for evaluating building sustainability, so that measures can be taken in advance to safeguard cultural heritage.

The effects of climate change, manifested in successive alterations in temperature, humidity, and wind intensity, together with a rise in sea level and extreme occurrences, have an impact on cultural heritage properties. Consequently, there is a rapidly expanding body of scientific research on the effects of climatic factors of distress on cultural heritage and on the impact analysis of climate change on cultural heritage properties (Sesana et al. 2021).

The position of cities in Serbia is defined by a number of challenges, including an absence of controlled and guided urbanization processes, an accelerated suburbanization trend, and the destruction of green spaces in cities. Additionally, there is a need for enhanced connectivity between cities. The destruction of the green hinterland of cities, resulting in the creation of heat islands, is especially emphasized. This further leads to the problems of large urban settlements dominated by concrete. In these conditions, the rehabilitation of cultural heritage objects is difficult, which will become an increasing problem in the future if drastic measures are not taken to stabilize the impact of evident climate changes in overpopulated city centers with too many concreted surfaces, such as Belgrade.

Belgrade, situated within the humid continental climate region, has a climate characterized by relatively high summer temperatures, low winter temperatures, and transitional seasons between them (spring and fall). In the ecological Atlas of Belgrade, five topographic-climate zones are identified: the city center, Novi Beograd, Krnjača, Surčin Plateau, and the hilly periphery (Environmental Quality in Belgrade (EQB 2017). To date, the urban heat island (UHI) and outdoor thermal comfort (OTC) indices in Belgrade have been identified in relation to cities in neighboring countries (Milovanović et al. 2020) or based on official meteorological station datasets (Lukić et al. 2021).

A review of historical data reveals that between the years 1951 and 2023, there was a discernible increase in the average temperature. This is evidenced by the fact that the ten warmest years on record occurred between the years 2000 and 2023 (RHMZ Annual Bulletin for 2020).

2 The Case of Belgrade

2.1 A Brief Introduction to Belgrade and to the Large-Scale Urban Destructions of the Twentieth Century

Belgrade is one of the oldest cities in Europe according to archeological excavations, dating from the Paleolithic period. Although Belgrade has not changed its physical position, it has been the capital of a number of states over time. For example, it was the capital of the Kingdom of Yugoslavia from 1918 to 1945, before becoming the capital of the Socialist Federal Republic of Yugoslavia from 1945 to 1992, and then from 1992 to 2003 of the Federal Republic of Yugoslavia. Belgrade then went on to become the capital of Serbia and Montenegro from 2003 to 2006. Since 2006, it has been the capital of Serbia (Arandelovic 2020a). The city has experienced a complicated and ever-changing history over the past two centuries, which has unfortunately led to the loss of a significant portion of its cultural heritage due to multiple instances of destruction. This is particularly true with regard to the wars of the twentieth century, in which Belgrade experienced five major devastating attacks and suffered enormous losses, not only in terms of people, but also in terms of the irreparable destruction of the urban and cultural heritage.

The first two attacks were by Austro-Hungarian forces, in 1914 and 1915, during which around 50% of the existing housing stock was completely demolished. Then, on April 6, 1941, Belgrade underwent the third attack, as Nazi bombers dropped 360 tons of bombs within two days. They demolished 714 houses and severely damaged 22,000 others, along with countless infrastructure and commercial premises. The complete loss of priceless cultural heritage that was forever destroyed on this occasion is irretrievable. Air raids resulted in the destruction of the National Library, which housed 350,000 books and a large number of rare manuscripts that originated in the twelfth century. In 1944, the fourth bombing attack on Belgrade in the twentieth century happened when the Allies demolished numerous business and manufacturing premises, as well as all six state hospitals and a few existing bridges. The Allied objectives were the Germans and their military structures. Regardless of the goal and reason, the city of Belgrade suffered destruction once again. The fifth, the NATO bomb attack, carried out in 1999, demolished 76 buildings and damaged 4489 others (Perović and Žegarac 2000).

Following the Second World War, Belgrade underwent a period of significant urban expansion, driven by a substantial influx of rural residents into the city, which was subsequently accompanied by the construction of numerous large-scale residential complexes. In the period between 1950 and 1991, the total number of dwelling units in Belgrade rose from approximately 107,000 to 512,000. These units were constructed in the form of large residential towers, which were designed as an urban housing solution for the growing population. This approach differed from the traditional model of housing that those new city residents were used to, which consisted of individual houses with private gardens (Hirt and Petrović 2010).

2.2 A Brief Overview of the Destruction of Cultural Heritage Buildings

When considering the impact of war on architecture, two types of damage can be identified. Firstly, there is the destruction resulting directly from military actions to conquer territory. Secondly, there is the targeted destruction of culture achieved through the annihilation of artifacts. This implies the intentional targeting of particular building typologies or architectural practices, leading to the loss of heritage, tradition, and the identity associated with architecture (Bevan 2007). There is a long list of buildings of cultural importance that have been destroyed. However, the Generalštab complex is the most famous among them because of its historical significance, location, and the continuous controversy it has brought with it over the years regarding its potential rehabilitation.

The Generalštab complex, situated at the intersection of Nemanjina Street and Kneza Miloš Street in Belgrade, suffered extensive damage during the 1999 NATO bombing. The Generalštab building, the sole edifice in Serbia designed by architect Nikola Dobrović, constructed between 1954 and 1963, exemplifies the legacy of the modern movement in architecture. It is probable that the building suffered significant damage that is not immediately apparent. This may have been caused by mechanical impacts, vibration, and explosions during the bombing. It is also likely that the lack of protection, and exposure to various chemical and physical agents after the building sustained extensive damage in 1999, contributed to the deterioration (Nikolić and Delić-Nikolić 2012).

Such architectural destruction refers to the deliberate destruction of a particular building typology or architectural legacy, in this case a cultural heritage that has shaped an architectural epoch of modern national history. The memory of a particular aspect of national history is inextricably linked to the physical structure in which it occurred. The role of this event in the formation of the identity of Belgrade after the Second World War can be understood by analyzing the political, economic, and ideological notions of power that were at play. In order to gain a full appreciation of the meaning and significance of the Generalštab Complex, it is essential to understand its historical context and the circumstances surrounding the construction of the complex in the wider context of the space on which it was built, as well as at the time of its creation (Popović et al. 2021).

Fig. 1 Contemporary ruins of Generalštab Complex, Photography: Biljana Arandelovic

The demolition of this edifice is simultaneously a manifestation of the collective memory of a nation and a subject of special cultural protection as an object of exceptional significance. Its destruction is a symbol of a broader cultural issue. The ruins (Fig. 1) have remained in the same place since 1999, and they have been the subject of repeated discussions by all local and state administrations. This is due to the building's location in the very center of the city, on the street where all state administration facilities are located. A decision regarding the future of the building has not yet been made. It is unclear whether it will be reconstructed according to the original plan or whether a new building will be constructed on its ruins. Over the years, various proposals have been put forth, but the final decision has consistently been postponed due to the controversy surrounding any proposed solution.

2.3 Culture Heritage in Belgrade—Selected Examples in the Savamala Area

All of Belgrade's other cultural heritage buildings serve as a testament to the city's rich historical development. For example, evidence of Austria's urban heritage from the early eighteenth century is evident in Belgrade's city center, particularly in the neoclassical facades that were constructed following the influence of the Habsburg capital, Vienna. The city of Belgrade is home to numerous structures that were directly influenced by the Viennese Secession style and were constructed at the beginning of the twentieth century. These buildings were designed by Serbian architects who had received their education in Austria, and subsequently returned to their native country (Samardžić 2015).

Fig. 2 Belgrade cooperative building—Geozavod building, Photography: Biljana Arandelovic

At the location within the area of Savamala, there are several cultural heritage buildings, which today belong to the area known as Belgrade Waterfront project. **The Belgrade Cooperative Building** (Fig. 2), situated at 48 Karadjordjeva Street in the Savamala district, was built between 1905 and 1907 in the style of Academism by Serbian architects Andra Stevanović and Nikola Nestorović. The building, influenced by French architecture of the early twentieth century, served as an administrative office for the Belgrade Cooperative, the first Serbian insurance company, formed in 1882 (Aleksić 2012). It played a significant role in the economic history of the city, and of Serbia as a whole, providing financial support for industrial development, trade and manufacturing. However, the building lost its original character and ceased operations in 1946 after the Cooperative was relocated during World War I. The Belgrade Cooperative was constructed on the site of the former Little Market, which is regarded as the heart of Savamala, and it marks the beginning of the modernization of Belgrade along the Sava coast. This authentic building is believed to be one of the most architecturally significant in Belgrade, and the first to feature a lavish hall and a monumental staircase leading to the upper floor of its banquet halls. The construction process used was modern and innovative, introducing reinforced concrete for the first time in Belgrade. After World War II, it served as the headquarters for various institutions, including the Geozavod. Despite being assigned to the Institute of Geophysics, it is still commonly referred to as the Geozavod building. Since 1966, the Belgrade Cooperative Building has been protected, recognized as a significant cultural heritage site in 1979 (Arandelovic 2020b).

This significant cultural and historical object was a part of Serbian culture and history until the second decade of the twenty-first century, when the Serbian government transferred a significant area along the Sava River to an Arab firm from Dubai for the development of the Belgrade Waterfront project.

The Hotel Bristol in Belgrade (Fig. 3) was a significant symbol of the city's transformation from an oriental town to a modern European capital. However, due to economic and political crises, the hotel deteriorated over time. After World War II, it was taken over by the Ministry of Defense, specifically the Military Institute Dedinje, and used to accommodate displaced military families. The hotel's future is now influenced by the urban redevelopment of Savamala, as it is part of the Belgrade Waterfront development project along with other heritage buildings (Arandelovic 2020b).

Fig. 3 Hotel Bristol in Savamala before renovation. Photography: Biljana Arandelovic

3 Problem Statement and Study Aim

The aim of the research is to explore and identify the main challenges and obstacles in solving the problem of rehabilitating Belgrade's cultural heritage buildings and to determine the role of climate change in this process. In order to establish the findings, it is first and foremost necessary to indicate which cultural heritage objects have been included in this observational research and the reason for their selection, as well as why they are considered to be useful examples of heritage objects to be observed. The results will indicate whether human factors, such as management, lack of adequate financial resources, or failure to comprehend the actual reality of climate change, pose a larger threat to Belgrade's cultural heritage than climate change directly. The paper also questions the approach of placing profit above the restoration of cultural heritage buildings in Belgrade, which can be perceived as being unregulated and investment-oriented.

4 Discussion

Today, the city of Belgrade faces significant challenges in the rehabilitation of cultural heritage buildings. Climate change, which is a global concern, necessitates a re-evaluation of construction practices for new buildings, complexes, and urban units. This is particularly relevant in the context of cultural architectural heritage, which is subject to the same environmental pressures and requires special measures for its rehabilitation. The city center is home to numerous historical facades in a state of deterioration. This is due to a number of factors, including unresolved ownership, lengthy restitution processes, and disputes over ownership. Many heirs are unable to reclaim property that was confiscated by the state during the communist era and subsequently appropriated after the Second World War. Furthermore, the misallocation of state funds toward investments that do not directly benefit the cultural heritage, which is essential for the well-being of society as a whole, is a de facto concern.

Fig. 4 Belgrade Waterfront district in Savamala. Photography: Biljana Arandelovic

Society is still in the process of transition, while the laws adopted to protect cultural architectural heritage are not being consistently implemented due to the continuing high level of corruption in society. Some cultural heritage objects have been entirely devastated by bombing, while the remaining cultural heritage has not been a priority to decision-makers. The study identifies two main obstacles to the appropriate rehabilitation of cultural heritage properties: climate change, understood as a global problem, and the unregulated investor-friendly approach in Belgrade, which prioritizes profit over the preservation of cultural heritage buildings. In fact, in recent years, a large number of cultural heritage buildings have been handed over to foreign investors, as is the case in the Belgrade Waterfront (Figs. 4 and 5), or sold to investors who plan either to completely demolish them or to renovate them beyond recognition, regardless of whether they are buildings of cultural significance.

5 Policy Implications and Future Research Direction in Belgrade

It is evident that relevant institutions play a pivotal role in the rehabilitation of cultural heritage objects. Furthermore, they assume complete responsibility for the development of urban areas, the protection of public land, and the preparation of planning regulations aimed at limiting the expansion of urban areas. The lack of guidelines for dealing with uncertainties and disturbances in the post-socialist context is evident in Serbia, with the capital city of Belgrade being the most affected. One of the key uncertainties in the recent urban development of Belgrade is the extent to which foreign investors will participate in areas of the city that are of cultural and historical significance. The presence of buildings of significant cultural and historical value, which require rehabilitation, within these urban areas

Fig. 5 Belgrade Waterfront in Savamala, view inside the district. Photography: Biljana Arandelovic

represents a particularly acute challenge for the Belgrade urban system. Rehabilitation in accordance with the relevant regulations is of paramount importance.

It is possible for urban systems to become more adaptable to regulatory change and reshaping. Consequently, future research should focus on preparing cultural heritage objects for a range of potential outcomes, thereby reducing their sensitivity to short-term and long-term disturbances, uncertainties, and external shocks caused by natural disasters.

6 Conclusion

It is imperative that Belgrade formulates local policies and future directions of research and policy-making with regard to the sustainability agenda. In order to avert the imminent threat of architectural heritage collapse, it is imperative that local authorities assume the role of initiators and leaders in the enforcement of regulatory compliance. Without such decisive action, the potential for irreparable damage looms large. In light of the actions undertaken by the local authorities in Belgrade thus far, it is evident that profit is the primary consideration, even when it poses a threat to the architectural heritage of the city. Such practices are untenable and require immediate modification.

References

Aleksić, V.: Beogradska Zadruga i njena uloga u privrednom razvoju Srbije krajem XIX i početkom XX veka. Belgrade Cooperative and its role in the economic development of Serbia from the late 19th to the early 20th century, Bankarstvo 6, 108–133 (2012)

Almeida, C.P., Ramos, A.F., Silva, J.M.: Sustainability assessment of building rehabilitation actions in old urban centres. Sustain. Cities Soc. **36**, 378–385 (2018)

Arandelovic, B.: Architectural analysis of the first major rehabilitation in the 21st century of Olbrich's secession building in Vienna. Buildings **14**(5), 1229 (2024)

Arandelovic, B., Musil, R.: The renovation, rehabilitation and adaptation of historical heritage buildings in public ownership. The case of historical buildings of exceptional importance in Vienna. Build. Environ. 110937 (2023)

Arandelovic, B.: Creative city Berlin. In: Public Art and Urban Memorials in Berlin. The Urban Book Series. Springer, Cham, pp. 1–26 (2018)

Arandelovic, B.: The urban, political, and socioeconomic rise and fall of Belgrade through its history. In: Belgrade: The 21st Century Metropolis of Southeast Europe. The Urban Book Series. Springer, Cham, pp. 1–63 (2020a)

Arandelovic, B.: Reimaging Belgrade: the case of Savamala District. In: Belgrade: The 21st Century Metropolis of Southeast Europe. The Urban Book Series. Springer, Cham, pp. 169–221 (2020b)

Arfa, F.H., Zijlstra, H., Lubelli, B., Quist, W.: Adaptive reuse of heritage buildings: from a literature review to a model of practice. Hist. Environ.: Policy Pract. **13**(2), 148–170 (2022)

Bertolin, C.: Preservation of cultural heritage and resources threatened by climate change. Geosciences **9**(6), 250 (2019)

Bevan, R.: The Destruction of Memory: Architecture at War. Reaktion books (2007)
Brimblecombe, P., Grossi, C.M., Harris, I.: Climate change critical to cultural heritage. In: Survival and Sustainability: Environmental Concerns in the 21st Century, pp. 195–205 (2011)
Clark, K.: Policy review: valuing culture and heritage capital: a framework towards informing decision making. Hist. Environ.: Policy Pract. **12**(2), 252–258 (2021)
Environmental Quality in Belgrade (EQB): Belgrade: Secretariat for Environmental Protection (2017)
Ferretti, V., Bottero, M., Mondini, G.: Decision making and cultural heritage: an application of the multi-attribute value theory for the reuse of historical buildings. J. Cult. Herit. **15**(6), 644–655 (2014)
Gandini, A., Garmendia, L., San Mateos, R.: Towards sustainable historic cities: mitigation climate change risks. Entrep. Sustain. Issues **4**(3), 319 (2017)
Guzmán, P.C., Roders, A.P., Colenbrander, B.J.F.: Measuring links between cultural heritage management and sustainable urban development: an overview of global monitoring tools. Cities **60**, 192–201 (2017)
Hirt, S., Petrović, M.: The gates of Belgrade: safety, privacy, and new housing patterns in the post-communist city. Probl. Post-Communism **57**(5), 3–19 (2010)
Leissner, J., Kilian, R., Kotova, L., Jacob, D., Mikolajewicz, U., Broström, T., Ashley-Smith, J., Schellen, H.L., Martens, M., van Schijndel, J., Antretter, F., Vyhlídal, T.: Climate for Culture: assessing the impact of climate change on the future indoor climate in historic buildings using simulations. Herit. Sci. **3**, 1–15 (2015)
Lukić, M., Filipović, D., Pecelj, M., Crnogorac, L., Lukić, B., Divjak, L., Lukić, A., Vučićević, A.: Assessment of outdoor thermal comfort in Serbia's urban environments during different seasons. Atmos. Basel **12**(8), 1084 (2021)
Masini, N., Soldovieri, F. (eds.).: Sensing the Past: From Artifact to Historical Site, vol. 16. Springer (2017)
Medda, F., Lipparini, F.: Impact investment for urban cultural heritage. City Cult. Soc. **26**, 100413 (2021)
Milovanović, B., Radovanović, M., Schneider, C.: Seasonal distribution of urban heat island intensity in Belgrade Serbia. J. Geogr. Inst. "Jovan Cvijić" SASA **70**(2), 163–170 (2020)
Nikolić, E., Delić-Nikolić, I.: Facade renewal of Generalštab building in rehabilitation and adaptive reuse processes. In: Planning, Design, Construction and Renewal in the Civil Engineering: Proceedings, pp. 550–562 (2012)
Pereira Roders, A., Van Oers, R.: Bridging cultural heritage and sustainable development. J. Cult. Herit. Manag. Sustain. Dev. **1**(1), 5–14 (2011)
Perović, M., Žegarac, Z.: The destruction of an architectural culture: the 1999 bombing of Belgrade. Cities **17**(6), 395–408 (2000)
Petti, L., Trillo, C., Makore, B.N.: Cultural heritage and sustainable development targets: a possible harmonisation? Insights from the European perspective. Sustainability **12**(3), 926 (2020)
Popović, T., Marić, J., Vaništa Lazarević, E.: Reshaping approaches of architectural heritage devastated through bombing: case study of Generalštab, Belgrade. Urban Des. Int. **26**, 296–309 (2021)
RHMZ Annual Bulletin for 2020, Republički hidrometeorološki zavod Srbije: Godišnji bilten za Srbiju 2020. godina. Beograd, 14.januar 2021 (2021). https://www.hidmet.gov.rs/data/klimatologija/latin/2020.pdf
Samardžić, N.: Belgrade 1714–2014: Utopianism and Urbicide. In: Doytchinov, G., Đukić, A., Ioniță, C. (eds.) Planning capital cities: Belgrade, pp. 212–224. Verlag der Technischen Universität Graz, Bucharest, Sofia (2015)
Sesana, E., Gagnon, A.S., Ciantelli, C., Cassar, J., Hughes, J.J.: Climate change impacts on cultural heritage: a literature review. Wiley Interdiscip. Rev.: Clim. Chang. **12**(4), e710 (2021)
Tweed, C., Sutherland, M.: Built cultural heritage and sustainable urban development. Landsc. Urban Plan. **83**(1), 62–69 (2007)
Yates, J.F., De Oliveira, S.: Culture and decision making. Organ. Behav. Hum. Decis. Process. **136**, 106–118 (2016)

Curriculum Vitae

Editor name	: Dr. Shahfahad
Affiliation	: Department of Geography, Institute of Science
Institution	: Banaras Hindu University
Address	: Varanasi, Uttar Pradesh
Postal code	: 221005
City	: Varanasi
Country	: India
Telephone number	: (+91) 8800343539
E-mail	: fahadshah921@gmail.com
Homepage URL	: https://www.scopus.com/authid/detail.uri?authorId=57214364016
ORCID ID	: https://orcid.org/0000-0001-5868-1062

P. Mohammad et al. (eds.), *Geographic Approaches to Climate Change and Mitigation: Urban and Rural Perspectives (Volume 2)*,
Advances in Science, Technology & Innovation, https://doi.org/10.1007/978-3-031-92115-5

Dr. Shahfahad is Assistant Professor of Geography at Department of Geography, Institute of Science, Banaras Hindu University (BHU). Before joining BHU, Dr. Shahfahad worked as Assistant Professor of Geography at Department of Geography, Faculty of Sciences, Jamia Millia Islamia during 2023–24. He completed his B.A. (H) Geography in 2015, M.A. Geography in 2017 and Ph.D. in 2022 from the Department of Geography, Jamia Millia Islamia, New Delhi. He has also done Post-Graduate Diploma in Geoinformatics from Centre for Distance and Online Learning, Jamia Millia Islamia during 2022–23. He was awarded prestigious National Post Doctoral Fellowship (N-PDF) by Science and Engineering Research Board (SERB) of Department of Science and Technology (DST) in 2024. He was junior research fellow (JRF) during 2017–2019 and senior research fellow (SRF) during 2019–2022 in the Department of Geography, Jamia Millia Islamia. His core areas of research are urban expansion, climate change, urban hydrology as well as applications of geospatial techniques and artificial intelligence in urban studies and natural hazards. He has extensively studied the climate pattern and change over the metropolitan cities of Indian subcontinent. He has published two edited book volume and more than 80 scientific research articles in the journals of international repute such as Habitat International, Urban Climate, Journal of Environmental Management, Scientific Reports, Ecological Indicators, Remote Sensing, Geocarto International and Ecological Informatics. Beside this, he is currently editorial board member of two internal journals. Also, he has served as the reviewer for several reputed international journals.

Editor name : Dr. Pir Mohammad
Affiliation : Department of Land Surveying and Geo-Informatics, The Hong Kong Polytechnic University
Institution : The Hong Kong Polytechnic University
Address : Flat 6, 10/F, Po Sing Court, 25 Shek Ku Lung Road, Kowloon City, Kowloon, Hong Kong
Postal code : 00000
City : Kowloon
Country : Hong Kong
Telephone number : +852-55932487
E-mail : dr-pir.mohammad@polyu.edu.hk
Homepage URL : https://sites.google.com/view/pir-mohammad/home?authuser=0
ORCID ID : 0000-0002-0075-6372

Pir Mohammad is currently working as a postdoctoral fellow in the Department of Land Surveying and Geo-Informatics at The Hong Kong Polytechnic University, Hong Kong. He earned his Ph.D. in remote sensing, specializing in urban heat island studies over Indian cities from the Department of Earth Sciences, Indian Institute of Technology Roorkee, India, in 2021. He received the Excellence in Doctoral Research Award for his outstanding Ph.D. work in the science stream under the category of fundamental research. His area of research interest includes remote sensing, urban climate, urban heat island, and land use land cover change. He has authored a total of 20 SCI journal publications in the peer-reviewed journal of high repute, including GI Sciences and Remote Sensing, Science of the Total Environment, Urban Climate, Building and Environment etc., one of which was selected as a highly cited paper in Quaternary International journal as per the Web of Science. Pir also serves as the reviewer for more than 40 different SCI journals and reviewed more than 200 papers till now. He has also presented papers at various national and international conferences. Besides, his research has also been covered by various

Indian newspapers like The Times of India, and Bharat Times. Moreover, he is also able to contribute to the town planning policy document of the Ahmedabad city of India, published by the World Bank Group in 2020. He is also a life member of Indian Society of Remote Sensing (ISRS), Indian Society of Geomatics (ISG), and Indian Science Congress Association (ISCA).

https://www.scopus.com/authid/detail.uri?authorId=57210885586

Dr. Pir Mohammad
Research Assistant Professor
Department of Land Surveying and Geo-Informatics
The Hong Kong Polytechnic University, Hong Kong
E-mail: pir.mohammad@polyu.edu.hk, pirmohammad291@gmail.com
Links: Google Scholar / Personal Homepage /ORCID
Citations=**1392**, h-index=**19**, i10-index=**24**

Research Interests

My research focuses on the intricate nexus between urban environments, climate, and sustainability. I am particularly interested in leveraging remote sensing and geospatial techniques, alongside environmental and climate models, to tackle pressing issues related to urban environmental and climatic changes. Currently, my research encompasses the following themes:

- **Remote Sensing of Urban Environments**: Utilizing remote sensing and GIS techniques to monitor and analyze urban landscapes.
- **Extreme Weather Events and Climate Change**: Investigating the impacts of extreme weather events and their vulnerability in the phase of climate change scenarios.
- **Heat Mitigation and Thermal Comfort:** Explores strategies to reduce urban heat and enhance thermal comfort for residents, integrating climate-responsive design and sustainable practices.

Professional Background

03/2025 - continue — **Research Assistant Professor,** Department of Land Surveying and Geo-informatics, The Hong Kong Polytechnic University, Hong Kong

04/2022 – 02/2025 — **Postdoctoral Fellow,** Department of Land Surveying and Geo-informatics, The Hong Kong Polytechnic University, Hong Kong (Advisor: Prof. Qihao Weng)

07/2016 - 11/2021 — **Junior Research Fellow / Senior Research Fellow,** Department of Earth Sciences, Indian Institute of Technology Roorkee, India (Advisor: Prof. Ajanta Goswami)

Educational Qualification

07/2016 – 11/2021 — **PhD, Indian Institute of Technology Roorkee, India**
Department of Earth Sciences
Thesis: *Urban Heat Island study and modelling optimal urbanization of selected Indian cities*
Advisor: Prof. Ajanta Goswami

07/2013 – 06/2016 — **MTech, Indian Institute of Technology Roorkee, India**
Department of Earth Sciences
Thesis: *Urban Heat Island Impact Study in Ahmedabad City, Gujarat Using Earth Observation Techniques*
Advisor: Prof. Ajanta Goswami

07/2010 – 06/2013 — **BSc, Gauhati University, India**
Department of Geology, Cotton College

Fellowships and Awards

2023 — **Excellence in Doctoral Award** for PhD at IIT Roorkee

2022 - 2025 — Research Talent Hub of **Innovation and Technology Fund** of Hong Kong, **Postdoc**

2020	**Special Registration Support Program (SRSP)** by International Geological Congress Conference
2019	**JRF** and **Lectureship** by Council of Scientific & Industrial Research (**CSIR**), India
2016 – 2021	**JRF / SRF** Fellowship by Ministry of Human Resource Development (**MHRD**)
2013	**Chief Minister's Scheme** by Government of Assam
2013	**IIT-Join Admission for Masters (JAM)**
2013	**Dr Pabindranath Datta Memorial Silver Medal by Cotton College**
2008	**Arundoram Borooah Award** by Planning and Development Department of Assam

Project Grants

- **PI**: Resilience of Urban Slums to Heat Waves by Using Integrated Remote Sensing and GeoAI Approach, The Hong Kong Polytechnic University, **HKD 300,000**, 06/2025-05/2027.
- **Co-I**: Enhancing the Resilience of Rural Infrastructure and Communities, The Hong Kong Polytechnic University, **HKD 8,000,000**, 03/2025-02/2027.

Research Experience

The Hong Kong Polytechnic University (Postdoctoral Research Experience) **2022-2025**

- Importance of selecting suitable heat wave indices in identifying the dangerous heat wave condition outdoors. (Research Video: Link)
- Asian heat stress variation under historical and future climate change and mitigation-based scenarios and its heat exposure mapping in urban and rural perspective.
- Effect of agricultural practice and cropping patterns in mitigating soil temperature.

Indian Institute of Technology Roorkee (PhD Research Experience) **2016-2021**

- Investigating the surface urban heat island variation over 150 Indian cities and its relationship with different driving factors like vegetation, evapotranspiration, albedo, thermal inertia, soil moisture, etc. using remote sensing data.
- Spatial variation of land surface temperature and green space of selected Indian cities from different climatic zone along the urban-rural gradient.
- Evaluate the impact of land use land cover in thermal environment of a city, future urban growth prediction using cellular automata model, machine learning based future LST prediction.

Teaching and Mentoring Experience

Mentoring graduate students' thesis, The Hong Kong Polytechnic University

- Mingyu Jiang, Research Assistant, *The Hong Kong Polytechnic University* — 2025-
  Research: Global Urban Slum Mapping (Main Supervisor)
- On Wing Hin, B.Sc., *The Hong Kong Polytechnic University*
  Research: Urban Growth Impact assessment in the northern metropolis in Hong Kong (Main Supervisor) — 2025-2026
- Zheng Peiyuan, M.Sc, *The Hong Kong Polytechnic University*
  Research: Hydrodynamic-Deep Learning Coupled Model-Driven Methodology for Urban Flooding Prediction: Case Study of Wong Tai Sin District (Co- Supervisor with main supervisor as Dr. Zhiwei Li) — 2025-2026

Teaching Assistant, Indian Institute of Technology Roorkee (during PhD) 2016 - 2020

- Undergraduate and Postgraduation course on **Remote Sensing and GIS** (*Autumn & Spring Semester- 2016-17, 2017-18, 2018-19, and Autumn Semester 2019-20*)
  - Assisting the students during their practical hours related to different applications of remote sensing and GIS in Earth Sciences using different software (most frequently ArcGIS)
  - Assist supervisor for the upgradation of advanced remote sensing and GIS course structure
- Undergraduate course on **Basic Petrology** (*Autumn Semester 2016-17, 2017-18,*

2018-19, 2019-20)
  - Assisting supervisor in upgradation of course materials for the lectures (igneous, sedimentary and metamorphic rocks)
  - Assist the students during practical classes - describe both the hand specimen and microscopic description of basic types of rocks
- Undergraduate course on **Geomorphology** (*Spring Semester 2016-17, 2017-18, 2018-19*)
  - Assisting supervisor in taking tutorial classes and preparation of tutorial materials
  - Preparation of slides for the lectures including different aspects of geomorphology

Assisting graduate student's thesis (during PhD and Postdoc)

- Chung Cheuk Hei Cyrus, *The Hong Kong Polytechnic University* 2024 – 2025
  Research: Development of a modified thermal humidity index using integrated remote sensing and geospatial techniques (Advisor: Prof. Qihao Weng)
- Pragyanand, *Indian Institute of Technology Roorkee* 2016 – 2017
  Research: Study of Urban Heat Island Impact on Guwahati City Using Earth Observation Techniques (Advisor: Prof. Ajanta Goswami)
- Diksha Singh, *Indian Institute of Technology Roorkee* 2017 – 2018
  Research: Urban Heat Island Study of Raipur City, Chhattisgarh (Advisor: Prof. Ajanta Goswami)
- Prajwal Bhatt, *Indian Institute of Technology Roorkee* 2017 – 2018
  Research: Urban Heat Island Characterization Using Thermal Remote Sensing Data (Advisor: Prof. Ajanta Goswami)
- Manmohan S. Charan, *Indian Institute of Technology Roorkee* 2018 – 2019
  Research: Urban Sprawl Study Using Terrestrial Camera (Advisor: Prof. Ajanta Goswami)
- Nilesh Tanwar, *Indian Institute of Technology Roorkee* 2018 – 2019
  Research: Impact of Land use Landcover Changes on Urban Heat Island (Advisor: Prof. Ajanta Goswami)
- Atul Kumar, *Indian Institute of Technology Roorkee* 2017 – 2018
  Research: Temporal Alteration of Urban Climate in Gurugram: UHI Characterization (Advisor: Prof. Ajanta Goswami)
- Brian Basumatary, *Indian Institute of Technology Roorkee* 2018 - 2019
  Research: Impact Assessment of Change in LULC on Urban Heat Island (Advisor: Prof. Ajanta Goswami)

Edited Journal Special Issue

1. **Guest Editor**, Special Issue on "*Environmental Impacts and Consequences of Urban Sprawl*" in ***Environmental Sciences and Pollution Research*** (Springer), 30(49), October 2023. https://link.springer.com/journal/11356/volumes-and-issues/30-49

Book Published

1. **Mohammad, P.**; Shahfahad; Comino, J. R.; Matamanda, A. R. **2025**. "Geographic Approaches to Climate Change and Mitigation: Urban and Rural Perspectives (Volume 1)", Springer Cham. https://link.springer.com/book/9783031921186
2. Shahfahad; **Mohammad, P.**; Matamanda, A. R.; Comino, J. R. **2025**. "Geographic Approaches to Climate Change and Mitigation: Urban and Rural Perspectives (Volume 2)", Springer Cham. https://link.springer.com/book/9783031921148

Book Chapter

1. Mahdavi Estalkhsari, B.; **Mohammad, P**.; Karimi, A. **2023**. Land Use and Land Cover Change Dynamics and Modeling Future Urban Growth Using Cellular Automata Model Over Isfahan Metropolitan Area of Iran. In: Chatterjee, U., Akanwa, A.O., Kumar, S., Singh, S.K., Dutta Roy, A. (eds) *Ecological Footprints of Climate Change*. Springer Climate. Springer, Cham. https://doi.org/10.1007/978-3-031-15501-7_19
2. Khan F.; Das B.; **Mohammad P**. **2022.** Urban Growth Modeling and Prediction of Land Use Land Cover Change Over Nagpur City, India Using Cellular Automata Approach. In: Rai P.K., Mishra V.N., Singh P. (eds) *Geospatial Technology for Landscape and Environmental Management*. Advances in Geographical and Environmental Sciences. Springer, Singapore. https://doi.org/10.1007/978-981-16-7373-3_13

Journal Publications (*corresponding author)

2025

1. **Mohammad, P**.; Voogt, J.; Peng, J.; Garcia-Garcia, A.; Chen, J.M.; Zeng, Z.; Cescatti, A.; Weng, Q. **2025**. Global Evidence of Mitigating Soil Temperature through Sustainable Agricultural Practices. *Nature Communication* (Under Revision)
2. Cao, M.; **Mohammad, P**.; Rehman, M.M.; Weng, Q. **2025.** Differential Impacts of Soil Physical Properties on Root and Leaf Phenology in Cropland. *Remote Sensing of Environment* (Under Review)
3. Hui, S. P.; Liu, Y.; **Mohammad, P.**; Weng, Q. **2025**. Assessing the diurnal relationship between urban morphology and land surface temperature in a highly dense city: A case study in Hong Kong. *Urban Climate* (Under Review)
4. Tang, N., Farhan, M., **Mohammad, P.**, Abdullah-Al-Wadud, M., Hussain, S., Hamza, U., Zulqarnain, R. M., Rebouh, N. Y. **2025**. Enhancing Urban Heat Island Analysis Through Multi-Sensor Data Fusion and GRU-Based Deep Learning Approaches for Climate Modeling. *IEEE Journal of Selected Topics in Applied Earth Observations and Remote Sensing*, 18, 9279–9296. https://doi.org/10.1109/JSTARS.2025.3554529
5. **Mohammad, P.**; Weng, Q. **2025**. Asian heat stress variations in a changing climate: Implications for disproportionate urban and rural population exposure. *Habitat International, 156,* 103294. https://doi.org/10.1016/j.habitatint.2025.103294

2024

6. Shahfahad; Talukdar, S.; Naikoo, M.W.; Rihan, M.; **Mohammad, P**.; Rahman, A. **2024**. Seasonal dynamics of land surface temperature and urban thermal comfort with land use land cover pattern in semi-arid Indian cities: Insights for sustainable Urban Management. *Urban Climate, 57*, 102105. https://doi.org/10.1016/j.uclim.2024.102105
7. **Mohammad, P.**; Weng, Q. **2024**. Comparing existing heat wave indices in identifying dangerous heat wave outdoor conditions. *Nexus*, 1(3), 100027. https://doi.org/10.1016/j.ynexs.2024.100027

2023

8. Das B.; Khan F.; **Mohammad P. 2023.** Impact of urban sprawl on change of environment and consequences. *Environmental Science and Pollution Research,* 30(49), pp. 106894–106897. https://doi.org/10.1007/s11356-023-29192-3
9. Gohain, K.J.; Goswami, A.; **Mohammad, P.**; Kumar, S. **2023.** Modelling relationship between land use land cover changes, land surface temperature and urban heat island in Indore city of central India. *Theoretical and Applied Climatology, 151(3-4),* pp. 1981–2000. https://doi.org/10.1007/s00704-023-04371-x
10. Kafy, A. A.; Bakshi, A.; Saha, M.; Faisal, A. A.; Almulhim, A.I.; Rahaman, Z.A.; **Mohammad, P**. **2023**. Assessment and prediction of index based agricultural drought vulnerability using machine learning algorithms. *Science of the Total Environment, 867,* 161394. https://doi.org/10.1016/j.scitotenv.2023.161394
11. Mahdavi Estalkhsari, B.; **Mohammad, P.**; Razavi, N. **2023.** Change detection in a rural landscape : A case study of processes and main driving factors along with its response to thermal environment in Farim,

Iran. *Environmental Science and Pollution Research,* 30(49), pp. 107041–107057. https://doi.org/10.1007/s11356-022-24504-5

12. Karimi, A.; **Mohammad, P.**; Martínez, A.G.; Rangel, D.M.; Gachkar, D.; Gachkar, S. **2023**. New developments and future challenges in reducing and controlling heat island effect in urban areas. *Environment, Development and Sustainability,* 25(10), pp. 10485–10531. https://doi.org/10.1007/s10668-022-02530-0
13. **Mohammad, P.***; Goswami, A. **2023.** Exploring different indicators for quantifying surface urban heat and cool island together: A case study over two metropolitan cities of India. *Environment, Development and Sustainability,* 25(10), pp. 10857–10878. https://doi.org/10.1007/s10668-022-02509-x

2022

14. Karimi, A.; Kim, Y.J.; Mohammad, Z.N.; Martínez, A.G.; Delfani, S.; Brown, R.D.; Rangel, D.M.; **Mohammad, P**. **2022**. Assessment of Outdoor Design Conditions on the Energy Performance of Cooling Systems in Future Climate Scenarios—A Case Study over Three Cities of Texas, Unites States. *Sustainability, 14(22),* 14848. https://doi.org/10.3390/su142214848
15. **Mohammad, P.***; Goswami, A. **2022**. Predicting the impacts of urban development on seasonal urban thermal environment in Guwahati city, northeast India. *Building and Environment, 226*, 109724. https://doi.org/10.1016/j.buildenv.2022.109724
16. Halder, B.; Karimi, A.; **Mohammad, P.***; Bandyopadhyay, J.; Brown, R.D.; Yaseen, Z.M. **2022**. Investigating the Relationship between Land Alteration and the Urban Heat Island of Seville City Using Multi-temporal Landsat Data. *Theoretical and Applied Climatology, 150(1-2),* pp. 613–635. https://doi.org/10.1007/s00704-022-04180-8
17. Karimi, A.; **Mohammad, P.* 2022**. Effect of outdoor thermal comfort condition on visit of tourists in historical urban plazas of Sevilla and Madrid. *Environmental Science and Pollution Research, 29(40),* pp. 60641–60661. https://doi.org/10.1007/s11356-022-20058-8
18. **Mohammad, P**.; Goswami, A.; Chauhan, S.; Nayak, S. **2022.** Machine Learning algorithm based prediction of land use land cover and land surface temperature changes to characterize the surface urban heat island phenomena over Ahmedabad city, India. *Urban Climate, 42*, 101116. https://doi.org/10.1016/j.uclim.2022.101116
19. **Mohammad, P.***; Goswami, A. **2022.** Spatial variation of surface urban heat island magnitude along the urban-rural gradient of four rapidly growing Indian cities. *Geocarto International, 37(5)*, pp. 4269–4291. https://doi.org/10.1080/10106049.2021.1886338

2021

20. Karimi, A.; **Mohammad, P.**; Gachkar, S.; Gachkar, D.; Martínez, A. G.; Rangel, D. M.; Brown, R. D. **2021.** Surface Urban Heat Island Assessment of a Cold Desert City: A Case Study over the Isfahan Metropolitan Area of Iran. *Atmosphere, 12(10)*, 1368. https://doi.org/10.3390/atmos12101368
21. **Mohammad, P.***; Aghlmand, S.; Fadaei, A.; Gachkar, S.; Gachkar, D.; Karimi, A. **2021.** Evaluating the role of the albedo of material and vegetation scenarios along the urban street canyon for improving pedestrian thermal comfort outdoors. *Urban Climate, 40*, 100993. https://doi.org/10.1016/j.uclim.2021.100993
22. **Mohammad, P**.; Goswami, A. **2021.** Quantifying diurnal and seasonal variation of surface urban heat island intensity and its associated determinants across different climate zones over Indian cities. *GIScience & Remote Sensing, 58(7)*, pp. 955-981. https://doi.org/10.1080/15481603.2021.1940739
23. **Mohammad, P**.; Goswami, A. **2021.** A Spatio-Temporal Assessment and Prediction of Surface Urban Heat Island Intensity Using Multiple Linear Regression Techniques Over Ahmedabad City, Gujarat. *Journal of the Indian Society of Remote Sensing, 49(5)*, pp. 1091-1108. https://doi.org/10.1007/s12524-020-01299-x
24. Gohain, K.J.; **Mohammad, P**.; Goswami, A. **2021.** Assessing the impact of land use land cover changes on land surface temperature over Pune city, India. *Quaternary International, 575-576*, pp. 259-269. https://doi.org/10.1016/j.quaint.2020.04.052 *[Highly cited paper as per WoS]*

2020

25. Malik, S.; Pal, S.C.; Sattar, A.; Singh, S.K.; Das, B.; Chakrabortty, R.; **Mohammad, P. 2020.** Trend of

extreme rainfall events using suitable Global Circulation Model to combat the water logging condition in Kolkata Metropolitan Area. *Urban Climate, 32*, 100599. https://doi.org/10.1016/j.uclim.2020.100599

2019

26. **Mohammad, P.***; Goswami, A. **2019**. Temperature and precipitation trend over 139 major Indian cities: An assessment over a century. *Modeling Earth Systems and Environment, 5(4)*, pp. 1481–1493. https://doi.org/10.1007/s40808-019-00642-7
27. **Mohammad, P.***; Goswami, A.; Bonafoni, S. **2019**. The Impact of the Land Cover Dynamics on Surface Urban Heat Island Variations in Semi-Arid Cities : A Case Study in Ahmedabad City, India, Using Multi-Sensor/Source Data. *Sensors (Switzerland), 19(17)*, 3701. https://doi.org/10.3390/s19173701

Conference Publications

1. **Mohammad, P.**; Weng, Q. **2023.** Assessing the impact of historical and future climate change scenarios on diurnal heat stress in Asia. In Proceedings of *ISPRS Geospatial Week 2023*, Arab Academy for Science, Technology & Maritime Transport, Cairo, Egypt, paper ID: 1448.
2. **Mohammad, P.**; Weng, Q. **2023.** Historical and Projected socio-economic impacts of diurnal heat waves in Asia. In Proceedings of *Research Excellence for Societal Impacts*, PAIR Conference, The Hong Kong Polytechnic University, Hong Kong, abstract no: RILS 05.
3. **Mohammad, P.**; Goswami, A. **2022.** Surface urban heat island assessment over a humid subtropical city of India using remote sensing. In Proceedings of *36th International Geological Congress Abstracts*, New Delhi, abstract no: 4790–4277.
4. **Mohammad, P.**; Goswami, A. **2020.** Surface urban heat island variations over major Indian cities across different climatic zone. In Proceedings of *EGU General Assembly Conference Abstracts*, Vienna, abstract no 6444. https://ui.adsabs.harvard.edu/abs/2020EGUGA..22.6444M
5. **Mohammad, P.**; Goswami, A. **2019.** Temporal assessment of urban heat island over Lucknow city using thermal remote sensing. In Proceedings of the *Role of Remote Sensing and GIS in Natural Hazard Assessment for Sustainable Development: Present Scenario and Future Perspective*; Lucknow, p. 10.
6. **Mohammad, P.**; Goswami, A. **2018.** A Remote Sensing Based Evaluation of Urban Heat Island and Its Spatial Pattern over the City Of Raipur, Chhattisgarh, India. In Proceedings of the *AGU Fall Meeting Abstracts*, Washington, abstract #GH22A-10. https://ui.adsabs.harvard.edu/abs/2018AGUFMGH22A..10M/abstract
7. **Mohammad, P.**; Goswami, A.; Sattar, A.; Bonafoni, S. **2016.** A remote sensing approach to study the temporal growth of the urban limits for the city of Hyderabad from 2001-2015. In Proceedings of the *International Conference on Civil Engineering for Sustainable Development-Opportunities & Challenges*; Guwahati, pp. 388–391.
8. Goswami, A.; **Mohammad, P.**; Sattar, A.; Mahanta, R.; Bonafoni, S. **2016.** An Urban Heat Island magnitude study over the city of Guwahati-An integration of Remote Sensing and Ground based approach. In Proceedings of the *International Conference on Civil Engineering for Sustainable Development-Opportunities & Challenges*; Guwahati, pp. 383–387.
9. Goswami, A.; **Mohammad, P.**; Sattar, A. **2016.** A temporal study of Urban Heat Island (UHI) -A Evaluation of Ahmedabad city ,Gujarat. In Proceedings of the *International Conference on Climate Change Mitigation and Technologies for Adaptation*; Shillong, pp. 63–68.

Invited Speakers/Talks

1. Invited Resource Person Lecture of the refresher course on Disaster Management on topic "*Exploring Urban Heat Islands and Extreme Heatwaves: Trends and Impacts*" on 18th Dec, 2024 organised by Malaviya Mission Teacher Training Centre and Department of Political Science at Lucknow University, India

Contribution in Reports

1. Data Contribution in compiling an urban heat map of a policy document "*Ahmedabad, India Scaling Up*

with Contiguous Replication of Town Planning Schemes" published by Madhu Bharti and Shagun Mehrotra, **World Bank Group**, 2020. (Link) (Page 27: Urban Heat Map figure, see legend for credit, also see acknowledgement section)

Members of the Organizing Committee

1. Reviewer member of the **IGARSS** (International Geoscience and Remote Sensing Symposium) **2025 Scientific Committee**, schedule to be held on 3-8 August 2025 in Brisbane, Australia
2. Member of the **International Advisory Committee** of the *"Climate Change, Disaster Management and Environment Sustainability"* from 21-22nd Feb, 2024 organized by Department of Geography, Jamia Millia Islamia, New Delhi, India
3. Member of the **Organizing Committee** of the *"Workshop on GeoAI and Big data for Urban, Environment, and Sustainability"* from 9-10th Dec, 2023, organized by JC STEM Lab of Earth Observations and Research Center for Artificial Intelligence in Geomatics at the Hong Kong Polytechnic University

Selected Seminars/Conferences Attended

1. Attended two-day workshop on **"2025 Workshop and Tutorial in GeoAI Advancements and Applications"** from 14-15th July, 2025 organized by Research Center for Artificial Intelligence in Geomatics at the Hong Kong Polytechnic University.
2. Attended four-day workshop on "**GEO Symposium and ODOK Workshop**" from 23-26th Sept, 2024 organized by Group of Earth Observation at Hangzhou International Innovation Institute of Beihang University in Hangzhou, China.
3. Attended two days international conference on "**Nature-based Solutions for Urban Infrastructure**" from 25-26th April, 2024 organized by the Hong Kong Polytechnic University.
4. Participated in "**Workshop on GeoAI and Big data for Urban, Environment, and Sustainability**" from 9-10th Dec, 2023, organized by JC STEM Lab of Earth Observations and Research Center for Artificial Intelligence in Geomatics at the Hong Kong Polytechnic University.
5. Attended and presented a poster in "**ISPRS Geospatial Week 2023**" from 2-7th Sept, 2023, organized by Arab Academy for Science Technology & Maritime Transport at Cairo, Egypt.
6. Attended and presented a poster in "**Research Excellence for Societal Impacts**" from 8-11th May 2023, organised by PolyU Academy for Interdisciplinary Research (PAIR) at the Hong Kong Polytechnic University.
7. Participated in "**Advanced Urban Remote Sensing Workshop**" help from 3-4th Dec, 2022, organized by the University of Hong Kong at Hong Kong.
8. Successively completed the five days training program on "**Space-based information for ecosystem-based disaster reduction (Eco-DRR)**", between 25-29th November, 2019, jointly organized by United nations office for outer space affairs & Centre of excellence in disaster mitigation and management, IIT Roorkee, India.
9. Successively completed the one-week short course on "**Data Analytics with Python**", between 18-22th June, 2019, organized by Electronics & ICT Academy, IIT Roorkee.
10. Participated in "**Asia Pacific climate week**" held from 2-6th Sept, 2019, organized by United Nation at **Bangkok, Thailand**.
11. Presented paper in national conference *"Role of Remote Sensing and GIS in Natural Hazard Assessment for Sustainable Development: Present Scenario and Future Perspective"* in **Lucknow, India** in 2019.
12. Participated in **"IIRS User Interaction Meet 2018 (IUIM-2018)"** held on 27th Feb, 2018, at IIRS, Dehradun, India.
13. Successively completed the GIAN workshop on "**Urban Stormwater Management: Science, Policy and Engineering Training for SMART CITIES**", 27th Feb- 6th, Mar 2017, organized by Department of Hydrology, Indian Institute of Technology, Roorkee.
14. Participated in **"Three Days Training Programme on Geospatial Technologies**", 18-20th Feb, 2017, conducted by Department of Environmental Science, Central University of Rajasthan.

Peer Review Summary

Selected top 10 journal reviewers: Sustainable Cities and Society, Urban Climate, Remote Sensing of Environment, GIScience & Remote Sensing, IEEE Access, Habitat International, Building and Environment Atmospheric Research, Remote Sensing, Scientific Report

Media Coverage

1. Research findings from "Comparing existing heat wave indices in identifying dangerous heat wave outdoor conditions" Nexus, 2024 has been covered by various media worldwide including EurekaAlert, AAAS (Link), Science Magazine (Link), ScienceDaily (Link), The Telegraph (Link), The Economic Times (Link)
2. *70% Ahmedabad set to burn at 45°C by 2030* – covered by **The Times of India** on March 21, 2022 (Link)
3. *70% Ahmedabad set to burn at 45°C by 2030* – covered by **Ahmedabad News** on March 21, 2022 (Link)
4. *70% Ahmedabad set to burn at 45°C by 2030 | Ahmedabad News* – covered by **Bharat Times** on March 21, 2022 (Link)
5. *70% Ahmedabad set to burn at 45°C by 2030 | Ahmedabad Information* – covered by **The Times of Truth** on March 21, 2022 (Link)
6. Research output contributed in *"Ahmedabad, India Scaling Up with Contiguous Replication of Town Planning Schemes"* policy document published by Madhu Bharti and Shagun Mehrotra, **World Bank Group**, 2020. (Link)
7. *Land surface is getting warmer* – covered by **The Times of India** on Dec 23, 2019 (Link)

Technical Skills

Programming Skills: Python, MATLAB
Simulations and Software: Microsoft Office (Words, Excel, PowerPoint), Arc GIS, ERDAS IMAGINE, Q GIS, SAGA GIS, ENVI, Global Mapper, CorelDraw X7, Origin.

Member – Professional Body

1. Indian Society of Remote Sensing (ISRS), Dehradun, Life Member, **L-5002**
2. Indian Society of Geomatics (ISG), Ahmedabad, Life Member, **ISG-L-2238**
3. Indian Science Congress Association (ISCA), Kolkata, Life Member, **L-39591**

Personal Details

Date of Birth	04 May 1992
Nationality	Indian
Permanent Address	Ward No: 02, Bishunpur Lakhansen, Rampur Meena Dist: Muzaffarpur, Bihar-843127, India
Present and Correspondence Address	Department of Land Surveying and Geo-Informatics The Hong Kong Polytechnic University ZN418, 4/F, Block Z, 181 Chatham Road South, Hung Hom, Kowloon, Hong Kong

Editor name : Abraham R. Matamanda
Affiliation : Department of Urban and Regional Planning
Institution : University of the Free State
Address : 205 Nelson Mandela Drive, Park West
Postal code : 9301
City : Bloemfontein
Country : South Africa
Telephone number : +27 639557097
E-mail : matamandaa@gmail.com
ORCID ID : https://orcid.org/0000-0001-5260-5560

Abraham R. Matamanda (Ph.D. Urban and Regional Planning) is an NRF Y2-rated Urban and Regional Planner. Abraham lectures at the University of Free State (UFS) and is also the Deputy-Chair of the SARChI Chair on City Region Economies in the Faculty of Economic and Management Sciences at UFS. Currently, he serves as the co-editor of the *Town Planning Journal* published by UFS and serves on the academic editorial board of *Plos Water Journal*. Abraham is also the current Chairperson of the Free State Chapter of the South African Planning Institute (SAPI). He is a fellow of the Department of Higher Education and Training (DHET) Future Professoriate Programme Phase 1, third Cohort. Abraham is the South African PI for a global collaborative research project exploring how children and young people from monetary poor households have adapted to the COVID-19 pandemic focusing on food, education and play/leisure. His research focuses on urban governance and planning, climate change adaptation and resilience, informal Global South urbanism, urban land markets and housing studies. Abraham has authored and edited several books published with Springer Nature including:

- *Urban Geography in Postcolonial Zimbabwe: Paradigms and Perspectives for Sustainable Urban Planning and Governance*;
- *Housing and Technology: Special Focus on Zimbabwe*;
- *Urban Infrastructure in Zimbabwe: Departures, Divergences and Convergences*; and
- *Secondary Cities and Local Governance in Southern Africa.*

CURRICULUM VITAE FOR ABRAHAM R. MATAMANDA (Y2 NRF Rated Scholar & Fellow of the DHET Future Professoriate Programme Cohort 3 Phase 1)

Age: 37 (2023/2024)
Member of the Society of South African Geographers (Membership No. 19/21)

Member of the Regional Studies Association (Membership Number 21616)

Chairperson of the South African Planning Institute (SAPI) Free State Chapter (Jan 2024 – Present)

Member of the Reviewer Board for ***Frontiers in Sustainable Cities Journal***

Member of the Editorial Board for ***Environmental Science and Sustainable Development*** (ESSD) Journal

Member of the Academic Editorial Board for ***Plos Water Journal*** (June 2023-present)

Co-Editor of the ***Town and Regional Planning Journal*** Published with the University of the Free State (1 September 2022 - present)

Sits on the Executive Committee of Kovsie Journals at UFS since 2022

Deputy Chair on SARCHi Chair on City Economy Regions in Economic Management Sciences at UFS since June 2024

Executive committee member of the SARCHI Chair on city regions at the University of the Free State (January 2022-present)

ORCID: https://orcid.org/0000-0001-5260-5560
Google Scholar: https://scholar.google.co.za/citations?user=wNoCT1MAAAAJ&hl=en
Telephone: +27 (0)639557097
Physical address: 205 Nelson Mandela Drive, PO Box 339 Bloemfontein, 9301, South Africa
E-mail: matamandaa@gmail.com or MatamandaRA@ufs.ac.za
Date of this CV: 23 APRIL 2024

Summary

- PhD in Urban and Regional Planning, University of the Free State (2019). MSc Social Ecology, University of Zimbabwe (2015). BSc Hons Urban & Regional Planning, University of Zimbabwe (2013)
- Y2 NRF Rated Scholar (January 2022-)
- Member of the UFS Future Professoriate Programme second cohort, since January 2022
- Member of the South African Department of Higher Education and Training Cohort 3 Future Professoriate Since February 2023
- Researcher focusing on urban food systems, urban planning and governance, informal urbanism, climate change adaptation and resilience and housing studies.

- Over ten years of experience in research. More than 40 published academic articles, including multiple articles in high-impact journals. Google Scholar H-Index of 13. Scopus H-index of 7. Five publications are in press or accepted, and three forthcoming edited books.
- I have been involved with 3 research and policy projects as PI, Co-I, consultant or sub-consultant, with a cumulative value of more than ZAR1 000, 000.
- Currently lead PI of a 3-year NRF-funded collaborative research project with team members from Brazil and the United Kingdom. From the South African side, the project value is R 499,818 for 2022 and 2024.
- Was part of the consultant team that developed the Zimbabwe National Wetlands Guidelines in 2018-2019, contracted by the Environmental Management Agency.
- Was part of the consultant team contracted by the UN-HABITAT Zimbabwe to develop a policy framework for resilience and safe urban food markets during the COVID-19 pandemic, between November 2021 to January 2022
- Successfully supervised 3 PhD students; 8 Master's two passing with a distinction and all passing with an average of 73%; currently supervising and co-supervising five Masters, eight PhDs and one Honours students.

Skills

I am proficient in various programs, including Excel and other MS Office programs. I have basic competency in SPSS and NVivo for qualitative data analysis.

QUALIFICATIONS

1. *Registered for a Ph.D. in Urban and Regional Planning at the University of the Free State, July 2016 –December 2018. Passed March 2019, graduated 28 June 2019.*

Thesis title: *Exploring emerging human settlement forms and urban dilemmas nexus: Insights and challenges from Hopley, Harare, Zimbabwe.*

2. Masters of Science Degree in Social Ecology (September 2013 – June 2015) University of Zimbabwe **Degree Class**: Merit

Dissertation: *Stakeholders' awareness and perceptions on ecosystem services: implications for sustainable wetland use and management in Harare, Zimbabwe.*

3. Bachelor of Science Hons Degree in Rural and Urban Planning (January 2010 – July 2013) University of Zimbabwe

Degree Class: 2.1
Dissertation: *Approaches to Inner-city revitalization: Case of the Kopje area, Harare.*

SHORT COURSES

1. August 2021 – December 2021

CREST Online Training Course for Supervisors of Doctoral Candidates at African Universities, Stellenbosch University, South Africa

2. January 2021 – December 2021

Engaged Leadership Programme – University of the Free State
Modules covered this far: Discovering my authentic self; Igniting my emotional agility; Igniting my abstract thinking; Building social intelligence to create a high-performance team; Building cultural intelligence through diversity management.

MENTORSHIP PROGRAMS

1. January – December 2021, Emerging Scholars Accelerator Programs, University of the Free State
2. January 2022 – to date, moved from the ESAP cohort to join the Future professoriate Programme, University of the Free State as from January 2022
3. Selected to be part of the Third cohort of the DHET Future Professors Programme to be hosted by Stellenbosch University

---

PROFESSIONAL EXPERIENCE

Associate Professor in Urban & Regional Planning: (Associate Professor in the Department of Urban and Regional Planning, *1st of July 2024 - Present*)

Senior Lecturer in Human Geography: Department of Geography, UFS (*January 2021 – Present)*

Duties: Lecturing, supervising postgraduate students,

Writing consultant: Center for Postgraduate Studies, University of the Free State (January 2022-Present)

Duties:

Postdoctoral fellow: Urban & Regional Planning, University of the Free State (*July 2019 – Dec 2020)*

Research Assistant, Department of Urban & Regional Planning, UFS (*February 2018-June 2019*)

Duties for the research assistance included marking assignments, organizing an international conference hosted at UFS, conducting research and publishing.

<u>Lecturing</u>:

- Spatial Economics (2023)
- Applied Urban Development and Spatial Transformation (2023-24)
- Urban Geography for Honours (2021-2024)
- Introduction to Urban Geography for undergraduate second years (2021-2024)
- Philosophy of Geography for Honours (2021-2024)

My teaching philosophy is premised on the notion that the classroom environment must be intriguing and provide a learning environment where students are encouraged to express their ideas and participate in their education experience. While encouraging individual responsibility for learning in the classroom, I also assume the role of a facilitator by providing explanations and guidance relating to the course materials and subject matter. I also believe students have to interact and be challenged by the intellectual discussion of the course material. In this regard, I also strive to promote individuality where each student should have the chance to share their own experiences, which is especially critical in Urban Geography and planning considering the diversity of the students and their different orientations and socialization which would make the individual experiences of the respective students a critical part of understanding issues from their lived experiences. I also desire to encourage students to use technical tools and techniques in undertaking their work. This will be premised on encouraging students to learn to think logically, learn problem-solving methods and techniques and improve their writing skills.

RESEARCH GRANTS AWARDED

I was involved in the following funded research projects:

1. ***Adaptations of young people in monetary-poor households for surviving and recovering from COVID-19 and associated lockdowns*** (https://panexyouth.com/)

Years: 2022-2024
Role: Lead Principal Investigator for South African team and co-investigator for the global team with partners from Brazil and the United Kingdom (https://www.transatlanticplatform.com/adaptations-of-young-people-in-monetary-poor-households-for-surviving-and-recovering-from-covid-19-and-associated-lockdowns/)
Funder: National Research Foundation

- Value: ZAR498,000

2. ***Exploring the Situated Political Ecology and Economy of Agricultural Land Policies in BRICS: A Case Study of China and South Africa***

Years: 2020-2024
Role: Co-investigator
Funder: National Institute for the Humanities and Social Sciences and Chinese Academy of Social Sciences
Value: ZAR500,000

3. ***Awarded a research Grant by the Regional Studies Association International to host an international conference on Regional Science in southern Africa.***

Year of award: July 2023
Role: Co-investigator
Funder: Regional Studies Association International
Value: £3, 800

4. ***NRF Rating Grant***

Year of Award: 2023
Role: Principal investigator
Funder: National Research Fund, South Africa
Value: ZAR50, 000

5. ***Sustainable development Goals and Urban Health***

Year of Award: 2023
Role: Principal investigator
Funder: Department of Research, University of the Free State
Value: ZAR50, 000

6. ***Informal Rental Housing in Southern Africa***

Year of Award: 2023
Role: Principal investigator
Funder: SarChi Chair on City Regions, University of the Free State
Value: ZAR140, 000

POSTGRADUATE STUDENT SUPERVISION

HONOURS SUPERVISION			
STUDENT	**TITLE**	**SUPERVISOR**	**STATUS**
Palesa B Ngo	Analyzing the evolution of informal settlements: insights from Caleb Motshabi, Bloemfontein	**AR Matamanda**	Graduated April 2022 and passed with Distinction
Tshepiso E Tseladimitloa	An investigation of water pollution drivers in urban streams and strategies employed to address this pollution: a case study of Bloemspruit, Bloemfontein	**AR Matamanda** DR. E Rudolph	Graduated April 2022
Dr. Olugbenga A Ige	A systematic and meta-analytic review of homelessness in South Africa for 1994-2020	Ms E Kruger **AR Matamanda**	Graduated April 2022
Nthabiseng Erica	Alternative forms of urban agriculture in poor neighbourhoods: Case of Ekangala, Bronkhorstspruit, Gauteng, South Africa	**AR Matamanda**	(Jan- Dec 2022) Passed with Distinction
Theo Khutxo	Impacts of COVID-19 on migrants in the informal economy of Bloemfontein, South Africa	**AR Matamanda** DR JI Bhanye	(Jan – December 2022) Passed with Distinction
Xolani Makhubo	Safety and Security in a middle-income suburb: Case study of Phase 2 Bloemanda, Bloemfontein	Mrs T Mehlomakhulu **AR MATAMANDA**	(Jan – December 2022)

MASTERS DISSERTATION STUDENTS SUPERVISED TO COMPLETION			
NAME	**TITLE**	**SUPERVISOR**	**STATUS**
1. Tokelo Regina Nche	Exploring peripheral settlements and public transport nexus: insights from Naledi Trust, Kroonstad, Free State, South Africa	**AR MATAMANDA**	Graduated 2021
2. Rets'epile Charity Kalaoane	Implications of land use planning on urban mobility: case analysis of the Main North 1 Road in Maseru, Lesotho	**AR MATAMANDA**	Graduated 2021
3. Conwell Beukes	Evaluating the impact of participative planning in the upgrading of informal settlements in Windhoek Namibia: A case study approach	**AR MATAMANDA**	Graduated 2021
4. Phinda Nogcinisa	Spatial Planning for Green Infrastructure: The Case of Mayfield Township, Makhanda (Grahamstown)	**AR MATAMANDA**	Passed with Distinction, Graduated April 2022
5. Malebona Makoetje	Exploring Spatial Planning In Marginalised Rural Communities: The Case Study Of Makhotlong, Lesotho	**AR MATAMANDA** Mrs Rouve Bingle	Graduated April 2022

6. Sibongile Moloise (June 2021-Nov 2023)	Investigating the role of Traditional Ecological Knowledge in managing and conserving ecosystem services among rural communities in Capricorn District Municipality, Limpopo, South Africa	**AR Matamanda**	Passed with Distinction, Graduated in April 2024
7. Moleboheng Pherane (Jan – Dec 2023)	Stakeholder Perceptions of the use and value of urban wetlands: Case of the Seven Dams Conservancy (SDC), Bloemfontein, South Africa.	**AR MATAMANDA** DR E RUDOLPH	Graduated in April 2024
8. Tshifihwa Mudau (Feb 2022- Nov 2023)	Challenges and possible solutions for managing Water Infrastructure Assets to improve service delivery: A case study of Rustenburg Local Municipality	DR M DUNN **AR Matamanda**	Graduated in April 2024

ONGOING RESEARCH MASTERS SUPERVISION		
NAME	**TITLE**	**SUPERVISOR**
1. Chipo Marecha (January 2021-)	Investigating the impacts of mixed-use development on the liveability of the city: Case of Nkulumane, Bulawayo	**AR Matamanda** Prof V Nel
2. Palesa Ngo (February 2022-)	The aftermath of colonialism: Analysing the post-colonial urban expansion in Bloemfontein	**AR Matamanda** Prof V Nel
3. Erica Mashanye (February 2023 -)	Exploring Nature Based solutions for sustainable urban agriculture in Bloemfontein, South Africa	**AR Matamanda**
4. Maria Dawa Gazuwa	The environmental and socio-economic impacts of iron ore mining in the Gamagara Local Municipality, South Africa: Unpacking local stakeholders' perspectives	**AR MATAMANDA**
5. Bopane Lephoto	The Rise Of Informal Settlements In Lesotho: A Case Study Of Koalabata, Berea	**AR MATAMANDA** DR K MOCWAGAE
6. Henry Gurajena	Examining Incremental Backyard Housing Developments In The Face Of Land Value Capture In Gaborone	Prof A Scheba **AR Matamanda** Prof L Marais

COMPLETED PHDS		
NAME	**TITLE OF THESIS**	**SUPERVISOR**
1. Patronellah E Hadebe	A comparative assessment of spatial planning of the City of Johannesburg Metropolitan Municipality and the City of Tshwane Metropolitan Municipality in Gauteng (***PhD in Geography, University of the Free State)***	Prof S Adelabu Prof T Okelo **AR MATAMANDA**
2. Natal Buthelezi (January 2021 – April 2023)	Morphological Characterisation of Informal Settlements and the Adaptation to Natural Disasters in the Face of Climate Change: A Case Study of Buffalo City, South Africa (***PhD in Urban & Regional Planning, University of the Free State)***	**AR MATAMANDA** Prof V Nel
3. Willoughby Zimunya (December 2019- December 2023)	Spatiality, Planning and Policy Framework in the Mapping of Urban Outcomes in Zimbabwe: The Case of Mutare (***PhD in Urban & Regional Planning, University of the Free State)***	**Matamanda AR** Prof. MM Campbell Prof I Chirisa

ONGOING PHD SUPERVISION		
NAME	**TITLE**	**SUPERVISOR**
1. Abongile Mgwele (February 2021 -)	Power and Political Dynamics in Spatial Governance through the lens of informal settlements: Challenges and Perspectives from Mthatha, South Africa (**PhD in Urban & Regional Planning, University of the Free State**)	**Matamanda AR** Prof MM Campbell Prof J Chakwizira
2. Busisiwe Nkonki (January 2021 -)	Assessing climate vulnerability for building community resilience and sustainability: A case of Green Farm, Port St. Johns	**AR Matamanda** DR. A Van Der Walt Prof A Adelabu
3. Halleluha Chirisa (January 2022 -)	The Spatial-Temporal Patterns of Epidemiological Incidences in the Urban Landscape in Zimbabwe: Analysis and Evidence from Greater Harare	**AR Matamanda** DR. A Van Der Walt DR M Mulondo
4. Siphokazi Rammile (August 2022 -)	Toponyms, identity and urban liveability in poor neigbhourhoods of South Africa (**PhD in Urban & Regional Planning, University of the Free State**)	**AR Matamanda**
5. Tapiwa Nyamukapa (January 2023-)	Nature, Processes And Outcomes Of Land Governance In Harare: A Case Study Of Caledonia Peri-Urban Settlement	**AR MATAMANDA** Prof L Marais DR M Napier
6. Leemisa S Matooane	Land use planning and urban food systems nexus in Maseru, Lesotho (**PhD in Urban & Regional Planning, University of the Free State**)	**AR Matamanda** DR JI Bhanye Prof V Nel
7. Tebello Putsoane	Spatial planning for ecologically-sensitive sites in Maseru, Lesotho (**PhD in Urban & Regional Planning, University of the Free State**)	**AR MATAMANDA** DR JI Bhanye
8. Ndumiso Mbuthuma	Micro-enterprises and economic risks associated with climate change-induced disasters in eThekwini Municipality	Prof A Scheba Prof L Marais **AR MATAMANDA**
9. Shirley Letono	Impacts Of Climate Change On Small- Scale Maize Production Farmers In The Peri-Urban Areas In Free State Province, South Africa	**AR MATAMANDA** Dr A Van Der Walt
10. Samson Mutarisi	Land-based financing for service delivery in Harare, Zimbabwe	**AR MATAMANDA** Prof I Chirisa DR T Maramura

POSTDOCTORAL SUPERVISION	
NAME	**TITLE OF RESEARCH**
DR Johannes I Bhanye (1 January 2022 – 30 June 2024)	Dynamics of peri-urban land governance in Zimbabwe
Dr Shamiso Mafuku (1 August 2024 - present)	Housing and Infrastructure Development in Southern Africa

ASSESSMENTS AND MODERATIONS

1. 2023 November was part of the technical team assigned to assess the programmes offered in the Faculty of Engineering at Eswatini University. My task was to assess the BSc Degree in Geography, Environmental Sciences and Planning as a subject matter expert.

2. Internal moderator for the University of Free State Honours and Masters in Urban and Regional Planning programme, for 2016.
3. 2022, external moderation for University of Johannersburg, Department of Town and Regional Planning. MSc module – Environmental Planning and GIS Application

EXTERNAL & INTERNAL ASSESSMENT OF POSTGRADUATE THESES

PhD Examination

1. March 2024 – University of Johannesburg, Department of Urban and Regional Planning
2. January 2024 – University of Johannesburg, Department of Urban and Regional Planning
3. July 2023 – Wisdom Sibanda, Department of Geography, University of the Free State, Qwaqwa Campus
4. May 2023 – Kowanai Mhlanga, Department of Geography, University of the Free State, Qwaqwa Campus
5. April 2023 – Internal examination of PhD thesis in the Department of Geography, University of the Free State
6. February 2022 – Examined Ph.D. thesis internally in the Department of Geography, University of the Free State, South Africa
7. February 2022 – External examiner for a Ph.D. thesis from Rhodes University, South Africa
8. February 2020 - Examined Ph.D. thesis for Central University of Technology, Department of Civil Engineering

Masters Thesis Examination

1. February 2024 – University of Johannesburg, Department of Urban & Regional Planning
2. February 2024 – University of Mpumalanga, MSc in Environmental Management Sciences
3. May 2023 – University of Zululand, MSC thesis in Department of Geography
4. March 2023 – Internal examination of MSc thesis for Department of Geography, University of the Free State
5. December 2021 – External examiner for a Masters Thesis for the Department of Architecture and Spatial Planning, Namibia University of Science and Technology, Namibia
6. December 2021 – Examined 2 Masters thesis for the Department of Urban & Regional Planning, University of the Free State, South Africa
7. November 2021 – Examined Masters thesis internally in the Department of Geography, University of the Free State, South Africa
8. February 2021 – External examiner for Masters Dissertation, Department of Urban and Regional Planning, University of Johannesburg
9. January 2020; January 2021 - Internal assessor for Masters mini-dissertations, Department of Urban and Regional Planning, University of the Free State
10. July 2020 – Examined Masters thesis for Central University of Technology, Department of Civil Engineering

Consultant

Development of the Zimbabwe National Wetlands Utilisation Guidelines
Year: 2018-2019
Role: Sub-consultant
Client: Environmental Management Agency, Zimbabwe
This project was on undertaking a situational analysis of the wetlands in Zimbabwe in the rural and urban areas, focusing on identifying what was and was not working concerning achieving the national goals of preserving the wetland ecosystems in the country. The project comprised of a team of different experts that resulted in a sectoral and interdisciplinary approach in preparing the guidelines. My role was to coordinate the teams and disseminate information with the Principal investigator. I was also responsible in facilitating the fieldwork that involved focus group discussions with communities, in-depth interviews and document analysis.

Preparation of the Master Plan for Gokwe South Rural District Council
Year: 2024

Role: Lead Researcher
Client: Gokwe South Rural District Council

SELECTED RECENT PUBLICATIONS

Peer-Reviewed Research Articles

2024

1. **Matamanda AR**, Chirisa I, Mazanhi P, Toriro P, 2024. The Interface Between Politics, Ethics and Urban Planning: The Case of Land and Space Barons in Harare, Zimbabwe. *Environment and Planning C: Politics and Space.*
2. **Matamanda AR**. (2024). Genesis of Hopley Farm Informal settlement in independent Harare, Zimbabwe. *African Geographical Review*, 43(1), 144-157. Doi: 10.1080/19376812.2022.2117216
3. Ramogwebo T, Hlongwane JS, Bhanye J, **Matamanda AR,** 2024. Doing fieldwork among hard-to-reach populations: an account of local female researchers studying foreign migrants in Downtown Bloemfontein, South Africa. *International Journal of Qualitative Methods,* 23, 1-12
4. Kalaoane RC, **Matamanda AR,** Bhanye JI. (2024) Analysis of urban mobility along the Main North 1 Road in Maseru, Lesotho. *Discover Sustainability,* 5(40), 1-18. https://doi.org/10.1007/s43621-024-00226-1
5. Bhanye J, Kloeffel T, Beyers M, Mabaso M, Rajadurai K, Winklmaier, **Matamanda AR (**2024) Decentralised water-energy-food (WEF) systems in Africa: space analysis, least-cost modelling of sack farming and establishment of renewable energy technologies in the Diepsloot slums of Johannesburg, South Africa. *Discover Global Society*, 2(24), 1-17
6. Musosa L, Shekede MD, Gwitira I, Chirisa I, Tevera D, **Matamanda AR** (2024) Auditing the spatial and temporal changes in urban cropland in Harare Metropolitan Province, Zimbabwe. *African Geographical Review*, 43(1), 170-185. Doi: https://doi.org/10.1080/19376812.2022.2128834
7. Kalaoane RC, Musakwa W, Kibangou A, Gumbo, Musonda I, **Matamanda AR**. (2024). Bibliometric analysis of quality of service in public transportation: current and future trends. *Scientific Africa*.

2023

8. Bhanye J, Shayamunda RH, Mpahlo RI, **Matamanda AR**, Kachena L (2023). Land politics and settlers' responses to land tenure under threat in emerging peri-urban spaces in Zimbabwe. *Land Use Policy*, 135. https://doi.org/10.1016/j.landusepol.2023.106945
9. Moloise SD, **Matamanda AR,** Bhanye JI. (2023) Traditional ecological knowledge and practices for conservation and management of Savanna ecosystem services in Capricorn District, Limpopo, South Africa. *International Journal of Sustainable Development and World Ecology.* Doi: https://doi.org/10.1080/13504509.2023.2249856.
10. Leboto-Khetsi L, Mangara F, Dunn M, Manana K, **Matamanda AR**, Chirisa I. (2023) Off-campus student housing supply and utilization: perspectives of landlords and students in Chinhoyi, Zimbabwe. *South African Geographical Journal.* https://doi.org/10.1080/03736245.2023.2227150.
11. **Matamanda AR,** Kalaoane RC, Chakwizira J. (2023) "Leave us alone": 'Right to the city' of street vendors along Main North 1 Road, Maseru, Lesotho. *GeoJournal* 88: 4473-4491. https://doi.org/10.1007/s10708-023-10881-y

2022

12. **Matamanda AR,** Mafuku SH, Bhanye JI. (2022). The potential of Chinhoyi as a fast-growing secondary city in addressing urban challenges in Zimbabwe. *Journal of Asian and African Studies*. Doi: 10.1177/00219096221137660.
13. **Matamanda AR,** Nel V, Leboto-Khetsi L, Dunn M. (2022). Risk communication in an informal settlement during COVID-19: case of Dinaweng, Bloemfontein South Africa. *Urban Governance*. Doi: https://doi.org/10.1016/j.ugj.2022.10.002
14. **Matamanda AR**, Mphambukeli TN. (2022). Urban (in) security in an emerging human settlement: Perspectives from Hopley Farm Settlement, Harare, Zimbabwe. Frontiers in Sustainable Cities. 4:933869. doi: 10.3389/frsc.2022.933869

15. **Matamanda AR**, Mphambukeli TN, Chirisa I. (2022). Exploring water-gender-health nexus in human settlements: The burden on women and the girl-child. *Cities & Health*. Doi: 10.1080/23748834.2022.2136557.
16. **Matamanda AR**. (2022). Gatekeeper politics and urban planning research in the contested space of an emerging settlement: reflection on experiences in Hopley farm settlement, Harare. *International Journal of Social Research Methodology*. Doi: 10.1080/13645579.2022.2097393
17. **Matamanda AR,** Mandebvu-Chaora C, Rammile S (2022) The interplay of urban agriculture and spatial (in) justice in Harare, Zimbabwe. *Land Use Policy*. 105, 106029. Doi: https://doi.org/10.1016/j.landusepol.2022.106029
18. **Matamanda AR,** Dunn M, Nel V (2022). Broken bridges over troubled waters: COVID-19 and the urban poor residing in Dinaweng informal settlement, Bloemfontein, South Africa. *South African Geographical Journal*, 104(3): 309-327. Doi: 10.1080/03736245.2022.2028669
19. Chirisa I, Mutambisi T, Chivenge M, Mabaso E, **Matamanda AR,** Ncube R. (2022). The urban penalty of COVID-19 lockdowns across the globe: manifestations and lessons for Anglophone sub-Saharan Africa. *GeoJournal,* 87: 815–828. https://doi.org/10.1007/s10708-020-10281-6
20. Filho WL, Ternova L, Fayyaz MM, Abubakar IR, Kovaleva M, Donkor FK, Anuga SW, **Matamanda AR**, Djekic I, Umar IA, Olooto FM, Meirelles M, Nagy GJ, May J, May M, Ebhuoma E and Begum H. (2022). An analysis of climate change and health hazards: results from an international study. *International Journal of Climate Change Strategies and Management*. https://doi-org.ufs.idm.oclc.org/10.1108/IJCCSM-08-2021-0090

2021

21. **Matamanda AR** (2021). Mugabe's urban legacy: a postcolonial perspective on urban development in Harare, Zimbabwe. *Journal of Asian and African Studies*. 56(4) 804–817. doi:.org/10.1177/0021909620943620
22. **Matamanda AR**, Chirisa I, Rammile S. (2021). Elitist Domination and Its Import: Survey of Four Decades of Perpetuation of Inequities in Zimbabwe, *Politikon*, 48(3): 450-467. DOI: 0.1080/02589346.2021.1952736
23. **Matamanda AR,** Chirisa I, Leboto L, Dunn M, Kwangwama AN. (2021). The Planning Profession Questioned: Evidence from the Role and Practice of Planners in Zimbabwe. *Journal of Planning Education and Research*, 0739456X211028295.
24. **Matamanda AR,** Chirisa I, Kanonhuhwa TN, Mhlanga DD. (2021). Politics and the legacy of street renaming in postcolonial Zimbabwe. *Journal of Cultural Geography*, 38(3), 336-353, DOI: 10.1080/08873631.2021.1927348
25. Mashingaidze N, Chirisa I, Mutambisi T, **Matamanda AR. (**2021). Paradoxes surrounding carbon credits and local area development: the case of Mbire District, Zimbabwe. *Local Environment*, 26(10), 1175-1185. DOI: 10.1080/13549839.2021.1967899
26. Ibrahim KO, Gomo M, Oke SA, **Matamanda AR.** (2021). Hand-dug wells in rural areas of developing countries. *Sustainable Water Resources Management*, 7(3), 1-7.

2020

27. **Matamanda AR (**2020). Living In An Emerging Settlement: The Story Of Hopley Farm Settlement, Harare Zimbabwe. *Urban Forum*, 31(4): 473-487. Doi: 10.1007/S12132-020-09394-5
28. **Matamanda AR**, Chinozvina QL. (2020). Driving forces of citizen participation in urban development practice in Harare, Zimbabwe. *Land Use Policy*, *99*, 105090. https://doi.org/10.1016/j.landusepol.2020.105090
29. **Matamanda RA** (2020) Battling the Informal Settlement Challenge through Sustainable City Framework: Experiences and Lessons from Harare, Zimbabwe. *Development Southern Africa*, 37(2), 217-231. https://doi-org.ufs.idm.oclc.org/10.1080/0376835X.2019.1572495
30. **Matamanda AR,** Mphambukeli TN, Chirisa I (2020). Applied Systems Analysis In Water Access For Emerging Human Settlements: A Case Study Of Hopley Farm, Harare, Zimbabwe. *Urban Water Journal*, 17(9), 763-773. Doi: 1811882 DOI:10.1080/1573062X.2020.1811882.
31. **Matamanda AR,** Chirisa I, Dzvimbo MA, Chinozvina Q. (2020). The political economy of Zimbabwean Urban informality since 2000 – A contemporary governance dilemma. *Development Southern Africa*. 37(4), 694-707, DOI: 10.1080/0376835X.2019.1698410.
32. Chirisa I, **Matamanda AR**, Mazanhi P. (2020). Resisting, frustrating or embracing the urban agenda: Chieftaincies in Southern Africa examined constitutionally and statutorily. *Land Use Policy* 95(2020). Doi: https://doi.org/10.1016/j.landusepol.2020.104618.

2019-2014

33. Chirisa I, **Matamanda, AR**. (2019). Forces Shaping Urban Morphology in Southern Africa Today: A reflection on the interplay among people, practice and policy. *Journal of Urbanism: International Research on Placemaking and Urban Sustainability*, 12(3): 354-372. DOI:10.1080/17549175.2019.1626262.
34. Chirisa I, Chivambe A, Mukarwi L, **Matamanda AR**. (2019). Spatializing Class Struggles of Post-Colonialism and Post Modernism: Of the Messing and Cleaning in the City of Harare. *Critical Planning Journal*, 24: 134-149
35. **Matamanda AR**, Chirisa I, Mangara F, Dzvimbo MA. (2019). The ecological politics surrounding the downsizing and downgrading of public park: A reflection on the history of change of the Harare Gardens in Zimbabwe. *Case Studies in the Environment,* 3(1): 1-7.
36. **Matamanda AR,** Chirisa I, Mukamuri BB, Kaduwo P, Mhlanga M. (2018). Harare's deteriorating wetlands: why sound policies and legislations are not enough. *Case Studies in the Environment*, 2(1): 1-11.
37. Chirisa I, Simba-Mlambo E, **Matamanda AR**. (2018). Leadership, governance and the political economy of the African urban systems: searching for a robust policy framework. *Journal of Public Administration and Development Alternatives*, 3(2): 1-14.
38. **Matamanda AR**, Dzvimbo MA, Kadebu RT. (2017). Climate change and infrastructure planning: Implications for sustainable urban management in Harare, Zimbabwe. *Journal of Public Administration and Development Alternatives*, 2(1), 92-108
39. Chirisa I, **Matamanda AR.** (2016). Addressing Urban Poverty in Africa in the Post-2015 Period: Perspectives for Adequate and Sustainable Housing. *Journal of Settlements and Spatial Planning*, 7(1): 79 – 87.
40. Chirisa I, Dirwai SO, **Matamanda A**. (2015). A Review on the Systems Approach for Cities in Africa in Solving the Environmental Problematique. *Future, Cities and Environment*, 1:11.
41. **Matamanda A**, Chirisa I (2014). Ecological planning as an Inner-city revitalization planning agenda for Harare, Zimbabwe. *City, Territory and Architecture,* 1:14. https://doi.org/10.1186/s40410-014-0014-1.
42. Chirisa I, Bandauko E, **Matamanda AR,** Mandisvika G. (2014). Decentralised domestic wastewater systems in developing countries: the case study of Harare (Zimbabwe). *Applied Water Science*, 7(3), 1069-1078.
43. Chirisa I**, Matamanda AR**, Bandauko E. (2014). Ruralised urban areas vis-à-vis urbanised rural areas in Zimbabwe: implications for spatial planning. *Regional Development Dialogue*, 35: 65-80.

Selected Articles under Review

1. **Matamanda AR,** Scheba A, Matooane LS, Ayob F, Lephoto B, Turok I, Malaene Matters: Small scale rental housing in Lesotho, Maseru, *Habitat International*
2. Bhanye J., Dzingirai V., **Matamanda A.**, Emerging land inheritance dynamics among migrants in foreign peri-urban spaces. An ethnographic study among Malawian diaspora in Zimbabwe. *Land Use Policy*
3. Bhanye J., **Matamanda A.,** Bandauko E., Kohima JM. Implications of land reform programs on urban land markets: A case study of Harare, Zimbabwe. *GeoJournal*

Book Chapters

1. Chirisa I, Bobo T, Mpahlo R, **Matamanda AR**. 2024. History, Practices, challenges and opportunities of urban planning in Zimbabwe. In PB Cobbinah and E Gaisie, *Reimagining urban planning in Africa*, Cambridge University Press: Cambridge.

2. **Matamanda AR**, Chanza N, Nyamugadza E, Chinozvina QL. 2024.Land use planning for climate change adaptation in secondary cities: Insights from Chinhoyi, Zimbabwe.

3. Mugumbate H, Selemani W, **Matamanda AR.** 2024. Governing urban food systems in secondary cities: Contestations and struggles from Chitungwiza, Zimbabwe.

4. Nel V, Hugo M, **Matamanda AR**, Oranje M. 2024. Sasolburg: A town built around the chemical industry suffering under poor governance and its environmental legacy.

5. Chakwizira J, **Matamanda AR**, Nel V, Chatiza K. 2024. The future of secondary cities in (Southern) Africa: concluding remarks and research agenda.

6. Kalaoane RC, **Matamanda AR.** 2024. Women in the urban informal economy and pathways towards inclusive cities. In A Lauren, JR Bryson, A Ersoy, L Reardon, Pandemic Recovery? Reframing and rescaling societal challenges. Edward Elgar, 78-93

7. Bhanye J, Hugo M, **Matamanda A** & Mocwagae K (2023). Reimagining airport travellers' experiences in the post-COVID-19 pandemic era: current reality and practical solutions. In the 2nd Book in ICCCASU Springer Book Series, "Making Sense of Planning and Development for the post-Pandemic Cities."

8. Karakadzai T, **Matamanda AR,** Chirisa I. (2022). Rainwater harvesting for water security in informal settlements: techniques, practices, and options. In Leal Filho W, Azul AM, Doni F, Salvia AL, In: *The Palgrave Encyclopedia of Urban and Regional Futures*. Palgrave Macmillan, Cham. https://doi.org/10.1007/978-3-030-51812-7_84-1

9. **Matamanda AR**, Kohima JM, Nel V, Chirisa I, 2022. Climate Change Adaptation And Planning Education In Southern Africa. In Gebregiorgis GA, Greiving S, Namangaya AH, Kombe W, *Planning cities in Africa*. Springer Nature, Cham. Doi: https://doi.org/10.1007/978-3-031-06550-7_6

11. Matamanda AR, Nel V, Dunn M, Mgwele A, Rammile S, Leboto-Khetsi L, Kohima JM, Ngo PB, (2022) Sustainable development goals and urban health challenges in informal settlements of Mangaung Municipality, South Africa. In WL Filho, Azul AM, Doni F, Salvia AL, *Handbook of Sustainability Science in the Future: Policies, Technologies and Education by 2050.* Springer Nature, Cham. Doi: https://doi.org/10.1007/978-3-030-68074-9

12. **Matamanda AR**, Nel V, Chanza N, Leboto-Khetsi L, Mangara F, Paradza P, 2022. The political economy of COVID-19 pandemic: lessons learned from the responses of local government in sub-Saharan Africa. In C. Nunes Silva (eds.) *Local Government and the COVID-19 pandemic*. Cham: Springer Nature, pp. 103-128

13. **Matamanda AR,** 2022. Are we there yet? Prospects and barriers to implementing smart city initiatives in Harare, Zimbabwe. In Chavunduka C, *Sustainable and smart spatial planning in Africa: Case studies and solutions*. CRC Press, New York

14. Dzvimbo MA, **Matamanda AR,** Mawonde A, Magijani F, 2022. Exploring climate change impacts on smallholder farmers in Mhondoro-Ngezi District, Zimbabwe. In, Filho WL, Djekic I, Smetana S, Kovaleva M, *Handbook of climate change across the food supply chain*, pp. 381-402. Cham: Springer Nature

15. **Matamanda AR,** Nel V, Chirisa I, 2021. Introduction to the Urban Geography Scape of Zimbabwe. In, **Matamanda AR**, Verna N, Chirisa I (2021) (eds). *Urban Geography in Postcolonial Zimbabwe: The Paradigms and Perspectives for Sustainable Urban Planning and Governance*. Cham: Springer Nature. pp. 1-15

16. Nel V, **Matamanda AR,** Chirisa I, 2021. Social Justice in Spatial Governance. In, **Matamanda AR**, Verna N, Chirisa I (2021) (eds). *Urban Geography in Postcolonial Zimbabwe: The Paradigms and Perspectives for Sustainable Urban Planning and Governance*. Cham: Springer Nature. pp. 17-35

17. **Matamanda AR**, Dube T, Campbell M, 2021. Studentification and Its Interplay on Urban Form and Urban Policy: Reflection from Bulawayo, Zimbabwe. In, **Matamanda AR**, Verna N, Chirisa I (eds). *Urban Geography in Postcolonial Zimbabwe: The Paradigms and Perspectives for Sustainable Urban Planning and Governance*. Cham: Springer Nature. pp. 57-75

18. **Matamanda AR,** Nel V, 2021. (Un)Healthy Cities: Reflections on Urban Public Health in Poor Neighbourhoods. In, **Matamanda AR**, Verna N, Chirisa I (2021) (eds). *Urban Geography in Postcolonial Zimbabwe: The Paradigms and Perspectives for Sustainable Urban Planning and Governance*. Cham: Springer Nature. pp. 93-119

19. Nel V, **Matamanda AR,** Chirisa I, 2021. Putting Together the Broken Pieces: Rewiring the Urban Geography in Zimbabwe. In, **Matamanda AR**, Verna N, Chirisa I (2021) (eds). *Urban Geography in Postcolonial Zimbabwe: The Paradigms and Perspectives for Sustainable Urban Planning and Governance*. Cham: Springer Nature. pp. 307-317

20. **Matamanda AR**, Nel V, Leboto L, 2021. Ecological Risks of the Postcolonial City: Experiences from Harare, Zimbabwe. In, **Matamanda AR**, Verna N, Chirisa I (2021) (eds). *Urban Geography in Postcolonial Zimbabwe: The Paradigms and Perspectives for Sustainable Urban Planning and Governance*. Cham: Springer Nature. pp. 141-161.

21. Mutambisi T, Chanza N, **Matamanda AR,** Ncube R, Chirisa I, 2021. Climate Change Adaptation in Southern Africa: Universalistic Science or Indigenous Knowledge or Hybrid. In: Leal Filho W., Oguge N., Ayal D., Adeleke L., da Silva I. (eds) *African Handbook of Climate Change Adaptation*. Springer, Cham. https://doi.org/10.1007/978-3-030-42091-8_8-1

22. Mandaza-Tsoriyo W, Usingarawe G, **Matamanda AR,** Chirisa I, 2021. Making of Smart and Intelligent Cities Challenges and Options for Africa. In R. Brears (ed.), *The Palgrave Encyclopedia of Urban and Regional Futures*. Cham: Springer. https://doi.org/10.1007/978-3-030-51812-7_75-1

23. Chirisa I, **Matamanda AR**, Maphosa E, Ncube R, 2021. 'Land Barons' as the Elephant in the Room: Planning and the Management of Urban Space in Chitungwiza and Harare. In Chirisa I and Chigudu A, *Resilience and Sustainability in Urban Africa Context, Facets and Alternatives in Zimbabwe* pp. 171-188, Cham: Springer
24. Chirisa I, Mazanhi P, **Matamanda AR**, 2021. Computational urban planning. Gaps, Challenges, and Strides in Africa. In: Brears R (eds.) The Palgrave Encyclopedia of Urban and Regional Futures. Cham: Palgrave Macmillan https://doi.org/10.1007/978-3-030-51812-7_73-1
25. **Matamanda AR,** Nel V, 2020. "Sustainable Urbanization in Africa: The Critical Enablers and Disablers." In *Sustainable Cities and Communities. Encyclopedia of the UN Sustainable Development Goals* edited by Leal, F. W., Azul, A., Brandli, L., Özuyar, P., Wall, T. pp. 738-751, Cham: Springer.
26. **Matamanda AR,** Mafuku SH, Mangara F, 2020. "Informal Settlement Upgrading Strategies: The Zimbabwean Experience." In *Sustainable Cities and Communities. Encyclopedia of the UN Sustainable Development Goals* edited by Leal, F. W., Azul, A., Brandli, L., Özuyar, P., Wall, T. pp. 316-327, Cham: Springer.
27. Chirisa I, Mukarwi L, **Matamanda AR**, 2019. Resilience, Resistance and Risings in African Cities: Cases of Civil Responses to the Non-Responsiveness and Unilateralism by the State to Urban Challenges. Chapter 20 in IO Albert and T Lawanson, Urban Crisis and Management in Africa: A Festschrift for Prof. Akin Mabogunje. Texas: Pan-African University Press. pp. 335-349
28. Chirisa I, Mukarwi L, **Matamanda AR**, 2018. Sustainability in Africa: The service delivery issues of Zimbabwe. In: S. J. Garren (Ed.), *The Palgrave handbook of sustainability* (pp. 699–715). Cham: Palgrave Macmillan.
29. Chirisa I, Mukarwi L, **Matamanda AR**, 2018. Sustainability in Africa: The Service Delivery Issues of Zimbabwe. Chapter 37 in Brinkmann, R and Garren, S (ed.) *The Palgrave Handbook of Sustainability: Case Studies and their Practical Solutions*. Palgrave Macmillan: New York.
30. Chirisa I, **Matamanda AR,** Mutambwa J, 2018. Africa's Dilemmas in Climate Change Communication: Universalistic Science Versus Indigenous Technical Knowledge. In: Leal Filho W., Manolas E., Azul A., Azeiteiro U., McGhie H. (eds) *Handbook of Climate Change Communication: Vol. 1. Climate Change Management*. Springer, Cham

Technical Reports

Andres, L., Moawad, P., Kraftl, P., Denoon-Stevens, S., Marais, L., **Matamanda, A.,** Bizzotto, L., Giatti, L. (2023), The Impact of COVID-19 on Education, Food & Play-Leisure and Related Adaptations for Children and Young People: International and National Overviews. PANEX-Youth WP2 Full Report, London; https://panexyouth.com/

Andres, L., Moawad, P., Kraftl, P., Denoon-Stevens, S., Marais, L., **Matamanda, A.,** Bizzotto, L., Giatti, L. (2023), The Impact of COVID-19 on Education, Food & Play-Leisure and Related Adaptations for Children and Young People: International Overview. PANEX-Youth WP2 Short Report, London; https://panexyouth.com/

Conference Proceedings

Matamanda AR, Mphambukeli T. (2019). *Planning for Co-production through Innovative Technology in Informal and 'City-Edge' Human Settlements in the Global South: The Case of Hopley Farm, Harare, Zimbabwe. The Proceedings of the* South Africa Sweden University Forum 2019 Symposium. Nelson Mandela University, Port Elizabeth

Skiyi V, **Matamanda AR**, Chirisa I. (2016). *Constraints and opportunities in integrating wetland ecosystem services in urban land use planning, Harare, Zimbabwe.* The Proceedings of the 2016 ILASA Conference 55-63, Institute for Landscape Architecture in South Africa (ILASA). University of Pretoria (UP), Pretoria

Books: Monographs

1. Bhanye J, Mangara F, **Matamanda AR,** Kachena L (2023) *Covid-19 Lockdowns And The Urban Poor In Harare, Zimbabwe: Emerging Perspectives And The Morphing Of A Sustainable Urban Future*. Springer Nature: Cham
2. **Matamanda AR**, Chirisa I, Rammile S, Marais M (2022*) Housing and technology: Special focus on Zimbabwe.* Springer Nature, Cham. https://doi.org/10.1007/978-3-031-09098-1

3. Chirisa I, **Matamanda AR**. 2018. *Housing and the Environment: Emerging Paradigms for Urban Zimbabwe*. University of Zimbabwe Publications, Harare.

Edited Books

1. **Matamanda AR**, Chakwizira J, Chatiza K, Nel V (eds.) (2024) *Secondary cities and local governance in Southern Africa*. Springer Nature, Cham. https://link.springer.com/book/10.1007/978-3-031-49857-2.
2. Alberti F, Gallo P, **Matamanda AR**, Strauss EJ (eds.) (2024) Resilient planning and design for sustainable cities. Springer Nature: Cham. https://doi.org/10.1007/978-3-031-47794-2
3. Chirisa I, **Matamanda AR** (eds.) (2023). Urban Infrastructure in Zimbabwe: Departures, *divergences and convergences*. Springer Nature, Cham
4. Alberti F, **Matamanda AR,** He B, Galderisi A, Smol M, Gallo P (eds.) (2023) *Urban and Transit Planning (3rd Edition) – city planning: Urbanization and circular development*. Springer Nature, Cham. https://doi.org/10.1007/978-3-031-20995-6
5. **Matamanda AR**, Verna N, Chirisa I (2021) (eds). *Urban Geography in Postcolonial Zimbabwe: The Paradigms and Perspectives for Sustainable Urban Planning and Governance*. Cham: Springer Nature. https://link.springer.com/book/10.1007/978-3-030-71539-7

CONFERENCES

1. **Matamanda AR**, Selemani W, Mugumbate H (2023) Urban governance and food systems in secondary cities: contestations and struggles from Chitungwiza, Zimbabwe. *International Conference Paradigm Shifts in Local and Urban Governance*, IGU Commission Geography of Governance Annual Conference – 2023, Budapest, Hungary 4 - 6 September 2023
2. Hugo M, Bhanye J, **Matamanda AR** (2023) Urban planning and public transportation: a historical overview of South Africa. *International Conference Paradigm Shifts in Local and Urban Governance*, IGU Commission Geography of Governance Annual Conference – 2023, Budapest, Hungary 4 - 6 September 2023
3. Mgwele A, **Matamanda AR** (2023) 'Democracy spatial legacies': Implications on South African spatial governance and planning. *International Conference Paradigm Shifts in Local and Urban Governance*, IGU Commission Geography of Governance Annual Conference – 2023, Budapest, Hungary 4 - 6 September 2023
4. **Matamanda AR**, Dzvimbo MA, Paradza P (2023) Universities As Nodes For Regional Development? Insights From Chinhoyi University Of Technology, Zimbabwe. *Conference on Rethinking the Regional Development Opportunities and Challenges in Southern Africa.* 5-7 July 2023, University of Zimbabwe, Harare, Zimbabwe
5. Chionwa-Gaza M, **Matamanda AR** (2023) The place of Civil society organisations in community infrastructure and regional development. *Conference on Rethinking the Regional Development Opportunities and Challenges in Southern Africa.* 5-7 July 2023, University of Zimbabwe, Harare, Zimbabwe
6. Hugo M, Bhanye J, **Matamanda AR** (2023) Spatial inequalities in transport systems and implications for regional development. A South African perspective. *Conference on Rethinking the Regional Development Opportunities and Challenges in Southern Africa.* 5-7 July 2023, University of Zimbabwe, Harare, Zimbabwe
7. Pherane M, **Matamanda AR**, Rudolph E (2023) Urban wetlands and human environmental footprint and conflict: Case of the Seven Dams Conservancy, Bloemfontein. Global Change Conference, Bloemfontein
8. Rammile SN**, Matamanda AR** (2022) A place called home: understanding toponyms. 58th ISOCARP world planning congress "From wealthy to healthy cities", hosted by Brussels capital region on 3-6 October 2022, Brussels.
9. Buthelezi N, **Matamanda AR**, Nel V. (2022). Climate change disaster risk reduction and management systems in low-income areas: A case study of Buffalo City, South Africa. *International Geographical Union Commission on Geography of Governance (IGU CGoG) webinar 'Local Governance and Climate Emergency'.* https://sites.google.com/view/lgce2022/abstracts-powerpoints
10. Buthelezi N, **Matamanda AR**, Nel V (2022). Climate Change disaster risk reduction and management systems in vulnerable urban communities of South Africa. *International Geographical Union Commission on Geography of Governance (IGU CGoG): 'Local Governance in a time of global emergencies' Conference,* 29-31 August, Mexico City, Mexico. https://sites.google.com/view/geogov2022/abstracts

11. Buthelezi N, **Matamanda AR**, Nel V (2022). Climate change adaptation policies in Southern Africa and inequality disparity in urban informal settlements. *Development Days Conference: Urban infrastructure as key to achieving SDG 11: conditions, experiences and perspectives from the Global South,* 17-18 February 2022, Helsinki, Finland. https://www.swedev.dev/event/development-days-2022-infrastructures-technologies-and-vulnerabilities-in-global-development/
12. Ngo PB, **Matamanda AR,** Nel V, Owolabi S, Ayantokun AS (2022) Is segregation a thing of the past or is it here to stay: analysing the decolonisation process through the spatial transformation of the post-colonial city of Bloemfontein, South Africa. Society of South African Geographers Biennial Conference, 12-14 September 2022, University of Pretoria, Pretoria, South Africa
13. Moloise SD, **Matamanda AR (**2022). *Is there a place for traditional ecological knowledge in South Africa's legislation? Navigating Complex Pluriversal Relations: Indigeneity, Natural Resources Governance and Intercontinental Relations in the 21st century*. University of Pretoria, South Africa, 4-7 July 2022
14. Moloise SD, **Matamanda AR (**2022). *Factors affecting the use and value of Traditional Ecological Knowledge in Conservation and Management of Savanna Ecosystem among the rural communities of Capricorn District Municipality, Limpopo, South Africa.* 28th International Sustainable Development Research Society Conference, Stockholm, (Virtual Presentation), 15-17 June 2022.
15. **Matamanda AR**, Kohima JM, Nel V, Chirisa I (2021). *Climate Change Adaptation And Planning Education In Southern Africa*. Presented at the 5th AAPS-Conference Urban Africa in the 21st Century, Current Issues and Future Prospects of urban governance and planning, Virtual Conference, 18-20 November 2021
16. Buthelezi N**, Matamanda AR,** Nel V (2021) Climate change induced hazard impact awareness on urban informal settlement communities in South Africa. Presented at the Human Settlements and Housing African Regional Student Conference, University of Fort Hare, 27-29 October 2021
17. **Matamanda AR**, Nel V (2021) *The Public Transport and Land Use Planning Nexus: The Case Harare, Zimbabwe and Bloemfontein, South Africa.* Presented at IGC 34th International Geographical Congress, Geography: Bridging the Continents, Virtual Conference, 16-20 August 2021
18. **Matamanda AR**, Mphambukeli TN (2019) *Planning for Co-production through Innovative Technology in Informal and 'City-Edge' Human Settlements in the Global South: The Case of Hopley Farm, Harare, Zimbabwe.* Presented at the South Africa Sweden University Forum 2019 Symposium. Nelson Mandela University, Port Elizabeth, 6 May 2019.
19. **Matamanda AR** (2018) Exploring Water-Gender-Health Nexus in emerging Human settlements: The Burden on Women and Girl-Child in Hopley, Harare. Presented at the Water, Food and Health Nexus in BRICS Plus: Problems, Progress and Prospects BRICS-Plus International Conference, University of the Free State, Bloemfontein, South Africa. 2-5 August, 2018.
20. **Matamanda AR**, Dzvimbo MA (2016) *Urbanisation, Environmental Protection and the Sustainable Development Goals: Sifting Relevant Policies and Strategies for Harare*. Presented at the First Biennial Symposium on Human Settlements Planning, Development and Sustainability in Zimbabwe, University of Zimbabwe, Harare, 17-19 August 2016.
21. Chirisa I, Maphosa A, **Matamanda AR** (2016) *Terrain Unevenness as a Determinant of Rural-Urban Migration Patterns in Zimbabwe: The Past and the Present.* Presented at an International Conference on the Migration Phenomenon and its Impact on Sustainable Development in Africa and Beyond: Emphasizing African Unity, Peace and Prosperity at University of Zimbabwe on 29 July 2016.
22. **Matamanda AR** (2015) *Battling informality and housing challenges through sustainable cities framework: Lessons from experiences in Harare, Zimbabwe***.** Presented at Conference for Planning Students and Young Graduates at the University of Johannesburg from 30 October 2015 – 1 November 2015.

PEER REVIEW

I have conducted blind peer-review for the following journals:

Journal	Publisher	Date of Peer review (s)
City Architecture and Territory	Springer	May 2024
Journal of Planning Education and Research	SAGE	June 2023
Planning Practice and Research	Taylor & Francis	July 2023
Plos Global Public Health	Plos One	July 2023
Politikon	Taylor & Francis	April 2023

Antipode	Wiley	November 2022
Journal of Urban Research and Practice	Taylor & Francis	October 2022
Open House International	Emerald Publishing	May 2022
Urban Governance	Elsevier	March 2022
Journal of Housing and the Built Environment	Springer Nature	March 2022
International Journal of Urban and Regional Research	Wiley	March 2022; September 2021
African Geographical Review	Taylor & Francis	February 2022
Cities & Health	Taylor & Francis	May-June 2020, reviewed 4 manuscripts for a special issue on COVID-19; March 2022; February 2024
Journal of Asian and African Studies	Sage	December 2021
Cogent Social Sciences	Taylor & Francis	July 2023; August 2021; September 2021
Urban Forum	Springer	June 2021; May 2024
International Social Science Journal	Wiley	February 2021
CODESRIA journal Africa Development	CODESRIA	January 2021
Journal of Energy in Southern Africa		October 2020
GeoJournal	Springer	July 2020; October 2020; May 2023
Town and Regional Planning	UFS	September 2020; March 2022
Progress in Planning	Elsevier	March 2020
Africa Review Journal	Taylor & Francis	March 2020
Land Use Policy	Elsevier	September 2019
Case Studies in the Environment	California University Press	June 2019

- Reviewed manuscripts submitted to the Society of South African Geographers Student Conference 2021and 2023
- Occasionally assigned to stand in for the Editor of Journal of Town and Regional Planning and conduct pre-assessment of manuscripts in 2021

GUEST INVITATIONS

1. Invited to be a guest presenter at the inaugural workshop for the Department of Public Health and Administration, University of the Free State. The workshop was on 23 August 2023. Presented on how postgraduate students can integrate the Sustainable Development Goals in their research.
2. Presented at a seminar in Sao Paulo, Brazil May 2023 “Adaptability of Vulnerable communities and youth in post-pandemic cities” Youtube link https://www.youtube.com/watch?v=Z5ramzQrdjY
3. Invited to the University of Zimbabwe to present on “Writing a Journal article” in the Department of Rural and Urban Planning on 26 November 2020
4. Invited to the University of Zimbabwe to present on writing the PhD thesis, in the Department of Rural and Urban Planning

REFERENCES

1. **Professor James Chakwizira**

Full Professor, Department of Urban and Regional Planning
University of Venda, University Road, Thohoyandou, Limpopo, 0950
South Africa
Cell: +27 763877814
Email: jameschakwizira@yahoo.com

2. **Professor Lochner Marais**

Full Professor, Center for Development Support
University of the Free State, PO Box 339, Bloemfontein 9300
South Africa
Cell: +27 826787735
Email: MaraisJGL@ufs.ac.za

3. **Professor Verna Nel**

Full Professor, Department of Urban & Regional Planning
University of the Free State, PO Box 339, Bloemfontein 9300
South Africa
Phone: 051 4012499
Email: NelVJ@ufs.ac.za

4. **Professor Innocent Chirisa**

Pro Vice Chancellor, Academic Affairs
Zimbabwe Ezekiel Guti University
1901 Barrassie Road, Bindura
Cell: +263733307130
Email: chirisa.innocent@gmail.com

Editor name : Jesús Rodrigo-Comino
Affiliation : Departamento de Análisis Geográfico Regional y Geografía Física
Institution : University of Granada
Address : Facultad de Filosofía y Letras, Campus Universitario de Cartuja, Universidad de Granada
Postal code : 18071
City : Granada
Country : Spain
E-mail : jesusrc@ugr.es
Homepage URL : www.egemap.eu
ORCID ID : https://orcid.org/0000-0002-4823-0871

MINISTERIO DE CIENCIA E INNOVACIÓN

Cofinanciado por la Unión Europea

AGENCIA ESTATAL DE INVESTIGACIÓN

CURRICULUM VITAE (CVA)

CV date	3/10/2024

Part A. PERSONAL INFORMATION

First name	Jesús		
Family name	Rodrigo Comino		
Gender (*)	Male	Birth date	22/02/1989
ID number	76429998D	Age	35
e-mail	jesusrc@ugr.es	URL Web:	
WoS Researcher ID L-4535-2016 SCOPUS Author ID 56178653000 ORCID 0000-0002-4823-0871			

A.1. Current position

Position	Associate Professor (Profesor Titular)		
Initial date	25/06/2024		
Institution	University of Granada		
Department/Center	Departamento de Análisis Geográfico Regional y Geografía Física, Facultad de Filosofía y Letras, Campus Universitario de Cartuja		
Country	Spain	Teleph.	637937558
Key words	Land degradation; regional geography; soil geography		

A.2. Previous positions (research activity interuptions, art. 14.2.b)

Period	Position/Institution/Country/Interruption cause
07/2013-11/2013	University of Trier (DAAD-Fellowship)
09/2014-08/2016	University of Trier (La Caixa-DAAD Fellowship)
09/2016-07/2018	University of Málaga, FPU Grant (Spanish Goverment)
08/2018-06/2019 (*first son)	University of Trier (Postdoc research stay)
06/2019-05/2021 (*second daughter)	University of Trier (Smart Light-HUB project; postdoc)
09/2019-05/2021	University of Valencia (FIRElinks COST Action; postdoc)
05/2021-To date (**third son)	University of Granada (Assistant and Associate professor)

***Interruptions (2 months -Parental leave in Germany-); ** Interruption (5 months - Parental leave Spain).**

A.3. Education

PhD, Licensed, Master	University/Country	Year
Lic. of Geography	Málaga	2007-2012
Master in Geographical Analysis in the Planning of the Territory. Technologies of the Geographic Information	Granada/Málaga	2012-2013
PhD Geography	Málaga	2014-2018
DEng Geomatics and Topography	Politécnica de Valencia	2021-2023

Part B. CV SUMMARY

PhD DEng Jesús Rodrigo Comino, a graduate in Geography, currently works as an Associate Professor at the University of Granada and has been included in the World's Top 2% Scientists ranking by Stanford University since 2021. He got a Master in Territorial Planning and Geographic Information Systems (2013) at the University of Málaga/Granada, whose final work was divided into 3 national publications and a monograph. During his predoc stage, he obtained three scholarships for doctoral studies: DAAD (German Academic Exchange Service), La Caixa Foundation and FPU (Ministry of Education, Spain). During this period, he completed his first doctoral thesis in Geography between the University of Trier (Germany -2 years-) and Malaga (2 years) in 2018: "Actual geomorphological processes in sloping vineyards. A comparison between Ruwer-Mosel Valley (Trier, Germany) and Montes de Málaga (Málaga, Spain)". In 2023, he defended his second doctorate in engineering in Geomatics and Topography at the Polytechnic University of Valencia. His research career consists of four complete monographs (Nova, Springer, etc.) and some edited books (Elsevier, CRC, Nova), >200 indexed publications in Scopus and WoS, leading international collaborations with research teams over the world. Regular reviewer in more than 150 indexed international journals, a member of 2 doctoral theses committees, and an evaluator of projects for the Ministries of Science of Chile, Peru, the United States, Serbia, Switzerland, Kazakhstan or Poland, and postgraduate scholarships for DAAD. He has organized several international scientific meetings and congresses (for example COST Actions, Biohydrology, Fire in the Earth, etc.), sessions at international conferences (EGU, TerraEnvision, Conference of the Arabian Journal of Geosciences, etc.), oral presentations and conferences master classes (Germany, Bulgaria, Norway, Algeria, etc.). Editor-in-chief of the indexed journal (Scopus and ESCI; Q2) Air Soil and Water Research (SAGE) and Euro-Mediterranean Journal for Environmental Integration (Springer). In addition, he was Associate Editor at Scientific Reports (Nature) and Hydrological Science Journal (Taylor & Francis) and continues in the Arabian Journal of Geosciences (Springer), and Journal of Mountain Science (Springer). He has participated as a researcher in R+D+I projects on social issues related to housing or the census, or transfer and knowledge at a European level, such as the INTERREG Smart-Light HUB project (light pollution) or COST FIRElinks (fires). He has been invited to give lectures on agriculture, sustainable management and erosion. He has supervised 5 final degree projects and 3 master's degrees. He has taught regulated and certified teaching at the Universities of Granada, Valencia, Málaga, León, Oviedo, Trier (in German), Quito (Ecuador) or La Habana (Cuba) on geopolitics, geomorphology, Geographic Information Systems, remote sensing and statistical techniques. He acted as PI of some international (FIRElinks COST Action), national (BBVA Foundation) and regional projects and supervise several PhD theses. Nowadays, he coordinate one European project H2030 (Marie Curie Staff Exchange), national project (Plan Nacional) and regional transference project (Grupos Operativos, Junta de Andalucía). Moreover, participated in other two European projects coordinating activities in different WPs (Erasmus +: Urban Inprint, and Soil Living Labs: SOILCRATES).

Part C. RELEVANT MERITS

C.1. Publications

1. RODRIGO-COMINO, J., Seeling, S., Seeger, M.K., Ries, J.B., 2021. Light pollution: A review of the scientific literature: The Anthropocene Review. https://doi.org/10.1177/20530196211051209
2. RODRIGO-COMINO, J., Giménez-Morera, A., Panagos, P., Pourghasemi, H.R., Pulido, M., Cerdà, A., 2020. The potential of straw mulch as a nature-based solution for soil erosion in olive plantation treated with glyphosate: A biophysical and socioeconomic assessment. Land Degradation & Development 31, 1877–1889. doi: 10.1002/ldr.3305
3. RODRIGO-COMINO, J., Keshavarzi, A., Zeraatpisheh, M., Gyasi-Agyei, Y., Cerdà, A., 2019. Determining the best ISUM (Improved stock unearthing Method) sampling point number to model long-term soil transport and micro-topographical changes in vineyards. Comput. Electron. Agric. 159, 147–156. DOI: 10.1016/j.compag.2019.03.007
4. RODRIGO-COMINO, J., Seeger, M., Iserloh, T., Senciales González, J.M., Ruiz-Sinoga, J.D., Ries, J.B., 2019. Rainfall-simulated quantification of initial soil erosion processes in sloping and poorly maintained terraced vineyards - Key issues for sustainable management systems. Science of the Total Environment 660, 1047–1057. DOI: 10.1016/j.scitotenv.2018.12.451
5. RODRIGO-COMINO, J., Novara, A., Gyasi-Agyei, Y., Terol, E., Cerdà, A., 2018. Effects of parent material on soil erosion within Mediterranean new vineyard plantations. Engineering Geology 246, 255–261. doi: 10.1016/j.enggeo.2018.10.006
6. RODRIGO-COMINO, J., Keesstra, S.D., Cerdà, A. 2018. Connectivity assessment in Mediterranean vineyards using Improved Stock Unearthed Method, LiDAR and soil

erosion field surveys. Earth Surface Processes and Landforms 43 (10), 2193-2206. doi: 10.1002/ esp.4385

7. RODRIGO-COMINO, J. 2018: Five decades of soil erosion research in "terroir". The State-of-the-Art. Earth-science Review 179, 436-447. DOI: 10.1016/j.earscirev.2018.02.014
8. RODRIGO-COMINO, J.; Senciales, J.M., Cerdà, A., Brevik, E.C. 2018. The multidisciplinary origin of soil geography: A review. Earth-Science Review 177, 614-623. doi: 10.1016/j.earscirev.2017.11.008
9. RODRIGO-COMINO, J., Cerdà, A. Improving Stock Unearthing Method to measure soil erosion rates in vineyards. Ecological Indicators 85, 509-517. doi: 10.1016/j.ecolind.2017.10.042
10. RODRIGO-COMINO, J., Iserloh, T., Lassu, T.,Cerdà, A., Keestra, S.D., Prosdocimi, M., Brings, C., Marzen, M.; Ramos, M.C.; Senciales, J.M., Ruiz Sinoga, J.D., Seeger, M., Ries, J.B. 2016. Quantitative comparison of initial soil erosion processes and runoff generation in Spanish and German vineyards. Science of the Total Environment 565, 1165-1174. doi: 10.1016/j.scitotenv.2016.05.163.

C.2. Workshops and meetings (oral/poster presentations, organisation of events)

1. 80 (15 oral presentations, 55 posters: EGU, AK Geomorphologie, Young Geomorphologists, etc.). **2. 5 invited keynotes** (TerraEnvision, Facing Fire, Ministry of Agriculture, Algeria, National Congress of Water, etc.). **3. Organisation of 3 international meetings: Biohydrology 2019 and Fire in the Earth 2021 (Valencia, Spain) and 2023 (Granada, Spain). 4. Organisation of a several COST meetings (Connecteur; FIRElinks) since 2016.**

C.3. Projects:

C.3.1. Participation in european R & D & I knowledge management and transfer projects

1. Cost Action: CA18135 FIRElinks. Description: Grant holder. Skills: Management, administration and organisation of meetings with scientists, stakeholders and policymakers about wildfires, working with social media and communication/transference with participatory agents such as politicians, employees and scientists. Entity: European Uniion (COST office)/Universitat de València Type: International EU. From: 07/09/2019 To: Today. Duration: 4 years. Access system: Concurrence.

2. INTERREG: Smart Light-HUB. Description: Project manager. Skills: Management and administration of the project for German partner, preparing transfer materials/activities for participatory agents such as politicians and educators about light pollution. Entity: European Union/ Universität Trier Type: International EU. From: 15/06/2018 To: Today. Duration: 3 years. Access system: Concurrence.

3. Grazing Effects on Runoff and Soil Erosion over Grassland – A field Plot-Scale Hydrologic Experimental Approach. Reference: PN-III-P1-1.1-TE-2019-1180; Contract number: TE 140/2020; Starting date: 01/2021; Finishing date: 06/2022; Entity: The Executive Agency for Higher Education, Research, Development and Innovation Funding – UEFISCDI; Romania). Quantity: 319,400 lei.

4. Efectos de los cambios de uso del suelo en la dinámica eco-geomorfológica en ambientes mediterráneos, a diferentes escalas, en el contexto del cambio global (ECUDES). PI: José Damián Ruiz Sinoga. Entity: Ministerio de Economía y Competitivad. Plan Estatal de Investigación Científica y Técnica y de Innovación 2017-2020. Starting date: 1/6/2020 Finishing date: 30/6/2023. Quantity: 36,600€.

5. **Urban Inprint** (Erasmus+): Connecting Universities and local governments to implement urban agendas: IP. Esteban Romero Frías. Desde: 01/09/2023 a 31/08/2025. Cantidad: 250,000€.

6. **SOIL Innovation Labs**: Co-Regenerating and Transforming European Soils (SOILCRATES). H2030. IP. UGR Partner: Esteban Romero Frías. Tarea: líder grupo de trabajo 3. Duración 1/10/2024 a 1/10/2026. Cantidad: 16,000,000€.

C.3.2. Research project as PI

1. Creación de una base nacional de erosión en viñedos para potenciar la protección del suelo

fértil (Creation of a national base for erosion in vineyards to enhance the protection of fertile soil). Entity: Fundación BBVA, Becas Leonardo. Quantity: 40,000€. Starting date: 30/6/2021 Finishing date: 31/12/2022.

2. Plan propio University of Granada (4 regional projects): Visiting Scholar (15000€), Precompetitive projects (1500€), young researchers (2500€) and consolidated researchers (7000€). Total: 26000€. Starting: 09/2022 Finishing: 06/20234.

3. Cost Action: CA18135 FIRElinks. IP European Union (COST office)/Universidad de Granada; Type: International EU. From: 01/09/2022 to 01/11/2023. Budget: 210,990.1€.

4. **Already accepted projects (2024):** Grupos operativos Junta de Andalucía (300000€); Marie Curie Staff Exchange (UE, 1.59 million €); Plan Nacional (65000€).

C.4. CHIEF EDITOR AND ASSOCIATE EDITOR

1. Editor in Chief: Air, soil and water research, SAGE (Scopus and Q2, ESCI) and Euro-Mediterranean Journal for Environmental Integration, Springer (Scopus and ESCI).

2. Associate Editor (indexed journals): Journal of Mountain Science (Springer); La Rivista di Economia, Demografia e Statistica; Arabian Journal of Geosciences (Springer). In the past: Scientifi Reports (Nature), Hydrological Science Journal (Taylor and Francis) and Land (mdpi).

C.5. Project evaluations

- **Since 2019:** Evaluating proposals for open calls of the National Science Centre of Poland (Narodowe Centrum Nauki), ul. Twardowskiego 16, 30-312 Kraków, NIP 6762429638; REGON 121361537. https://ncn.gov.pl/?language=en
- **Since 2019:** Evaluating proposals for Consejo de Ciencia concurso Nacional de Proyectos FONDECYT regular 2020 CONICYT – Chile. https://www.conicyt.cl/fondecyt/
- **09/2020:** Evaluating proposals for open calls of the Geomorphology and Land-use Dynamics Program at the National Science Foundation (NSF), United States of America: https://www.nsf.gov/funding/pgm_summ.jsp?pims_id=13690
- **Since 2020**: Evaluating proposals for National Centre of Science and Technology Evaluation", 404 Office, Bogenbay batyr street 221, 050026, Almaty, Kazakhstan: http://www.ncste.kz/
- **Since 2021**: E067-2021-02 "Proyectos Especiales para la Reactivación. Unidad de Evaluación y Selección del Fondo Nacional de Desarrollo Científico, Tecnológico y de Innovación Tecnológica – FONDECYT, Perú.
- **Since 2021:** Swiss National Science Foundation. Agency for the evaluation of national projects in the area of Geomorphology.
- **Since 2021**: Science Fund of the Republic of Serbia. Andricev venac 2/III, Belgrade, www.fondzanauku.gov.rs

C.6. Teaching and supervision of final degrees/master

1. Universidad de Málaga (2017-2018); Universidad de Valencia (2017-2020); Universität Trier (2013-2021); University of Granada, University of León (2020-2021); Quito University (Ecuador) 2021-2022, University of La Habana (2022).
2. **5** final degrees y **4** master thesis at the Universität Trier, Pontificia de Chile and Granada
3. **4 PhD students**.

C.7. OTHERS

Coordinating Terra-Lab Group 2 (EGEMAP: Environmental Geography and Mapping): www.egemap.eu

Professional and official pilot of drones.

Official Environmental Judicial Expertise.

GPSR Compliance

The European Union's (EU) General Product Safety Regulation (GPSR) is a set of rules that requires consumer products to be safe and our obligations to ensure this.

If you have any concerns about our products, you can contact us on ProductSafety@springernature.com

In case Publisher is established outside the EU, the EU authorized representative is:

Springer Nature Customer Service Center GmbH
Europaplatz 3
69115 Heidelberg, Germany

Batch number: 10706089

Printed by Printforce, the Netherlands